AN INTRODUCTION TO
PROBABILITY THEORY

AN INTRODUCTION TO
PROBABILITY THEORY

P. A. P. MORAN

CLARENDON PRESS · OXFORD

Oxford University Press, Walton Street, Oxford OX2 6DP
Oxford New York Toronto
Delhi Bombay Calcutta Madras Karachi
Petaling Jaya Singapore Hong Kong Tokyo
Nairobi Dar es Salaam Cape Town
Melbourne Auckland
and associated companies in
Beirut Berlin Ibadan Nicosia

Oxford is a trade mark of Oxford University Press

Published in the United States
by Oxford University Press, New York

© Oxford University Press 1968

First published 1968
First published in paperback, with corrections 1984
Reprinted 1986

British Library Cataloguing in Publication Data
Moran, P. A. P.
An introduction to probability theory.
1. Probabilities.
I. Title
519.2   QA273
ISBN 0–19–853242–3

Printed in Hong Kong

CONTENTS

PREFACE

A NUMBER of excellent books on probability theory have recently been published. Nearly all of these are either elementary introductions to the subject or else are written only for those who are interested in deep problems of modern abstract analysis. The present book has been written to provide an outline of the main facts of probability theory which, on the one hand, can be generalized in a highly abstract manner and, on the other, used to describe many complicated phenomena in natural science.

Such a treatment must necessarily include both measure theory and random processes. No attempt has been made to give a complete account of measure theory, and in particular the theory of conditional probability measures is not considered in full generality. Enough of measure theory has, however, been given to provide the necessary basis for the treatment of infinite sums and sequences and to introduce the reader to the more detailed theory which can be found in the books of Doob, Halmos, and Loève. The book also contains an introduction to random processes which, although far from covering all the main problems of the subject, serves to illustrate the theory of the rest of the book and to show how many special probability distributions arise in practice.

The book is designed to be read by honours students in about their third year. Holding, as I do, the view that however important abstraction and generalization are, no mathematical science can remain vigorous unless it draws some of its inspiration from the natural sciences, an attempt has been made to illustrate the subject by some of the many attractive problems in what has now come to be known as 'applied probability theory'.

An attempt has also been made to provide enough references to enable the reader to follow up particular questions in the literature and also to find the main tables of probability distributions.

I am indebted to a number of friends who have read parts of the manuscript and in particular to Dr. A. W. Davis, Mr. R. Davis, Mrs. S. Ohlsen, and Dr. C. Pearce. I am also very indebted to Mrs. B. Cranston, who typed successive drafts of the manuscript.

Canberra
April 1967

P. A. P. M.

1. The Probabilities of Events

1.1. The idea of probability

THE theory of probability is a branch of mathematics which deals with the combinations of certain numbers, or measures, which are used to measure the frequencies of events in the real world. Its origins are to be found in the interest taken by certain mathematicians of the seventeenth century in games of chance and it is therefore natural to illustrate the beginning of the theory by examples such as that of tossing a penny. Such an event may have one of two outcomes, either a head being thrown or a tail. If the penny is very nearly symmetrical then we can 'reasonably' expect that if the penny is tossed a large number of times the relative frequency with which a head is thrown, compared with the total number of throws, will be close to $\frac{1}{2}$. It is then natural to say that the probability of throwing a head in a single toss is $\frac{1}{2}$, this number being applied to the single event and being equal to what we reasonably expect the relative frequency of such an event to be in a large number of repetitions.

We cannot be certain that such a relative frequency will be near $\frac{1}{2}$ in a large number of trials since it is clearly possible that no heads at all occurred, but using the word 'probable' in an everyday non-numerical sense we can say that when the series of events is long, it is highly probable that the relative frequency of 'heads' is near $\frac{1}{2}$. Later we shall (in the Law of Large Numbers) make this statement more precise in a mathematical sense.

To generalize the above situation we therefore suppose that some particular event occurs in one of a number of different manners, as for example if we carry out any experiment which can turn out in a number of different ways. In the first three chapters of this book we suppose that the number of ways in which the event can occur is either finite (for example in tossing a penny, or drawing a card at random from a pack of cards), or 'enumerable', i.e. that the events can be numbered off, or put into one-to-one correspondence with the integers. The latter case may be exemplified by the number of times it is necessary to toss a penny in order to obtain a head for the first time. The restriction to events which may be classified in either a finite or enumerable number of ways rules out the type of problem exemplified by choosing a real number 'at random' lying between 0 and 1. This type of problem requires a more elaborate mathematical analysis which is deferred until later in the book.

We can, however, relate such a situation to the present case in the following manner. Suppose that the outcome of the experiment is a number X which may take a continuous range of values. If we define two events depending on X to be

$$X \leqslant x,$$

$$X > x,$$

where x is some number, we can talk of the probabilities that these events take place. Thus we can write $F(x)$ for the probability that $X \leqslant x$, this being the relative frequency with which this event can be reasonably expected to occur in a large number of repetitions of the experiment. $F(x)$ depends on x and we say that $F(x)$ is a 'cumulative distribution function' or simply 'distribution'. In most circumstances we rule out the possibility of infinite values of X, and since $F(x)$ is clearly non-decreasing, any function $F(x)$ for which:

(1) $0 \leqslant F(x) \leqslant 1$,

(2) $\lim\limits_{x \to \infty} F(x) = 1$,

(3) $\lim\limits_{x \to -\infty} F(x) = 0$,

(4) $F(x)$ is non-decreasing,

can serve as a distribution. We shall find this concept useful even in the first three chapters of this book where we deal only with discrete, i.e. finite or enumerable, probabilities.

1.2. The space of events

Suppose therefore that the experiment can turn out in one of a finite or enumerable number of exclusive ways which we denote as E_1, E_2, E_3,..., and which we call the 'elementary events'. The set of all such events is called the 'space' of elementary events and we write it as Ω. We can now define events, which we denote by A_1, A_2,..., B_1,..., and which consist of all, some, or none of the E's. For example we could define an event which is to consist of either E_1 or E_2. In particular the set of all events, Ω, is such an event, and it is also convenient to define an event, O, which consists of none of the E_i's, and therefore cannot happen.

As an illustration consider an experiment which consists of a random choice of one card from a well-shuffled pack of 52 cards. The 52 elementary events, E_1,..., E_{52} will consist of the choice of the 52 different cards in the pack, Ω will be the event that consists of the choice of any card, and we can define other events such as the choice of a card of a particular suit or of a particular rank.

Given any two events, A_1 and A_2, we define a 'product' event denoted by $A_1 A_2$ which occurs if any events E_i occur which belong to both A_1 and A_2. This operation corresponds to the 'multiplication' of sets, and $A_1 A_2$ occurs if and only if both A_1 and A_2 occur. $A_1 A_2$ is also sometimes called the 'intersection' of A_1 and A_2. Thus if A_1 is the event which consists of the choice of a card which is a spade, A_2 the choice of a card which is a king, $A_1 A_2$ will consist of the choice of the king of spades. Since we have agreed to denote by O an impossible event we can write $A_1 A_2 = 0$ when A_1 and A_2 have no common E's.

Clearly the operation of combining two events A_1 and A_2 into an event $A_1 A_2$ is associative, and commutative, in the sense that $(A_1 A_2) A_3 = A_1 (A_2 A_3)$ and $A_1 A_2 = A_2 A_1$. We can therefore define the operation of taking the common part of a finite or enumerable number of sets $A_1 A_2, \ldots$ and in the enumerable case we write this

$$\prod_1^\infty A_i.$$

Such a set always exists, although it may be the null set O.

Similarly we can define the 'sum' of events, $A_1 + A_2$, to consist of all those events E_i which belong to A_1, to A_2, or to both. This is also associative and commutative and we can define

$$\sum_1^\infty A_i$$

which always exists. It is also clear that the distributive law

$$A_1(A_2 + A_3) = A_1 A_2 + A_1 A_3$$

holds.

Next we write $A_1 \subset A_2$ (or $A_2 \supset A_1$) if all the events E_i in A_1 are also in A_2, and say that A_1 is 'included' in A_2. This will happen if A_1 is the event of choosing the king of spades, and A_2 is the event of choosing a king. In particular any A is included in Ω, the sum of all sets E_i, and so we can always define $\Omega - A$ which we write as $\bar{A}$, and call the 'complement' of A. It is then clear that

$$\bar{\Omega} = O, \quad \bar{O} = \Omega,$$

$$A + \bar{A} = \Omega \text{ for any } A,$$

$$A_1 - A_2 = A_1 \bar{A_2} \text{ for any } A_2 \text{ such that } A_2 \subset A_1,$$

$$\overline{A_1 A_2} = \bar{A_1} + \bar{A_2},$$

$$\overline{A_1 + A_2} = \bar{A_1} \bar{A_2}, \text{ for any } A_1, A_2.$$

Furthermore these relationships also hold when finite 'sums' and 'products' are replaced by infinite sums and products. It follows from the above that any relationship between events of the above kind involving sums, products, complements, and the inclusion relationship '$\subset$', implies a similar relationship between these sets which is obtained by transforming all such terms as $A_1 A_2$, $A_1 + A_2$ into $\bar{A_1} + \bar{A_2}$, $\bar{A_1} \bar{A_2}$ respectively, $\supset$ into $\subset$, and leaving '=' and the complementary symbol unchanged.

It is convenient to call the events A_1, A_2,... 'sets' of 'points' E_1, E_2,..., the set, Ω, of all points E_i being referred to as a 'space'.

If we have a class of such sets with the property that it is closed under the operations of forming finite sums, finite products, and complements, i.e. for any A_1, A_2, $A_1 + A_2$, $A_1 A_2$ and $\overline{A_1}$ belong to the class, we say that such a class is a 'field'. This will be sufficient for all problems in which there are only a finite number of possible outcomes but we wish to include also the cases where there are at least an enumerable number. If therefore we have a field in which all enumerable sums and products,

$$\sum_1^\infty A_i, \quad \prod_1^\infty A_i$$

are defined and belong to the field we say that we have an enumerable 'σ-field'. Such a generalization is necessary, for example, for the problem considered above in which a penny was repeatedly tossed until a head appeared. The elementary events E_i ($i = 1, 2,...$) may be taken to be the events that exactly i tosses are required to obtain a head for the first time. The probability that the number of tosses required to obtain a head is odd is then the probability of the infinite sum of events

$$E_1 + E_3 + E_5 + ...$$

or we can regard it as the probability of the event defined by the infinite product

$$(\Omega - E_2)(\Omega - E_4)(\Omega - E_6)....$$

Thus given any finite or enumerable set of elementary events E_1, E_2,... we can define our field of possible events to consist of all sets which consist of none, some, or all of the events E_i and from now on we shall always regard this as our basic field. Clearly if the number of E_i is finite, the number of sets A_i is finite, but if the number of E_i is enumerably infinite, the number of A_i is non-enumerably infinite. This is not, of course, the only field than can be defined using an enumerable set of elementary events E_i for we might, for example, consider the σ-field consisting of all selections of the E_i with the restriction that E_1 and E_2 must always be both included or both excluded from the sets A_i. The number of different such fields which can be formed out of a finite set of the E_i is finite and has been studied by Szekeres and Binet (1957).

Given the E_i the construction of the A_i is obvious. If, conversely, we are given the A_i, and the fact that they form a field, the question arises as to whether they can be regarded as equivalent to the field formed by subsets of a collection of objects. The affirmative answer (in a certain sense) to this question is given by Stone's Representation Theorem (see Halmos 1950, Stone 1936).

1.3. Probability defined on the space of events

We now have to elucidate the meaning of probability as applied to the sets (events) A_i. We want the probability of an event A_i to measure the relative proportion of times the event A_i will be reasonably expected to occur in an indefinitely long series of trials. Thus if we write $p(A_i)$ for the probability of A_i this will be a numerically valued function of A_i which satisfies the condition $0 \leqslant p(A_i) \leqslant 1$. Suppose that A_1, A_2,... are a finite or enumerable set of events which are exclusive so that no pair can occur together. Then the event ΣA_i will occur when exactly one of the A_i occurs and vice versa so that the relative frequency of ΣA_i in a long series of trials must equal the sum of the relative frequencies of the A_i. Hence $p(A_i)$ must satisfy

$$p(\Sigma A_i) = \Sigma p(A_i)$$

whenever the A_i are a finite or enumerable sequence of exclusive events. Since the event O is impossible we must have $p(O) = 0$, and similarly $p(\Omega) = 1$. These conditions are sufficient to define probability on an enumerable system of the above kind.

DEFINITION. Let the sets A_i (including O and Ω) form a σ-field. Then any numerical valued function $p(A)$ of the sets A_i is called a probability distribution if

(1) $0 \leqslant p(A_i) \leqslant 1$ for all A_i.

(2) $p(\Sigma A_i) = \Sigma p(A_i)$ if the sets A_i are mutually exclusive, and ΣA_i is a finite or enumerable sum.

This definition is phrased in terms of the sets A_i forming the field. In the problems of discrete probability considered in the first three chapters of this book these A_i will always be defined as sums of elementary events E_i.

A 'probability' in this definition is necessarily real and non-negative. Negative and complex valued probabilities have been defined (Bartlett 1945, Cox 1955), but have found very few applications.

An elementary consequence of the above definition is worth noting. If A_1,..., A_n are a set of events which may or may not be exclusive we have

$$p\left(\sum_1^n A_i\right) = p(A_1) + p(\overline{A_1} A_2) + p\{(\overline{A_1} \overline{A_2}) A_3\} + \ldots + p\{(\overline{A_1} \ldots \overline{A_{n-1}}) A_n\},$$

$$(1.1)$$

and since

$$p\{(\overline{A_1} \ldots \overline{A_j}) A_{j+1}\} \leqslant p(A_{j+1})$$

it follows that

$$p\left(\sum_1^n A_i\right) \leqslant \sum_1^n p(A_i). \qquad (1.2)$$

This is known as Boole's inequality. Later we shall extend and strengthen this result (§ 1.18).

1.4. Random variables and indicators

A useful concept in many problems involving discrete probabilities is that of an 'indicator function'. This is a special case of the more general concept of a 'random variable' and we therefore define the latter first, leaving its detailed study to later. A random variable is also sometimes called a 'variate'.

A random variable, X, is defined to be a numerically valued function of the elementary events E_i so that it can be written $X(E_i)$. Since the E_i are regarded as 'points', a random variable is a function of a point, and not, like probability, a function of a set consisting of one or more points. As an example we might take the suffix i of E_i so that $X(E_i) = i$, or in the case of the choice of a card at random from a pack we might put $X = 1, 2, 3, 4$ according as the card is diamond, heart, club, or spade. Since the E_i are enumerable, the number of different values which $X(E_i)$ can take is at most enumerable. It may, of course, be much less than the number of E_i as for example in the card case where we might define X to be equal to 1 if the card is a red king, and equal to zero otherwise. If we number off the possible values of X as $x_1, x_2, x_3,...$ we could define a new sample space consisting of events $E_1', E_2',...$ corresponding to the cases where X takes the values $x_1, x_2,...$ and each E_i' would correspond to the set of all E_i for which $X(E_i) = x_i$. It is often more convenient, however, not to do this, especially when we generalize the procedure in Chapter 4.

We define the expectation of a random variable X (denoted by $E(X)$) by

$$E(X) = \Sigma X(E_i)p(E_i) \tag{1.3}$$

whenever this sum exists in the absolutely convergent sense, i.e. whenever

$$\Sigma |X(E_i)|p(E_i) \tag{1.4}$$

is finite. The condition (1.4) is introduced to take care of the cases where the number of events is enumerable so that we avoid the cases where the series (1.3) is either infinite or has a sum depending on the order in which the terms are taken. It is, however, convenient to say that $E(X)$ exists and is positively or negatively infinite when the sum (1.3) is infinite (and of fixed sign) whatever the order of the terms.

If X and Y are two random variables for which $E(X)$ and $E(Y)$ both exist and are finite, it is clear that $X + Y$ is a random variable and

$$E(X + Y) = E(X) + E(Y). \tag{1.5}$$

A similar result naturally does not hold in general for the random variable XY.

If X is a random variable, so also are $X^2, X^3,...$, and we write

$$m_r = E(X^r), \tag{1.6}$$

for $r = 1, 2,...$, and call m_r the 'rth moment' of X whenever it exists. Clearly $m_2, m_4,...$ cannot be negative, and must either be positive or positively infinite, whilst $m_1, m_3,...$ may have either sign.

A particular type of random variable is the 'indicator function'. The indicator function of a set A_i belonging to the σ-field of events is defined to be a random variable which takes the values 1 for any event E_i contained in A_i, and zero for any other E_i. We write it as $I(A_i)$. In pure mathematics this is sometimes called the 'characteristic function' of the set A_i, but in probability theory this phrase is used in another sense. Then it is clear that

$$E\{I(A_i)\} = p(A_i).$$

As a consequence of the set operations defined before we have

$$I(A + \bar{A}) = I(\Omega) = 1, \tag{1.7}$$

$$I(\bar{A}) = 1 - I(A), \tag{1.8}$$

$$I(AB) = I(A)I(B) \quad \text{for any } A, B, \tag{1.9}$$

$$I(A + B) = 1 - \{1 - I(A)\}\{1 - I(B)\}. \tag{1.10}$$

Similarly

$$I\left(\prod_1^\infty A_i\right) = \prod_1^\infty I(A_i) \tag{1.11}$$

and

$$I(\textstyle\sum A_i) = I(A_1) + I(\bar{A_1}A_2) + I(\bar{A_1}\bar{A_2}A_3) + \dots$$
$$= I(A_1) + \{1 - I(A_1)\}I(A_2)$$
$$+ \{1 - I(A_1)\}\{1 - I(A_2)\}I(A_3) + \dots. \tag{1.12}$$

These results will later be found to be useful in deriving combinatorial theorems.

1.5. Independence

Suppose that we have a basic finite or enumerable set of events $E_1, E_2,...$ and that we have defined as before the finite or enumerable σ-field of events A_i consisting of all finite or enumerable sums of events E_i. We then say that two events A_1 and A_2 are independent if

$$p(A_1 A_2) = p(A_1)p(A_2). \tag{1.13}$$

In particular any event will be taken as independent of the events Ω and O. More generally, we say that any finite or enumerable sequence of events $A_1, A_2, A_3,...$ are jointly independent if for every selection $i_1, i_2,...$ of the suffices of the A's we have

$$p(A_{i_1} A_{i_2}...) = p(A_{i_1})p(A_{i_2}).... \tag{1.14}$$

Thus every subset of the A's are then independent also.

We now show that if any set of events are independent, the same is true of any selection of events out of this set and the set of their complements. Thus in particular for two events we must show that if

$$p(A_1 A_2) = p(A_1)p(A_2)$$

then

$$p(A_1 \overline{A_2}) = p(A_1)p(\overline{A_2}),$$

and

$$p(\overline{A_1} \, \overline{A_2}) = p(\overline{A_1})p(\overline{A_2}).$$

To show this we have

$$
\begin{aligned}
p(A_1 \overline{A_2}) &= p(A_1 - A_1 A_2) \\
&= p(A_1) - p(A_1 A_2) \\
&= p(A_1)\{1 - p(A_2)\} \\
&= p(A_1)p(\overline{A_2}),
\end{aligned}
$$

by using the rules for the probability of the sum of exclusive events. Similarly

$$
\begin{aligned}
p(\overline{A_1} \, \overline{A_2}) &= p(\overline{A_1} - \overline{A_1} A_2) \\
&= p(\overline{A_1}) - p(\overline{A_1} A_2) \\
&= p(\overline{A_1}) - p(\overline{A_1})p(A_2) \\
&= p(\overline{A_1})p(\overline{A_2}),
\end{aligned}
$$

since A_1 and A_2 are independent. Notice also that it follows from the definition that if A_1, A_2,... are jointly independent

$$EI(A_1 A_2 \ldots) = \{EI(A_1)\}\{EI(A_2)\}\ldots.$$

If a set of events A_1, A_2,... are independent in pairs it does not necessarily follow that they are independent as a whole. As an example suppose we toss a penny three times and consider only those cases in which an even number of heads occurs, i.e. in an obvious notation

$$E_1 \quad \text{T T T},$$
$$E_2 \quad \text{T H H},$$
$$E_3 \quad \text{H T H},$$
$$E_4 \quad \text{H H T}.$$

If we ascribe to each of these the probability $\frac{1}{4}$ and define A_1, A_2, A_3 to be the events that the first, second, and third toss gives a head, i.e. $A_1 = E_3 + E_4$, $A_2 = E_2 + E_4$, $A_3 = E_2 + E_3$, we obviously have

$$p(A_1 A_2) = p(A_1)p(A_2),$$
$$p(A_1 A_3) = p(A_1)p(A_3),$$
$$p(A_2 A_3) = p(A_2)p(A_3),$$

but
$$p(A_1 A_2 A_3) = 0,$$
whereas
$$p(A_1)p(A_2)p(A_3) = \tfrac{1}{8}.$$

Indicator variables are said to be independent when the events to which they refer are independent and this idea can be extended to random variables in general. To do this we first notice that a random variable X defined on an enumerable field can only take an enumerable set of different values. Let these be denoted by $x_1, x_2,\ldots$ and the corresponding disjoint sets (which must belong to the field) by $A_1, A_2,\ldots$. Then X can be written as

$$X = \sum_i x_i I(A_i).$$

Another random variable, Y, with the representation

$$Y = \sum_j y_i I(A_j'),$$

will be said to be independent of X (and X independent of Y) if each event A_i is independent of every event A_j'.

It then follows that if X and Y are independent so are $f(X)$, $g(Y)$ where f and g are arbitrary functions, and also that

$$E(XY) = \sum_{ij} x_i y_j p(A_i A_j')$$
$$= \sum_i x_i p(A_i) \sum_j y_j p(A_j')$$
$$= E(X)E(Y). \tag{1.15}$$

Thus the expectation of a sum of random variables is always equal to the sum of their expectations, but the expectation of a product of random variables is only equal, in general, to the product of the expectations when they are independent.

If two events A_1 and A_2 are independent in the above sense the relative frequency with which one of them turns up in the long run will not be affected in any way by any stipulation we make about the other. This gives us a method of combining the probabilities of two or more experiments whose outcomes are independent in the usual sense of independence. In particular in this way we can ascribe probabilities to the outcome of a set of independent repeated trials of the same experiment.

Thus suppose we have two experiments which are independent of each other. If the results of the first are denoted by $E_1, E_2,\ldots$ as before we can set up the σ-field of events A_i consisting of sums of the E_i, together with the associated probabilities, $p(A_i)$. Similarly if the results of the second experiment are denoted by $E_1', E_2',\ldots$ we can set up a σ-field of sets B_i consisting of sums of the E_i' together with their probabilities $p(B_i)$. We denote the space of all events E_i by Ω, and the space of all events E_i' by Ω'.

2

The possible results of both experiments considered together can then be denoted by (E_i, E'_j), and if the experiments are to be independent we must ascribe the probability $p(E_i)p(E'_j)$ to this event. We can now set up a σ-field consisting of all sets of events which are sums of events (E_i, E'_j) with their corresponding probabilities. The space of all such events is known as the product space of the two spaces formed by the E_i and E'_i, and the corresponding probability field is known as the product probability field.

It is now easily seen that events defined only by what happens in one of the fields are 'independent' of events defined by what happens in another field, in the sense in which events were defined to be independent at the beginning of this section. Thus in the example given the events (A_i, Ω') and (Ω, B_j) are independent because

$$(A_i, B_j) = (A_i, \Omega')(\Omega, B_j)$$

and

$$p(A_i, B_j) = p(A_i, \Omega')p(\Omega, B_j)$$

$$= p(A_i)p(B_j).$$

A similar method can be applied to set up the product probability field formed by any finite number of probability fields. The case of an enumerable number of probability fields requires rather more discussion and is deferred until Chapter 4.

As an example suppose a penny is tossed three times. Then the product space will consist of all sums of elementary events defining by a sequence of three letters each of which is H or T. There are eight of the latter and by the rule above we ascribe to each the probability

$$\tfrac{1}{8} = \tfrac{1}{2} \times \tfrac{1}{2} \times \tfrac{1}{2}.$$

It is then obvious that the occurrence of a head at the first trial is independent of the occurrence of a head at the second or third trials.

1.6. Conditional probability

Suppose we have a finite or enumerable σ-field of events, A_i, whose probabilities are defined. Consider an event A_1 whose probability is non-zero. Then

$$p(A_2 | A_1) = \frac{p(A_1 A_2)}{p(A_1)} \qquad (1.16)$$

is called the conditional probability of A_2 given A_1. It clearly corresponds to the relative frequency with which A_2 will be expected to occur in the long run when we restrict our attention to all cases in which A_1 also occurs. In a similar way we can write

$$p(A_3 | A_1 A_2) = \frac{p(A_1 A_2 A_3)}{p(A_1 A_2)}$$

$$= \frac{p(A_1 A_2 A_3)}{p(A_2 | A_1) p(A_1)}$$

and

$$p(A_1 A_2 A_3) = p(A_3 | A_1 A_2) p(A_2 | A_1) p(A_1),$$

with an obvious generalization to more than three events.

From (1.16) if B_1, B_2,... are a class of exclusive sets $\sum_j B_j = \Omega$ then

$$\sum_j p(B_j | A_1) = 1,$$

so that for any σ-field of sets B_j belonging to the field $\{A_i\}$, the conditional probability defines a probability field so long as $p(A_1) \neq 0$. Moreover, we also have

$$p(A) = \sum_j p(B_j) p(A | B_j),$$

where we take $p(B_j) p(A | B_j) = 0$ if $p(B_j) = 0$. From this we can deduce the following result.

THEOREM 1.1 (Bayes's theorem). *If $p(A) > 0$, and the B_j are as above,*

$$p(B_j | A) = \frac{p(AB_j)}{p(A)}$$

$$= \frac{p(B_j) p(A | B_j)}{\sum_j p(B_j) p(A | B_j)}. \tag{1.17}$$

In a similar way we can define conditional expectations.

The independence of two or more events can be defined in terms of conditional probabilities. Thus for two events, A and B, condition (1.13) is equivalent to asserting that either

$$p(A_2 | A_1) = p(A_2)$$

or

$$p(A_1) = 0.$$

Any event is independent of any other event with zero probability or with probability one.

1.7. Repeated trials

In section 1.5 we have shown how probabilities of independent trials can be combined. Consider a sequence of n independent trials in each of which an event E may or may not occur with probabilities p and $q = 1 - p$. There are then 2^n possible sequences. The probability of any one of these in which E occurs r times, and does not occur $n-r$ times, is $p^r q^{n-r}$. Since there are $\binom{n}{r}$ such sequences the probability of E occurring exactly r times is

$$\binom{n}{r} p^r q^{n-r} \quad (r = 0, 1, ..., n), \tag{1.18}$$

these terms arising in the expansion of the binomial expression

$$(q+p)^n. \tag{1.19}$$

This is known as the binomial distribution of Bernoulli (for tables see §2.8).

The number of times E occurs is a random variable which we may write as X. There are 2^n elementary sets. We can also represent X in another way. Define X_i to be equal to 1 or 0 according as the ith trial results in E or not. Then each X_i is a random variable and $X = X_1 + ... + X_n$.

The set of all sequences of n events is the product space of the n spaces formed by E and $\bar{E}$ at the ith trial, and X_i can be regarded as a random variable defined on the ith of these. Clearly

$$E(X_i) = p,$$

$$E(X_i^2) = p.$$

Then using (1.5) we have

$$E(X) = E(X_1) + ... + E(X_n)$$

$$= np, \tag{1.20}$$

and

$$E(X^2) = E(X_1 + ... + X_n)^2$$

$$= nE(X_1^2) + n(n-1)E(X_1 X_2)$$

$$= np + n(n-1)p^2 = n^2p^2 + np(1-p). \tag{1.21}$$

X is said to follow the 'binomial distribution with probability p and index n'.

If we write p_r for the probability that $X = r$ we can regard the whole experiment as a single trial in which the outcome is one of $n+1$ events, namely $X = 0, 1, ..., n$. The probabilities of these events are p_r. Since

$$\frac{p_{r+1}}{p_r} = \frac{n-r}{r+1} pq^{-1},$$

p_r will increase as r increases (so long as $npq^{-1} > 1$) until $(r+1)(n-r)^{-1}$ exceeds pq^{-1} when p_r will start to decrease again. Thus the distribution has a single hump and is said to be 'unimodal'. It is easy to see that the point at which the largest value of p_r is attained is the integer nearest to np. It may also happen that there are two such points in which case both are within unity of np.

1.8. Moments about the mean

Another type of moment is of great importance. Suppose we have a random variable X taking the values $x_1, x_2,\ldots$. We define the rth moment *about the mean* to be

$$\mu_r = E\{X - E(X)\}^r$$
$$= E(X - m_1)^r. \tag{1.22}$$

In particular

$$\mu_2 = E(X - m_1)^2$$
$$= E(X^2 - 2m_1 X + m_1^2)$$
$$= m_2 - m_1^2, \tag{1.23}$$

where μ_2, which is sometimes written 'Var(X)', is known as the 'variance' of the distribution and its square root as 'the standard deviation'.

This notation for moments using m_r for moments about the origin, and μ_r for moments about the mean, is different from the notation usual in statistical theory where *roman* letters are used for sample values and Greek letters for population values.

For the binomial distribution

$$\mu_2 = n^2 p^2 + np(1-p) - (np)^2$$
$$= np(1-p). \tag{1.24}$$

One important property of the variance is that it is additive for independent random variables. Thus if X and Y are independent we have

$$E\{X + Y - E(X + Y)\}^2 = E(X + Y)^2 - \{E(X + Y)\}^2$$
$$= E(X^2) + 2E(XY) + E(Y^2)$$
$$\quad - \{E(X)\}^2 - 2E(X)E(Y) - \{E(Y)\}^2$$
$$= E\{X - E(X)\}^2 + E\{Y - E(Y)\}^2 \tag{1.25}$$

because $E(XY) = E(X)E(Y)$. This result is reflected in formula (1.24) since X can be regarded as the sum of n independent random variables each with variance $p(1-p)$.

1.9. Tchebychev's inequality and the weak law of large numbers

If X is a random variable and

$$m_1 = E(X),$$
$$\mu_2 = E\{X - E(X)\}^2,$$

are both finite it is possible to put a bound on the probability of X being different from $E(X)$ in absolute value by more than any specified amount. If X takes the values x_1, x_2,... with probabilities p_1, p_2,... we have

$$\mu_2 = \sum_i p_i \{x_i - E(X)\}^2$$

$$\geqslant \sum_i' p_i \{x_i - E(X)\}^2,$$

where the sum $\sum'$ is taken over all i such that

$$|x_i - E(X)| \geqslant x, \quad \text{say.}$$

Thus

$$\mu_2 \geqslant x^2 \sum_i' p_i$$

$$\geqslant x^2 \operatorname{pr}\{|x_i - E(x_i)| \geqslant x\}$$

and thus

$$\operatorname{pr}\{|x_i - E(x_i)| \geqslant x\} \leqslant \mu_2 x^{-2}. \tag{1.26}$$

This is usually known as Tchebychev's inequality but was also discovered by Bienaymé and possibly others. It provides a partial clarification of the relationship of the frequency of an event in a large number of independent trials to its probability.

Suppose that n independent trials are made in each of which the event E has probability p. (1.18) gives the probability that r of these trials result in E. rn^{-1}, the proportion of trials in which E occurs, can then be shown to be near p with high probability since the random variable $X = rn^{-1}$ has mean p and variance $n^{-1}p(1-p)$. Hence we have the following result.

THEOREM 1.2
$$\operatorname{pr}(|X - p| \geqslant x) \leqslant n^{-1}p(1-p)x^{-2}. \tag{1.27}$$

If x is fixed this can be made as small as desired by increasing n. This result could, of course, be proved by estimating the sum of terms in (1.18) corresponding to $\operatorname{pr}(|X - p| > x)$ (the sum of the 'tails' of the distribution of X), but the use of Tchebychev's inequality is much simpler and generalizes more easily. On the other hand, the bound given by (1.27) is very crude.

Theorem 1.2 is the simplest expression of the 'Weak Law of Large Numbers'. It shows that in a large number of independent trials of the same experiment, the probability that the relative frequency with which a particular result eventuates will differ from the probability of this result by more than a small given amount is small and can be made as small as we please by making n large. The relative frequency is said to 'converge in probability' to p. The result therefore links the frequency with which an event occurs with its probability, but it cannot be used to provide a definition of probability in terms of frequency since the Weak Law of Large Numbers is in itself a probability statement.

(1.27) is a statement about the relative frequency of the event E in a sequence of n trials, where n is a fixed number. Suppose that the sequence of trials is indefinitely prolonged and that the number of times E occurs in the first n trials is r_n. Then it is natural to ask whether a sense can be ascribed to the statement that the sequence of numbers

$$r_n n^{-1}$$

converges. Clearly it does not necessarily converge in the ordinary sense, and hence we must define the probability of its convergence. In order to do this it is necessary to define probability in a space of events each of which is an enumerable sequence of failures and successes. Such a space is not enumerable, since it is in one-to-one correspondence with the set of all enumerable sequences of zero's and one's, and hence it cannot be treated by the definition of probability used in the first three chapters of this book.

We shall consider the problem of such a definition in Chapter 8, and we shall then prove a theorem (Theorem 8.36) which shows that the sequence $r_n n^{-1}$ converges 'almost certainly' in a natural sense. The result is said to be the 'Strong Law of Large Numbers'. Theorem 8.36 is a difficult theorem but for the particular case of binomial trials can be quite simply proved in a direct manner.

1.10. Some other schemes of successive trials

Suppose that there are n successive independent trials as before but that the probability of a particular result varies from trial to trial. Let these probabilities be $p_1, \ldots, p_n$, and write $q_i = 1 - p_i$. Thus the probability of the event occurring r times will be equal to the sum of $\binom{n}{r}$ terms each of which is the product of r different p's chosen out of the above set, multiplied by the product of all the q's corresponding to p's which were not chosen. This distribution, sometimes known as Poisson's binomial, is awkward to handle directly and we therefore use another approach to find the mean and variance of the random variable defined to be the number of successes.

Define indicator variables X_i such that X_i is equal to 1 when the event occurs at the ith trial and is zero otherwise. As before these variables are independent and the total number of times the event occurs is

$$X = \sum_1^n X_i.$$

Then arguing as for (1.20) and (1.21) we have

$$E(X) = \sum_i E(X_i) = \sum_i p_i, \tag{1.28}$$

$$E(X^2) = E\left(\sum_i X_i\right)^2$$

$$= E\left(\sum_i X_i^2\right) + E\left(\sum_{i \neq j} X_i X_j\right)$$

$$= \sum_i p_i + \sum_{i \neq j} p_i p_j$$

$$= \left(\sum_i p_i\right)^2 + \sum_i p_i - \sum_i p_i^2. \tag{1.29}$$

The variance is then

$$\text{var}(X) = E(X^2) - \{E(X)\}^2$$

$$= \sum_i (p_i - p_i^2), \tag{1.30}$$

which might have been seen directly since the variance is additive.

These results may be compared with the case of n independent trials each with probability equal to the mean of the p_i, i.e. with $p = n^{-1}\sum p_i$. In the two cases the expected number of successes is the same. It is plausible that in the second case the fact that the p_i are unequal will make the variance of X larger than in the Bernoulli case, but the reverse is in fact true.

To show this we have

$$\sum_i p_i(1-p_i) = \sum p_i - \sum p_i^2$$

$$= np - np^2 - (\sum p_i^2 - np^2)$$

$$= np(1-p) - \sum(p_i - p)^2. \tag{1.31}$$

The sum on the right-hand side is greater than zero if the p_i are not all equal and thus the variance in the Poisson case is less than the variance in the Bernoulli case.

Another variation of Bernoulli's binomial distribution is known as the Lexian distribution after W. Lexis. Here again we have n repeated independent trials with the same probability p but p is here chosen before to be one of a set of values $p_1, \ldots, p_k$, each of these being chosen with probability k^{-1}. The sample space for the whole experiment is then the product space of the set of k alternatives $(p_1, \ldots p_k)$, and the set of all sequences of successes or failures in n trials.

To find the moments we use the obvious fact that the expectation of a random variable is the expectation of its conditional expectation. Writing E_1 for any expectation conditional on the value of p, and E_2 for the expectation over all values of p, we have

$$E(X) = E_2 E_1(X)$$

$$= E_2 np$$

$$= nk^{-1}\sum p_i. \tag{1.32}$$

$$E(X^2) = n^2 E_2(p^2) + nE_2\{p(1-p)\}$$

$$= n(n-1)k^{-1}\sum p_i^2 + nk^{-1}\sum p_i.$$

Then if $\hat{p} = k^{-1}\sum p_i$,

$$\text{var}(X) = E(X^2) - E(X)^2$$

$$= n\hat{p}(1-\hat{p}) + n(n-1)k^{-1}\sum(p_i - \hat{p})^2. \tag{1.33}$$

In this case the variance is clearly increased. This is an example of a situation which occurs frequently in which the 'parameter' of a probability distribution is itself supposed to follow a probability distribution. The resulting overall distribution is then said to be a 'compound distribution'.

1.11. The hypergeometric distribution

Many elementary problems in probability can be represented as problems in the choice of balls of different kinds out of an urn or box, and hence are known as 'urn problems'. Thus if we have an urn containing m white balls and $M-m$ black balls and choose a ball at random the probability that it is white is mM^{-1}. If we replace the ball and choose again, the probabilities of black or white at the second trial are independent of the first. Thus if we choose a ball n times, replacing it after each choice, the distribution of the total number of white balls chosen will be a binomial distribution with probability mM^{-1} and index n.

The hypergeometric distribution arises when the ball chosen is not replaced after each choice, so that successive choices do not have independent probabilities.

To obtain the probability of choosing exactly k white balls in n choices $(0 \leqslant k \leqslant m)$ we suppose that the balls are all distinguishable in respect of some characteristic. The number of ways of choosing n balls in sequence, taking account of the order in which they are chosen, is

$$M(M-1)...(M-n+1).$$

The number of ways of choosing k white balls and $n-k$ black balls, again taking account of their individual order within the two sequences of white and black, is

$$m(m-1)...(m-k+1)(M-m)(M-m-1)...(M-m-n+k+1).$$

In the n successive choices the k white balls chosen could have occurred in

$$\binom{n}{k}$$

different orders in the sequence of n. Thus the overall probability of choosing k white balls can be written

$$\binom{M}{n}^{-1}\binom{m}{k}\binom{M-m}{n-k}. \tag{1.34}$$

This is the hypergeometric distribution.

We could also have obtained this result more directly by noticing that

$$\binom{M}{n}$$

is the number of different ways of choosing n balls out of M whilst

$$\binom{m}{k} \quad \text{and} \quad \binom{M-m}{n-k}$$

are the numbers of different ways of choosing k white balls out of m, and $n-k$ black balls out of $M-m$.

If we define a random variable X equal to the number of white balls chosen, p_k, the probability that $X = k$ is given by (1.34). The first two moments of X can be found by direct summation using (1.34), but it is easier to proceed as follows using indicator variables as before.

We write $X = X_1 + \ldots + X_n$ where $X_i = 0, 1$ according as the ith ball chosen is black or white. Then the probability that $X_i = 1$, considered by itself, is mM^{-1}, and the probability that $X_i = 1$, $X_j = 1$ $(i \neq j)$ is

$$\frac{m(m-1)}{M(M-1)}.$$

Hence

$$E(X) = nE(X_i)$$
$$= nmM^{-1}. \tag{1.35}$$

$$E(X^2) = E(\Sigma X_i)^2$$
$$= nE(X_1) + n(n-1)E(X_1 X_2)$$
$$= nmM^{-1} + n(n-1)m(m-1)\{M(M-1)\}^{-1}. \tag{1.36}$$

From this it follows that

$$\text{Var } X = E(X^2) - E(X)^2$$
$$= \frac{n(M-n)m(M-m)}{M^2(M-1)}. \tag{1.37}$$

Clearly if M gets very large and

$$mM^{-1} \to p \neq 0, \quad (M-m)M^{-1} \to 1-p = q \neq 0,$$

we would expect the hypergeometric distribution to be well approximated by a binomial distribution with index n and probability p. We can verify this by approximating to the factorials in (1.34) which involve M and m by Stirling's formula (see 1.12) and letting M and m tend to infinity. It is obvious that the moments given by (1.35), (1.36), and (1.37) also tend to the moments of the binomial distribution, (1.37) being smaller than the variance for a binomial with index n and probability mM^{-1}. This is what we should accept since if the proportion of the whole population taken increases to unity, the variance should tend to zero.

We consider this distribution in more detail in § 2.12.

1.12. Stirling's approximation for n!

We have stated above that Stirling's formula can be used to show the hypergeometric distribution is approximated by the binomial and since it is constantly used in combinatorial problems it is convenient to discuss it in some detail here. Stirling's formula asserts that n! is asymptotically equal to

$$(2\pi)^{\frac{1}{2}}n^{n+\frac{1}{2}}e^{-n}, \tag{1.38}$$

in the sense that the ratio of these two expressions tends to unity. Although it is useful to know that two expressions are asymptotically equal it is often helpful to be able to go further and put exact bounds on their ratio. We shall therefore prove a stronger result, the proof of which is due to Darmois (1928) and Robbins (1955). For a still stronger result see Feller (1957b).

THEOREM 1.3.

$$n! = (2\pi)^{\frac{1}{2}}n^{n+\frac{1}{2}}e^{-n+r_n}, \tag{1.39}$$

where

$$(12n+1)^{-1} < r_n < (12n)^{-1}. \tag{1.40}$$

If we put $r_n = (12n)^{-1}$, (1.39) is remarkably accurate, as is shown by the following table in which

$$R_1 = n!\{(2\pi)^{\frac{1}{2}}n^{n+\frac{1}{2}}e^{-n}\}^{-1},$$

and

$$R_2 = n!\{(2\pi)^{\frac{1}{2}}n^{n+\frac{1}{2}}e^{-n+(12n)^{-1}}\}^{-1}.$$

TABLE 1.1

n	R_1	R_2
1	1·084438	0·997731
2	1·042207	0·999674
3	1·028065	0·999900
4	1·021008	0·999957
5	1·016784	0·999998

To prove the above result we write

$$S_n = \log(n!) = \sum_{p=1}^{n-1} \log(p+1).\tag{1.41}$$

Now consider the integral

$$A_p = \int_p^{p+1} \log x\, dx$$

$$= (p+1)\log(p+1) - p\log p - 1.$$

Log x is an increasing function which is convex above and so we write

$$A_p = \log(p+1) - B_p + C_p,\tag{1.42}$$

where

$$B_p = \tfrac{1}{2}\{\log(p+1) - \log p\},$$

$$C_p = \int_p^{p+1} \log x\, dx - \tfrac{1}{2}\{\log(p+1) + \log p\}.\tag{1.43}$$

Here C_p will be small and positive since it is the area between the graph of the function $\log x$ and the line joining the points on this curve at which $x = p$ and $x = p+1$. Then from (1.41)

$$S_n = \sum_{p=1}^{n-1}(A_p + B_p - C_p)$$

$$= \int_1^n \log x\, dx + \tfrac{1}{2}\log n - \sum_{p=1}^{n-1} C_p$$

$$= (n+\tfrac{1}{2})\log n - n + 1 - \sum_{p=1}^{n-1} C_p.\tag{1.44}$$

From (1.43) we have

$$C_p = (p+\tfrac{1}{2})\log\!\left(\frac{p+1}{p}\right) - 1.$$

Now

$$\log\!\left(\frac{1+x}{1-x}\right) = 2(x + \tfrac{1}{3}x^3 + \tfrac{1}{5}x^5 + \ldots)$$

when $|x| < 1$, and using this with $x = (2p+1)^{-1}$ we get

$$C_p = \sum_{n=1}^{\infty}\{(2n+1)(2p+1)^{2n}\}^{-1}.$$

Then

$$C_p < \frac{1}{3(2p+1)^2}\left\{1 + \frac{1}{(2p+1)^2} + \frac{1}{(2p+1)^4} + \ldots\right\}$$

$$< \frac{1}{12p(p+1)} = \frac{1}{12}\left(\frac{1}{p} - \frac{1}{p+1}\right),$$

and (since $p \geqslant 1$)

$$C_p > \frac{1}{3(2p+1)^2}\left\{1 + \frac{1}{3(2p+1)^2} + \frac{1}{9(2p+1)^4} + \ldots\right\}$$

$$> \frac{1}{12}\left\{\left(p + \frac{1}{12}\right)^{-1} - \left(p + 1 + \frac{1}{12}\right)^{-1}\right\}.$$

If we put

$$D = \sum_1^\infty C_p, \quad r_n = \sum_n^\infty C_p$$

we have

$$\frac{1}{13} < D < \frac{1}{12}, \quad \frac{1}{12n+1} < r_n < \frac{1}{12n},$$

and

$$S_n = (n + \tfrac{1}{2})\log n - n + 1 - D + r_n$$

so that if $K = e^{1-D}$, it follows that

$$n! = K n^{n+\frac{1}{2}} e^{-n+r_n}. \tag{1.45}$$

This is the required result except that we have still to show that

$$K = \sqrt{(2\pi)}.$$

To prove this we first prove a theorem of Wallis. Integrating by parts we have

$$I_{2n} = \int_0^{\frac{1}{2}\pi} \sin^{2n}\theta \, d\theta = \frac{1.3.5. \ldots .(2n-1)}{2.4.6. \ldots .2n} \frac{\pi}{2},$$

and

$$I_{2n+1} = \int_0^{\frac{1}{2}\pi} \sin^{2n+1}\theta \, d\theta = \frac{2.4.6. \ldots .2n}{3.5. \ldots .(2n+1)}.$$

Since

$$I_{2n+1} < I_{2n} < I_{2n-1}$$

we have

$$\frac{(2.4.6. \ldots .2n)^2}{(3.5. \ldots .2n-1)^2(2n+1)} < \frac{1}{2}\pi < \frac{(2.4.6. \ldots .2n-2)^2(2n)}{(3.5. \ldots .2n-1)^2},$$

and the ratio of the extreme terms tends to unity so that

$$\pi = \lim_{n\to\infty} \frac{2^{4n}(n!)^4}{n(2n!)^2}.$$

This is known as Wallis's formula; and inserting (1.45) it follows that $K = \sqrt{(2\pi)}$, so that (1.39) is proved.

By considering the asymptotic behaviour of their ratios to (1.39), it is easy to see that the expressions

$$(2\pi)^{\frac{1}{2}}(n + \tfrac{1}{2})^{n+\frac{1}{2}} e^{-n-\frac{1}{2}} \tag{1.46}$$

and

$$(2\pi)^{\frac{1}{2}}(n + 1)^{n+\frac{1}{2}} e^{-n-1} \tag{1.47}$$

are also asymptotically equal to $n!$.

The factorial function, $n!$, is the value for $x = n+1$, as positive integer, of the gamma function,

$$\Gamma(x) = \int_0^\infty e^{-z} z^{x-1} \, dx,$$

which satisfies the equation $\Gamma(x+1) = x\Gamma(x)$, so that $n! = \Gamma(n+1)$. By a number of methods, differing from that used above in the proof of Theorem 1.3, it is possible to show that for x large and real, $\Gamma(x)$ is asymptotically equal to

$$(2\pi)^{\frac{1}{2}} e^{-x} x^{x-\frac{1}{2}} \tag{1.46}$$

which agrees with (1.47) for $x = n+1$, and formulae similar to (1.38) and (1.48) are then also easy to establish.

The ratio of $\Gamma(x)$ to (1.48) is usually expressed as an asymptotic series but a convergent series due to Binet is given in Whittaker and Watson (1935), p. 253.

The following useful theorem now easily follows.

THEOREM 1.4. *If $a > 0$, $b > 0$, $c > 0$,*

$$\frac{a(a+c)\ldots(a+nc)}{b(b+c)\ldots(b+nc)}$$

is asymptotically equal to

$$\frac{\Gamma(bc^{-1})}{\Gamma(ac^{-1})} n^{(a-b)c^{-1}}, \tag{1.49}$$

as n increases.

In particular if n and m are integers such that $n-m > 0$, $n > 0$, the ratio

$$(n+m)!/n!$$

is asymptotically equal to

$$n^m$$

when m is kept fixed and n increases. For further work on asymptotic formulae of this kind see Wise (1954a, b).

We shall always use the convention that $0! = 1$, and never use $n!$ for n negative. Similarly it is convenient to define the symbol

$$\binom{n}{r}$$

to mean

$$\frac{n(n-1)\ldots(n-r+1)}{r!}$$

where r is any positive integer and n is any real number. In addition we take

$$\binom{n}{0} \quad \text{and} \quad \binom{n}{n}$$

to be equal to unity. When r is not an integer

$$\binom{n}{r}$$

is not defined but is taken as zero when r is a negative integer. It is easy to verify that if n is a negative integer,

$$\binom{n}{r} = (-1)^r\binom{r-n-1}{r},$$

a result which is often useful.

In many problems it is sufficient to use Stirling's formula for numerical work but occasionally the exact values of $n!$ are needed. *Chambers's Six-figure Mathematical Tables* give the first seven significant figures of $n!$, and the common logarithm of $n!$ to six decimal places, for $n = 1(1)\ 1000$. Salzer (1951) gives $n!$ for $n = 1(1)\ 1000$ to sixteen significant figures. Fry (1928) gives the common logarithm of $n!$ to 10 decimal places for $n = 1(1)\ 1200$. Exact values of $n!$ for $n = 1(1)\ 60$ are given in the appendix to Peters (1922), and the common logarithm of $n!$ to 33 decimals is given by Duarte (1927) for $n = 1(1)\ 3000$. The binomial coefficients $\binom{n}{r}$ are given exactly for $n = 1(1)\ 19$, and to eight significant figures for $n = 20(1)\ 100$ in Fry (1928), and for exact values for $n = 1(1)\ 60$ in the appendix to Peters (1922). Values for $n = 1(1)\ 200$ (all r), and for selected values of r and n up to 5000 are given in the *Royal Society Mathematical Tables*, Vol. 3 (1954), edited by J. C. P. Miller. For other tables see Fletcher, Miller, Rosenhead, and Comrie (1962), and Greenwood and Hartley (1962).

1.13. The multinomial distribution

This is a generalization of the binomial distribution that is obtained when we consider n independent trials in each of which there are $k > 2$ possible outcomes. If these are denoted by $E_1,..., E_k$, and have probabilities $p_1,..., p_k$, where $p_1 + ... + p_k = 1$, the probability that they occur $r_1,..., r_k$ times $(r_1 + ... + r_k = n)$ is clearly

$$\frac{n!}{r_1!...r_k!}p_1^{r_1}...p_k^{r_k}, \tag{1.50}$$

by the same kind of argument as before. This is the relevant term in the expansion of

$$(p_1 + ... + p_k)^n.$$

Using a different notation from the binomial case and writing $X_1,..., X_r$ for the numbers of times each type of event occurs we clearly have

$$E(X_i) = np_i, \tag{1.51}$$

$$\text{Var}(X_i) = np_i(1 - p_i), \tag{1.52}$$

as before by regarding E_i as one type of event and

$$\bar{E}_i = E_1 + \ldots + E_{i-1} + E_{i+1} + \ldots + E_k$$

as the other. The main other quantity of interest is the cross-moment $E(X_i X_j)$. This can be found directly by using the indicator argument as before or by writing

$$E(X_i X_j) = \tfrac{1}{2}E(X_i + X_j)^2 - \tfrac{1}{2}E(X_i^2) - \tfrac{1}{2}E(X_j^2)$$
$$= n(n-1)p_i p_j. \tag{1.53}$$

In general the largest term in a multinomial distribution is not that for which the r_i satisfy $|r_i - np_i| < 1$.

Just as the hypergeometric distribution is related to the binomial, so there exists a multinomial hypergeometric distribution analogous to the multinomial distribution. Suppose we have N objects which are of k different types whose numbers are $N_1, \ldots, N_k$, where $N_i = Np_i$. If we choose M objects at random out of these, replacing each object after each choice, the distribution of the numbers of the various types is clearly given by the multinomial distribution (1.50) with $p_i = N_i N^{-1}$. If the objects are not replaced after each choice the probability of obtaining m_1 of type 1, m_2 of type 2, and so on ($\Sigma m_i = M$), is clearly

$$\binom{N_1}{m_1} \cdots \binom{N_k}{m_k} \binom{N}{M}^{-1}, \tag{1.54}$$

which is the multivariate hypergeometric distribution.

1.14. Chain binomials, an example of chain dependence

Another interesting generalization of the binomial distribution was devised by Greenwood (1931) to study the progress of infectious diseases in small groups such as families. (For an extensive discussion see Bailey 1957 and Sugiyama 1961.) This was a simplification of earlier work by L. J. Reed and W. H. Frost.

Consider a family of n individuals one of whom becomes infected from the outside by an infectious disease. The disease is supposed to remain latent for a unit length of time after which the infected individual is capable of infecting other members of the family. The $n-1$ other members of the family are then each supposed to have independently the probability p of becoming infected so that the chance of r new infections is

$$\binom{n-1}{r} p^r q^{n-1-r}, \tag{1.55}$$

where $q = 1 - p$. At the end of a further unit period of time the individuals who became infected at the end of the previous period will have passed through their latent period of non-infectibility, and if there is at least one such individual a further crop of newly infected individuals will arise. Thus if we start with one infected individual in a family of n followed by $s_1, s_2,..., s_k$ infections at time intervals 1, 2,..., k, and the process then stops, the probability of this occurring is (putting $s_0 = 1$)

$$\prod_{t=1}^{k} \binom{n - s_0 - s_1 - ... - s_{t-1}}{s_t} p^{s_t} q^{n - s_0 - ... - s_t}, \tag{1.56}$$

where $s_0 + s_1 + ... + s_k \leqslant n$, all $s_i > 0$. Thus for a family of four the probability of sequences such as (1), (11), (111), (12),... are q^3, $3pq^4$, $6p^2q^4$, $3p^2q^2$,.... In the Reed–Frost model the assumptions are modified. If at time t there are r individuals which are capable of being infected, and s individuals whose latent period has just come to an end, it is supposed that each susceptible individual has an independent probability p of having a contact with each of the s infecting individuals which transfers the infection. In order to become infected one such contact is adequate and the probabilities of any susceptible individual becoming infected or not is therefore $1 - q^s$, and q^s, where $q = 1 - p$. The chance of t new infections is therefore

$$\binom{r}{t}(1 - q^s)^t q^{s(r-t)}. \tag{1.57}$$

This model leads to the same results as the Greenwood model for $n = 3$ and a single initial infective. For $n = 4$, however, some of the probabilities become different. Thus for $n = 4$, and a sequence of type (121) we get

$$\binom{3}{2} p^2 q \binom{1}{1} p = 3p^3 q$$

for the Greenwood model, and

$$\binom{3}{2} p^2 q \binom{1}{1} (1 - q^2) = 3p^3 q(1 + q)$$

for the Reed–Frost model. These models have been extensively used in the study of infections in households and are instructive as an example of a sequence or chain of events whose probabilities are dependent on what has already happened. For a further generalization see Irwin (1954).

1.15. Probabilities of combined events

Suppose that in some field there are defined k events $A_1,..., A_k$ about which we make no further assumptions. An important class of combinatorial formulae enable us to write down expressions for the probabilities that exactly m $(0 \leqslant m \leqslant k)$ of these events occur. These formulae have been

3

known since at least the early eighteenth century and have a very wide application in problems of discrete probability.

Consider first the probability that at least one of the events occurs. Define X_i to be an indicator variate which is equal to 1 when A_i occurs and equal to zero otherwise. Then

$$(1-X_1)(1-X_2)...(1-X_k)$$

will equal zero only when at least one of the X_i is equal to 1, and will equal 1 otherwise. Thus

$$1-(1-X_1)(1-X_2)...(1-X_k) \tag{1.58}$$

will be the indicator variable of the event that at least one of $A_1,..., A_k$ occurs. The probability that this happens, which we write as P_1, will be the expectation of (1.58), so that

$$P_1 = E\{1-(1-X_1)...(1-X_k)\}$$

$$= E\left\{\sum_i X_i - \sum_{i \neq j} X_i X_j + ... + (-1)^{k+1} X_1 ... X_k\right\}.$$

Write $p_{ij...l}$ for the probability that events $A_i A_j ... A_k$ happen, and S_m for the sum of all $p_{ij...l}$ which involve m different events ($m = 1,...,k$). Then

$$p_{ij...l} = E(X_i X_j ... X_l),$$

so that

$$P_1 = \sum p_i - \sum p_{ij} + \sum p_{ijl} - ... \tag{1.59}$$

$$= S_1 - S_2 + S_3 - ... + (-1)^{k+1} S_k, \tag{1.60}$$

where the sums are always taken over unequal suffices. This result is easily checked by applying it to various simple cases. Thus if the A_i are all independent events we have

$$p_{ij...l} = p_i p_j ... p_l$$

so that

$$P_1 = 1-(1-p_1)(1-p_2)...(1-p_k),$$

which is obviously correct. On the other hand, if the A_i are exclusive events $p_{ij} = p_{ijl} = ... = 0$ and the probability that at least one event occurs is equal to the probability that exactly one event occurs, which is $\sum p_i$.

(1.60) can be generalized in various ways. Using the same assumptions write P_m for the probability that at least m of the events occur and $P_{[m]}$ for the probability that exactly m of the events occur. We first prove that

$$P_m = S_m - \binom{m}{1} S_{m+1} + \binom{m+1}{2} S_{m+2} + ... + (-1)^{k-m} \binom{k-1}{k-m} S_k \tag{1.61}$$

by verifying that the right-hand side is equal to the left. Define a random variable Y dependent on the outcome of the whole set of events and such that

$$Y = Y_1 + ... + Y_n \quad (n = 2^k - 1),$$

where the Y_i are defined in the following way. Each Y_i corresponds to one of the $2^k - 1$ ways in which a selection of at least one of the events $A_1, \ldots, A_k$ can be made. For each such selection Y_i is equal to zero if the number of events selected is less than m, and equal to

$$(-1)^s \binom{m+s-1}{s}$$

if the number selected is $m+s$ ($s = 0, \ldots, k-m$) and all of them happen, and equal to zero if not all of these $m+s$ events happen. Suppose that r events happen. If $r < m$ all the Y_i are zero. If $r = m+t$ ($t = 0, 1, \ldots, k-m$) the sum of all the non-zero Y_i is

$$\sum_{s=0} (-1)^s \binom{m+s-1}{s} \binom{m+t}{m+s}.$$

This is equal to the coefficient of x^m in

$$(1+x)^{m+t}(1+x^{-1})^{-m} = (1+x)^t x^m$$

which is unity. Hence Y is equal to unity if m or more events occur and equal to zero otherwise, and the expectation of Y must equal P_m. On the other hand, $\sum EY_i$ is clearly equal to the right-hand side of (1.61).

The derivation of a formula for $P_{[m]}$ is easier. Denote a selection of m out of the k events by $(i_1, \ldots, i_m)$ and the remaining suffices by $(j_1, \ldots, j_{k-m})$. Then the expression

$$\sum (X_{i_1} \ldots X_{i_m})(1 - X_{j_1}) \ldots (1 - X_{j_{k-m}}), \tag{1.62}$$

where the sum is taken over all such selections, will equal unity if exactly m events occur, and zero otherwise, so that its expectation is $P_{[m]}$. Multiplying this out we get a sum of terms $S_m, S_{m+1}, \ldots, S_k$ each multiplied by a constant. For S_{m+s} the constant is $(-1)^s$ multiplied by the number of ways each term in S_{m+s} occurs in (1.62) which is clearly $\binom{m+s}{s}$ so that finally we obtain

$$P_{[m]} = S_m - \binom{m+1}{1} S_{m+1} + \binom{m+2}{2} S_{m+2} - \ldots + (-1)^{k-m} \binom{k}{k-m} S_k.$$

$$\tag{1.63}$$

From this (1.61) could be derived by observing that

$$P_m = P_{[m]} + P_{[m+1]} + \ldots + P_{[k]}.$$

Another way of proving such formulae is to consider the elementary events E_i if these can be constructed from the field. When the number of A_i in the field is finite it is always easy to construct the elementary events E_i from them (see remarks in §1.2). $P_{[m]}$ will be equal to the sum of the probabilities of all those E_i which occur in the representation of exactly m of the A_i and no more. Each S_{m+j} consists of a sum of terms which are themselves sum of selections of probabilities associated with elementary events E_i. An event E_i which belongs to less than m of the A's will not occur on the right hand. If it belongs to exactly m of the A_i it will occur in just one of the terms which make up S_m, and in none of the S_{m+j} for $j > 0$. Suppose it occurs in $m+j$ (> 0) of the events A_i. Then it will occur in S_m, S_{m+1},..., S_{m+j} and the sum of coefficients by which it will be multiplied will be

$$\binom{m+j}{m} - \binom{m+1}{1}\binom{m+j}{m+1} + \ldots + (-1)^j\binom{m+j}{j}\binom{m+j}{m+j}$$

$$= \binom{m+j}{m}\left\{1 - \binom{j}{1} + \binom{j}{2} - \ldots + (-1)^j\binom{j}{j}\right\}$$

$$= 0.$$

Thus the right- and left-hand sides of (1.63) are verified to be equal.

It is possible (Fréchet 1940, 1943; Broderick 1937) to extend such formulae to cases where we have several sets of events. Thus supposing we have a discrete Borel probability field and two sets of events $(A_1,..., A_k)$, $(B_1,..., B_l)$, it is possible to find formulae for the probability that exactly m of the A_i and exactly n of the B_i occur whether or not the sets are independent. For further work on these formulae see Chung (1941, 1942, 1943a, b, c, 1945), Geiringer (1940), and Takacs (1958).

1.16. We now consider an example of the application of (1.61) and (1.63) to a particular problem.

Suppose that all permutations of the set of numbers 1, 2,..., N are equally probable so that there are $N!$ events E_i, each with probability $(N!)^{-1}$. Define the events $A_1,..., A_N$ so that A_i is the event that the number i occurs in the ith place in the permutation of the sequence 1, 2,..., N. These events are clearly not independent, since the probability of each is clearly N^{-1} but the probability that i is in the ith place given that j ($\neq i$) is in the jth place is $(N-1)^{-1}$. Clearly the probability that a specified set of m events A_i occur is

$$(N-m)!(N!)^{-1}.$$

Then the probability that at least one of the events A_i occurs is

$$S_1 - S_2 + S_3 - \ldots + (-1)^{N-1}S_N$$

$$= \binom{N}{1}\frac{N-1!}{N!} - \binom{N}{2}\frac{N-2!}{N!} + \ldots + (-1)^{N-1}\binom{N}{N}\frac{1}{N!}$$

$$= 1 - \frac{1}{2!} + \frac{1}{3!} - \ldots + (-1)^{N-1}\frac{1}{N!} \qquad (1.64)$$

This is very nearly equal to $1 - e^{-1}$, the approximation being very good even for N small.

This problem is well known under the name of the *problème de recontres* or the 'matching distribution'. A solution was given by Montmort in 1708, and various generalizations since then have resulted in an extensive literature to which references will be found in Irwin (1955) and Barton (1958). See also David and Barton (1962), who give a detailed account of the theory, and Krishna Iyer (1954).

1.17. From (1.63) we can go further and calculate the probability that there are exactly m numbers whose place is equal to their value. This is

$$P_{[m]} = S_m - \binom{m+1}{1} S_{m+1} + \ldots + (-1)^{N-m} \binom{N}{N-m} S_N$$

$$= \frac{1}{m!} - \binom{m+1}{1} \frac{1}{m+1!} + \ldots + (-1)^{N-m} \binom{N}{N-m} \frac{1}{N!}$$

$$= \frac{1}{m!} \left\{ 1 - \frac{1}{1!} + \frac{1}{2!} - \frac{1}{3!} + \ldots + (-1)^{N-m} \frac{1}{N-m!} \right\}. \qquad (1.65)$$

This is very nearly equal to

$$e^{-1} \frac{1}{m!}$$

which is the $(m+1)$th term of a Poisson distribution which will be studied in the next chapter. Some numerical comparisons are given in Feller (1957b).

This suggests that we should study the random variable defined to be equal to the number of integers which are in their correct places. Write X_i for the indicator variable of the event A_i in the present definition. Then the number of integers in their correct places is a random variable X given by

$$X = X_1 + \ldots + X_N,$$

and

$$E(X) = NE(X_1)$$

$$= 1. \qquad (1.66)$$

Furthermore,

$$E(X^2) = E(X_1 + \ldots + X_N)^2$$

$$= NE(X_1^2) + N(N-1)E(X_1 X_2)$$

$$= 1 + N(N-1)\{N(N-1)\}^{-1} \qquad (1.67)$$

so that $\text{var}(X) = 1$.

1.18. Further results on the combination of events

We have already seen (§ 1.3) that

$$P_1 \leqslant S_1,$$

which is another way of writing (1.2) and this is a special case of a series of inequalities known as Bonferroni's.

THEOREM 1.5. *Using the notation of* (1.61) *and* (1.63), *and taking r as an even non-negative integer,*

$$P_m \leqslant S_m - \binom{m}{1}S_{m+1} + \ldots + \binom{m+r-1}{r}S_{m+r} \tag{1.68}$$

$$\geqslant S_m - \binom{m}{1}S_{m+1} + \ldots - \binom{m+r}{r+1}S_{m+r+1}, \tag{1.69}$$

and similarly

$$P_{[m]} \leqslant S_m - \binom{m+1}{1}S_{m+1} + \ldots + \binom{m+r}{r}S_{m+r} \tag{1.70}$$

$$\geqslant S_m - \binom{m+1}{1}S_{m+1} + \ldots - \binom{m+r+1}{r+1}S_{m+r+1}. \tag{1.71}$$

Thus the sequence of terms in the expressions for P_m and $P_{[m]}$ are 'enveloping' series in the sense that their partial sums are alternately greater and less than the true value. These inequalities would therefore be easy to prove if successive terms were decreasing in absolute magnitude but this is in fact not true in general.

Proof. We first express the S_i in terms of the P_i and $P_{[i]}$, and this is possible in a unique fashion since, for example, $S_k = P_k$ and S_{k-r} can be expressed uniquely in terms of P_{k-r} and $S_{k-r+1}, \ldots, S_k$. We shall show that

$$S_m = \sum_{r=m}^{k} \binom{r}{m} P_{[r]} \tag{1.72}$$

$$= \sum_{r=m}^{k} \binom{r-1}{m-1} P_r. \tag{1.73}$$

This could be done by direct substitution but a more elegant method is to use generating functions. Write

$$P(z) = \sum_{r=0}^{k} P_{[r]} z^r. \tag{1.74}$$

Then substituting for the $P_{[r]}$ by means of (1.63) we get

$$P(z) = \sum_{r=0}^{k} S_r (z-1)^r, \tag{1.75}$$

where we make the convention that $S_0 = 1$. Similarly defining

$$Q(z) = \sum_{r=1}^{k} P_r z^r, \tag{1.76}$$

we get, using (1.61),

$$Q(z) = z \sum_{r=1}^{k} S_r (z-1)^{r-1}. \tag{1.77}$$

Putting $z-1 = w$ in (1.74) we get

$$\sum_{r=0}^{k} S_r w^r = \sum_{r=0}^{k} P_{[r]}(1+w)^r,$$

and equating powers of w we get (1.72). Similarly from (1.77) we get (1.73). The use of generating functions thus enables us to write the relations (1.61), (1.63), (1.72), and (1.73) in a very simple form.

Now consider (1.70) and (1.71). These are equivalent to asserting that the sum of the terms omitted from (1.63) to obtain (1.70) has the same sign as the first of the omitted terms. Thus we want to show that for every integer $l \geqslant m$,

$$\sum_{r=l}^{k} (-1)^{r-l} \binom{r}{r-m} S_r \geqslant 0.$$

Substituting for S_r by means of (1.72) the right-hand side becomes

$$\sum_{r=l}^{k} (-1)^{r-l} \binom{r}{r-m} \sum_{t=r}^{k} \binom{t}{r} P_{[t]},$$

and the coefficient of $P_{[t]}$ in this for any fixed t in the range $l \leqslant t \leqslant k$ is

$$\sum_{r=l}^{t} (-1)^{r-l} \binom{r}{r-m} \binom{t}{r}$$

which is equal to the coefficient of z^{t-l} in

$$\binom{t}{m} (-1)^{t-l}(1-z)^{t-m-1}(1-z^{t-l-1})$$

and is therefore positive. A similar argument can be used to prove (1.68) and (1.69).

A number of other inequalities of a similar kind are given in Fréchet (1940, 1943).

1.19. Occupancy problems

Suppose that there are N different balls which are to be placed in n different boxes which we call 'cells'. The number of different ways of doing this is clearly n^N, since each ball can be placed in a box in n different ways. A large number of problems, known as problems of 'occupancy', arise when we ascribe probabilities to the various ways of doing this and calculate the resulting probabilities of various resulting events. These various ways of ascribing probabilities to different ways of distributing the balls are known by the names of physicists who applied such distributions in statistical physics.

The most obvious distribution is obtained when we suppose that each of the above n^N arrangements are equally probable. This is what happens if each ball has a chance n^{-1} of going into each box independently of the other balls. The resulting model is said to exhibit 'Maxwell–Boltzmann' statistics. If the boxes are denoted by the numbers $1, \ldots, n$, and the numbers of balls which go into these are $N_1, \ldots, N_n$ we have

$$N_1 + \ldots + N_n = N,$$

and the probability of this particular distribution is, by applying the multinomial distribution,

$$\frac{N!}{N_1! \ldots N_n!} n^{-N}. \tag{1.78}$$

The distribution may be written $(N_1, N_2, \ldots, N_n)$. The number of different such distributions may be found as follows. Represent each ball by a letter X and the n boxes by the spaces between $n + 1$ vertical 'rules' $|$. Then to each different distribution corresponds a different sequence of the form

$$| XXX | X || XX | \ldots,$$

where one rule occurs at each end. Between them there are N X's and $n - 1$ rules, and the number of different ways in which these can be arranged in order is clearly

$$\frac{N + n - 1!}{N! n - 1!} = \binom{N + n - 1}{n - 1}. \tag{1.79}$$

Since to each different symbol of this form corresponds a different distribution (1.79) also gives the number of different distributions. If the balls are indistinguishable, some distributions of the balls will appear the same and for some purposes it is necessary to assume that all different distributions have the same probability, which must therefore equal

$$\binom{N + n - 1}{n - 1}^{-1}. \tag{1.80}$$

In this case we say that the balls follow 'Bose–Einstein' statistics and this is the kind of thing which occurs in quantum theory when different particles are regarded as indistinguishable in principle.

Another case is obtained if not more than one ball is allowed in each box. Then we must have $N \leqslant n$, and the number of different distributions is equal to the number of ways of choosing N out of n which is

$$\binom{n}{N}. \tag{1.81}$$

These are to have the same probability which must therefore equal

$$\binom{n}{N}^{-1}, \tag{1.82}$$

and the balls are said to follow 'Fermi–Dirac' statistics.

There is clearly a fourth situation in which the balls are distinguishable but not more than one can lie in each box. The number of ways of putting the balls into the boxes is then

$$\frac{n!}{n-N!}$$

and each distribution in the Fermi–Dirac case corresponds to $N!$ of these so that the distributions have the same probability as before. Such a situation does not seem to occur in physics.

In statistical mechanics the balls correspond to particles and the boxes to energy states so that a particle in state i has energy e_i. Assuming suitable values for the e_i it is of interest to calculate the probability distribution of the total energy

$$E = \sum N_i e_i$$

and also to consider the distribution of the N_i conditional on a fixed value of E. These distributions and the ones above are also of interest as the stationary distributions in certain random processes (see exercises 3.4 and 3.5).

1.20. Occupancy problems with Maxwell–Boltzmann statistics

We have already seen that the number of different distributions in this case is given by (1.79). From this it follows that the number of different distributions in which there is at least one ball in each box (which necessitates $N \geqslant n$) is

$$\binom{N-1}{n-1}, \tag{1.83}$$

since we may obtain each distribution by first putting one ball in each box and then distributing the remaining $N - n$ balls in any way. This is the number of different distributions ignoring any distinctions between the individual balls. However, the probability that no box is empty is a special case of the probability that exactly m boxes are empty and to obtain this we use the methods of § 1.15, since we now have to distinguish the individual balls.

We define A_i to be the event that the ith box is empty. Then the probability of A_i is clearly

$$\left(1 - \frac{1}{n}\right)^N$$

and the probability of m different events $A_i, \ldots, A_j$ is

$$\left(1 - \frac{m}{n}\right)^N$$

so that

$$S_m = \binom{n}{m}\left(1 - \frac{m}{n}\right)^N.$$

Hence the probability of exactly m boxes being empty, and the others occupied is, using (1.63),

$$P_{[m]} = S_m - \binom{m+1}{1}S_{m+1} + \dots + (-1)^{n-m}\binom{n}{n-m}S_n$$

$$= \binom{n}{m}\sum_{r=m}^{n}\binom{n-m}{r-m}(-1)^{r-m}\left(1 - \frac{r}{n}\right)^N$$

$$= \binom{n}{m}n^{-N}\sum_{s=0}^{n-m}\binom{n-m}{s}(-1)^{n-s-m}s^N. \tag{1.84}$$

Notice that this is true for $N < n$. (1.84) is conveniently written in finite difference notation. If $f(n)$ is a function of n we write

$$\Delta f(n) = f(n+1) - f(n),$$

$$\Delta^2 f(n) = \Delta\{\Delta f(n)\}$$

$$= f(n+2) - 2f(n+1) + f(n),$$

and so on, the general formula being

$$\Delta^r f(n) = \sum_{k=0}^{r}\binom{r}{k}(-1)^{k+r}f(n+k). \tag{1.85}$$

When $f(n) = n^s$, where s is a positive integer we write this, for the value $n = 0$, as

$$\Delta^r 0^s, \tag{1.86}$$

which is known as a 'difference of zero'. Tables of

$$\frac{\Delta^r 0^s}{r!}$$

for $s = 1(1)\ 25, r = 1(1)\ s$, are given in Fisher and Yates (1938 and subsequent editions). In the first three editions of this book there are twelve errors which are listed by Miller (1950), but which do not occur in later editions. Gupta (1950) has given a table for $s = 1(1)\ 50, r = 1(1)\ s$. Asymptotic formulae are given by David and Barton (1962), and by Good (1961). These quantities are closely related to the Bernoulli numbers.

Using this notation, (1.84) becomes

$$P_{[m]} = \binom{n}{m}n^{-N}\Delta^{n-m}0^N. \tag{1.87}$$

We could also have proved (1.84) by observing that the m empty cells can be chosen in $\binom{n}{m}$ ways and finding the probability that $n-m$ given cells are not empty. In a similar way we can show that

$$P_m = \binom{n}{m} \sum_{r=m}^{n} (-1)^{r-m} \binom{n-m}{r-m} \frac{m}{r} \left(1-\frac{r}{n}\right)^N \tag{1.88}$$

but this expression does not seem to be simply expressible in terms of differences of zero.

Formulae (1.84) and (1.88) can also be derived by using generating functions. This is done for (1.84) by Domb (1952), who applies the result to a cosmic ray counter problem discussed by Schrödinger (1951).

Tables of the distribution $P_{[m]}$ for $N = n$ when $n = 3(1)\ 20$ are given by David (1950), and unpublished tables for $N = 1(1)\ 30$, $n = 2(1)\ 20$ have been calculated by W. L. Nicholson, who has published some of the probability levels at the tails of the distribution (Nicholson 1961) for use in statistical tests of significance. Many generalizations of the above results are given in Chapter 14 of David and Barton (1962), who also give approximations for the differences of zero. Tukey (1949) obtains the moments of the numbers of boxes with given numbers of balls and considers some practical applications, and Weiss (1958) discusses limiting distributions. Occupancy problems are also considered by Riordan (1958) and Irwin (1955).

1.21. Bose–Einstein and Fermi–Dirac statistics

In the Bose–Einstein case we have

$$\binom{N+n-1}{n-1}$$

different possibilities with equal probabilities. To find the probability that exactly m cells are empty we observe first that these m cells can be chosen in

$$\binom{n}{m}$$

ways. The probability that the remaining $n-m$ cells are not empty is then, by using (1.79) and (1.83),

$$\binom{N-1}{n-m-1}\binom{N+n-1}{n-1}^{-1},$$

so that the required result is

$$\binom{n}{m}\binom{N-1}{n-m-1}\binom{N+n-1}{n-1}^{-1}. \tag{1.89}$$

This could also have been obtained by using (1.63).

In the case of Fermi–Dirac statistics only one ball can be put in each cell so that there are always exactly $n - N$ empty cells and the above problem does not arise.

It is possible to generalize the above occupancy models so that they can all be included in one scheme (Loève 1963, p. 43).

1.22. Permutations

In §1.16 we considered a problem in which the basic set of events consisted of all the $N!$ permutations of the N numbers 1, 2,..., N. Many probability problems arise for this model, mainly when the probability ascribed to each permutation is $(N!)^{-1}$, and we have already considered one in §1.16 where we obtained the probability that none of the numbers is in its correct place.

Two distributions based on permutations have been extensively studied in connexion with the theory of 'rank correlation' in statistics (see M. G. Kendall 1948). In this subject some quantity is used as a measure of agreement between a ranking or permutation and the natural order 1, 2,..., n.

One of these is the number of inversions of order in the permutation. Consider the set 123 and the particular permutation 312. Of the three pairs two, (31), (32) are in inverse order, and one, (12) is not. It is easy to see that in the six possible permutations one has no inversions, two have one inversion, two have two inversions, and one has three. Feller (1945a) has shown how to obtain the corresponding result for general n. Suppose that $P_{n,i}$ is the probability that there are i inversions in the $n!$ permutations of the n integers 1, 2,..., n. Given any such permutations we can obtain $n+1$ permutations of $n+1$ integers by putting the integer '$n+1$' in any of the $n-1$ places between the n integers of the given permutation or on the extreme left, or extreme right. Counting from the right these positions result in 0, 1,..., $n+1$ inversions, so that $P_{n+1,i}$ is the coefficient of z^i in

$$(P_{n,0} + P_{n,1}z + ... + P_{n,j}z^j)(1 + z + ... + z^n)(n+1)^{-1},$$

where $j = \tfrac{1}{2}n(n-1)$, the maximum number of inversions possible in a sequence of n. It is therefore natural to use generating functions and if we write

$$P_n(z) = P_{n,0} + P_{n,1}z + ... + P_{n,j}z^j,$$

we obtain

$$P_{n+1}(z) = \frac{1 - z^{n+1}}{(n+1)(1-z)} P_n(z)$$

$$= \frac{1}{(n+1)!(1-z)^{n+1}} \prod_{i=1}^{n+1} (1 - z^i). \qquad (1.90)$$

Tables of the coefficients in the expansion of $P_n(z)$ are given by M. G. Kendall (1938), with a slightly different notation, for $n = 1(1)\,10$. Tables of the probability in the tails of this distribution are given by L. Kaarsemaker and A. van Wijngaarden (1953) to three decimals for $n = 1(1)$ 40. An earlier table of the coefficients was given by Bourget (1871). We shall see later that these coefficients can be approximated in terms of a continuous distribution (the 'normal' distribution).

Alternatively, suppose that in a permutation of 1, 2,..., n the number i occurs in the $p(i)$th place from the left. The quantity

$$S = \sum_{i=1}^{n} \{p(i) - i\}^2 \qquad (1.91)$$

can take the values 0, 2, 4,..., $\frac{1}{3}n(n^2 - 1)$, and in the space of all permutations, each with probability $(n!)^{-1}$, has a probability distribution which is much more awkward to obtain than the number of inversions. Its distribution is given by Kendall, Kendall, and Babington Smith (1939) for $n = 1(1)$ 8, and by David, Kendall, and Stuart (1951) for $n = 9, 10$, but cannot be given in a simple analytic form. Suitably scaled, its distribution also tends to that of the normal distribution (Kendall 1948).

Another class of permutation problems arises when the permutations of n objects which are not all different are considered. Suppose the objects are of only two different kinds which we denote by a and b. Let there be n_1 a's and n_2 b's ($n = n_1 + n_2$). Then there are

$$\frac{n!}{n_1! n_2!} \qquad (1.92)$$

possible permutations, all equally likely.

Each permutation will consist of sequences of a's and sequences of b's alternately. Each of these is called a 'run'. Thus in the sequence

$$aabbbabbaa$$

there are three runs of a's and two of b's. Suppose the number of runs of a's and b's are r_1 and r_2. We first consider the joint distribution of r_1 and r_2. Clearly $r_1 = r_2 - 1$, r_2, or $r_2 + 1$. Suppose first that $r_1 = r_2$ so that the total number of runs, $2r_1$, is even.

We must first find the number of ways of dividing n_1 objects into r_1 groups, in order, such that each group has at least one object. We again use the method of generating functions. The first group can contain 1, 2,... objects and similarly for the others so that the required answer is the coefficient of x^{n_1} in

$$(x + x^2 + \ldots)^{r_1} = x^{r_1}(1 - x)^{-r_1}$$

and the required coefficient is

$$\binom{n_1 - 1}{r_1 - 1},$$

which is otherwise obvious by applying (1.83). Since the runs of a's and of b's must alternate and the whole sequence must start with either a or b, the total number of permutations with r_1 runs of a's and r_1 runs of b's is

$$2\binom{n_1-1}{r_1-1}\binom{n_2-1}{r_1-1},\qquad(1.93)$$

and the probability is obtained by dividing this by (1.92). Similarly if $r_1 = r_2 - 1$, the probability is

$$\binom{n_1-1}{r_1-1}\binom{n_2-1}{r_2-1}n_1!\,n_2!(n!)^{-1}.\qquad(1.94)$$

The mean and variance of $r_1 + r_2$ can be obtained by summation but as we shall see later are more easily obtained in another manner.

The distribution of runs has been very extensively studied and the above results can be generalized in various ways. Thus instead of considering the joint distribution of the total number of runs of a's and b's it is not hard to obtain the joint distribution of the runs of various specified lengths, and more than two different kinds of objects may also be considered. David and Barton (1962) give a detailed account of the theory, and Mood (1940) should also be consulted.

Instead of studying the distribution of runs we may consider the distribution of successive pairs. For example, we may consider the number of successive pairs of the form aa, or ab. These may be found from the results (1.93) and (1.94), for if s_1, s_2, and t are the numbers of pairs aa, bb, and (ab or ba) we have

$$s_1 = n_1 - r_1,$$

$$s_2 = n_2 - r_2,$$

$$t = r_1 + r_2 - 1,$$

which checks with the fact that $s_1 + s_2 + t$ must equal $n_1 + n_2 - 1$.

The moments of the distribution of s_1 (or s_2) and of t could be found from (1.93) and (1.94) but it is easier to use again the method of indicator function. To illustrate this we find the moments of s_1. We have

$$s_1 = X_1 + \ldots + X_{n-1},$$

where $n = n_1 + n_2$, and $X_i = 1$ if the ith and $(i+1)$th elements are a's, and zero otherwise. Then

$$E(s_1) = (n-1)E(X_1)$$

$$= (n-1)\frac{n_1(n_1-1)}{n(n-1)} = n^{-1}n_1(n_1-1),\qquad(1.95)$$

and

$$E(s_1^2) = E(X_1 + \ldots + X_{n-1})^2$$

$$= (n-1)E(X_1^2) + 2(n-2)E(X_1 X_2) + (n^2 - 5n + 6)E(X_1 X_3)$$

$$= \frac{n_1(n_1-1)}{n} + \frac{2n_1(n_1-1)(n_1-2)}{n(n-1)} + \frac{n_1(n_1-1)(n_1-2)(n_1-3)}{n(n-1)}. \quad (1.96)$$

By a similar argument we can find the moments of t. In this problem the universe to which the probabilities refer consists of all the $n!$ possible permutations of the n given objects, regarded as different. Another type of problem closely related to this involves finding the distribution of the number of aa, bb, and (ab or ba) joins in a sequence of n letters each of which is independently a or b with probabilities p and q.

The distribution of the number of ab (and ba) joins has been obtained in this case by Wishart and Hirschfeld (1936) (see also Wishart 1954). Here we obtain the first and second moments to illustrate again the method of indicator functions which is of very wide application in many generalizations of this problem. Thus defining

$$t = X_1 + \ldots + X_{n-1},$$

where $X_i = 1$, or 0, according as the letters in the ith and $(i+1)$th place are ab (or ba), or not. Then

$$E(t) = (n-1)E(X_1)$$

$$= 2(n-1)pq, \quad (1.97)$$

and

$$E(t^2) = (n-1)E(X_1^2) + 2(n-2)E(X_1 X_2) + (n-2)(n-3)E(X_1 X_3)$$

$$= 2(n-1)pq + 2(n-2)pq + 4(n-2)(n-3)p^2q^2. \quad (1.98)$$

The moments of the distribution of the number of aa joins could be found just as easily. These results could have been obtained by multiplying the results for a fixed number of a's and b's by binomial probabilities and adding. This type of problem has been generalized in a number of different directions.

1.23. Distributions in contingency tables

Another combinatorial problem with wide applications arises when we consider the classification of a number of objects by two or more systems of classification. Suppose that N objects each belong to one of m classes $A_1, \ldots, A_m$ and also to one of n classes $B_1, \ldots, B_n$. Let the number belonging to both A_i and B_j be N_{ij}. For simplicity we shall use the notation

$$N_{i.} = \sum_j N_{ij},$$

$$N_{.j} = \sum_i N_{ij},$$

$$N_{..} = \sum_{ij} N_{ij}. \qquad (1.99)$$

The numbers N_{ij} may be arranged in a rectangular array

$$N_{11}, \ldots, N_{1n}$$

$$N_{21}, \ldots, N_{2n}$$

$$\ldots\ldots\ldots\ldots$$

$$N_{m1}, \ldots, N_{mn},$$

for which the row sums are $N_{i.}$, and the column sums $N_{.j}$. This is known as a contingency table. If each of the N objects was independently chosen to belong to each of the N_{ij} possibilities $A_i B_j$ with probabilities p_{ij} the joint distribution of the N_{ij} would be a multinomial distribution of the form

$$N! \prod_{ij} (N_{ij}!)^{-1} p_{ij}^{N_{ij}}, \qquad (1.100)$$

and the moments and cross-moments of the N_{ij} could be found by using the formulae of 1.13. However, in this case it is of more interest to obtain the moments of the N_{ij} conditional on the values $N_{i.}$ $(i = 1, \ldots, m)$ and $N_{.j}$ $(j = 1, \ldots, n)$, and given that

$$p_{ij} = p_{i.} p_{.j},$$

where

$$p_{i.} = \sum_j p_{ij}, \quad p_{.j} = \sum_i p_{ij}. \qquad (1.101)$$

The distribution of the N_{ij} conditional on $N_{i.}$ and $N_{.j}$ is therefore

$$\frac{N!}{\prod_{ij} N_{ij}!} \prod_{ij} p_{ij}^{N_{ij}} \left(\frac{N!}{\prod_i N_{i.}!} p_{i.}^{N_{i.}} \frac{N!}{\prod_j N_{ij}!} p_{.j}^{N_{.j}} \right)^{-1}. \qquad (1.102)$$

In this expression the p's cancel so that the required probability of the observed N_{ij} given the row and column totals is

$$\frac{\prod_i N_{i.}! \prod_j N_{.j}!}{N! \prod_{ij} N_{ij}!}. \qquad (1.103)$$

It therefore follows that the sum of such terms, taken over all possible values of N_{ij} satisfying the row and column conditions, is unity. We may write this

$$\sum_{N_{ij}} \frac{\left(\prod_i N_{i.}!\right)\left(\prod_j N_{.j}!\right)}{N! \prod_{ij} N_{ij}!} = 1, \tag{1.104}$$

where the summation is taken over all values of N_{ij} satisfying the conditions. To find the moments of the quantities N_{ij} we first find the 'factorial moments'. Just the sth moment of N_{ij} is defined to be the average value of N_{ij}^s taken over the whole probability distribution so the sth factorial moment is the average value of

$$N_{ij}(N_{ij}-1)\ldots(N_{ij}-s+1). \tag{1.105}$$

We write this quantity as $N_{ij}^{(s)}$ and similarly for any number x we put

$$x^{(s)} = x(x-1)\ldots(x-s+1). \tag{1.106}$$

We can write (1.104) in the form

$$\sum_{N_{ij}} \prod_i \left\{ N_{i.}! \left(\prod_j N_{ij}!\right)^{-1} \right\} = N! \left(\prod_j N_{.j}!\right)^{-1}. \tag{1.107}$$

The sth factorial moment of N_{kl} can be written

$$E(N_{kl}^{(s)}) = \sum_{N_{ij}} N_{kl}^{(s)} \frac{\prod_i N_{i.}! \prod_j N_{.j}!}{N! \prod_{ij} N_{ij}!} \tag{1.108}$$

which can be put in the form

$$\frac{N_{k.}^{(s)} \sum_{M_{ij}} \prod_i \left\{ M_{i.}! \left(\prod_j M_{ij}!\right)^{-1} \right\}}{N! \left(\prod_j N_{.j}!\right)^{-1}},$$

where

$$M_{i.} = N_{i.} \quad \text{for} \quad i \neq k,$$
$$\quad\; = N_{k.}-s \quad \text{for} \quad i = k,$$
$$M_{.j} = N_{.j} \quad \text{for} \quad j \neq l,$$
$$\quad\; = N_{.l}-s \quad \text{for} \quad j = l,$$

and

$$M_{ij} = N_{ij} \quad \text{for} \quad i \neq k \text{ or } j \neq l,$$
$$\quad\; = N_{kl}-s \quad \text{for} \quad i = k \text{ and } j = l.$$

The summation in the numerator is taken over all values of M_{ij} with the prescribed $M_{i.}$ and $M_{.j}$. From (1.107) the second part of the numerator is

4

equal to

$$\frac{(N-s)!}{\prod_j M_{.j}!}.$$

Hence

$$E(N_{kl}^{(s)}) = \frac{N_{k.}^{(s)} N_{.l}^{(s)}}{N^{(s)}}, \qquad (1.109)$$

which is the required result. In particular we find

$$E(N_{ij}) = N_{i.} N_{.j} N^{-1}, \qquad (1.110)$$

and

$$E(N_{ij}^2 - N_{ij}) = \frac{N_{i.}^{(2)} N_{.j}^{(2)}}{N^{(2)}}, \qquad (1.111)$$

so that

$$E(N_{ij}^2) = \frac{N_{i.}^{(2)} N_{.j}^{(2)}}{N^{(2)}} + \frac{N_{i.} N_{.j}}{N}. \qquad (1.112)$$

Similarly for the cross-moments we find

$$E(N_{ij}^{(s)} N_{kl}^{(t)}) = \frac{N_{i.}^{(s)} N_{.j}^{(s)} N_{k.}^{(t)} N_{.l}^{(t)}}{N^{(s+t)}} \quad \text{for} \quad i \neq k, j \neq l,..., \qquad (1.113)$$

$$= \frac{N_{i.}^{(s+t)} N_{.j}^{(s)} N_{.l}^{(t)}}{N^{(s+t)}} \quad \text{for} \quad i = k, j \neq l,... \qquad (1.114)$$

and a similar result when $i \neq k, j = l$.

When $m = n = 2$ the distribution is particularly simple since if the row and column are prescribed the choice of any one number N_{ij} fixes all the rest. The probability (1.103) then becomes

$$\frac{N_{1.}! N_{2.}! N_{.1}! N_{.2}!}{N! N_{11}! N_{12}! N_{21}! N_{22}!}. \qquad (1.115)$$

Comparing this with (1.34) and writing

$$M = N,$$

$$n = N_{1.}, \quad M-n = N_{2.},$$

$$m = N_{.1}, \quad M-m = N_{.2},$$

$$k = N_{11}, \quad m-k = N_{21},$$

$$n-k = N_{12}, \quad M-m-n+k = N_{22},$$

we see that N_{11}, for example, is distributed in a hypergeometric distribution on the range

$$n - \min(M-n, n), \min(m, n). \qquad (1.116)$$

A useful discussion of contingency tables is given in Wilks (1946).

1.24. Random numbers

Random digits are not useful solely for the purpose of providing illustrations of the application of probability theory but are also extensively used in statistics for the purpose of introducing unbiased procedures into methods of sampling and experimental design. Their most important application, however, is to 'Monte Carlo' methods, which are methods of experimental sampling used to provide answers to probability problems which are too difficult to solve analytically. Furthermore, their construction and use lead to some interesting mathematical problems.

What is needed in practice is a sequence of decimal (or sometimes binary) digits such as 2519648 ..., which have been obtained in some manner which is such that we can feel considerable assurance that there is no fundamental reason for systematic divergence from the two basic assumptions:

(1) that the probability of any particular digit occurring at a particular place in the sequence, is $\frac{1}{10}$;

(2) that events at different places in the sequence are completely and jointly independent of each other.

Various tables of such sequences have been published and their construction raises some interesting questions. The first such table was that of Tippett (1927), who gave 41 600 decimal digits. These were said to have been obtained by selecting numbers from census reports. There are a number of reasons why such a method of construction should not be satisfactory but statistical tests made on this table show few signs of bias (Kendall and Babington Smith 1938).

Another method of construction which would appear at first sight to be satisfactory is to take a table of some mathematical function to a large number of digits and pick out the digit in some intermediate place, not far enough to the right to be affected by the systematic effect of small differences, and not far enough to the left to be unaffected by the change from one table entry to the next. Surprisingly enough this method cannot be justified theoretically in general. Thus consider the logarithms to base 10 of the integers 1, 2,... and let $v_g(n)$ be the number of times the integer g ($g = 0, 1,..., 9$) occurs in the kth place in the logarithms of 1, 2,..., n. Then

$$\frac{v_g(n)}{n}$$

does not tend to $\frac{1}{10}$ as n increases and in fact does not tend to a limit at all. However, if the logarithm is replaced by the square root the corresponding ratio does tend to $\frac{1}{10}$. These results were discovered by J. Franel (1919), and more accessible proofs can be found in the solutions to problems 178 and 181 in §II of Polya–Szego (1925).

A table of 15 000 decimal digits obtained in this way was given by Fisher and Yates (1938, and later editions). These were taken from the fifteenth to nineteenth places of tables of logarithms to twenty places, and were found on examination to be biased. They were therefore subsequently adjusted by methods which are described in the first edition only (1938).

Kendall and Babington Smith (1939b) gave 100 000 decimal digits which were obtained by a 'randomizing machine' (described in Kendall and Babington Smith, 1938). This consisted of a disc divided into ten equal segments numbered 0 to 9 and illuminated by a neon lamp which was flashed at random instants. In spite of the possibility of bias being introduced by the well-known psychological effect of 'number preferences' (Yule 1927) the numbers passed a battery of statistical tests satisfactorily.

The largest table of all is that published by the Rand Corporation (1955) and gives 1 000 000 decimal digits. The construction of these tables is described by Brown (1951). They were produced on a randomizing machine described as an 'electronic roulette wheel'. An electric source producing random impulses at the rate of about 100 000 per second was gated (switched on and off) once every second. The pulses allowed through, which averaged about 3000, were fed into a 5-place binary counter which therefore corresponds to a 32-place (2^5) roulette wheel. They were then converted to decimals, throwing away 12 of the digits, and the last decimal digit punched on a card. These digits were then 'compounded' by adding each digit to that on the next card modulo 10. Statistical tests on the resulting series showed it to be satisfactory (see in particular the review of these tables by Tompkins 1956).

If a series of digits departs from randomness in certain ways it is possible to process them in such a way that the result is a more satisfactory series. Suppose for simplicity that we are dealing with a sequence of binary digits (0's and 1's). Such a series can depart from randomness in two ways:

(1) if the probability of occurrence of a 0 (or 1) at each stage is not $\frac{1}{2}$;

(2) if there is serial dependence between successive members of the sequence.

If the successive members are certainly independent but the probabilities are not quite $\frac{1}{2}$ improvements can be made by compounding.

If the probability of '1' at each stage is really $\frac{1}{2} + \varepsilon$ we consider successive distinct pairs and replace 00 or 11 by 0, 01, or 10 by 1. The probability of 0 is then

$$(\tfrac{1}{2} - \varepsilon)^2 + (\tfrac{1}{2} + \varepsilon)^2 = \tfrac{1}{2} + 2\varepsilon^2$$

which is much closer to $\frac{1}{2}$ if ε is small. In doing this we have reduced the total number of binary digits by $\frac{1}{2}$. However, this is not necessary and we could have taken successive pairs such that the last digit of one is the first of the next, so that from a sequence such as

$$01101011\ldots$$

we get

$$1011110\ldots.$$

If the original series was independent the new series has a slight serial dependence when $\varepsilon \neq 0$.

We could have written the last sequence as

$$\{X_1 + X_2(\mathrm{mod}\,2)\},\ \{X_2 + X_3(\mathrm{mod}\,2)\},\ldots$$

which recalls the method of compounding used on the Rand decimal digits. It is easy to show that if

$$X_1,\ X_2,\ldots$$

is a sequence of decimal digits and successive pairs are added modulo 10 the resulting numbers have probabilities diverging from $\frac{1}{10}$ by quantities of order ε^2 if the original probabilities diverged from $\frac{1}{10}$ by quantities of order ε.

A better method is, in the case of binary digits, to take successive disjoint pairs, ignore those which are 00 and 11, and write 1 for 01, 0 for 10. Whatever the probabilities are, this results in a series having probabilities of 0 and 1 exactly equal to $\frac{1}{2}$ so long as there is no serial dependence. This results in a new sequence whose total number is less than the original by a fraction $p(1-p) \leqslant \frac{1}{4}$, where p is the true probability of 1 in the original sequence. Thus this process is also more expensive.

Similar processes of compounding will remove, in part or in whole, serial dependence of simple kinds. A number of papers on this subject have been written and a list is given in Hull and Dobell (1962).

1.25. Pseudo-random numbers

Instead of producing random numbers by an empirically random process it is often very useful to be able to produce a series of numbers by a mathematical process which is such that the rule of generation is so obscurely related to the actual values that the results behave as genuinely random numbers would usually be expected to behave when statistical tests are applied to them. Such numbers are said to be 'pseudo-random'.

An example of such a method is the 'middle of the square' process suggested originally by von Neumann. We begin with some number, n_1, of N digits. We take these as decimal digits for the purposes of illustration. This is squared and a zero added on the left if the result has $2N-1$ digits. Of the $2N$ digits thus obtained the middle N digits are picked out to form the next random number, n_2, in the sequence, and the process then repeated (it is usual to take N as even). The resulting sequence $n_1, n_2,\ldots$ is then regarded as a sequence of random digits. The two great advantages of such methods are, first, that when using high-speed computers it is not necessary to feed random numbers in from an outside source or store and, secondly, that the sampling procedure is always exactly repeatable.

The midsquare method has, however, been shown to be unsatisfactory. Better methods have been devised, which are based on congruences. For a survey of these see Hull and Dobell (1962) and Hammersley and Handscomb (1964) (both references contain an extensive bibliography).

Random permutations are also useful. Appearances notwithstanding, it is difficult to obtain such permutation by shuffling cards (Kosambi and Rao 1958, Joseph 1959). The theory of shuffling in general is considered by Feller (1957b) and by Borel and Chéron (1940) (see also Moran 1962b and Golomb 1961).

Mathematical techniques for constructing random permutations are given by Rao (1961), Beckenbach (1964), and Sandelius (1962), and tables of random permutations have been published by Moses and Oakford (1963).

1.26. Theory of gambling. Martingales

Consider a gambler playing a series of games against a banker who possesses an infinitely large capital. We suppose that the gambler starts with an initial capital X_0. The series of games played is denoted by the integers $1, 2, ...$, and we suppose that X_0 and the amounts won or lost at each stage are integral multiples of some monetary unit. The gambler is not allowed at any time to play the game in any way which might involve him in losing more than his capital.

If his capital after the first game is X_1, $X_1 - X_0$ is his winnings (which may be negative) and is a discrete random variable. If the expected value of X_1 is equal to X_0 so that

$$E(X_1 - X_0) = 0, \tag{1.117}$$

the game is said to be 'fair'. This is the classic definition of 'fairness' and is the most natural mathematically but its limitations must be understood, for the fairness of the game can be consistent with an overwhelming probability that the gambler will lose rather than win or vice versa. Thus a person who buys a lottery ticket costing £1 in a lottery with 1 000 000 tickets and a single prize of £1 000 000 is engaged in a fair game since his expected gain is £$(10^{-6} \times 10^6) - £1$, which is zero, but his chance of losing is $1 - 10^{-6}$. If there are several prizes his expected winnings is the sum of those associated with each prize (since expectations are additive). In practice this sum is less than the cost of a ticket since all commercial gambling games must be 'unfair'. However, most people are willing to pay something for the undoubted pleasure of uncertain possibilities of great gain.

Alternatively, the taking out of insurance exchanges the small possibility of a large loss for the certainty of a small loss. Since administration costs something this is again an unfair game but most people are also willing to pay something for the assurance of not going bankrupt. Thus it is quite rational for a man to take out insurance and gamble at the same time.

Returning to the gambler playing a sequence of games we may suppose that his capital after the second game is X_2, and X_n after the nth game. We say the game is 'fair' if, for all n,

$$E(X_{n+1}|X_n) = X_n. \qquad (1.118)$$

It is useful to interpret (1.118) in a slightly more general way. The gambler may make certain choices at each game, e.g. of how much to stake, of which of a number of gambling games to play, and so on. These choices may depend on the past history of the series of games from 1 to n and we shall then suppose that the expectation in (1.118) is calculated not only conditionally on the value of X_n but also on the whole previous history of the sequence.

Such a system is said to be a 'martingale', a word coined by J. Ville (1939) who, together with P. Lévy, first studied such situations in general. Historically this word had a variety of meanings among which was a system of gambling.

Since we may naturally define an 'unfair' game by the inequality

$$E(X_{n+1}|X_n) \leqslant X_n, \quad \text{all } n, \qquad (1.119)$$

we say that a sequence of games is 'unfair' if (1.119) holds for all of them. Such a system is said to be a 'semi-martingale' and the theory of such systems is due to Doob (1953). Strictly speaking, a game is in fact 'unfair' only if there is strict inequality in (1.119) but this definition is more convenient.

Gamblers have spent much time in devising gambling systems. These may be defined in general as rules by which at each stage $i+1$ the gambler may decide, on the basis of his experience in the games $1, 2,..., i$, to choose some particular type of game, to skip the game $i+1$, or to stop altogether. Clearly so long as the particular games he may choose at each stage are fair, or unfair, the same will be true of the whole system. Doob developed a theory which shows that optional skipping or stopping, based on the previous experience cannot convert a fair system into an unfair one, nor an unfavourable system into a fair or favourable one. We consider here solely the case of optional stopping.

THEOREM 1.6. *Suppose $\{X_n\}$ is an 'unfair' sequence of games (and thus a semi-martingale) defined in the above manner. If the decision to go on after the ith game is an event whose probability is dependent solely on the history of the sequence up to i, and if X'_{i+1} is defined to be equal to X_{i+1} if the sequence is not terminated by the $i+1$th game, and equal to X_m if the sequence was terminated at $m \leqslant i$, then $\{X'_i\}$ is also a semi-martingale.*

Let m be the number of games played. Let A_i represent the outcome of the ith game. Then the set $(A_1,..., A_i)$ includes the sequence $X_1,..., X_i$, if $i \leqslant m$, or $X_1,..., X_m$, if $i = m$, and in the latter case the information that the sequence is now terminated. Write a_i for a random variable dependent on the sequence $A_1,..., A_i$, and equal to 0, 1 with specified probabilities dependent on $A_1,..., A_i$. a_i is taken equal to 1 if the game is continued from the ith game to the $(i+1)$th, and zero otherwise. Then

$$m = 1 + a_1 + a_2 + ...$$

We want to show that

$$E(X'_{n+1}|A_1,..., A_n) \leqslant X'_n. \tag{1.120}$$

Suppose that $m < n$. Then there exists $i < n$ so that

$$a_i = a_{i+1} = a_{i+2} = ... = 0,$$

and X'_{n+1} is equal to X'_n since both are equal to X'_i. If, on the other hand, $m \geqslant n$ the game goes on if $a_n = 1$ and stops if $a_n = 0$. In the first case

$$E(X'_{n+1}|A_1,..., A_n) \leqslant X_n = X'_n,$$

and in the second case

$$E(X'_{n+1}|A_1,..., A_n) = X_n = X'_n.$$

Thus in all cases

$$E(X'_{n+1}|A_1,..., A_n) \leqslant X'_n.$$

Since the same argument applies for a semi-martingale with the inequality reversed, if $\{X_n\}$ is a martingale so is $\{X'_n\}$.

Similar theorems are proved by Doob (1953) for optional 'sampling' and 'skipping'. Thus no gambling system can alter the 'fairness' or 'unfairness' of a sequence of games as defined by expectation. All that can be done is to transform a sequence so that the probability of winning is increased or decreased while at the same time the amount won is decreased or increased. Such an exchange may, as we have seen, increase the pleasure of gambling.

The system of gambling which consists of doubling the stakes at each trial and stopping at the first win (the probability of winning being $\frac{1}{2}$) does not come into the scope of the above theorem since it requires that the gambler, like the banker, has an infinite capital. If his capital is finite and he continues doubling his stake until he can go no further, his overall expectation remains zero. If, on the other hand, he has an infinite capital his freedom of optional stopping enables him to force a win of any previously prescribed amount.

Any gambling system therefore does not alter the 'fairness' or 'unfairness' of a game, but only enables the gambler to alter the probability distribution of his final gain (positive or negative) in such a way that although its mean is unaltered, high probabilities of small gains may be exchanged for low probabilities of large gains.

For further discussions of games see Feller (1945*b*, 1957*b*), Robbins (1952), Breiman (1961), Kendall and Murchland (1964), and Midas (1964).

1.27. Entropy

A mathematical quantity, defined in terms of a distribution, which has some very curious properties and applications, is 'entropy'. Consider a trial which can result in one of a series of different events E_1, E_2,... which occur with probabilities p_1, p_2,.... The entropy of this probability situation is defined to be

$$E = -\sum p_i \log p_i, \qquad (1.121)$$

where $p_i \log p_i$ is taken equal to zero when $p_i = 0$.

If we had associated with each event E_i a quantity X_i, the mean value of X_i is

$$\sum p_i X_i,$$

and if X_i is chosen to be minus the logarithm of the probability of E_i when p_i is not zero, and zero otherwise, we obtain (1.121). $-\log p_i$ is necessarily non-negative and is large when p_i is small. It may therefore be regarded as a measure of the unlikelihood or unexpectedness of E_i, and E as a measure of the mean unexpectedness of whatever eventuates from the trial.

For a finite set E_1,..., E_n the entropy can also be regarded as a measure of the inequality of the probabilities p_i and in fact we can show that

$$-\sum_1^n p_i \log p_i, \qquad (1.122)$$

where $\sum p_i = 1$, is greatest when all the p_i are equal. The function $-x \log x$, for $0 < x \leqslant 1$ (and conventionally defined as zero when $x = 0$), is a non-negative function whose derivative is $-1 - \log x$, which is decreasing as x increases. $-x \log x$ is therefore a 'convex' function, i.e. the line joining the points $(x_1, -x_1 \log x_1)$ and $(x_2, -x_2 \log x_2)$ $(x_1 < x_2)$ lies entirely below the curve between these points. Any function $\phi(x)$ for which $\phi''(x) \leqslant 0$ throughout some range is convex in this sense throughout this range, as is easily shown by using Taylor's theorem.

In fact the usual definition of convexity is somewhat weaker than that given above and states that a function $\phi(x)$ is convex in some interval if for any two points of this interval

$$2\phi\left(\frac{x_1 + x_2}{2}\right) \geqslant \phi(x_1) + \phi(x_2). \qquad (1.123)$$

(Sometimes strict inequality is required.) Clearly (1.123) is true if $\phi''(x) \leqslant 0$ between x_1 and x_2. A more general result is the following.

THEOREM 1.7. *If $\phi(x)$ is continuous, and convex in the sense of* (1.123), *and if $p_i \geqslant 0$ ($i = 1,...,n$), $\sum p_i = 1$, then*

$$\phi(\sum p_i x_i) \geqslant \sum p_i \phi(x_i). \tag{1.124}$$

Proof. Since $\phi(x)$ is continuous it is sufficient to prove this when the p_i are rational fractions with the same denominator. Furthermore, we can if we wish regard some of the x_i as identical. It is therefore sufficient to prove that

$$\phi\left(\frac{1}{n} \sum x_i\right) \geqslant \frac{1}{n} \sum \phi(x_i). \tag{1.125}$$

By repeated application of (1.123) it follows that (1.125) is true whenever n is a power of 2, 2^k say. If n is not a power of 2, let k be such that $2^k > n$. Then for any number X in the range,

$$\phi[2^{-k}\{\sum x_i + (2^k - n)X\}] \geqslant 2^{-k}\{\sum \phi(x_i) + (2^k - n)\phi(X)\}.$$

If we now put $X = n^{-1}\sum x_i$, which is certainly in the range, the result follows. Theorem 1.7 has applications in other probability problems.

Since $-x \log x$ is a convex function the entropy of a finite set of events is a maximum when their probabilities are equal. This would also be true if the entropy were defined as $\sum \phi(p_i)$ for any other convex function and we have to inquire just what makes this particular function so useful. We therefore look for further conditions which can be imposed on the definition of entropy which will necessitate the use of $-x \log x$. Although the idea of entropy is still at present very obscure at least two such characterizations have been obtained. One of these (Moran 1961) arises from the use of entropy in the kinetic theory of gases and involves the properties of certain types of differential equations. The other, which we consider here, is based on the use of entropy as a measure of 'information' (Khintchine 1957).

Consider a finite set of possibilities $E_1,..., E_n$, with probabilities $p_1,..., p_n$, where $\sum p_i = 1$. We wish to construct a function, $H(p_1,...,p_n)$ which will represent a measure of the expected amount of 'information' conveyed by the knowledge that one of these events has occurred. $H(p_1,...,p_n)$ will therefore be a symmetric function of the p_i and should be a maximum when all the p_i are equal, since in that case the amount of initial uncertainty is greatest. Furthermore, we should expect such a measure to be a continuous function of the p_i, and since also it is to be defined for all n, we should have

$$H(p_1,...,p_n,0) = H(p_1,...,p_n).$$

We therefore impose the following conditions:
(1) $H(p_1,...,p_n)$, defined for $n = 1$, $2,...$, is a continuous symmetric function of the p_i which attains its maximum when all $p_i = n^{-1}$.
(2) $H(p_1,...,p_n,0) = H(p_1,...,p_n)$.

Now consider a finite set of events E_{ij} $(i = 1,...,m$ and $j = 1,...,n)$ whose probabilities are p_{ij}. We call A_i the event that one of the E_{ij} $(j = 1,...,n)$ has occurred, and B_j the event that one of the E_{ij} $(i = 1,...,m)$ has occurred.

Given two successive experiments it is natural to insist that the expected amount of information conveyed by both together should be the sum of the expected amounts of information conveyed by each. If the second experiment depends on the first, we take the information conveyed by it to be the expected value of the information conveyed by it conditional on the outcome of the first experiment.

Now if A_i occurs (an event with probability $\sum\limits_j p_{ij}$) the conditional probability of B_j occurring is

$$p(B_j \mid A_i) = p_{ij}\left(\sum_j p_{ij}\right)^{-1}$$

$$= P_{ij}, \quad \text{say.} \tag{1.126}$$

We can therefore define the information in the second experiment as

$$H_i(B) = H(P_{i1},..., P_{in})$$

when A_i has occurred, and the mean value of this is the mean information supplied by the second experiment,

$$H_A(B) = \sum_{ik} p_{ik} H(P_{i1},..., P_{in}). \tag{1.127}$$

We now impose the condition:

(3) $H(AB) = H(A) + H_A(B),$ \hfill (1.128)

or

$$H(p_{11}, p_{12},..., p_{mn}) = H\left(\sum_k p_{1k},..., \sum_k p_{mk}\right) + \sum_{ik} p_{ik} H(P_{i1},..., P_{in}).$$
$$\tag{1.129}$$

We can then state the following characterization theorem.

THEOREM 1.8. *A function satisfying the above conditions* (1), (2), *and* (3) *is a constant multiple of*

$$-\sum_1^n p_i \log p_i.$$

Proof. Write $H(n^{-1},..., n^{-1}) = L(n)$, when there are n possible events. Then by conditions (1) and (2)

$$L(n) = H(n^{-1},..., n^{-1}) = H(n^{-1},..., n^{-1}, 0)$$
$$\leqslant H\{(n+1)^{-1},..., (n+1)^{-1}\} = L(n+1).$$

Consider n^k events $A_{ij}...$ where each of the n suffices take the values $1,..., n$, and suppose that each of these has probability n^{-k}. Then using (1.128), and the fact that the set of events defined by any one of the suffices is independent of the others, we have

$$L(n^k) = kL(n).\tag{1.130}$$

This holds for any pair of integers k and n. Choose some other pair l and m so

$$L(m^l) = lL(m).\tag{1.131}$$

Choose k so that $n^k \leqslant m^l \leqslant n^{k+1}$, and thus $k\log n \leqslant l\log m \leqslant (k+1)\log n$. It follows that

$$L(n^k) \leqslant L(m^l) \leqslant L(n^{k+1}),$$

and therefore

$$kL(n) \leqslant lL(m) \leqslant (k+1)L(n),$$

so that

$$\frac{k}{l} \leqslant \frac{L(m)}{L(n)} \leqslant \frac{k+1}{l}.\tag{1.132}$$

This implies that

$$\left|\frac{L(m)}{L(n)} - \frac{\log m}{\log n}\right| \leqslant \frac{1}{l}.$$

But the left-hand side of this inequality is independent of l which can be chosen arbitrarily large so that

$$\frac{L(m)}{L(n)} = \frac{\log m}{\log n},$$

which, in turn, can be written

$$L(m) = \lambda \log m,$$

where λ is a positive constant since $L(m)$ is monotonic increasing.

We have thus found $H(p_1,\ldots,p_n)$ when all the p_i are equal to n^{-1}. Now suppose the p_i are arbitrary rational numbers satisfying $p_i \geqslant 0$, $\Sigma p_i = 1$. We may suppose the p_i all to have the same denominator d, and numerators $a_1,\ldots,a_n$ so that $\Sigma a_i = d$. Suppose that $\{A_i\}$ is a scheme of n events with probabilities p_i, and that B_{ij} $(j = 1,\ldots,a_i)$ is a scheme of a_i equally probable events which may occur conditional on A_i having occurred. Then the set of events $A_i B_{ij}$ consists of d equally probable events. Using (1.128) we have

$$\lambda \log d = H(p_1,\ldots,p_n) + \lambda \sum_i p_i \log a_i$$

so that

$$H(p_1,\ldots,p_n) = -\lambda \sum_i p_i \log(a_i d^{-1})$$

$$= -\lambda \sum_i p_i \log p_i,\tag{1.133}$$

as required. Since $H(p_1,\ldots,p_n)$ is a continuous function of its arguments, the same result holds when the p_i are irrational numbers, and the theorem is proved.

The idea of the entropy of a distribution which is either discrete or, as later in this book, continuous finds its main applications in communication theory, and in statistical physics. In statistical theory 'information' is usually 'information about a particular parameter' of a probability distribution, and is measured by the reciprocal of the square of the standard deviation of some estimator of that parameter. It therefore refers to a particular parameter and a particular method of estimation. Entropy, on the other hand, is a measure of the 'information' contained in the observed value of a quantity with a specified probability distribution, and does not refer to any parameter. The use of information in this sense in statistical inference is discussed at length by Kullback (1959). The best general introduction to 'entropy' in probability theory is Renyi (1962).

1.28. Applications of the idea of entropy

It is a remarkable fact that both in the elementary theory of probability and in the classical analytical theory of probability, as opposed to its applications, very little use of this concept has been made, and furthermore what applications have been discussed are more of a logical and combinatorial character than probabilistic. An account of some interesting problem of this type is given in Yaglom and Yaglom (1959) and we consider here some of the simplest of their examples.

The inhabitants of the south side of Canberra are well known to be incapable of giving an untruthful answer to a question, but the reverse is the case for inhabitants of the north side. A tourist lost in Canberra wishes to find out which side he is in by asking an inhabitant of the town, who may have come from either side, questions which will settle the problem, and which require answers 'Yes' or 'No' only.

We suppose that the probability that the tourist is in either side is $\frac{1}{2}$. The knowledge that he requires has an information content measured by the entropy

$$E = -\tfrac{1}{2}\log\tfrac{1}{2} - \tfrac{1}{2}\log\tfrac{1}{2}$$
$$= \log 2.$$

A question may have the answer 'Yes' or 'No', and if the probability of these are p and $q = 1 - p$ the measure of the information in the answer is

$$-p\log p - q\log q.$$

This is always less than $\log 2$ unless $p = q = \frac{1}{2}$. This does not show that the tourist can ask a question which will give him the required answer but it does show that if he can, the answers must have equal probability. In fact a suitable question does exist and is 'Do you live on this side of Canberra?' If the answer is 'Yes' the tourist must be on the south side, and if 'No' on the north side.

A more instructive example concerns the parlour game in which a person, A, thinks of an object and a panel of questioners attempt to discover the object by asking questions which require the answers 'Yes' or 'No' only. For simplicity suppose the object thought of is one of the numbers 1, 2,..., N. If the probabilities of choosing these are $p_1,..., p_N$ the information contained in the answer is

$$-\sum p_i \log p_i,$$

which has a maximum equal to $\log N$ when all the p_i are equal. Suppose to begin with that this is so, i.e. that $p_i = N^{-1}$ for all i.

We have already seen that the information contained in the answer to one question is at most $\log 2$. It is then easy to see that the information contained in n questions is at most $\log 2^n = n \log 2$. Hence in order to find what number was thought of, at least

$$\log N (\log 2)^{-1} \tag{1.134}$$

questions must be asked. When this quantity is not an integer the number of questions required is the least integer greater than (1.134).

It is easy to see that it is in fact possible to find the number thought of by asking this number of questions. For simplicity suppose $N = 32 = 2^5$, so that at least five questions are necessary. The first will be 'Is the number greater than 16?' If the answer is 'Yes', the next question is 'Is the number greater than 24?' whilst if the answer is 'No' the next question is 'Is the number greater than 8?' and the process is similarly continued, so that five questions are sufficient for an exact determination.

So far the use of probability in such problems is a little artificial. However, if $-\sum p_i \log p_i$ is not equal to its maximum and this fact is known to the questioner we should expect that in some sense fewer questions will be required. Clearly it may be necessary to ask all five questions to obtain an answer but we will show that on the average fewer are required. To simplify the problem, suppose that A thinks of one of the four numbers 1, 2, 3, 4 with probabilities $\frac{1}{2}$, $\frac{1}{4}$, $\frac{1}{8}$, $\frac{1}{8}$ respectively. Since the number of cases is 4 the information is not greater than $2\log 2$ and the questioner can certainly find the number by two questions. However, the information is in fact equal to

$$-\tfrac{1}{2}\log\tfrac{1}{2} - \tfrac{1}{4}\log\tfrac{1}{4} - \tfrac{1}{4}\log\tfrac{1}{8} = \tfrac{7}{4}\log 2.$$

We can show that by a suitable choice of questions the answer can be obtained in n questions where n is a random variable whose mean is $\frac{7}{4}$. The first question is 'Is the number 1?' If the answer is 'No' the second question is 'Is the number 2?' and if the answer to this is 'No', the third question is 'Is the number 3?' The probabilities of the number of questions necessary being 1, 2, or 3 are easily seen to be $\frac{1}{2}$, $\frac{1}{4}$, $\frac{1}{4}$, so that the mean number required is

$$\tfrac{1}{2} + 2(\tfrac{1}{4}) + 3(\tfrac{1}{4}) = \tfrac{7}{4},$$

and the mean information extracted is $\frac{7}{4}\log 2$ as expected. Notice that the questioner has paid for the advantage of reducing the mean number of questions required by allowing it to be possible that sometimes he will have to ask three questions.

This example can obviously be generalized to much more elaborate situations, and one application of this kind is to the design of filing systems (MacDonald 1952). Another is to the well-known problem of weighing a set of pennies one of which is defective (Yaglom and Yaglom 1959).

For other interesting discussions of this subject see Renyi (1962), Kullback, Kupperman, and Ku (1962) and Barnard (1951).

1.29. The aim of this book is to describe the mathematical theory of probability regarded as a theory of measure applied to events in the real world, the values of the measure being chosen so as to bear a relation, in some sense, to the relative frequencies with which such events occur. Many papers and books have been written which attempt to clarify what 'probability', in a quantitative sense, can mean, and there has been much resulting controversy which can be considered only briefly here, but is of the greatest importance in relation to statistical inference.

Generally speaking we can classify theories of probability into three classes, in the first of which probability is regarded as a numerical measure of belief in a proposition rather than a likely relative frequency of an event. We then have numerically valued functions of propositions instead of functions of events, and hence such theories are sometimes described as 'subjective'. The main originator of this type of theory was W. E. Johnson, whose *Logic* appeared in three volumes in 1921–4. J. M. Keynes's *A Treatise on Probability* appeared in 1921 and was influenced by Johnson's teaching at Cambridge. Ramsey (1931), Jeffreys (1961), and Good (1950a) have developed this approach further, principally in connexion with statistical inference, for there appear to be problems in the latter which are more easily resolved by this approach.

However, from the point of view of the scientist concerned with problems in physical and biological worlds, subjective theories of probability suffer from disadvantages. In the first place it is not very obvious that 'degrees' of belief are readily quantifiable. It is clearly a different thing to assert that the proposition 'α-Centauri is a star which has planets' has probability $\frac{1}{2}$ (because we have no evidence one way or the other), and to assert that a penny, carefully made to be symmetrical, has probability $\frac{1}{2}$ of landing head upwards when tossed in a properly random manner. In the first case we ascribe the probability $\frac{1}{2}$ to the proposition because we have no knowledge adequate for a decision, while in the second case we have positive knowledge about the symmetry of the penny, and what we might describe as the highly chaotic character of the system of causes which determine which way the penny falls.

It is worth noticing, however, that if 'degrees of belief' can be put into an order so that we can set up an ordering relationship, then they can also be given a cardinal number, for if beliefs in two exclusive propositions A, B, are ordinally equal, we can ascribe the cardinal number $\frac{1}{2}$ to the conditional probability of A given that either A or B is true. (I owe this remark to Professor D. G. Champernowne.)

Secondly, even if degrees of belief can be consistently given numerical measures it is not clear how these measures can be related to the relative frequencies with which events actually occur, and for this reason they do not provide the natural approach in studying physical and biological phenomena.

Another approach to probability theory is due to von Mises, who attempted to start from the study of mathematical objects which he called 'collectives'. These are sequences of digits the relative frequencies of which among the first n not only converge as n increases, but are also such that this is also true for any subsequences selected from the sequence by any method not dependent on what the actual digits are. Such a definition leads to considerable difficulties and, moreover, even if a proper and consistent theory of such collectives could be constructed it would probably be equivalent to a theory of measure.

Measure as a basis for probability theory will be considered in detail in Chapter 4, but the basic ideas have already been exemplified in the particular case of a finite discrete set of events already considered in. this chapter. Suppose that we have n events $A_1,..., A_n$, and know that one and only one of them must occur at a particular 'trial'. Then we ascribe numbers $p_1,..., p_n$ to them satisfying the conditions $p_i \geqslant 0$, $\sum p_i = 1$, and we do this in a way which will make the relative frequencies with which these will occur in the long run equal to these values. In order to do this we need not only a mathematical theory but positive empirical knowledge.

The numbers, p_i, are then regarded as the probability measures ascribed to the individual events, and to any sum of events we ascribe a measure equal to the sum of corresponding p_i. In Chapter 4 we shall see how to generalize this scheme to situations in which the outcome of each trial is more complicated than the occurrence of one of a series of discrete events.

It remains to consider in what way the Weak and Strong Laws of Large Numbers relate probability to observed frequency. This is best exemplified by the simple case of a binomial sequence of independent trials in each of which an event E can occur with probability p $(0 < p < 1)$, or fail with probability $q = 1 - p$.

The Weak Law of Large Numbers asserts that in a sequence of N trials the observed relative frequency of occurrence of E will diverge from p, by more than any prescribed quantity $\varepsilon < 0$, with a probability, P say, which converges to zero as N increases. We therefore see that this law cannot be used to define 'probability' in terms of 'frequency' in general because it relates p to the observed relative frequency only by means of another probability P. However, if we have any consistent system of probabilities applied to events, and if we once admit that events with very small probabilities are very unlikely to occur (and the smaller the probability, the less likely) then the Weak Law of Large Numbers shows that observed relative frequencies will be highly likely to be near to the corresponding probabilities, and that this approximation will be better if the number of trials is large.

The Strong Law of Large Numbers is a statement of quite a different kind (Theorem 8.36). In it we consider an infinite sequence of trials in each of which the event may or may not occur, and we are concerned with the convergence of the relative frequency to p in the ordinary mathematical sense of convergence. To define a 'probability' for such an event requires a more elaborate theory than we have so far considered and this will be discussed in detail in Chapter 4. The resulting probabilities are necessarily determined in their values by the values of the probabilities of events on every finite section of the sequence. The theorem then asserts that the probability that the relative frequency of E in the first N trials converges to p in the ordinary sense is unity.

This theorem relates p to the relative frequency but again cannot be used to define probability unless it is previously agreed that events with zero probability do not happen, an assertion which already involves the idea of probability itself. Furthermore, the result cannot be tested in practice since it is impossible to carry out an infinite series of trials.

These results show that neither of the Laws of Large Numbers can be used to provide a definition of probability. It is also to be noticed that the two laws differ in their empirical nature in that the first is empirically testable whilst the second is not.

Bibliographical notes

An adequate history of the theory of probability has never been written. Todhunter (1949) is a very detailed synopsis of early work up to and including that of Laplace. David (1962) gives an interesting history of gambling games and their theory. See also a series of articles on the history of probability and statistics published in *Biometrika* (David 1955; M. G. Kendall 1956, 1957, 1961, 1963; Thatcher 1957; Plackett 1958; Bayes 1958).

References to the various interpretations of what 'probability' means have already been given in § 1.29.

Combinatorial probability problems have a large literature and so does the related theory of combinatorial analysis. For the latter see McMahon (1915–16), Riordan (1958), Beckenbach (1964), and David and Barton (1962), which is a very detailed survey of combinatorial probability theory but does not give adequate references. Whitworth (1897, 1901) contain a very large number of special problems, many of which are very elementary, but also deals with some general theorems which are important. The proofs of these can be much simplified by the use of generating and indicator functions. Difficult but artificial special problems will be found in the examinations printed in the *Annual Year Book of the Institute of Actuaries* (see also Bizley 1957).

In many combinatorial problems, especially in the determination of moments, it is useful to consider symmetric functions of a set of variables $x_1,..., x_n$. For the theory of these see David and Barton (1962). Extensive tables are given in David and Kendall (1949–58), a series of papers which will shortly be published in book form.

Exercises

Exercise 1.1. A large number, N, of persons are to be tested for a certain disease by a simple and sensitive test on a sample of their blood. If all are tested individually the required number of tests is N. Suppose that they are divided into groups of size n, where $N = kn$, and the blood of each group pooled before testing. Any group which shows a positive response then has all its members tested. Prove that the expected number of tests is then

$$k + N\{1 - (1-p)^n\},$$

where p is the probability of an individual showing a positive response. (See Dorfman 1943 and Sterett 1957. This technique can be modified in various ways. How?)

Exercise 1.2. Consider N events, $A_1,..., A_N$, which may or may not be independent, and which need not exhaust all possibilities. Let X be the number of these events which occur and k a positive integer. Using indicator functions or otherwise, prove that the expected value of

$$X(X-1)...(X-k+1)$$

is $k!$ multiplied by the sum of the probabilities that k specified events out of the N occur.

Exercise 1.3. $N = mn$ points are arranged in a rectangular array of m rows of n points. $M (\leqslant mn)$ of these points are chosen at random and marked. Prove that the first and second moments of the number of neighbouring (in a vertical or horizontal direction) marked points are

$$m_1 = (2mn - m - n)\frac{M(M-1)}{N(N-1)},$$

and

$$m_2 = (2mn - m - n)\frac{M(M-1)}{N(N-1)}$$
$$+ 4(3mn - 3m - 3n + 2)\frac{M(M-1)(M-2)}{N(N-1)(N-2)}$$
$$+ (4m^2n^2 - 4m^2n - 4mn^2 + m^2 + n^2 - 12mn + 13m + 13n - 8)$$
$$\times \frac{M(M-1)(M-2)(M-3)}{N(N-1)(N-2)(N-3)}.$$

Exercise 1.4. Two lists of names are considered. The names in each list are all different and there are M and N of them respectively. D names are common to both lists, and it is required to estimate D by comparing two samples. Suppose that m names are chosen at random from the first list, and n from the second. Let d be the number of names common to the two samples and put $p = DM^{-1}$. Show that the probability distribution of d is (with the usual convention)

$$P(d) = \binom{D}{d}\binom{M}{m}^{-1}\binom{N}{n}^{-1} \sum_{k=d}^{D} \binom{D-d}{k-d}\binom{M-D}{m-k}\binom{N-k}{n-d},$$

and that

$$E(\hat{p}) = M^{-1}D,$$

and

$$\text{var}(\hat{p}) = Np\, m^{-1}n^{-1}\left\{1+\left(\frac{m-1}{M-1}\right)\left(\frac{n-1}{N-1}\right)(D-1)\right\}-p^2,$$

where

$$\hat{p} = m^{-1}n^{-1}Nd. \quad \text{(Deming and Glasser 1959.)}$$

Exercise 1.5. a members of a population of N animals are captured, marked, and released. The animals are then recaptured one by one until m ($\leqslant a$) marked animals have been recaptured. Prove that P_n, the probability that it is necessary to capture n animals to obtain m which are marked, is given by

$$P_n = \frac{a}{N}\binom{a-1}{m-1}\binom{N-a}{n-m}\binom{N-1}{n-1}^{-1} \quad (m \leqslant n \leqslant N+m-a)$$

(negative hypergeometric distribution).

Exercise 1.6. In the above example show that P_n for a population of $N+1$ animals in greater than or less than P_n for a population of N according as

$$N \lessgtr anm^{-1}-1.$$

This suggests that $n(a+1)m^{-1}-1$ be used to estimate N. Prove that the mean of this quantity is N, and that its variance is

$$\frac{(a-m+1)(N+1)(N-a)}{m(a+2)}.$$

This is an example of estimation by capture–recapture methods. For general surveys of this subject see Bailey (1951) and Chapman (1954).

Exercise 1.7. A population of N animals in an isolated region are exposed to the danger of being trapped on k successive occasions. Assuming that on each occasion each animal has independently the probability q of being trapped ($p = 1-q$), prove that the probability that the numbers caught on the k successive occasions are $r_1,\ldots, r_k$ is

$$\frac{N!}{r_1! \ldots r_k!\, (N-s_k)!}\, q^{s_k} p^{kN-\sum_1^k s_i},$$

where $s_i = \sum_{j=1}^{i} r_j$.

This provides a method of estimating populations by removal. See Moran (1951) and Zippin (1956).

Exercise 1.8. A man carries two boxes of matches in his left and right pockets. Initially they contain N matches each, and are emptied, a match at a time, the box having a match taken from it being chosen at random. Let r be the number of matches remaining in the other box when one of the boxes is first found to be empty. Prove that the probability distribution of $r = 0, 1, ..., N$ is given by

$$p_r = \binom{2N-r}{N} 2^{-2N+r},$$

and that the mean of r is

$$2^{-2N}(2N+1)\binom{2N}{N} - 1.$$

(Banach's match-box problem—see Feller 1957b.)

Exercise 1.9. N individuals form a group and each chooses at random d other individuals in the group. X is the number of individuals who are chosen by nobody. Prove that the probability that X takes the value K is

$$\sum_{j=k}^{N-1-d} (-)^{k+j} \binom{j}{k}\binom{N}{j}\binom{N-j}{d}^j \binom{N-j-1}{d}^{N-j}\binom{N-1}{d}^{-N}$$

and obtain the mean and variance of this distribution (Katz 1952).

Exercise 1.10. s animals which may be one of a number of particular genetic types $A_1, ..., A_N$ are bred. Each animal is, independently, one of these types with probability N^{-1}, and is male or female with probability $\frac{1}{2}$. Each type, A_i, is associated in a one-to-one correspondence with another type A_j. Prove that the expected number of females amongst the s animals which are such that there also exists at least one male of the corresponding type is

$$\tfrac{1}{2}s\left\{1 - \left(1 - \frac{1}{2N}\right)^{s-1}\right\},$$

and that the probability that there is no such female is

$$(2N)^{-s} \sum_{r=0}^{N} \binom{N}{r}(-1)^r 2^{N-r}(N-r)^s. \quad \text{(Fisher 1949.)}$$

Exercise 1.11. One version of the game of craps played with two dice is as follows. The player throws the two dice and if the sum of the points obtained is 7 or 11 he wins outright. If he obtains 2, 3, or 12 he loses. If neither of these cases occur at the first throw he throws repeatedly until he obtains a 7 and then loses, or obtains the same score as in his first throw when he wins. Prove that his chance of winning is 244 $(495)^{-1}$.

2. Discrete Distributions

2.1. Generating functions

IN CHAPTER 1 we have already considered many cases in which the outcome of an experiment is one of a finite or enumerable sequence of possible events E_1, E_2,.... If associated with each of these events we have a numerical value, X_j, the quantity X equal to X_j when E_j occurs is a 'random variable'. In nearly all cases of interest the values X_j are the non-negative integers 0, 1,..., or, more rarely, the integers 0, ± 1, ± 2,.... It is then natural to use generating functions, and if p_n is the probability that $X = n$, we write

$$P(z) = \sum_0^\infty p_n z^n, \tag{2.1}$$

when $p_n = 0$ for $n < 0$. Since $P(1) = \sum p_n = 1$, this series is convergent for $|z| \leqslant 1$. When $p_n > 0$ for $n < 0$ the similar series is not in general convergent unless $|z| = 1$, and in such cases it is best to put $z = e^{i\theta}$ and consider the function

$$\phi(\theta) = P(e^{i\theta}) = \sum_{-\infty}^\infty p_n e^{in\theta},$$

on the circle $0 \leqslant \theta < 2\pi$. If, however, p_n tends to zero fast enough as $n \to \pm\infty$, $P(z)$ will be convergent in a ring-shaped area of the form $r_1 \leqslant |z| \leqslant r_2$, where $0 < r_1 < 1 < r_2$. From now on we shall, unless otherwise specified, consider only the case where $p_n = 0$ for $n < 0$, so that the series for $P(z)$ is necessarily convergent for all z such that $|z| \leqslant 1$.

$P(z)$ is said to be the probability generating function for the distribution of X since the successive coefficients are the probabilities p_n.

As an example we may consider the binomial distribution (1.18) with probability p and index n. X is the number of successful events, and equals r with probability

$$\binom{n}{r} p^r q^{n-r},$$

where $q = 1 - p$ and $r = 0$, 1,..., n. $P(z)$ is then obviously $(q + pz)^n$.

Since the probability that $X = n$ is p_n, we can define the probability that $X > n$ as q_n, and we have

$$q_n = p_{n+1} + p_{n+2} \cdots. \tag{2.2}$$

We can define a generating function which generates the quantities q_n by

$$Q(z) = q_0 + q_1 z + q_2 z^2 + \dots, \tag{2.3}$$

the series being always convergent for $|z| < 1$ since $q_i \leqslant 1$. Taking $|z| < 1$, and multiplying by $(1-z)$ we have $(1-z)Q(z) = 1 - P(z)$, and therefore

$$Q(z) = \frac{1 - P(z)}{1 - z}. \tag{2.4}$$

Similarly the series

$$p_0 + (p_0 + p_1)z + (p_0 + p_1 + p_2)z^2 + \dots$$

is equal to

$$(1-z)^{-1}P(z),$$

which is therefore a generating function for the probabilities that $X \leqslant n$.

2.2. Moments

The mean, m_1, is $\sum_0^\infty np_n$, and we similarly define the sth moment, m_s, about the origin by

$$m_s = \sum_0^\infty n^s p_s. \tag{2.5}$$

When these moments are finite we can find them in terms of $P(z)$. For example,

$$m_1 = \sum_0^\infty nP_n = \frac{d}{dz}P(z) \quad (\text{at } z = 1),$$

and we denote the expression on the right by $P'(1)$. Similarly,

$$\sum n(n-1)p_n = P''(1),$$

so that $m_2 = \sum n^2 p_n = P''(1) + P'(1)$.

To obtain the variance,

$$\sum_0^\infty (n - m_1)^2 p_n,$$

we have

$$\text{var}(X) = m_2 - m_1^2$$
$$= P''(1) + P'(1) - \{P'(1)\}^2.$$

It is convenient to define the 'factorial moments', $m_{(k)}$, by the expression

$$m_{(k)} = \sum_0^\infty n(n-1)\dots(n-k+1)p_n$$
$$= P^{(k)}(1), \tag{2.6}$$

where the right-hand side is the kth derivative of $P(z)$, with z set equal to 1. It is clear that the factorial moment of order k can be expressed in terms of

$m_1, \ldots, m_k$ and similarly m_k in terms of $m_{(1)}, \ldots, m_{(k)}$. In fact

$$m_1 = m_{(1)},$$
$$m_2 = m_{(2)} + m_{(1)},$$
$$m_3 = m_{(3)} + 3m_{(2)} + m_{(1)},$$
$$m_4 = m_{(4)} + 6m_{(3)} + 7m_{(2)} + m_{(1)} \tag{2.7}$$

so that m_k can be found in terms of $P'(1), \ldots, P^{(k)}(1)$. The general formula for such a representation can be found in the following way.

Put $z = e^t$ in (2.1) and expand in powers of t (which is certainly permissible if $t < 0$). Then

$$P(e^t) = \sum_{n=0}^{\infty} p_n e^{nt}$$
$$= \sum_{k=0}^{\infty} t^k \sum_{n=0}^{\infty} p_n n^k / k!$$
$$= \sum_{k=0}^{\infty} \frac{m_k}{k!} t^k. \tag{2.8}$$

The expansion of $P(e^t)$ generates the quantities $m_k(k!)^{-1}$. $P(e^t)$ is therefore somewhat loosely called the 'moment generating function'. In a similar way we find

$$P(1+t) = \sum_{n=0}^{\infty} p_n (1+t)^n$$
$$= \sum_{k=0}^{\infty} \frac{m_{(k)}}{k!} t^k, \tag{2.9}$$

so that $P(1+t)$ is the 'factorial moment generating function'. It follows that $m_k(k!)^{-1}$ is the coefficient of t^k in

$$\sum_{n=0}^{\infty} \frac{m_{(n)}}{n!} (e^t - 1)^n, \tag{2.10}$$

and $m_{(k)}(k!)^{-1}$ is the coefficient of t^k in

$$\sum_{n=0}^{\infty} \frac{m_n}{n!} \{\log(1+t)\}^n, \tag{2.11}$$

from which we can derive the relations (2.7). In fact using (2.10) and (2.11) it is possible to write down general formulae for $m_{(k)}$ in terms of the m_k, and conversely (Frisch 1926). The coefficients involved are 'Stirling numbers' and are related to the 'differences of zero' defined in § 1.20. These formulae are useful since from (2.6) $m_{(k)}$ is often simpler to find than m_k. Extensive tables are given in David and Barton (1962). See also Riordan (1958).

The moments do not in general exist but if, as often happens, $p_n \to 0$ faster than some quantity ρ^n, where $0 < \rho < 1$, they all must exist. In this case the series for $P(z)$ is convergent inside a circle $|z| < \rho^{-1}$ and hence is analytic at $z = 1$, so that $P(e^t)$ and $P(1+t)$ can be expanded in series convergent inside circles around the point $t = 0$, and with radii equal to $-\log \rho$, and $\rho^{-1}-1$ respectively.

Just as we can express $m_{(k)}$ in terms of the $\{p_n\}$, so we can find p_n in terms of the $m_{(k)}$ by an 'inversion formula'. Consider (2.9) and put $1+t = u$. Then

$$\sum_{n=0}^{\infty} p_n u^n = \sum_{k=0}^{\infty} \frac{m_{(k)}}{k!} (u-1)^k.$$

If these series are analytic in a circle around $u = 1$, i.e. around $t = 0$, we identify powers of u and obtain

$$p_n = \sum_{k=n}^{\infty} (-1)^{k-n} \frac{m_{(k)}}{n!\,k-n!}. \tag{2.12}$$

This formula gives p_n explicitly but involves a knowledge of *all* the moments. It assumes, moreover, that the series are analytic in a circle about $u = 1$. This raises the question of whether a probability distribution is determined uniquely when its moments are known. We shall discuss this problem briefly in Chapter 6 and it suffices here to say that in general the answer is no. An example of a *discrete* distribution which has all its moments finite but is not determined by them is given by Wintner (1949).

We also need to find simple expressions for central moments, i.e. the moments about the mean defined by

$$\mu_k = \sum_{n=0}^{\infty} (n-m_1)^k p_n. \tag{2.13}$$

Their generating function is clearly

$$\sum_{k=0}^{\infty} \frac{\mu_k}{k!} t^k = \sum_{k=0}^{\infty} \frac{t^k}{k!} \sum_{n=0}^{\infty} (n-m_1)^k p_n$$

$$= \sum_{n=0}^{\infty} p_n e^{t(n-m_1)}$$

$$= e^{-tm_1} P(e^t). \tag{2.14}$$

The values of μ_k in terms of m_k, and vice versa, for $k = 1,\ldots, 4$ are given in equations (5.61) and (5.62).

Some results are more easily expressed in terms of 'cumulants' which are the coefficients, κ_i, defined by the expansion

$$K(t) = \sum_{s=0}^{\infty} \frac{\kappa_s}{s!} t^s = \log P(e^t), \tag{2.15}$$

when this expansion is convergent in some circle about $t = 0$. It is easy to verify that

$$m_1 = \kappa_1,$$

$$m_2 = \kappa_2 + \kappa_1^2.$$

Tables of $\kappa_1,..., \kappa_{10}$ in terms of $m_1,..., m_{10}$, and of $\mu_2,..., \mu_{10}$, and conversely, are given in Kendall and Stuart, Vol. II (1958). We use all these moments and coefficients again in Chapter 6 for more general distributions.

2.3. Multivariate discrete distributions

We can similarly construct generating functions for the joint distribution of several discrete random variables. Consider the case where there are two only, since three or more can be dealt with in an exactly similar way. Suppose then that the random variables X and Y take the values 0, 1, 2,..., and that the probability that $X = m$, $Y = n$ is p_{mn}. Then the joint probability generating function is

$$P(w, z) = \sum_{m,n} P_{mn} w^m z^n.$$

If the random variables are independent we can put

$$p_{mn} = p_m^{(1)} p_n^{(2)},$$

where $\{p_m^{(1)}\}$ is the probability distribution of X, and $\{p_n^{(2)}\}$ that of Y. Then the generating function $P(w, z)$ factorizes, and may be written

$$P(w, z) = P_1(w) P_2(z), \tag{2.16}$$

where

$$P_1(w) = \sum_m p_m^{(1)} w^m, \quad P_2(z) = \sum_n p_n^{(2)} z^n.$$

Even if the random variables are not independent the marginal distributions, $\mathrm{pr}(X = m)$ and $\mathrm{pr}(Y = n)$, are given by $P(w, 1)$ and $P(1, z)$, and the moments and joint moments can be found by differentiation as before.

As an example suppose there are N successive independent trials each of which may result in one of k events $E_1,..., E_k$, with probabilities $p_1,..., p_k$ where $\sum p_i = 1$. Let $n_1,..., n_k$ be the numbers of times each of these occur. Then the joint probability generating function of $n_1,..., n_k$ is

$$P(z_1,..., z_k) = (p_1 z_1 + ... + p_k z_k)^N$$

$$= \sum p_{n_1,..., n_k} z_1^{n_1} ... z_k^{n_k}, \tag{2.17}$$

where the coefficient of

$$z_1^{n_1} ... z_k^{n_k}$$

is the probability that E_1 occurs n_1 times, E_2 occurs n_2 times and so on. This is the multinomial distribution (1.50).

2.4. Convolutions

Consider again the case where there are two random variables, X and Y, which are independent. The probability that $X + Y$ is equal to an integer n is

$$p_n^{(1)}p_0^{(2)} + p_{n-1}^{(1)}p_1^{(2)} + \cdots + p_0^{(1)}p_n^{(2)},$$

and this is the coefficient of z^n in $P_1(z)P_2(z)$. Hence the probability distribution of $X + Y$, which is called the 'convolution' of the distributions of X and Y, has a probability generating function

$$P_1(z)P_2(z). \qquad (2.18)$$

This convolution distribution is often written

$$\{p_m^{(1)}\} * \{p_n^{(2)}\},$$

and clearly its cumulants are the sums of the cumulants of X and Y. The convolution operation, $*$, has some of the properties of multiplication in that it is associative,

$$\{p_m^{(1)}\} * [\{p_n^{(2)}\} * \{p_s^{(3)}\}] = [\{p_m^{(1)}\} * \{p_n^{(2)}\}] * \{p_s^{(3)}\},$$

and commutative,

$$\{p_m^{(1)}\} * \{p_n^{(2)}\} = \{p_n^{(2)}\} * \{p_m^{(1)}\},$$

but division is not in general possible. With this operation the set of all discrete distributions on $(0, 1, \ldots)$ is therefore said to form a 'semi-group'.

The reason why such distributions do not form a 'group' is that $P_1(z)P_2(z)^{-1}$ is not in general the generating function of a distribution.

It is also convenient to define the 'convolution power' of a distribution. If $P(z)$ is the probability generating function of a distribution $\{p_n\}$,

$$\{p_n\}^{n*} \quad (n = 1, 2, \ldots) \qquad (2.19)$$

is defined to be the 'nth convolution power' if it is generated by $P(z)^n$. This is the distribution of the sum of n independent quantities each with the distribution $\{p_n\}$.

A simple example of this is the binomial distribution whose variate is the sum of n independent variates whose generating functions are each $q + pz$.

2.5. The convergence of discrete distributions

Consider the convergence of a sequence of discrete distributions to a discrete distribution. For each integer $m = 1, 2, \ldots$ let $\{p_{mn}\}$ $(n = 0, 1, \ldots)$ be a discrete distribution, i.e.

$$p_{mn} \geq 0, \quad \sum_n p_{mn} = 1.$$

Then we say that the distributions $\{p_{mn}\}$ converge to a distribution $\{p_n\}$ if

$$\lim p_{mn} = p_n,$$

for every value of n. We can relate such convergence to the convergence of the corresponding generating functions by the following theorem which is a particular case of a much more general result which we will prove in Chapter 6.

THEOREM 2.1. *In the above circumstances a necessary and sufficient condition that p_{mn} converges to p_n for each n is that*

$$P_m(z) = \sum_n p_{mn} z^n$$

converges to

$$P(z) = \sum_n p_n z^n$$

for every value of z in $0 \leqslant z < 1$.

Notice that in this theorem the sufficiency part assumes that the limiting function, $P(z)$, is the generating function of a probability distribution. This is essential, for it is possible that p_{mn} should converge to a number $p_n \geqslant 0$ for each n without $\{p_n\}$ being a probability distribution. This happens, for example, for the binomial distribution, $(q+pz)^n$, if p and q are kept fixed and $n \to \infty$. Then each p_{mn} converges to zero. However, if p_{mn} converges to p_n for every n, and $\{p_{mn}\}$ is always a probability distribution, $P_m(z)$ will converge to $P(z)$ whether the latter is the generating function of a distribution or not.

Proof. First suppose that p_{mn} converges to p_n for every n, and consider a fixed value of z in the interval $0 \leqslant z < 1$. Given any $\varepsilon > 0$ choose N so that

$$z^N(1-z)^{-1} < \varepsilon.$$

Then

$$|P_m(z) - P(z)| < \sum_{n=0}^{N} |p_{mn} - p_n| z^n + \varepsilon, \qquad (2.20)$$

because

$$\left| \sum_{n=N+1}^{\infty} p_{mn} z^n - \sum_{N+1}^{\infty} p_n z^n \right| \leqslant \sum_{n=N+1}^{\infty} z^n \leqslant z^{N+1}(1-z)^{-1} < \varepsilon.$$

We can now choose m so large that each term in the sum on the right-hand side of (2.20) is arbitrarily small, and hence for each fixed z, $P_m(z)$ converges to $P(z)$.

Now suppose $P_m(z)$ converges to $P(z)$ for each z in $0 \leqslant z < 1$. For any fixed value of n consider the set of numbers p_{mn} ($m = 1, 2, ...$). If these do not converge we must have

$$\underline{\lim_m} \, p_{mn} < \overline{\lim_m} \, p_{mn} \leqslant 1, \qquad (2.21)$$

and if there is not convergence for every n there must be strict inequality in (2.21) for at least one value of n. Let n be the lowest integer for which this is true. Then we can choose a sequence

$$m(1, n), \ m(2, n),\ldots$$

such that

$$P_{m(i,n),n} \quad (i = 1, 2,\ldots)$$

converges to the upper limit. Out of this we can choose a subsequence

$$m(1, n+1), \ m(2, n+1),\ldots$$

for which

$$P_{m(i,n+1),n+1}$$

converges to some quantity. The process is continued by constructing sequences $m(i, n+j)$ $(j = 2, 3,\ldots)$, each a subsequence of the previous one, and such that

$$P_{m(i,n+j),n+j}$$

converges. Taking the 'diagonal' sequence

$$m' = m(1, n), \ m(2, n+1), \ m(3, n+2),\ldots$$

and using the first part of the theorem, $P_{m'}(z)$ must converge to a generating function whose coefficients are the above limits. But this generating function must be $P(z)$ and is therefore a probability generating function. Since the same argument could be applied using

$$\varlimsup_{m} P_{mn},$$

the upper and lower limits must be the same and the theorem is proved.

　　In the theory of discrete distributions frequent use is made of this theorem, and in Chapter 6 we shall prove its analogue for arbitrary distributions.

2.6. Compound distributions

　　In the first three chapters of this book we are concerned only with discrete distributions. However, as we shall see, there are several ways in which the theory of such distributions already requires the use of continuous distribution. In §1.1 we have already defined the distribution of a random variable X in general to be a non-decreasing function, $F(x)$, such that the probability that $X \leqslant x$ is $F(x)$. In the applications of this idea in this chapter $F(x)$ will be the integral of a non-negative integrable function $f(x)$, so that we can write

$$F(x) = \int_{-\infty}^{x} f(t) \, dt.$$

When $f(t)$ is continuous, $f(t) \, dt$ can be regarded as the probability that X lies in the range $(x, x+dt)$ when dt becomes small. $f(t)$ is then known as the probability density of the distribution $F(x)$ and $F(x)$ is said to be a continuous distribution. To avoid the possibility of infinite values of X we must have

$$\int_{-\infty}^{\infty} f(t)\, dt = 1, \qquad (2.22)$$

and thus any non-negative integrable function satisfying (2.22) represents a continuous distribution.

One application to discrete distributions arises in the idea of a 'compound distribution'. Suppose that $\{p_n(\theta)\}$, $n = 0, 1,\ldots$, is a discrete distribution depending on a parameter θ which lies in some range R. If we suppose that $f(\theta)$ is the probability density of some distributions defined on the range R, the set of quantities

$$p_n = \int_R p_n(\theta) f(\theta)\, d\theta, \quad n = 0, 1,\ldots, \qquad (2.23)$$

is a discrete probability distribution which is said to be 'compounded' and its generating function is

$$P(z) = \Sigma p_n z^n = \int_R P(z, \theta)\, d\theta, \qquad (2.24)$$

where $P(z, \theta)$ is the generating function of $\{p_n(\theta)\}$.

The idea of compounding can also be applied when θ is itself a discrete random variable and in particular when it is the index n in the convolution power of a distribution (2.19). Thus suppose that

$$Y = X_1 + \ldots + X_N, \qquad (2.25)$$

where each X_i is a discrete random variable with the distribution $\{p_n\}$ and generating function $P(z)$. If N is fixed and the X_i are independent, Y will have the generating function $P(z)^N$. Suppose, however, that N is a random variable with a distribution $\{\pi_N\}$ and generating function $\pi(z)$. Then Y will clearly have the generating function

$$\sum_N \pi_N P(z)^N = \pi\{P(z)\}. \qquad (2.26)$$

From this relationship we can easily find the moments of Y in terms of those of X_i and N.

We are now in a position to consider systematically a wide variety of discrete distributions, which often occur in practice (for a useful list and bibliography see Haight 1961).

2.7. The uniform discrete distribution

Here the random variable X assumes the values $1, 2,\ldots, k$ with probabilities k^{-1}. The probability generating function is

$$P(z) = k^{-1}(z + z^2 + \ldots + z^k) = k^{-1}z(1 - z^k)(1 - z)^{-1}. \qquad (2.27)$$

The moment generating function is

$$P(e^t) = \frac{e^t(1-e^{kt})}{k(1-e^t)}$$

$$= \frac{e^{\frac{1}{2}(k+1)t}\sinh\frac{1}{2}kt}{k\sinh\frac{1}{2}t.} \tag{2.28}$$

Expanding this it is possible to obtain the moments and a recurrence relation which they satisfy. In particular,

$$m_1 = \frac{1}{2}(k+1),$$

$$m_2 = \frac{1}{6}(k+1)(2k+1),$$

$$\mu_2 = \frac{1}{12}(k^2-1).$$

The first ten moments and some other results are given by Pierce (1940). General expressions for the moments and cumulants in terms of Bernoulli numbers are given in David and Barton (1962). Cumulants are given by Stuart (1950).

From (2.27) we can find the generating function of the probability distribution of the sum of N independent random variables each having the above distribution. This is clearly

$$\left\{\frac{z(1-z^k)}{k(1-z)}\right\}^N. \tag{2.29}$$

Such a sum can take any value between N and Nk. By expanding the numerator, and the reciprocal of the denominator by using the binomial theorem, and picking out the coefficient of z^n, we see that the probability of the sum equalling n ($N \leqslant n \leqslant Nk$) is

$$k^{-N}\sum_{s=0}^{s=u}(-1)^s\binom{N}{s}\frac{n-sk-1!}{n-N-sk!\,N-1!}, \tag{2.30}$$

where $n = N+uk+v$, u, v being non-negative integers and $v < k$.

This result was discovered by Montmort and De Moivre early in the eighteenth century (see Todhunter 1949). If N is kept fixed whilst k is made large, and the tail of the above distribution is considered when nk^{-1} is kept fixed, we can obtain as a limit the distribution of the sum of N random variables each uniformly distributed on the interval $(0, 1)$. We shall consider this problem again in §7.1.

Notice that in connexion with the distribution of the number of inversions in a permutation (§1.22) we have already obtained the generating function of a sum $X_1+\ldots+X_N$, where X_1 has a uniform distribution on the integers $(0, 1)$, X_2 a uniform distribution on the integers $(0, 1, 2)$, and so on.

2.8. The binomial distribution

We recall from § 1.7 and § 2.1 that the binomial distribution with probability $p = 1-q$ and index n is a distribution on the integers $0, 1,..., n$ given by

$$\binom{n}{i} p^i q^{n-i} \tag{2.31}$$

and generating function $(q+pz)^n$.

A table of individual terms and the cumulative sums to seven decimals for $n = 2(1)\ 49$, $i = 0(1)\ n-1$, $p = 0\cdot01(0\cdot01)\ 0\cdot50$, is given in Vol. 6 (1950) of the *Applied Mathematics Series* of the National Bureau of Standards. Romig (1953) extends this to $n = 50(5)\ 100$, $p = 0\cdot01(0\cdot01)\ 0\cdot50$ with six decimals. The *Harvard Computation Laboratory Tables* (1955) of the cumulative binomial probability distribution gives cumulative sums to five decimals for $n = 1(1)\ 50(2)\ 100(10)\ 200(20)\ 500(50)\ 1000$ for $p = 0\cdot01(0\cdot01)$ $0\cdot50$ and $p = \frac{1}{16}, \frac{1}{12}, \frac{1}{8}, \frac{3}{16}, \frac{5}{16}, \frac{1}{3}, \frac{3}{8}, \frac{5}{12}$, and $\frac{7}{16}$. The Ordnance Corps (1952) gives tables of the cumulative sums for $n = 1(1)\ 150$, and $p = 0\cdot01(0\cdot01)\ 0\cdot50$, to seven decimal places. The *Biometrika Tables for Statisticians*, Vol. I (1956), give individual terms to five decimals for $n = 5(5)\ 30$, $p = 0\cdot01$, $0\cdot02(0\cdot02)$ $0\cdot10(0\cdot10)\ 0\cdot50$.

The cumulative sums in these tables are defined by

$$B(k,n,p) = \sum_{i=0}^{k} \binom{n}{i} p^i q^{n-i}. \tag{2.32}$$

These sums satisfy the interesting equation

$$B(k,n,p) = (n-k)\binom{n}{k} \int_0^q t^{n-k-1}(1-t)^k \, dt. \tag{2.33}$$

To prove this integrate the integral on the right by parts to obtain

$$\binom{n}{k}\{t^{n-k}(1-t)^k\}_0^q + \binom{n}{k} k \int_0^q t^{n-k}(1-t)^{k-1} \, dt$$

$$= \binom{n}{k} p^k q^{n-k} + (n-k+1)\binom{n}{k-1} \int_0^q t^{n-k}(1-t)^{k-1} \, dt,$$

and the result then follows by induction. Formula (2.33) can be given another interpretation in that the expression on the right can be regarded as a cumulative distribution function of a continuous random variable defined on the interval $(0, 1)$. This is the beta distribution which we will discuss in § 7.21, and which can be expressed in terms of another important distribution, Fisher's F-distribution, which is extensively tabulated.

The moments are most easily found from the generating function $(q+pz)^n$, the moment generating function $(q+pe^t)^n$, and the factorial moment generating function $(1+pz)^n$. In this way we find

$$m_1 = np,$$

$$m_2 = n(n-1)p^2 + np,$$

$$m_3 = n(n-1)(n-2)p^3 + 3n(n-1)p^2 + np,$$

$$m_4 = n(n-1)(n-2)(n-3)p^4 + 6n(n-1)(n-2)p^3 + 7n(n-1)p^2 + np,$$

$$(2.34)$$

and

$$\mu_2 = np(1-p),$$

$$\mu_3 = np(1-p)(1-2p),$$

$$\mu_4 = 3n^2p^2(1-p)^2 + np(1-p) - 6np^2(1-p)^2. \qquad (2.35)$$

Higher moments can be found in Kirkham (1935). They are complicated but satisfy a simple recurrence relation given in Exercise 2.3. The cumulants $\kappa_1, \ldots, \kappa_{12}$ are given by Haldane (1940). Some distributions have particularly simple forms for some kinds of moments and this is illustrated in the case of the binomial distribution by the fact that the factorial moments are given by

$$m_{(k)} = p^k n_{(k)}, \qquad (2.36)$$

which may be easily verified from the factorial moment generating function $(1+pz)^n$, or proved by the method of Exercise 1.2. Formulae for the numerical computation of the binomial moments in powers of p are given by Larguier (1936).

A problem which is sometimes important in practice is to find the moments (especially the mean) of expressions such as $(a+bX)^{-1}$, where $a > 0$, $b > 0$, and X has a binomial distribution with parameter p and index n. Clearly this expectation does not exist when $a = 0$. Assume that $a > 0$. Then we have

$$E\{(a+bX)^{-1}\} = \sum_{i=0}^{n} (a+bi)^{-1}\binom{n}{i}p^i q^{u-i}$$

$$= \int_0^1 x^{a-1}(q+px^b)^n \, dx, \qquad (2.37)$$

but this is not of much help. Alternatively, an approximation can be obtained by expanding $(a+bX)^{-1}$ by the binomial series and taking a sufficient number of terms. A better approximation is found by expanding $\{a+bnp+b(X-np)\}^{-1}$ in powers of $b(X-np)(a+bnp)^{-1}$. See also Thionet (1963).

When $a = 0$, the expectation cannot exist because X^{-1} has a non-zero probability of having an infinite value but, nevertheless, when npq is not small the second expansion may still give a good measure of the centre of the distribution of $(bX)^{-1}$. To avoid the difficulty of infinite moments it is often usual to modify the distribution so that there is no zero class thus obtaining a 'truncated binomial' (see §2.11).

The sum of two binomial variates with the same p is clearly again a binomial variate but if the p's differ the individual probabilities are more complicated. The difference of two binomial variates with the same p also has a complicated expansion but can be expressed in a more concise form in terms of Legendre functions (de Castro 1952).

2.9. The normal approximation to the binomial distribution

We have already seen that the mean and standard deviation of the binomial distribution are np and $\sqrt{\{np(1-p)\}}$. If we keep p fixed and let n increase we readily see from Tchebychev's inequality (§1.9) that

$$\mathrm{pr}\{|X-np| > \alpha n\},$$

where α is any fixed positive number, tends to zero. This means that the distribution of Xn^{-1}, which can take the values 0, n^{-1}, $2n^{-1}$,..., $(n-1)n^{-1}$, 1, tends to become closely concentrated about the value p. We shall show that the shape of the distribution around the value p, when suitably scaled, tends to a fixed shape which is that of a certain continuous distribution known as the normal distribution (see §7.2). Write

$$\phi(x) = (2\pi)^{-\frac{1}{2}}e^{-\frac{1}{2}x^2}. \tag{2.38}$$

Then it is easy to show that

$$\int_{-\infty}^{\infty} \phi(x)\,dx = 1.$$

To do this observe that the square of the left-hand side can be written

$$(2\pi)^{-1}\int_{-\infty}^{\infty}\int_{-\infty}^{\infty} e^{-\frac{1}{2}(x^2+y^2)}\,dx\,dy = (2\pi)^{-1}\int_{0}^{\infty}\int_{0}^{2\pi} e^{-\frac{1}{2}r^2}r\,dr\,d\theta = 1.$$

Thus

$$\Phi(x) = (2\pi)^{-\frac{1}{2}}\int_{-\infty}^{x} e^{-\frac{1}{2}u^2}du \tag{2.39}$$

is a function which continuously increases from 0 to 1 as x increases from $-\infty$ to ∞. It is therefore the 'distribution function' (§1.1) of a random variable. Put

$$b(n, p, j) = \binom{n}{j}p^j q^{n-j} \quad (q = 1-p), \tag{2.40}$$

and

$$B(n, p, j) = \sum_{i=0}^{j}\binom{n}{i}p^i q^{n-i}. \tag{2.41}$$

6

We attempt to approximate to (2.40) by a function of the form (2.38), and to (2.41) by a function of the form (2.39). These two approximations are of a different character, for the variate of the binomial distribution is essentially discrete, whereas that of the normal distribution is continuous. On the other hand, we can regard (2.41) as defining a distribution function, $F(x)$, where i is the integral part of x, and after suitable scaling we shall show that $B(n,p,j)$ converges to $\Phi(x)$.

It is natural to use a scaling of the binomial variate, X say, so that its mean and standard deviation are those of the normal variate defined by $\phi(x)$. Since it is easy to show that

$$\int_{-\infty}^{\infty} x\phi(x)\,dx = 0, \qquad\qquad (2.42)$$

$$\int_{-\infty}^{\infty} x^2\phi(x)\,dx = 1, \qquad\qquad (2.43)$$

we shall consider, instead of X, the random variable

$$Z = \frac{X-np}{\sqrt{(npq)}}, \qquad\qquad (2.44)$$

and we first show that as n increases, p being fixed, $b(n,p,j)$ converges to

$$n^{-1}\phi\left(\frac{j-np}{\sqrt{(npq)}}\right). \qquad\qquad (2.45)$$

To do this we start from (2.40) and write $x = j-np$ so that $j = x+np$, $n-j = nq-x$. From (2.45) we are interested in convergence for cases where j and n increase in such a way that

$$\frac{j-np}{\sqrt{(npq)}}$$

remains constant or is at least bounded. This implies that $j-np = O(n^{\frac{1}{2}})$. We can relax this a little, and suppose that in the limit

$$n^{-\frac{2}{3}}|j-np| \to 0, \qquad\qquad (2.46)$$

so that for n sufficiently large we can ensure that

$$\frac{x}{np}, \frac{x}{nq}$$

are both less than $\frac{1}{2}$. Using Stirling's approximation (§ 1.12) we have

$$b(n,p,j) = \binom{n}{j}p^j q^{n-j},$$

$$= (2\pi)^{-\frac{1}{2}}\frac{n^{n+\frac{1}{2}}}{j^{j+\frac{1}{2}}(n-j)^{n-j+\frac{1}{2}}}p^j q^{n-j}\exp\left\{\alpha\left(\frac{1}{n}+\frac{1}{j}+\frac{1}{n-j}\right)\right\},$$

where $|\alpha| < 1$. This can be written as

$$(2\pi)^{-\frac{1}{2}}\left\{\frac{n}{j(n-j)}\right\}^{\frac{1}{2}}\left(\frac{np}{j}\right)^{j}\left(\frac{nq}{n-j}\right)^{n-j}\beta, \qquad (2.47)$$

where $\beta \to 1$ uniformly in the range (2.46). Using (2.46) again we see that the second factor in (2.47) tends uniformly to

$$(npq)^{-\frac{1}{2}}.$$

We also know that for $|x| < \frac{1}{2}$, say,

$$\log(1+x) = x - \tfrac{1}{2}x^2 + K_2 x^3, \qquad (2.48)$$

where K_2 is bounded. We can therefore write

$$\log\left\{\left(\frac{np}{j}\right)^{j}\left(\frac{nq}{n-j}\right)^{n-j}\right\} = -j\log\left(1+\frac{x}{np}\right) - (n-j)\log\left(1-\frac{x}{nq}\right)$$

$$= -(np+x)\left\{\frac{x}{np} - \frac{1}{2}\left(\frac{x}{np}\right)^2 + K_2\left(\frac{x}{np}\right)^3\right\}$$

$$+(x-nq)\left\{-\frac{x}{nq} - \frac{1}{2}\left(\frac{x}{nq}\right)^2 - K_2\left(\frac{x}{nq}\right)^3\right\}$$

$$= -\frac{1}{2}\frac{x^2}{npq} + K_3\frac{x^3}{n^2}, \qquad (2.49)$$

where $K_3 > 0$ is a bounded function of x, p, q, n in the given range. Taking exponentials we have the following theorem.

THEOREM 2.2. *If* $|x|n^{-\frac{2}{3}} = |j-np|n^{-\frac{2}{3}}$ *converges to zero as* n *tends to infinity,* $b(n,p,j)$ *is asymptotically equal to*

$$(2\pi npq)^{-\frac{1}{2}}\exp\left\{-\frac{1}{2}\frac{(j-np)^2}{npq}\right\}, \qquad (2.50)$$

and the convergence of the ratio of these expressions is uniform.

In particular this convergence is uniform for any region in which

$$\frac{j-np}{\sqrt{(npq)}}$$

is bounded.

Since the convergence is uniform we can immediately derive a corresponding theorem for $B(n,p,j)$, which concerns the convergence of the cumulative distribution represented by $\Phi(x)$.

THEOREM 2.3. *If* $|j-np|n^{-\frac{2}{3}}$, *and* $|k-np|n^{-\frac{2}{3}}$ *both converge to zero as* n *increases to infinity,* $j < k$, *then*

$$B(n,p,k) - B(n,p,j)$$

is asymptotically equal to

$$(2\pi npq)^{-\frac{1}{2}} \int_{j-\frac{1}{2}}^{k+\frac{1}{2}} \exp\left\{-\frac{1}{2}\frac{(u-np)^2}{npq}\right\} du, \qquad (2.51)$$

so long as $k-j$ *also tends to infinity.*

The proof is immediately obvious on considering the representation of an integral as the limit of a sum. Formulae (2.50) and (2.51) provide approximations which are adequate for many purposes, the largest term in the error usually arising when $p \neq \frac{1}{2}$. The limits of integration, $j-\frac{1}{2}$ and $k+\frac{1}{2}$, in (2.51) can naturally be replaced by j and k but the use of the correction $\frac{1}{2}$ results in a closer numerical approximation and is known as Yates's correction (Yates 1934).

Theorems 2.2. and 2.3 are of a different nature. Theorem 2.3 is a theorem about the convergence of distribution functions. Most limiting theorems in probability are of this type. Theorem 2.2, on the other hand, is a result on the limiting behaviour of the ordinate of a discrete distribution and such 'local' theorems are often harder to prove. They do not usually follow from the corresponding theorems about the distribution function unless further conditions are used.

Much research has been devoted to the problem of determining bounds for the error in the two above theorems. S. Bernstein devoted a series of rather inaccessible papers to this subject (see Bernstein 1943), and Feller (1945c) contains an improvement of Bernstein's results, in particular the remarkable theorem that a much better approximation to $b(n,p,j)$ than (2.50) is

$$\{(n+1)pq\}^{-\frac{1}{2}}\phi(x) \qquad (2.52)$$

where

$$x = \frac{j-(n+1)p+\frac{1}{2}}{\{(n+1)pq\}^{\frac{1}{2}}}. \qquad (2.53)$$

The use of $n+1$ instead of n, and the addition of $\frac{1}{2}$ is not at all the obvious sort of approximation to use since the mean and variance of j are np and npq, and not $(n+1)p-\frac{1}{2}$, $(n+1)pq$. This illustrates the devious cunning necessary in finding highly accurate approximations. (A simple discussion is given in Feller 1957b, 2nd edn only, pp. 163 and 182.) Other approximations for the tail of a binomial are given by Bahadur (1960).

2.10. The compound binomial

The idea of 'compounding' a distribution can be usefully applied to the binomial distribution by letting p have a continuous distribution on some range in $(0, 1)$. The most natural distribution to use is the beta distribution which will be discussed in §7.21 and which we have already used for a different purpose in (2.33).

Suppose that p has the continuous distribution defined by

$$f(p) = \frac{\Gamma(\alpha+\beta)}{\Gamma(\alpha)\Gamma(\beta)} p^{\alpha-1}(1-p)^{\beta-1}, \tag{2.54}$$

in the range $0 \leqslant p \leqslant 1$, so that

$$\int_0^x f(u)\,du$$

is the probability that $p \leqslant x$. This is known as the beta distribution and α and β can be any numbers greater than zero. Then the probability that the random variable X takes the value j is

$$\binom{n}{j} \frac{\Gamma(\alpha+\beta)}{\Gamma(\alpha)\Gamma(\beta)} \int_0^1 p^{j+\alpha-1}(1-p)^{n-j+\beta-1}\,dp$$

$$= \binom{n}{j} \frac{\Gamma(\alpha+\beta)\Gamma(j+\alpha)\Gamma(n-j+\beta)}{\Gamma(\alpha)\Gamma(\beta)\Gamma(n+\alpha+\beta)}. \tag{2.55}$$

This is a discrete distribution extending over the values $j = 0, 1, ..., n$, and is sometimes known as the negative hypergeometric distribution (see also Skellam 1948, and Exercises 1.5 and 1.6). Its generating function is clearly

$$\frac{\Gamma(\alpha+\beta)}{\Gamma(\alpha)\Gamma(\beta)} \int_0^1 (1-p+pz)^n p^{\alpha-1}(1-p)^{\beta-1}\,dp, \tag{2.56}$$

and from this the moments may be obtained. Alternatively, they can be found by considering $E(X^s)$ ($s = 1, 2, ...$), conditional on the value of p, and then taking the expectation over the distribution of p.

In particular we see that

$$E(X) = n\alpha(\alpha+\beta)^{-1}, \tag{2.57}$$

$$E(X^2) = n\alpha(\alpha+\beta)^{-1} + n(n-1)\alpha(\alpha+1)(\alpha+\beta)^{-1}(\alpha+\beta+1)^{-1}, \tag{2.58}$$

so that

$$\mathrm{var}(j) = \frac{n\alpha\beta(\alpha+\beta+n)}{(\alpha+\beta)^2(\alpha+\beta+1)}. \tag{2.59}$$

Compare this with a binomial distribution with the same mean and range. If such a distribution has index N and probability P we have $N = n$ and $P = \alpha(\alpha+\beta)^{-1}$. Its variance is $nP(1-P)$ which is smaller than (2.59) as we expect. This property is sometimes very useful in applications where we require a distribution on the integers $0, 1, ..., n$ with a larger variance than that of the binomial with the same mean. Conversely, if we require a distribution with a variance smaller than that of a binomial we can use a hypergeometric distribution.

2.11. Distributions related to the binomial distribution

It sometimes happens that we have to deal with random variables which have binomial distributions except for the fact that the zero value cannot appear. This occurs frequently in human genetics where we may wish to collect all families having at least one known offspring which shows a rare genetic character (for example albinos). Such families usually have both parents normal in appearance. We may then wish to test the hypothesis that the probability of an offspring, the brother or sister of the known person with the rare character, being an albino is $\frac{1}{4}$, say (which would be the case if it is controlled by a single recessive genetic factor), and to do this we must consider the binomial distribution with probability $\frac{1}{4}$, and the zero class excluded.

The general form of such a distribution is

$$p_0 = 0,$$

$$p_i = \binom{n}{i} p^i q^{n-i} (1 - q^n)^{-1}, \quad i = 1, \dots, n. \tag{2.60}$$

The moments of this distribution about the origin are then equal to the moments of the corresponding binomial multiplied by $(1-q^n)^{-1}$.

Since p_0 is zero, $E(X^{-1})$ is finite, and as it is occasionally useful in statistical theory to know its value, a number of writers have studied its calculation by suitable series (Stephan 1945). Tables have been published by Grab and Savage (1954) for $n = 2(1)\ 20$, $p = 0\cdot01$, $0\cdot05(0\cdot05)\ 0\cdot95$, $0\cdot99$, and for $n = 21(1)\ 30$, $p = 0\cdot01$, $0\cdot05(0\cdot05)\ 0\cdot50$ (see also Exercise 2·4).

It is also sometimes useful to exclude more than one of the possible values of i, and in this case it is more often necessary to truncate at the other end, thus obtaining the distribution

$$p_i = \frac{\binom{n}{i} p^i q^{n-i}}{\sum\limits_{j=0}^{N} \binom{n}{j} p^j q^{n-j}}$$

$$= \frac{\binom{n}{i} \lambda^i}{\sum\limits_{j=0}^{N} \binom{n}{j} \lambda^j} \quad (\lambda = pq^{-1}), \tag{2.61}$$

where $N < n$, and $i = 0, 1, \dots, N$. This is known as the Engset distribution and occurs in telephone congestion theory. The values, p_i, are easily found from tables of the binomial distribution.

We have already defined the multinomial distribution by (1.50). The moment generating function is clearly

$$(p_1 z_1 + \ldots + p_k z_k)^n,$$

from which all the moments can be derived. A table of cumulants to the fourth order is given in Kendall and Stuart (1958), Vol. I, and an extensive discussion of such cumulants with generalizations and tables is given by Wishart (1949).

Just as the ordinates and cumulative distribution of the binomial can be approximated by the univariate normal distribution, so the multinomial distribution can be approximated by the multivariate normal distribution considered in Chapters 5 and 7. The multinomial is the joint distribution of the numbers $r_1, \ldots, r_k$ of events $E_1, \ldots, E_k$ in n trials each of which must eventuate in one of the E_i. Hence $\Sigma r_i = n$, and the joint distribution of the r_i is 'singular', i.e. they satisfy an identity. In making a multivariate normal approximation we must therefore choose $k-1$ of the r_i and approximate by a multivariate normal distribution for $k-1$ random variables. This is in fact what we have done for the binomial where $k = 2$.

Other approximations for the terms of the multinomial distribution are given by Johnson (1960), Johnson and Young (1960), and Wise (1963). The compounding of the multinomial is considered by Mosimann (1962), and another generalization is given by Tallis (1962).

2.12. The hypergeometric distribution

We have already introduced this distribution in §1.11. There are three parameters M, m, and n, where $0 \leqslant m \leqslant M$, $0 < n \leqslant M$, and the probability that the random variable X takes the value k is (contrast (2.55))

$$\binom{M}{n}^{-1} \binom{m}{k} \binom{M-m}{n-k} \tag{2.62}$$

using the convention, always convenient in dealing with binomial coefficients, that if a, b are integers,

$$\binom{a}{b} = 0,$$

when $a > 0$ and $b < 0$, or $b > a$. Then k must lie in the range

$$\max(0, n-M+m), \quad \min(m, n).$$

The first two moments are given in (1.35), (1.36), and (1.37). To discuss the moments generally we consider the probability generating function obtained by multiplying (2.62) by z^k and summing over the possible values. Then, remembering the above convention,

$$P(z) = \binom{M}{n}^{-1} \sum_{k=0}^{n} \binom{m}{k}\binom{M-m}{n-k} z^k. \tag{2.63}$$

This is the coefficient of x^n in $(1+zx)^m(1+x)^{M-m}\binom{M}{n}^{-1}$, a result apparently first noticed by Soper (1922).

The hypergeometric function, from which this distribution derives its name, is defined (Whittaker and Watson 1935) by the series

$$F(a,b;c,z) = 1 + \frac{ab}{c}z + \frac{a(a+1)b(b+1)}{2!\,c(c+1)}z^2 + \dots, \tag{2.64}$$

which is analytic for $|z| < 1$ if c is not a negative integer. By substitution it satisfies the differential equation

$$z(1-z)\frac{d^2F}{dz^2} + \{c-(a+b+1)z\}\frac{dF}{dz} - abF = 0. \tag{2.65}$$

If we put $a = -n$, $b = -m$, $c = M-m-n+1$, we see that

$$P(z) = \frac{(M-m)!\,(M-n)!}{(M-m-n)!\,M!} F(-n, -m; M-m-n+1, z). \tag{2.66}$$

Substituting in (2.65) and putting $z = e^\theta$, $P(e^\theta) = \phi(\theta)$, we find that $\phi(\theta)$ satisfies

$$(1-e^\theta)\{\phi''(\theta) - (n+m)\phi'(\theta) + nm\phi(\theta)\} - nm\phi(\theta) + M\phi'(\theta) = 0. \tag{2.67}$$

Equating powers of θ to zero we can find the moments. m_1 and m_2 have already been given in (1.35) and (1.36), and μ_3, μ_4 are given in Kendall and Stuart (1958), Vol. I, who also give a recurrence relation from which the higher moments can be found. They are very complicated but the factorial moments are simpler (Exercise 2.1). A useful representation of $P(z)$ as a contour integral is given by Wise (1954a).

If we put $p = mM^{-1}$, $q = (M-m)M^{-1}$, and let M, m tend to infinity whilst keeping p, q, and n fixed, it is easily seen by using Stirling's formula that (2.62) tends to a binomial distribution with index n and probability p. It can also be shown that (2.67) then tends to the differential equation satisfied by the moment generating function of the binomial distribution, namely

$$(1-p+pe^\theta)\phi'(\theta) = npe^\theta\phi(\theta). \tag{2.68}$$

Since the binomial can be approximated in terms of the normal distribution the same is also true of the hypergeometric. It is also known that if np is kept fixed in the binomial distribution whilst n increases the limiting form is the 'Poisson distribution' considered in the next section, and the same is true of the hypergeometric when M, m are allowed to tend to infinity whilst the mean, nmM^{-1}, is kept fixed. An improved normal approximation to the hypergeometric distribution, along the lines of Feller's approximation to the binomial (2.52), has been given by W. L. Nicholson (1956).

Liebermann and Owen (1961) give tables of the hypergeometric distribution. These give individual terms and cumulative sums to six decimals for $M = 2(1)\ 50(10)\ 100,\ 1000,\ m = 1(1)\ 50,\ 500$, and $0 < n < m$.

The tails of the distribution may be calculated directly from tables of factorials but this may be a lengthy process. Asymptotic formulae have been given by Wise (1954a).

Generalizations of the hypergeometric distribution are given by Kemp and Kemp (1956).

2.13. The Poisson distribution

This is a distribution over all the positive integers defined by

$$p_n = e^{-\lambda}\lambda^n(n!)^{-1}, \quad n = 0, 1, 2,\dots \quad (\lambda > 0). \tag{2.69}$$

It is usually introduced as the limiting form of the binomial when $np = \lambda$ is kept fixed, and $n \to \infty$, $p \to 0$. Keeping j fixed, the jth term is clearly the asymptotic limit of

$$\frac{1}{j!}\left(\frac{n!}{n-j!}\right)p^j\left(1-\frac{\lambda}{n}\right)^{n-1},$$

which converges to

$$\frac{1}{j!}\lambda^j e^{-\lambda}.$$

However, the Poisson distribution occurs in a more fundamental way as the most naturally occurring distribution of the number of 'events' or 'points' in some region of space or time, when the distribution and occurrence of such events is 'random'. To illustrate this in the simplest case suppose that points are distributed at random on the whole line $(-\infty < x < \infty)$ in such a way that:

(1) the numbers of points occurring in two or more disjoint intervals, I_1 and I_2, are distributed independently of each other;

(2) the expected number occurring in any finite interval is finite, and proportional to the length of the interval (equal to $\lambda|I|$, say);

(3) if the length of the interval I is $|I|$, the probability of more than one point occurring in I tends to zero faster than $|I|$ when $|I| \to 0$.

From these assumptions it follows that the number of points occurring in any interval I, of length $|I|$, has a Poisson distribution with the λ of (2.69) replaced by $\lambda|I|$. For suppose the interval divided into n intervals of equal size. The expected number occurring in any interval of length l is λl. Then the probability of just one point occurring in one of these intervals will be $\lambda|I|n^{-1} + o(n^{-1})$, and the probability of more than one will be $o(n^{-1})$. The probability of exactly j points occurring in j of the intervals and none else-where will be

$$\binom{n}{j}\{\lambda|I|n^{-1}+o(n^{-1})\}^j\{1-\lambda|I|n^{-1}+o(n^{-1})\}^{n-j},$$

and this tends to

$$e^{-\lambda|I|}\{\lambda|I|\}^j(j!)^{-1}.$$

A similar discussion applies to the consideration of random points in spaces of higher dimension such as would arise, for example, if stars are distributed at random in space.

The probability generating function of (2.69) is

$$P(z) = \sum_{j=0}^{\infty} e^{-\lambda}(\lambda z)^j(j!)^{-1}$$

$$= e^{\lambda(z-1)}. \tag{2.70}$$

The convergence of the binomial to the Poisson could therefore also have been demonstrated by showing that (2.70) is the limit of the generating function of the binomial distribution when $np \to \lambda$, $n \to \infty$, $p \to 0$, and using Theorem 2.1.

We similarly have

$$P(e^\theta) = \exp \lambda(e^\theta - 1)$$

as the moment generating function, and thus in particular

$$m_1 = \lambda,$$

$$m_2 = \lambda + \lambda^2, \qquad\qquad \mu_2 = \lambda,$$

$$m_3 = \lambda + 3\lambda^2 + \lambda^3, \qquad \mu_3 = \lambda,$$

$$m_4 = \lambda + 7\lambda^2 + 6\lambda^3 + \lambda^4, \quad \mu_4 = \lambda + 3\lambda^2.$$

Since all cumulants are equal to λ, higher moments are easily found by substituting in the formulae giving the m_i in terms of the κ_i.

In (2.33) we have given a relationship between the tails of the binomial and those of the beta distribution. A similar relationship holds between the tails of the Poisson and the gamma distribution to be considered in §7.17. The latter is defined by the cumulative distribution

$$F(x) = \frac{\lambda^m}{\Gamma(m)} \int_0^x e^{-\lambda u} u^{m-1}\, du, \tag{2.71}$$

where $0 \leqslant x < \infty$, and m is a real positive number. Then if m is an integer, the equation

$$F(x) = e^{-\lambda x}\left\{\frac{(\lambda x)^m}{m!} + \frac{(\lambda x)^{m+1}}{m+1!} + \ldots\right\} \tag{2.72}$$

follows by repeated integration by parts. A generalization of this type of relationship, which gives rise to a generalized Poisson distribution, has been studied by Morlat (1952).

A distribution which occurs frequently in statistical theory, derived from the normal distribution, and described in §7.17, is that of a quantity known as χ^2. This distribution depends on a parameter n, known as the 'degrees of freedom of χ^2', and a random variable Z is said to have the χ^2 distribution with n degrees of freedom if it has the cumulative distribution (2.71) with $\lambda = \frac{1}{2}$, $m = \frac{1}{2}n$.

The tables of the χ^2 distribution from which (2.72) can be found are listed in §7.17. Here we describe only those tables of the terms and tails of the Poisson distribution which were calculated for this particular purpose. E. C. Molina (1942) gives individual terms and the tails to seven decimal places for $\lambda = 0.001(0.001)$ $0.01(0.01)$ 0.3, 0.4, and to six places for $0.5(0.1)$ $15(1)$ 100. *Biometrika Tables for Statisticians* (1954), Part I, gives a table of individual terms to six decimals for $\lambda = 0.1(0.1)$ 15 which is reproduced from the earlier *Tables for Statisticians and Biometricians* (1930) by K. Pearson. There is also a table to five significant places for $\lambda = 0.1(0.1)$ $1.0(1.0)$ 20 in Fry (1928).

It is clear from the probability generating function (2.70) that if X and Y are independently distributed in Poisson distributions with means λ and μ, $X+Y$ is distributed in a Poisson distribution with mean $\lambda + \mu$. Raikov (1938) has proved a remarkable converse to this result. If X and Y are independent random variables such that $X+Y$ has a Poisson distribution, then X and Y have Poisson distribution. In this theorem we may assume initially only that X and Y have general distributions and we therefore postpone the proof to Chapter 9 where we consider the general theory of decompositions of this kind.

The Poisson distribution can also be characterized in another way. If X and Y are independent random variables distributed in Poisson distributions with means λ and μ, their joint distribution is

$$e^{-\lambda-\mu}\lambda^X\mu^Y(X!)^{-1}(Y!)^{-1}, \tag{2.73}$$

and the distribution of their sum is

$$e^{-\lambda-\mu}(\lambda+\mu)^{X+Y}\{(X+Y)!\}^{-1}.$$

(2.73) can be written

$$\frac{(X+Y)!}{X!\,Y!}\left(\frac{\lambda}{\lambda+\mu}\right)^X\left(\frac{\mu}{\lambda+\mu}\right)^Y e^{-\lambda-\mu}(\lambda+\mu)^{X+Y}\{(X+Y)!\}^{-1},$$

and hence the conditional distribution of X, given $X+Y$, is binomial. This property can be shown to characterize Poisson distributions in the class of all discrete distributions on $0, 1, \ldots$ (Moran 1952, Daboni 1959, Lamperti 1959, Patil and Seshadri 1964).

The distribution of the difference, $X - Y$, of two independent Poisson random variables with means λ and μ is more complicated since it is a distribution on all the integers $(..., -1, 0, 1, 2, ...)$ and was first obtained by Skellam (1946) (Irwin (1937) gave the case $\lambda = \mu$). The probability that the difference is equal to k is the coefficient of z^k in

$$\exp\{\lambda(z - 1) + \mu(z^{-1} - 1)\}. \tag{2.74}$$

Now it is known that

$$\exp\{\tfrac{1}{2}z(t + t^{-1})\} = \sum_{-\infty}^{\infty} t^n I_n(z), \tag{2.75}$$

where $I_n(z) = i^{-n} J_n(iz)$ is a Bessel function which has been extensively tabulated. Using (2.75), (2.74) becomes

$$e^{-\lambda - \mu} \exp(\lambda z + \mu z^{-1}) = e^{-\lambda - \mu} \exp[\sqrt{(\lambda\mu)}\{z\sqrt{(\lambda\mu^{-1})} + z^{-1}\sqrt{(\mu\lambda^{-1})}\}]$$

$$= e^{-\lambda - \mu} \sum_{-\infty}^{\infty} \{z\sqrt{(\lambda\mu^{-1})}\}^n I_n\{2\sqrt{(\lambda\mu)}\},$$

so that

$$\mathrm{pr}(X - Y = k) = e^{-\lambda - \mu}(\lambda\mu^{-1})^{\frac{1}{2}k} I_k\{2\sqrt{(\lambda\mu)}\}. \tag{2.76}$$

When λ or μ is large, the ordinates of (2.76) can be well approximated by the normal distribution. This distribution provides a solution to D. G. Kendall's 'taxi-cab' queueing problem (§ 3.23).

From the manner in which we have approximated to the binomial distribution by the normal distribution, and the definition of the Poisson as the limit of the binomial, we expect that so long as λ is not too small we can approximate to

$$e^{-\lambda} \frac{\lambda^x}{x!}$$

by

$$(2\pi\lambda)^{-\frac{1}{2}} \exp\left\{\frac{-(x - \lambda)^2}{2\lambda}\right\},$$

and to

$$\sum_{x=a}^{b} e^{-\lambda} \lambda^x / x!$$

by

$$(2\pi\lambda)^{-\frac{1}{2}} \int_a^b \exp\left\{-\frac{(x - \lambda)^2}{2\lambda}\right\} dx.$$

This can be proved in the same way as for the binomial by using Stirling's formula. A bound for the error in the second of these approximations is given by Cheng (1949).

A number of interesting distributions have been derived by compounding the Poisson distribution with other distributions either by making its parameter λ have another distribution, or by using the random variable X

as the parameter of another distribution. Thus consider a binomial distribution with probability p and index n, where n is a random variable having a Poisson distribution with mean λ. Then the resulting generating function is

$$\sum_{n=0}^{\infty} \{(q+pz)^n e^{-\lambda}\}^n (n!)^{-1} = e^{-\lambda p(1-z)}$$

which is that of a Poisson distribution with mean λp.

This can be given an intuitive interpretation, for suppose that points are distributed at random on a line in such a way as to satisfy the assumptions (1)–(3) above, so that the number occurring in any unit interval is a Poisson variate with mean λ. If an observer counts the points in some unit interval in such a way that each point has independently a probability p of being observed, and q of being missed, the resulting number counted will have the above compound distribution. However, the points which occur, and are also counted, satisfy conditions (1)–(3) and therefore have a Poisson distribution which must have the mean λp.

If X is a Poisson variate with mean λ, and λ has some distribution on the positive real line ($0 \leqslant \lambda < \infty$), X will have a compound distribution. The most important case occurs when λ has a gamma or type III distribution and then X has the negative binomial distribution considered in §2.14.

A similar distribution arises when λ is itself proportional to a Poisson variate. A compound distribution of this kind has been studied by M. Thomas (1949) to deal with some distributions in plant and animal ecology. Suppose that plants are initially distributed at random on an area so that the number occurring in any region of unit area has a Poisson distribution with mean λ. Call these the 'parent' plants and suppose that each has a number of offspring distributed in a Poisson distribution with mean μ. We are interested in the distribution of the total number of plants in the area.

If there are X parents, the total number of offspring, Y, will have a Poisson distribution with mean μX, and the probability generating function, conditional on X, is

$$e^{\mu X(z-1)}.$$

The probability generating function of $X + Y$ is therefore

$$e^{-\lambda} \sum_{s=0}^{\infty} (s!)^{-1} \lambda^s z^s e^{\mu s(z-1)} = \exp\{-\lambda + \lambda z e^{\mu(z-1)}\}. \tag{2.77}$$

The probability of exactly k individuals, parents, or offspring is the coefficient of z^k in (2.77) which is

$$e^{-\lambda} \sum_{s=0}^{k} \frac{(\lambda e^{-\mu})^s}{s! \, k-s!} (\mu s)^{k-s}. \tag{2.78}$$

This is called the double Poisson distribution by Thomas, although this name is sometimes applied to mixtures of two Poisson distribution, i.e. distributions whose generating functions are of the form

$$\alpha e^{\lambda(z-1)} + \beta e^{\mu(z-1)},\qquad(2.79)$$

where α, $\beta > 0$, and $\alpha + \beta = 1$. The moments of (2.78) are easily found by differentiating (2.77), and thus we find

$$m_1 = \lambda(1+\mu),$$
$$\mu_2 = \lambda(1+3\mu+\mu^2),\qquad(2.80)$$

which shows that the variance is greater than that of a Poisson distribution with the same mean.

It would, perhaps, be more natural to apply the term 'double Poisson' to the distribution of offspring alone. The generating function of this is

$$e^{-\lambda} \sum_{s=0}^{\infty} (s!)^{-1}\lambda^s e^{\mu(z-1)s} = \exp\{\lambda(e^{\mu(z-1)} - 1)\}\qquad(2.81)$$

and the moments are

$$m_1 = \lambda\mu,$$
$$\mu_2 = \lambda(\mu+\mu^2).\qquad(2.82)$$

This is a particular case of the type of distribution defined by (2.26) and is known as Neyman's type A distribution.

Another less convenient distribution to use for λ is the log-normal distribution on $(0,\infty)$ whose cumulative distribution is

$$\frac{1}{\sigma\sqrt{(2\pi)}} \int_0^\lambda u^{-1}\exp-\frac{1}{2}\left\{\frac{1}{\sigma^2}(\log u)^2\right\} du.\qquad(2.83)$$

Explicit values of the probabilities must then be found by numerical integration (see Grundy 1951, Preston 1948) since the integrals giving p_n and the generating function have not been obtained explicitly.

It is sometimes also useful to compound the Poisson distribution with parameter λ by supposing that λ has the beta distribution on the finite interval $(0, 1)$. The resulting frequencies are then rather awkward since they are multiples of Whittaker's confluent hypergeometric function (see (6.23)). The F-distribution (§7.21) can also be used to compound the Poisson distribution.

Just as it is sometimes useful to consider a binomial distribution without its zero term so too it is often useful to consider a truncated Poisson. Removing the first term and rescaling the others to add to unity, we have, in analogy with (2.60),

$$p_n = e^{-\lambda}(1-e^{-\lambda})^{-1}\lambda^n/n!,\quad n = 1, 2,\dots.\qquad(2.84)$$

The moments of this distribution about the origin are then obviously equal to the moments of a Poisson distribution, with parameter λ, multiplied by

$(1-e^{-\lambda})^{-1}$, and the probability generating function is

$$\frac{e^{\lambda(z-1)}-e^{-\lambda}}{1-e^{-\lambda}}. \qquad (2.85)$$

We can also truncate the Poisson distribution on the other tail and obtain a bounded distribution

$$p_n = e^{-\lambda}\lambda^n \Big/ \Big\{ n! \sum_{j=0}^{N} e^{-\lambda}\lambda^j(j!)^{-1} \Big\}$$

$$= \lambda^n \Big/ \Big\{ n! \sum_{j=0}^{N} \lambda^j(j!)^{-1} \Big\}, \quad n = 0, 1,..., N. \qquad (2.86)$$

This is also known as 'Erlang's distribution' and arises in the theory of telephone traffic (compare Engset's distribution (2.61)). Palm (1954) gives p_n for $n = N$ to 6D for $N = 1(1)$ 150, $\lambda = 0.05(0.05)$ 1(0.1) 20(0.5) 30(1) 50, 52(4) 100, Jensen (1950) gives p_n for $n = N$, and various values of λ and N, and Brockmeyer, Halstrom, and Jensen (1948) give values of λ for which $p = 0.001(0.001)$ 0.005, 0.01, 0.02, 0.05, and $N = 1(1)$ 260. p_n is, however, very easily found from tables of the Poisson distribution.

$E(X^{-1})$ does not exist for the Poisson distribution but does exist for the Poisson without a zero class ((2.84)), and has been tabulated by Grab and Savage (1954) to five decimals for $\lambda = 0.01$, $0.05(0.05)$ 1.0(0.1) 2.0(0.2) 5.0(0.5) 7.0(1) 10(2) 20.

In statistical analysis it is also sometimes useful to know $E(X^{\frac{1}{2}})$. This arises because the distribution of $X^{\frac{1}{2}}$ can be more closely approximated by a normal distribution than that of X. Statistical data involving frequencies are therefore often transformed by a square-root transformation before further analysis. This has not been tabulated directly but $\text{var}(\sqrt{X}) = E(X)-\{E(\sqrt{X})\}^2$ has been tabulated by Irwin (1943) for $\lambda = 0(0.1)$ 15, 20, 50, and 100. It is easy to see that as λ gets large $\text{var}(\sqrt{x})$ tends to $\frac{1}{4}$.

For the multivariate Poisson distribution see Lukacs and Laha (1964).

2.14. The negative binomial distribution

Supposing that λ in the Poisson distribution has the gamma distribution

$$F(x) = \frac{a^m}{\Gamma(m)} \int_0^x e^{-au}u^{m-1}\, du,$$

where $a > 0$, $m > 0$, the compound distribution has the form

$$p_i = \frac{a^m}{\Gamma(m)} \int_0^\infty e^{-u-au}\frac{u^i}{i!}u^{m-1}\, du$$

$$= \left(\frac{a}{1+a}\right)^m \frac{\Gamma(m+i)}{\Gamma(m)\Gamma(i+1)}(1+a)^{-i}. \qquad (2.87)$$

Here i must be integral but m can be any positive real number, and $p_i z^i$ is then the ith term in the expansion of the generating function

$$\left(\frac{a}{1+a}\right)^m \left(1-\frac{z}{1+a}\right)^{-m}, \tag{2.88}$$

which gives rise to the name of 'negative binomial'. Putting $b = a^{-1}$, (2.88) becomes

$$P(z) = (1+b-bz)^{-m}, \tag{2.89}$$

where b can take any positive value.

The generation of a negative binomial distribution by compounding was first considered by Greenwood and Yule (1920) in order to account for the distribution of the number of accidents to workers in factories not being well fitted by a Poisson distribution. If the number of accidents occurring to any particular worker in some fixed period of time is a Poisson distribution with a mean which varies from worker to worker, the overall observed distribution will not be Poisson. The gamma distribution is an easily used distribution on $(0, \infty)$ whose mean and variance can be varied at will and thus provides a plausible distribution for the parameter in the Poisson distribution. This parameter may be regarded as a measure of 'accident proneness'. The resulting negative binomial fits observed accident distributions very well, but unfortunately, as we shall show in the next chapter, a negative binomial distribution of accidents can arise from a quite different but equally plausible hypothesis. (For surveys of the history of the application of this distribution to accident proneness see Greenwood 1950 and Irwin 1964.)

When m is an integer the distribution can arise in still another way. Consider first the case $m = 1$. Suppose that a series of independent trials is carried out of an experiment which has two possible outcomes, A and B, which occur with probabilities P and $Q = 1-P$. In particular this could be the sequential sampling of an infinite population consisting of two different kinds of object in the proportions P and Q. Let X be the number of trials up to and including the first trial at which A occurs. Then

$$\mathrm{pr}(X = s) = PQ^{s-1}, \quad s = 1, 2,\ldots, \tag{2.90}$$

and the generating function of the distribution of $X-1$ is

$$P(1-Qz)^{-1}. \tag{2.91}$$

(2.89) can be put in this form by taking $m = 1$ and $b = (1-P)P^{-1}$. (2.90) is known as the 'geometric distribution', and is an important special case of the negative binomial.

Now let X be the total number of trials up to and including the first trial at which A occurs for the kth time. Clearly $X-k$ has the generating function

$$P^k(1-Qz)^{-k}, \tag{2.92}$$

which is known as the 'Pascal distribution' for no very clear historical reason, and is again a negative binomial.

The distribution of X is the distribution of the number of coupons (such as those given away in packets of cigarettes) which need to be collected in order to obtain k copies of a particular coupon, when P is the probability of obtaining that coupon in any trial. The problem of the distribution of the number of coupons collected in order to obtain at least one example of s specified coupons, which occur in the population with specified probabilities $P_1, ..., P_s$, has a fairly large literature (see David and Barton 1962).

Returning to the generating function (2.89) we see that if X has a negative binomial distribution it can always be regarded as the sum of two independent random variables, Y, Z, which have non-degenerate distribution which are themselves negative binomial distributions (for example with parameters p and $\frac{1}{2}m$). This property which the negative binomial shares with the Poisson, but not the binomial, is related to the theory of infinite divisibility, and will be treated in detail in Chapter 9.

It is convenient to write (2.88) and (2.89) in the form

$$P(z) = P^m(1 - Qz)^{-m}, \tag{2.93}$$

by putting

$$P = 1 - Q = a(1 + a)^{-1} = (1 + b)^{-1}.$$

The moments which are then most easily found are the factorial moments about the origin. These are given by (formula (2.6))

$$m_{(s)} = p^{(s)}(1)$$
$$= \left(\frac{Q}{P}\right)^s m(m+1)...(m+s-1). \tag{2.94}$$

Using this result we find, putting $QP^{-1} = \alpha$ for convenience

$m_1 = \alpha m,$

$m_2 = \alpha m + \alpha^2 m(m+1),$

$m_3 = \alpha m + 3\alpha^2 m(m+1) + \alpha^3 m(m+1)(m+2),$

$m_4 = \alpha m + 7\alpha^2 m(m+1) + 6\alpha^3 m(m+1)(m+2) + \alpha^4 m(m+1)(m+2)(m+3),$

and

$\mu_2 = m(\alpha + \alpha^2),$

$\mu_3 = m(\alpha + 3\alpha^2 + 2\alpha^3),$

$\mu_4 = m(\alpha + 7\alpha^2 + 12\alpha^3 + 6\alpha^4) + m^2(3\alpha^2 + 6\alpha^3 + 3\alpha^4).$ \hfill (2.95)

The m_s could equally well have been found by averaging the moments of the Poisson distribution over the distribution of λ. This argument naturally does not apply to the μ_s.

7

If $\alpha = QP^{-1}$ is kept fixed and m increases, the distribution tends to normality both in its ordinates and in its tails, as can be shown by the same kind of argument as for the binomial.

Tables of the tails of the negative binomial are given by Grimm (1962). Rider (1962) gives $E(X^{-1})$, $E(X^{-2})$, and $\text{var}(X^{-1})$ for a negative binomial distribution, with the zero term removed, for $p = 0.01$, $0.05(0.05)$ $1.00(1)$ 5, and $m = 1(1)$ 30, where the generating function is taken in the form

$$(1 + p - zp)^{-m}.$$

Just as in the case of the Poisson distribution, it is possible to obtain a closed integral for the tail of a negative binomial. Taking the distribution in the form (2.87), we have

$$\text{pr}(X \geqslant N) = \left(\frac{a}{1+a}\right)^m \sum_{i=N}^{\infty} \frac{\Gamma(m+i)}{\Gamma(m)\Gamma(i+1)}(1+a)^{-i}. \qquad (2.96)$$

We can show that this is equal to (Fisher 1935)

$$\frac{a^m \Gamma(m+N)}{\Gamma(m)\Gamma(N)} \int_0^1 \frac{x^{N-1}}{(a+x)^{m+N}}\,dx, \qquad (2.97)$$

by writing the latter in the form

$$\frac{a^m \Gamma(m+N)}{\Gamma(m)\Gamma(N)} \int_0^1 \frac{(1-x)^{N-1}}{(a+1-x)^{m+N}}\,dx$$

$$= \left(\frac{a}{1+a}\right)^m \frac{\Gamma(m+N)}{\Gamma(m)\Gamma(N)} \int_0^1 \sum_{s=0}^{\infty} (1+a)^{-N} \frac{\Gamma(m+N+a)}{s!\,\Gamma(m+N)} \left(\frac{x}{1+a}\right)^s (1-x)^{N-1}\,dx,$$

which is equal to (2.96) on using the integral for the beta function. The equivalence of (2.96) and (2.97) could also be shown by averaging (2.72) over the distribution of λ.

(2.97) is known as a beta integral of the second kind and will occur again in §7.22 when we consider a continuous distribution known as Fisher's F-distribution which is very well tabulated. It follows that the values of the tails of a negative binomial can be found from those of Fisher's F-distribution.

In the last section we considered a biological problem in which clusters of individuals had a Poisson distribution, each cluster being made up of a number of individuals which itself had a Poisson distribution. A similar idea can be used here. If the number of clusters has a Poisson distribution with mean λ so the generating function is

$$P(z) = \exp\{\lambda(z-1)\},$$

and each cluster has a number of individuals with the negative binomial distribution whose generating function is

$$P^m(1 - Qz)^{-m},$$

the total number has a distribution with generating function

$$P(z) = \exp \lambda \left\{ \frac{P^m}{(1-Qz)^m} - 1 \right\}, \tag{2.98}$$

which is the generating function of the generalized Pólya–Aepli distribution (§2.18). Similarly, we might suppose the number of clusters to have the negative binomial distribution so that the resulting generating function is of the form

$$\left(\frac{1-q}{1-q\left(\frac{1-Q}{1-Qz}\right)^m} \right)^n, \tag{2.99}$$

where $0 < q$, $Q < 1$, and $m, n > 0$. Alternatively, we may suppose that the number of clusters has a negative binomial distribution and the number of individuals in each has a Poisson distribution so that we have a generating function of the form

$$\left[\frac{1-Q}{1-Q\exp\{\lambda(z-1)\}} \right]^m. \tag{2.100}$$

Clearly (2.98) and (2.100) can both be considered as limiting cases of (2.99) for the limit of

$$\left(\frac{1-Q}{1-Qz} \right)^m$$

as $Q \to 0$, $m \to \infty$ in such a way that $mQP^{-1} = \lambda$ is equal to

$$\exp\{\lambda(z-1)\}$$

so that in these circumstances the limit of a negative binomial is a Poisson distribution just as for the positive binomial distribution.

In the above cases we have compounded the distribution by making m proportional to the variate of a discrete distribution. However, m can take any positive value and could therefore be supposed to have a gamma distribution, an F-distribution, a beta distribution, or any other continuous distribution on $(0, \infty)$.

Bivariate Poisson and negative binomial distributions have also been considered. For the former see Aitken (1939) and Maritz (1952), and for the latter Arbous and Sichel (1954). A compound negative multinomial has been studied by Mosimann (1963). See also Lukacs and Laha (1964).

2.15. The logarithmic distribution

Let $0 < Q < 1$. Then

$$-\log(1-Q) = Q + \tfrac{1}{2}Q^2 + \tfrac{1}{3}Q^3 \dots$$

is a convergent series of positive terms. The coefficients of any such series can be taken as proportional to a set of probabilities on the integers. Here we take them as proportional to probabilities on the integers 1, 2,... so that $p_0 = 0$. Then the generating function is

$$P(z) = \frac{\log(1 - Qz)}{\log(1 - Q)}. \tag{2.101}$$

This distribution arose in ecology in the following way (Fisher, Corbet, and Williams 1943). It is a common practice to catch large numbers of species (such as insects) in a trap (such as a light trap). For any given species the numbers caught have, very plausibly, a Poisson distribution with a mean depending on the species. This mean varies from species to species and may be reasonably assumed to have a gamma distribution. Then if we choose a species at random the number of its representatives trapped will have a negative binomial distribution whose generating function we write as

$$P^m(1 - Qz)^{-m}. \tag{2.102}$$

However, species without representatives in the trap will be unobserved and hence the observed distribution of numbers trapped amongst the species which are actually observed will be a distribution on the positive integers 1, 2, 3,..., with probabilities proportional to the corresponding probabilities in (2.102), and will therefore have a generating function

$$\frac{P^m\{(1 - Qz)^{-m} - 1\}}{1 - P^m}. \tag{2.103}$$

The non-negative quantity m is the parameter in the gamma distribution which is inversely related to the variance of λ for a given mean. When m becomes large this variance tends to zero and (2.102) tends to a Poisson distribution. When m becomes small the distribution defined by (2.102) collapses into the value zero whilst (2.103), which is the remaining distribution when the zero value is removed, has as a typical term, p_k, the limit of

$$P^m(1 - P^m)^{-1}(1 - P)^k \frac{m(m+1)...(m+k-1)}{k!}.$$

Now $m(1 - P^m)^{-1}$ converges to $(-\log P)^{-1}$ as m tends to zero, and hence p_k is

$$-(1 - P)^k(k \log P)^{-1},$$

which is the typical term of a logarithmic distribution whose generating function is (2.101). This usually gives a good fit to observations because the variability of species is often well represented by a gamma distribution with a very small m. Applications of this kind occur in other subjects and now have a large literature (see, for example, Good 1953). A survey of the theory of the logarithmic distribution, with bibliography and tables, is given by Patil, Kamat, and Wani (1964).

The moments of (2.101) may be found by taking limits in (2.95), or from the generating function. The first four are (putting $\alpha = QP^{-1}$)

$$m_1 = \alpha(-\log P)^{-1},$$

$$m_2 = (\alpha + a^2)(-\log P)^{-1},$$

$$m_3 = (\alpha + 3\alpha^2 + 2\alpha^3)(-\log P)^{-1},$$

$$m_4 = (\alpha + 7\alpha^2 + 12\alpha^3 + 6\alpha^4)(-\log P)^{-1} \qquad (2.104)$$

and the moments around the mean are also easily found.

The tail probabilities can be found by taking the limit of (2.97) which becomes

$$-\frac{1}{\log P} \int_0^1 \frac{x^{N-1}}{(PQ^{-1}+x)^N} dx \qquad (2.105)$$

(which is easily expressed in terms of the F-distribution) or by observing that

$$(-\log P)^{-1}\left\{\frac{Q^N}{N} + \frac{Q^{N+1}}{N+1} + \ldots\right\} = (-\log P)^{-1} \int_0^Q \frac{x^{N-1}}{1-x} dx, \qquad (2.106)$$

an integral which is easily transformed into (2.105).

The logarithmic distribution is involved in a number of interesting compound distributions. The sum of n independent random variables, each with the distribution defined by (2.101), has generating function

$$\left\{\frac{\log(1-Qz)}{\log(1-Q)}\right\}^n,$$

which is a distribution on the integers n, $n+1$, $n+2$,... and is not of much interest in itself. However, if n is allowed to have a Poisson distribution with mean λ, the generating function becomes

$$\exp \lambda\left\{\frac{\log(1-Qz)}{\log(1-Q)} - 1\right\} = \left(\frac{1-Qz}{1-Q}\right)^{\{\lambda/\log(1-Q)\}}, \qquad (2.107)$$

which is a negative binomial. This result was obtained by Quenouille (1949b), and generalized by Gurland (1957) in the following way. If the n, instead of following a Poisson distribution, follows a negative binomial whose generating function is

$$\left(\frac{1-qz}{1-q}\right)^{-m}$$

the resulting compound distribution will have a generating function

$$\frac{1-q\left\{\frac{\log(1-Qz)}{\log(1-Q)}\right\}^{-m}}{1-q}. \qquad (2.108)$$

Next consider a compound negative binomial distribution whose generating function is

$$\left(\frac{1-Qz}{1-Q}\right)^{-X},$$

where Q has the same value as above, and X has the gamma distribution on $(0, \infty)$ defined by

$$F(X) = \frac{\alpha^m}{\Gamma(m)} \int_0^X e^{-\alpha x} x^{m-1}\, dx.$$

Let m have the same value as in (2.108) and put α equal to $(1-q^{-1})\log(1-Q)$. Then the resulting generating function is

$$\frac{\alpha^m}{\Gamma(m)} \int_0^\alpha \left(\frac{1-Qz}{1-Q}\right)^{-x} e^{-\alpha x} x^{m-1}\, dx = \left\{1 + \alpha^{-1}\log\left(\frac{1-Qz}{1-Q}\right)\right\}^{-m}.$$

Inserting the value of α, this is equal to (2.108) showing the equality of the results of the two methods of compounding.

2.16. The Pólya distribution

Let $p > 0$, $q = 1-p > 0$, and $\gamma > 0$. The Pólya distribution on the integers $0, 1,..., n$ is defined by

$$p_j = \binom{n}{j} \frac{p(p+\gamma)...\{p+\gamma(j-1)\}q(q+\gamma)...\{q+\gamma(n-j-1)\}}{(1+\gamma)(1+2\gamma)...\{1+(n-1)\gamma\}}, \qquad (2.109)$$

where $j = 0, 1,..., n$. If $\gamma = 0$ this is clearly the binomial distribution. If $\gamma \neq 0$ the fact that the sum of the p_j is equal to unity can be verified by equating the coefficients of t^n on both sides of the equation

$$(1-t)^{-a}(1-t)^{-b} = (1-t)^{-a-b},$$

where $a = p\gamma^{-1}$, $b = q\gamma^{-1}$.

If we put $p = \gamma\alpha$, $q = \gamma\beta$, and $\gamma = (\alpha+\beta)^{-1}$, (2.109) is equal to (2.55), the binomial distribution compounded with a beta distribution (the 'negative hypergeometric' distribution—see also Exercises 1.5 and 1.6).

A special case of (2.109) can be given an interpretation in terms of sampling from a finite population of objects, the numbers of which change with the removal of each individual unit. This is the way in which Pólya (1930) originally introduced this distribution. (See also Eggenberger and Pólya 1932, and Friedman 1949.)

Suppose an urn contains a white balls and b black balls. A ball is chosen at random and replaced, together with c balls of the same kind. If k successive drawings have already been made, of which l are white and $k-l$ black, the probability that the next ball drawn is white will then be

$$(a+lc)(a+b+lc)^{-1}.$$

The probability of extracting l white balls in a sequence of k drawings in a specified order is therefore

$$\frac{a(a+c)...\{a+c(l-1)\}b(b+c)...\{b+(k-l-1)c\}}{(a+b)(a+b+c)...\{a+b+c(k-1)\}},$$

whatever the order. Hence the overall probability of obtaining l white balls in a sequence of k is

$$p_l = \binom{k}{l} \frac{a(a+c)...\{a+c(l-1)\}b(b+c)...\{b+(k-l-1)c\}}{(a+b)(a+b+c)...\{a+b+c(k-1)\}}. \qquad (2.110)$$

This is equal to (2.109) if we put $n = k$, $j = l$, and

$$p = a(a+b)^{-1}, \quad q = b(a+b)^{-1}, \quad \gamma = c(a+b)^{-1}.$$

The remarkable thing about this distribution is that it has been obtained in two quite different ways, one using binomial type sampling with successively independent choices, but with p varying in a beta-type distribution, and the other by assuming serial dependence in the probabilities for each element selected. This duality, as already mentioned in §2.14, occurs also in the derivation of the negative binomial distribution. In fact the negative binomial distribution is a limiting case of Pólya's distribution. Taking (2.109) and letting $n \to \infty$, $p \to 0$, $\gamma \to 0$ in such a way that $np \to \lambda$, $n\gamma \to P^{-1}$, where $Q = (1+P)^{-1}$, we get

$$p_j = \frac{\lambda P(\lambda P+1)...(\lambda P+j-1)}{j!} \left(\frac{1}{1+P}\right)^j \left(\frac{P}{1+P}\right)^{\lambda P}$$

$$= \frac{\lambda P(\lambda P+1)...(\lambda P+j-1)}{j!} (1-Q)^{\lambda P} Q^j,$$

which is the general term of the negative binomial distribution whose generating function is (2.93) with $m = \lambda P$.

This limiting behaviour of the Pólya distribution can be looked at in another way. The Poisson distribution is the limit of the binomial when $p \to 0$, $n \to \infty$. It can be shown that the gamma distribution is the limit of the beta distribution (2.54) when $\alpha \to 0$, $\beta \to 0$ in a suitable manner and the variable is rescaled. Since the Pólya distribution is a compound of the binomial and the beta distribution we expect to obtain the negative binomial as a limiting distribution.

The first two moments and the generating function of the Pólya distribution have already been given in (2.57), (2.58), and (2.56) respectively.

In the derivation of (2.110), a and b were positive integers, and c a non-negative integer. If we had put $c = -1$ we would have a distribution which results when after the choice and replacement of each ball, a ball of the same colour is *removed* from the urn. This is just what happens in sampling without replacement, and (2.110) becomes $(k < a, b)$

$$p_l = \frac{k!\,a!\,b!\,a+b-k!}{k-l!\,l!\,a-l!\,b-k+l!\,a+b!}, \quad l = 0, 1,..., k, \qquad (2.111)$$

which is a hypergeometric distribution. We have therefore, in Pólya's distribution a distribution which has as special cases both the hypergeometric (which has a smaller variance than the binomial with the same mean) and the compound binomial (which has a larger variance than the binomial with the same mean).

2.17. The distribution of exceedances

The Pólya distribution also occurs in another way (Wilks 1942, Gumbel and v. Schelling 1950, Sarkadi 1957a). Suppose that $X_1,..., X_n, Y_1,..., Y_N$, are independent random variables all with the same distribution function, $F(x)$, which is a continuous function.

Let X_i be the mth largest of the set $X_1,..., X_n$ so that there are $m-1$ X's larger than X_i and $n-m$ smaller. Let p_j $(j = 0,..., N)$ be the probability that j of the $Y_1,..., Y_N$ are larger than X_i and $N-j$ smaller (clearly since $F(x)$ is continuous it is easy to show that the probability that any two of the X's and Y's are equal is zero). We shall show that

$$p_j = \frac{m\binom{n}{m}\binom{N}{j}}{(n+N)\binom{N+n-1}{m+j-1}}, \quad j = 0, 1,..., N. \qquad (2.112)$$

This probability is independent of the distribution function $F(x)$ so long as the latter is continuous. Clearly what matters is not the actual values of the X_i and Y_i, but their relative order. We therefore suppose that the sampling is carried out in the following manner. Let $Z_1,..., Z_{n+N}$ be a set of values of independent random variables with the distribution function $F(x)$. Let $g_1,..., g_{n+N}$ be the Z_i rearranged in increasing order. We choose n of these values at random and allocate them at random to the quantities $X_1,..., X_n$. The remaining N quantities are then allocated at random to the Y_i. We now prove that the distribution (2.112) holds in the probability space of all such choices and allocations, the set of values $g_1,..., g_{n+N}$ being held fixed. This will clearly be sufficient to prove (2.112).

The total number of ways of allocating the values $g_1,..., g_{n+N}$ to the quantities $X_1,..., X_n, Y_1,..., Y_N$ is clearly $(n+N)!$. Consider the number of ways of doing this conditional on the mth largest of the X_i taking the value Z_k. Clearly k must lie in the interval

$$n+1-m \leqslant k \leqslant N+n+1-m,$$

and given k, the number of Y's greater than Z_k has the fixed value

$$j = N+n+1-m-k.$$

The number of ways of allocating the X's so that the mth largest is Z_k will be equal to n times the number of ways of allocating the X's so that $X_1 = Z_k$, and X_1 is the mth largest. This is equal to n times the number of ways of allocating the set $(X_2, ..., X_n)$ to $m-1$ Z's greater than Z_k, and to $n-m$ Z's less than Z_k. This is clearly

$$n(n-1)! \binom{n+N-k}{m-1}\binom{k-1}{n-m}.$$

Having allocated the X's, the number of ways of allocating the Y's to the N remaining Z's is $N!$. Hence the overall probability is

$$\frac{n!\,n+N-k!\,k-1!\,N!}{m-1!\,n+N-k-m+1!\,n-m!\,k-1-n+m!\,n+N!}.$$

Putting $k = n+N+1-m-j$ we obtain (2.112). If we put N for k, j for l, $c = 1$, m for a, and $n+1-m$ for b in (2.110) we also get (2.112) so that the distribution of exceedances is a Pólya distribution. A table of this distribution for $n = N = 2(1)\,15(5)\,20$ is given in Epstein (1954). From (2.112) is clear that such tables are not hard to calculate so long as one starts from a known table of $\binom{n}{r}$.

The distribution of exceedances has been applied to some extreme value problems in an interesting way. Consider the particular case where $N = 1$, $m = 1$. The problem is then equivalent to finding the probability that x_{n+1} is the largest in a sample $(x_1, ..., x_{n+1})$ of independent random variables from some continuous distribution. Thus suppose the maximum daily flows of a river in n successive years have been observed and there are good reasons (for example a strong seasonal effect) for believing that these flows can be regarded as statistically independent. In such circumstances it is often important to know the probability that the flow, in some future year, exceeds any specified value. The usual way of doing this is to make some estimate of the probability distribution of flows by fitting some standard distribution to the observed values $x_1, ..., x_n$. It it then possible to estimate the probability that any future flow, x_{n+1} say, exceeds any particular value x which might, in particular, be taken as the observed value of $\max(x_1, ..., x_n)$. This is an assertion about the probability distribution of x_{n+1}, although it is not an exact assertion because the distribution cannot be estimated exactly from a finite sample. On the other hand, we can assert that

$$\mathrm{pr}\{x_{n+1} > \max(x_1, ..., x_n)\} = (n+1)^{-1}, \tag{2.113}$$

and this is an exact statement provided we interpret it as referring to the probability space consisting of the joint distribution of $x_1, \dots, x_{n+1}$. This distinction is important in the theory of inference. The same kind of argument applies for general values of N and m.

2.18. Contagious and compounded distributions in general

Having given a number of examples of discrete distributions and the ways in which they may be compounded we are now in a position to summarize the general results on the compounding of distribution.

Suppose first that some discrete distribution depends on a continuous parameter λ, so that its generating function may be written $P(z, \lambda)$. If λ can take a discrete set of values $\lambda_0, \lambda_1, \dots$ and does so with probabilities $\pi_0, \pi_1, \dots$ the overall compounded distribution has generating function

$$\pi_0 P(z, \lambda_0) + \pi_1 P(z, \lambda_1) + \dots, \qquad (2.114)$$

and this is sometimes called a 'mixture' of the distributions for $\lambda = \lambda_0, \lambda_1, \dots$. We have seen how this idea can be generalized by allowing λ to take a continuous range of values and thus averaging $P(z, \lambda)$ with some continuous distribution function. This is what was done to obtain the negative binomial from the Poisson distribution. In general if we average the Poisson parameter over any discrete or continuous distribution we shall say we have a 'compound' Poisson distribution. For example, suppose $a_1, a_2 > 0$, $a_1 + a_2 = 1$, $\lambda_1, \lambda_2 > 0$. Then the generating function

$$a_1 e^{\lambda_1(z-1)} + a_2 e^{\lambda_2(z-1)}$$

will define a compound Poisson distribution with (effectively) three parameters.

Any distribution may be convoluted with itself $n - 1$ times so that the result is its nth convolution power. In this distribution n may be regarded as a parameter and allowed to follow some discrete distribution. This special case of compounding has already been used in §2.13 to obtain distributions resulting from colonies. As a further example we may consider the Pólya–Aeppli distribution referred to above (2.98) (Pólya 1930, Aeppli 1924) and its generalization (Skellam 1952). Here the initial distribution is the geometric distribution whose generating function is

$$(1 - Q)(1 - Qz)^{-1},$$

and whose nth power is

$$(1 - Q)^n(1 - Qz)^{-n},$$

which generates a negative binomial. If n is allowed to follow a Poisson distribution we obtain the generating function of the Pólya–Aeppli distribution,

$$\exp\left\{ \lambda\left(\frac{1 - Q}{1 - Qz} - 1 \right) \right\}. \qquad (2.115)$$

If we had started from a negative binomial distribution

$$(1-Q)^m(1-Qz)^{-m},$$

where m is any positive number, the final result would be

$$\exp \lambda\left\{\left(\frac{1-Q}{1-Qz}\right)^m - 1\right\}, \qquad (2.116)$$

which is called the 'generalized Pólya–Aeppli' distribution by Skellam, and the 'Poisson–Pascal' distribution by Katti and Gurland (1961).

In these two cases we have used a Poisson distribution for the power, n, of the convolution. Clearly we can do this for any distribution. Confining ourselves to discrete distribution of the above kind whose generating functions can be written as $P(z)$, we obtain a generating function of the form

$$\exp[\lambda\{P(z)-1\}]. \qquad (2.117)$$

This is said to be a 'generalized' Poisson distribution in contrast with the 'compound' Poisson considered above.

Neyman (1939) gives a different classification of contagious distributions. His 'type A' is a generalized Poisson distribution of the form (2.117) in which $P(z)$ is either itself the generating function of a Poisson distribution, or is the mixture of a number (possibly infinite) of such distributions. Thus Neyman's type A is generated by (2.117) where $P(z)$ generates any compound Poisson distribution. In the particular case where $P(z)$ generates a simple Poisson distribution we get back to (2.81) which has two parameters, and is itself both a compound and a generalized Poisson distribution.

Neyman also defines 'type B' and 'type C' distributions which involve a somewhat different idea. Consider a random variable, X, following a Poisson distribution with generating function

$$P(z) = e^{\mu(z-1)}.$$

We may regard X as the number of adult individuals resulting from a colony of eggs laid at a point A on a plane. Consider a region, R, of the plane and suppose that each adult has independently a probability p of migrating to the region R. Then the probability that there are exactly j adults in R is

$$\sum_{n=j}^{\infty} e^{-\mu}\frac{\mu^n}{n!}\binom{n}{j}p^j(1-p)^{n-j} = e^{-\mu p}\frac{(\mu p)^j}{j!},$$

and therefore is a Poisson distribution with mean μp. This is just the result pointed out before (see §2.13) that the binomial distribution, with n compounded in a Poisson distribution with mean μ, is a Poisson distribution with mean μp. Since the position of the point A relative to the region R may vary, p will vary also. Suppose that it has a continuous distribution function $f(x)$

such that

$$\int_0^x f(u)\,du, \quad 0 \leqslant x \leqslant 1, \tag{2.118}$$

is the probability that $p \leqslant x$. Then the number of adults found in the region R and arising from the colony at A will have a distribution with generating function

$$\int_0^1 e^{\mu p(z-1)} f(p)\,dp. \tag{2.119}$$

Next suppose that the number of colonies which produce adults which can reach the region A is itself a random quantity with a Poisson distribution of mean λ, and that the positions of each of these colonies are independently distributed in such a way that the corresponding p has the distribution (2.118). Then the overall distribution of adults found in R will have a generating function

$$\exp \lambda \left\{ \int_0^1 e^{\mu p(z-1)} f(p)\,dp - 1 \right\}. \tag{2.120}$$

The type B distribution is obtained by supposing p is uniformly distributed over the interval $(0, 1)$. We then have

$$\int_0^1 e^{\mu p(z-1)}\,dp = \frac{e^{\mu(z-1)} - 1}{\mu(z-1)},$$

and the generating function is

$$\exp \lambda \left\{ \frac{e^{\mu(z-1)} - 1}{\mu(z-1)} - 1 \right\}. \tag{2.121}$$

For the type C distribution Neyman takes

$$f(p) = 2(1-p), \quad 0 \leqslant p \leqslant 1,$$

$$= 0, \qquad \text{elsewhere.}$$

Then

$$2\int_0^1 e^{\mu p(z-1)}(1-p)\,dp = \frac{e^{\mu(z-1)} - 1 - \mu(z-1)}{\tfrac{1}{2}\mu^2(z-1)^2},$$

and the generating function is

$$\exp \lambda \left\{ \frac{e^{\mu(z-1)} - 1 - \mu(z-1)}{\tfrac{1}{2}\mu^2(z-1)^2} - 1 \right\}. \tag{2.122}$$

For other work on contagious distributions see Arbous and Kerrich (1951), Beall and Rescia (1953), Feller (1943b), Gurland (1957, 1958), Maceda (1948), Skellam (1958), and Thompson (1954).

2.19. The Borel–Tanner distribution

As we have already pointed out, any powers series

$$\sum_0^\infty a_n z^n$$

is the probability generating function of a discrete probability distribution provided all the a_n are non-negative and they sum to unity. Borel (1942) constructed a remarkable distribution in this way, which was later (1953) generalized by Tanner. It arises out of the theory of queues, and this manner of derivation will be described in the next chapter (§ 3.17).

It is convenient to begin with the particular case considered by Borel. Let α satisfy $0 \leqslant \alpha < 1$. Then the Borel distribution is defined by:

$$p_0 = 0,$$

$$p_j = \frac{\alpha^{-1}j^{j-1}(\alpha e^{-\alpha})^j}{j!}, \quad j > 0. \tag{2.123}$$

The infinite series is convergent because

$$\frac{p_{j+1}}{p_j} = \left(1 + \frac{1}{j}\right)^{j-1} \alpha e^{-\alpha},$$

and this is asymptotically less than unity if $\alpha e^{-\alpha} < e^{-1}$, which is true so long as $\alpha < 1$.

We now have to show that the sum of the p_i is unity. To do this we use Lagrange's theorem on the inversion of series (Whittaker and Watson 1935, Chaundy 1935).

THEOREM 2.4. *If $f(y)$ and $g(y)$ are analytic functions of y in some interval about the point $y = a$, and if x is defined by the equation*

$$y = a + xg(y),$$

then in the same interval, provided $f(y)$ and $g(y)$ can be expanded in power series about $y = a$ which are convergent throughout it, and provided $1 - xg'(y)$ does not vanish,

$$f(y) = f(a) + \sum_{s=1}^\infty \frac{x^s}{s!} \frac{d^{s-1}}{da^{s-1}} \{g(a)^s f'(a)\}. \tag{2.124}$$

Proofs are given in the above references. To apply this theorem consider the equation

$$\alpha = a + \beta e^\alpha. \tag{2.125}$$

Then β is easily expressed as a function of α, and we wish to find α as a function of β. Take $y = \alpha$, $x = \beta$, $f(y) = y$, and $g(y) = e^y$ in (2.124) and we have for $a = 0$,

$$\alpha = \sum_{s=1}^{\infty} \frac{\beta^s}{s!} \frac{d^{s-1}}{da^{s-1}} (e^{sa})_{a=0}$$

$$= \sum_{s=1}^{\infty} \frac{\beta^s s^{s-1}}{s!}. \tag{2.126}$$

Inserting $\beta = \alpha e^{-\alpha}$ we obtain the required result on dividing by α. (2.126) is a formula which expresses α as a function, $\alpha(\beta)$ say, of β, when $\beta = \alpha e^{-\alpha}$. It therefore follows that the probability generating function of the Borel distribution is

$$P(z) = \alpha(\beta z)\alpha(\beta)^{-1}. \tag{2.127}$$

Tanner's generalization is a distribution defined on the integers $r, r+1,...$ only, where r is positive. It is given by

$$p_j = 0, \quad j = 0, i, ..., r-1,$$

and

$$p_j = \frac{re^{-\alpha j}\alpha^{j-r}j^{j-r-1}}{j-r!}, \quad j \geqslant r, \tag{2.128}$$

with the usual convention that $0! = 1$ so that $p_r = e^{-\alpha r}$. We use Theorem 2.4 with $f(y) = e^{ry}$, $g(y) = e^y$, so that we obtain

$$e^{\alpha r} = e^{ar} + \sum_{s=1}^{\infty} \frac{x^s}{s!} \frac{d^{s-1}}{da^{s-1}} (re^{sa+ra})$$

$$= e^{ar} + \sum_{s=1}^{\infty} \frac{x^s}{s!} r(r+s)^{s-1} e^{(s+r)a}.$$

Putting $a = 0$, and $x = \alpha e^{-\alpha}$, we find

$$1 = e^{-\alpha r} + \sum_{s=1}^{\infty} \frac{r(\alpha e^{-\alpha})^s}{s!} (s+r)^{s-1} e^{-\alpha r}$$

$$= \sum_{j=r}^{\infty} p_j. \tag{2.129}$$

Haight and Breuer (1960) gives tables of the cumulative sums of this distribution to five decimals for $\alpha = 0.01(0.01) 0.62$, $r = 1$, and values of j up to the point where the sum is not less than 0.999. They refer to a number of unpublished tables for other values of α and r.

The Borel distribution (2.123) can be convoluted with itself $r-1$ times to give its rth convolution power, and we now prove that this is equal to the Borel–Tanner distribution. (2.126) holds for all β in the range $(0 \leqslant \beta < 1)$, and from (2.129) we deduce that

$$\alpha^r = \sum_{j=r}^{\infty} \frac{r\beta^j j^{j-r-1}}{j-r!},$$

so that for these values of β we have the remarkable identity

$$\sum_{j=r}^{\infty} \frac{r\beta^j j^{j-r-1}}{j-r!} = \left\{ \sum_{j=1}^{\infty} \frac{\beta^j j^{j-1}}{j!} \right\}^r. \tag{2.130}$$

Replacing β by βz ($0 \leqslant z < 1$), and dividing by (2.130) we find

$$P_T(z) = P(z)^r, \tag{2.131}$$

where $P_T(z)$ is the generating function of the Borel–Tanner distribution (2.128), and $P(z)$, given by (2.127), that of the Borel distribution. In the next chapter where we will consider this distribution from a very different point of view, equation (2.131) will be given a direct and intuitive interpretation.

To obtain the moments we write (2.129) in the form

$$\alpha^r = \sum_{j=r}^{\infty} \frac{r\beta^j j^{j-r-1}}{j-r!},$$

which is an identity when $\beta = \alpha e^{-\alpha}$. Differentiating both sides with respect to β we get

$$E(X) = \beta \alpha^{-r} \frac{d}{d\beta}(\alpha^r),$$

and

$$E\{X(X-1)\} = \beta^2 \alpha^{-r} \frac{d^2}{d\beta^2}(\alpha^r).$$

From $\beta = \alpha e^{-\alpha}$ we find

$$\frac{d\alpha}{d\beta} = e^{\alpha}(1-\alpha)^{-1},$$

$$\frac{d^2\alpha}{d\beta^2} = e^{2\alpha}(2-\alpha)(1-\alpha)^{-3}.$$

Using these we find

$$E(X) = r(1-\alpha)^{-1}, \tag{2.132}$$

$$\text{var}(X) = E(X^2) - E(x)^2 = r\alpha(1-\alpha)^{-3}. \tag{2.133}$$

Higher moments can be found in a similar way.

2.20. Multivariate discrete distributions

We have already noted a few of these. Some others of the above distributions have been given bivariate or multivariate generalizations, mainly for statistical purposes. References to these for the binomial, Poisson, negative binomial, hypergeometric, and contagious cases will be found in Haight (1961).

One particular application of the multinomial distribution (§2.11) is so important in statistical physics that we discuss it here at some length. Suppose that n trials are made of an experiment which can result in one of N different cases which are equiprobable. The N different results are denoted by $A_1, ..., A_N$ and called 'states'. The probability that n_1 of the trials result in A_1, n_2 in A_2, and so on, where $\Sigma n_i = n$, is obviously

$$\frac{n! \, N^{-n}}{n_1! \dots n_N!}. \tag{2.134}$$

Let E be an integer satisfying $n \leqslant E \leqslant nN$ and consider only those sets of n trials for which we have the additional restriction

$$n_1 + 2n_2 + 3n_3 + \dots + Nn_N = E. \tag{2.135}$$

This corresponds in physics to the idea that each state A_i represents the energy level of a particle which has i units of energy, and that there are n particles whose total energy, E, is constant but which are otherwise distributed according to (2.134), i.e. according to 'Maxwell–Boltzmann statistics'. Then the probability of any observed distribution among the n particles defined by $(n_1, ..., n_N)$, and conditional on a given value of E, will be

$$\frac{Kn!}{n_1! \dots n_N!}, \tag{2.136}$$

where K is to be found by equating to unity the sum of all such terms over all possible sets $(n_1, ..., n_N)$ satisfying (2.135). Clearly K^{-1} will be the coefficient of w^E in

$$(w + w^2 + \dots + w^N)^n = \left\{ \frac{w(1 - w^N)}{1 - w} \right\}^n. \tag{2.137}$$

We have already evaluated an expression of this form in (2.30), and if we write $E = n + uN + v$ where u, v are integers satisfying $u \geqslant 0$, $0 \leqslant v < N$, we see that

$$K^{-1} = \sum_{s=0}^{s=u} (-1)^s \binom{n}{s} \frac{E - sN - 1!}{E - n - sN! \, n - 1!}. \tag{2.138}$$

What is of interest in the physical situation is to consider the asymptotic behaviour of (2.136) under the restriction (2.135) when n and N tend to infinity in such a manner that for each i, $n_i n^{-1}$ will converge in probability to a fixed non-zero quantity, i.e. it will tend to be nearly equal to this quantity except possibly in a set of cases of asymptotically small probability. To do this we first consider how to find $E(n_i)$. Let $P_i(z)$ be the probability generating function of n_i. Then

$$P_i(z) = \frac{\text{coefficient of } w^E \text{ in } (w + \dots + w^{i-1} + zw^i + w^{i+1} + \dots + w^N)^n}{\text{coefficient of } w^E \text{ in } (w + \dots + w^N)^n}.$$

The mean of n_i is therefore

$$P'_i(1) = \frac{\text{coefficient of } w^E \text{ in } nw^i \left\{ \dfrac{w(1-w^N)}{1-w} \right\}^{n-1}}{\text{coefficient of } w^E \text{ in } \left\{ \dfrac{w(1-w^N)}{1-w} \right\}^{n}}. \tag{2.139}$$

These coefficients are given by expressions of the form (2.138), and in fact these coefficients are proportional to the ordinates in the nth (or $(n-1)$th) convolution-power of discrete uniform distributions. The most important case occurs when $N \geqslant E$ and then the presence of the term in w^N in (2.139) has no effect on the value of the coefficients, so that we can put it equal to zero. $P'(1)$ is then

$$n \frac{E-i-1! \, n-1! \, E-n!}{n-2! \, E-i-n+1! \, E-1!}. \tag{2.140}$$

We suppose that E and n tend to infinity in such a way that En^{-1} tends to a positive constant $\alpha > 0$. Then using (1.49) we see that

$$n^{-1}P'_i(1) \to (\alpha-1)^{-1} \left(\frac{\alpha-1}{\alpha} \right)^i, \tag{2.141}$$

which clearly checks with the conditions that $\sum E(n_i) = n$, $\sum i E(n_i) = E$. This result can be expressed in another way by saying that if we consider only those sets of n trials which satisfy (2.135) the probability that one of these trials chosen at random results in A_i will have asymptotically a geometric distribution on the integers $i = 1, 2, \ldots$. It is also easy to obtain in the same way the second moment of n_i, and hence, using Tchebychev's inequality to show that $n_i n^{-1}$ 'converges in probability' (§ 1.9) to the right-hand side of (2.141).

Some early books on statistical physics attempt to prove this result by considering the distribution of the n_i subject to (2.135), and picking out the values of n_i which make the probability a maximum. This clearly does not prove the required result. Fowler and Darwin (Fowler 1936) gave the first exact discussion by finding asymptotic expressions for the coefficients in (2.139) by picking them out of the generating function, using a complex integral, and evaluating the latter by the method of steepest descents. This is useful in more complicated problems of the same kind.

The above results also hold if E and n tend to infinity in such a way that $En^{-1} \to \alpha$, and N also tends to infinity but N is not necessarily larger than E. This can be shown by proving that the sum of all the terms in (2.138) other than the first is asymptotically small in comparison with the first term.

In theoretical physics it is also important to extend the results of this case to situations where the basic distribution (2.134), which is 'Maxwell–Boltzmann', is replaced by a 'Bose–Einstein' or 'Fermi–Dirac' distribution. For further discussion of these problems see Fowler (1936, Chapter II), Khintchine (1950), Dingle (1940), and Münster (1950).

8

Bibliographical notes

Many miscellaneous results on particular discrete distributions will be found in David and Barton (1962). An index of discrete (and continuous) distributions, together with references, is given by Haight (1961). In this book the statistical theory of fitting distributions has been ignored, but many papers on the fitting of discrete distributions have been published in recent years in *Biometrics* and *Biometrika*. Some of the ways in which the distributions of this chapter can arise in the theory of random processes will be considered in the next chapter. An International Conference on Discrete Probability distributions was held in Montreal in 1963 and the proceedings have been published. Greenwood and Hartley (1962) is a most useful guide to tables of probability distributions.

Exercises

Exercise 2.1. Using the result of exercise 1.2 (Krishna–Iyer) show that the sth factorial moment of the hypergeometric distribution (2.62) is

$$s! \binom{n}{s}\binom{m}{s}\binom{M}{s}^{-1}.$$

Exercise 2.2. Obtain the factorial moments of the number of empty boxes in the problem discussed in § 1.19 by using the method of Exercise 1.2. Then use formula (2.12) to obtain the distribution (1.84).

Exercise 2.3. For the moments about the mean, μ_k, of the binomial distribution, prove the recurrence relation

$$\mu_{k+1} = pq\left(nk\mu_{k-1}+\frac{d\mu_k}{dp}\right),$$

where $q = 1-p$ is taken as dependent on p. Similarly, for the Poisson distribution with parameter λ, prove that

$$\mu_{k+1} = k\lambda_{k-1}+\lambda\frac{d\mu_k}{d\lambda}.$$

Exercise 2.4. For the truncated binomial distribution (2.60), prove that the negative moments,

$$m_{-k} = E(X^{-k}), \quad k = 1, \ldots,$$

$$m_0 = 1,$$

satisfy the equations

$$\frac{dm_{-k}}{dq} = \frac{nm_{-k}}{q(1-q^n)} - \frac{m_{1-k}}{q(1-q)}.$$

Hence obtain the lower moments, and establish a similar formula for the negative moments of the truncated Poisson distribution (Tiago de Oliveira 1952).

Exercise 2.5. If X has the binomial distribution (2.31) prove that

$$E\{(1+X)^{-1}\} = \frac{1-q^{n+1}}{(n+1)p}$$

and

$$E\{(1+X)^{-1}(2+X)^{-1}\} = \frac{1-q^{n+2}-(n+2)pq^{n+1}}{(n+1)(n+2)p^2}. \quad \text{(Thionet 1963.)}$$

Exercise 2.6. In the problem considered in § 2.20 obtain the second moment of $n_i\, n^{-1}$, and hence prove that the latter converges in probability to (2.141).

Exercise 2.7. With the probability distribution whose generating function is given by (2.29) prove that the probability that the variate is less than or equal to n can be written

$$k^{-N}\sum_{s=0}^{s=u}(-1)^s\binom{N}{s}\binom{n-sk}{N},$$

when $n = N+uk+v,\ 0 \leqslant v < k$.

3. Markov Processes

3.1. Introduction

IN CHAPTER 1 we have already dealt with some simple cases of successive trials in which the probabilities at each stage depend on what has happened before. The chain binomials considered in § 1.14 are examples of this kind. In this chapter we study such problems in a systematic way. As the successive stages can be regarded as successive instants of time the sequence of events may be regarded as a 'random' or 'stochastic' (from a Greek root meaning 'uncertain') process. The theory of such processes is now so large a subject that it would be impossible to cover it adequately in a single book, and all we can do here is to develop the main results in the more elementary parts of the theory for the purpose of showing some of the ways in which the distributions considered elsewhere in this book can arise, and how the theory of random processes is related to some of the other classical problems of probability theory.

3.2. Elementary theory of finite Markov chains

We consider events which can occur at successive discrete instants of time which we denote conventionally by a variable t which can take the values 0, 1,... or sometimes ..., -1, 0, 1,.... At each instant of time one of a finite number of events $E_1, E_2,..., E_n$ can occur. It is then sometimes convenient to regard these as the possible 'states' of some system and if the event E_i occurs at the time t we say that the system is 'in the ith state'.

At each stage $t+1$ we suppose that the events $E_1,..., E_n$ occur with certain probabilities which depend on which of the events occurred at time t, but if this is known, not on anything which had happened previously. We put p_{ij} for the probability that E_i occurs at time $t+1$ conditional on E_j having occurred at time t.† Such a system is said to define a 'finite Markov chain'.

The set of quantities, p_{ij}, $i = 1,..., n$, $j = 1,..., n$ are non-negative and satisfy the conditions

$$\sum_i p_{ij} = 1, \quad j = 1,..., n. \tag{3.1}$$

---

† Some authors use a notation in which the roles of i and j are interchanged.

It is convenient to arrange them in the form of a 'transition' matrix in such a way that i refers to the row and j to the column, and thus we write

$$\mathbf{P} = (p_{ij}). \tag{3.2}$$

If the probabilities of the events $E_1,\ldots, E_n$ at any time t are denoted by $p_1(t),\ldots, p_n(t)$, we have

$$p_i(t+1) = \sum_j p_{ij} p_j(t), \tag{3.3}$$

and the set of all such equations can be written in the matrix form

$$\mathbf{p}(t+1) = \mathbf{P}\mathbf{p}(t), \tag{3.4}$$

where $\mathbf{p}(t)$ is a column vector whose elements are $p_1(t),\ldots, p_n(t)$. By applying (3.4) repeatedly we see that

$$\mathbf{p}(t) = \mathbf{P}^t \mathbf{p}(0), \tag{3.5}$$

where t is an integer. A great deal of the theory of Markov chains is devoted to studying the asymptotic behaviour of (3.5) when t increases. We first consider some examples which illustrate what may happen.

(1) A man has four pairs of socks which he changes every day, choosing at random one of the three not worn the previous day. The event E_i is the wearing of pair i and the transition matrix is

$$\mathbf{P}_1 = \begin{bmatrix} 0 & \frac{1}{3} & \frac{1}{3} & \frac{1}{3} \\ \frac{1}{3} & 0 & \frac{1}{3} & \frac{1}{3} \\ \frac{1}{3} & \frac{1}{3} & 0 & \frac{1}{3} \\ \frac{1}{3} & \frac{1}{3} & \frac{1}{3} & 0 \end{bmatrix}. \tag{3.6}$$

Notice that in this case it is impossible to be in the same state on two successive occasions. If the pair used on the day corresponding to $t = 0$ is the first pair, the probabilities of using the pairs 1, 2, 3, and 4, on days 2, 3,... will be given by

$$(0, \quad \tfrac{1}{3}, \quad \tfrac{1}{3}, \quad \tfrac{1}{3})$$
$$(\tfrac{1}{3}, \quad \tfrac{2}{9}, \quad \tfrac{2}{9}, \quad \tfrac{2}{9})$$
$$(\tfrac{2}{9}, \quad \tfrac{7}{27}, \quad \tfrac{7}{27}, \quad \tfrac{7}{27})$$
$$(\tfrac{7}{27}, \quad \tfrac{20}{81}, \quad \tfrac{20}{81}, \quad \tfrac{20}{81})$$

and so on.

(2) As a second example suppose that we have the above situation modified so that each day the man has a probability $\frac{1}{2}$ of forgetting to change his socks. Then the transition matrix is

$$\mathbf{P}_2 = \begin{bmatrix} \frac{1}{2} & \frac{1}{6} & \frac{1}{6} & \frac{1}{6} \\ \frac{1}{6} & \frac{1}{2} & \frac{1}{6} & \frac{1}{6} \\ \frac{1}{6} & \frac{1}{6} & \frac{1}{2} & \frac{1}{6} \\ \frac{1}{6} & \frac{1}{6} & \frac{1}{6} & \frac{1}{2} \end{bmatrix}. \tag{3.7}$$

(3) Suppose that the system can be in one of four states E_1, E_2, E_3, and E_4 such that at each step it can only move, if at all, to a nearest neighbouring state, and also that once in the state E_1 it cannot move out of this state, while if in states E_2, E_3, E_4 it must move to another state at the next step. In particular we suppose the transition matrix is of the form

$$\mathbf{P}_3 = \begin{bmatrix} 1 & q & 0 & 0 \\ 0 & 0 & q & 0 \\ 0 & p & 0 & 1 \\ 0 & 0 & p & 0 \end{bmatrix}, \tag{3.8}$$

where $0 < p = 1-q < 1$. This is an example of a 'random walk'. For obvious reasons the state E_1 is said to be an 'absorbing state' and the state E_4 a 'reflecting state'.

(4) If in a system with four states the only permissible transitions are $E_1 \rightarrow E_2$, $E_2 \rightarrow E_3$, $E_3 \rightarrow E_4$, and $E_4 \rightarrow E_1$, the matrix is

$$\mathbf{P}_4 = \begin{bmatrix} 0 & 0 & 0 & 1 \\ 1 & 0 & 0 & 0 \\ 0 & 1 & 0 & 0 \\ 0 & 0 & 1 & 0 \end{bmatrix}, \tag{3.9}$$

and the process must continue to go through the states in a cyclic manner.

Define $p_{ij}^{(t)}$ as the probability of moving from state E_j to state E_i in t steps. Then it is obvious from (3.4) and (3.5) that the matrix $(p_{ij}^{(t)})$ is $\mathbf{P}^t$. If $p_{ij} > 0$, we say that state E_i is 'accessible from state j in one step', if $p_{ij}^{(t)} > 0$ we say state E_i is 'accessible from state E_j in t steps', and finally if there exists some t such that $p_{ij}^{(t)} > 0$ we simply say that state E_i is 'accessible from E_j'. In practice the determination of which states are accessible is often made easier by replacing all non-zero p_{ij} by unity, and powering up the matrix with the convention that at each stage each element is replaced by unity if it is non-zero.

In $\mathbf{P}_2$ all states are clearly accessible from all other states in one step and hence in any number of steps. In $\mathbf{P}_1$, on the other hand, all states are mutually accessible only for $t \geqslant 2$.

We now prove two theorems which determine the asymptotic behaviour of $\mathbf{P}^t$ for a large proportion of the simpler cases which occur in the applications of the theory of Markov chains.

THEOREM 3.1. *Suppose that all $p_{ij} > 0$, and consider the equation*

$$\mathbf{p}(t) = \mathbf{P}^t \mathbf{p}(0). \tag{3.10}$$

Then whatever the values of the elements of $\mathbf{p}(0)$, there exists a unique vector $\mathbf{p} = (p_1,...,p_n)$ such that each $p_i > 0$, and such that $\mathbf{p}(t)$ converges to $\mathbf{p}$ as t

increases. In fact

$$|p_{ij}^{(t)} - p_i| \leqslant (1 - n\delta)^t, \tag{3.11}$$

where

$$\delta = \min_{ij} p_{ij}.$$

Proof. Since $\sum_i p_{ij} = 1$, we have

$$\max_{ij} p_{ij} \leqslant 1 - (n-1)\delta.$$

Consider the equation

$$p_{ij}^{(l+1)} = \sum_k p_{ik}^{(l)} p_{kj}, \tag{3.12}$$

and write

$$m_i^{(t)} = \min_j p_{ij}^{(t)}, \tag{3.13}$$

$$M_i^{(t)} = \max_j p_{ij}^{(t)}. \tag{3.14}$$

Then

$$m_i^{(l+1)} = \min_j \sum_k p_{ik}^{(l)} p_{kj}$$

$$\geqslant m_i^{(l)} \min_j \sum_k p_{kj}$$

$$\geqslant m_i^{(l)}. \tag{3.15}$$

Thus the sequence $m_i^{(1)}, m_i^{(2)}, \ldots$ is non-decreasing, and similarly $M_i^{(1)}, M_i^{(2)}, \ldots$ is non-increasing. As t increases both sequences must therefore tend to limits which we prove to be the same.

In the equation, for given j and k,

$$\sum_i (p_{ij} - p_{ik}) = 0,$$

some terms will be such that $p_{ij} \geqslant p_{ik}$ and some may be such that $p_{ij} < p_{ik}$. Denote summation over the first set by $\sum^+$, and over the second set by $\sum^-$. Suppose that the first set corresponds to s values of i. Then

$$\sum_i^+ (p_{ij} - p_{ik}) = \sum_i^+ p_{ij} - \sum_i^+ p_{ik}$$

$$= 1 - \sum_i^- p_{ij} - \sum_i^+ p_{ik}$$

$$\leqslant 1 - (n-s)\delta - s\delta$$

$$\leqslant 1 - n\delta. \tag{3.16}$$

We can now write

$$M^{(l+1)} - m_l^{(l+1)} = \max_{jk} \sum_i p_{li}^{(l)}(p_{ij} - p_{ik})$$

$$\leqslant \max_{jk}\left\{\sum_i {}^+ p_{li}^{(l)}(p_{ij} - p_{ik}) + \sum_i {}^- p_{li}^{(l)}(p_{ij} - p_{ik})\right\}$$

$$\leqslant \max_{jk}\left\{\sum_i {}^+ M_l^{(l)}(p_{ij} - p_{ik}) + \sum_i {}^- m_l^{(l)}(p_{ij} - p_{ik})\right\}$$

$$\leqslant \max_{jk} \sum_i {}^+ (M_l^{(l)} - m_l^{(l)})(p_{ij} - p_{ik})$$

$$\leqslant (1 - n\delta)(M_l^{(l)} - m_l^{(l)}). \tag{3.17}$$

From this it follows that

$$M_l^{(l)} - m_l^{(l)} \leqslant (1 - n\delta)^l,$$

and hence $M_l^{(l)}$ and $m_l^{(l)}$ converge to the same non-zero constant as t increases. This proves (3.11). Notice in particular that the vector $\mathbf{p}$ with elements p_i satisfies the equation $\mathbf{p} = \mathbf{Pp}$. It is the only solution of this equation for which all the $p_i \geqslant 0$ and $\sum p_i = 1$. To see this suppose $\mathbf{p}_1$ is another such solution. Then from Theorem 3.1, $\mathbf{P}^t \mathbf{p}_1$ must converge to $\mathbf{p}$. But since it is identically equal to $\mathbf{p}_1$ for all t, $\mathbf{p}_1 = \mathbf{p}$.

Example (2) with the matrix (3.7) above satisfies the conditions of the theorem and $\mathbf{p}(n)$ must converge to $(\frac{1}{4}, \frac{1}{4}, \frac{1}{4}, \frac{1}{4})'$, which is the obvious solution of $\mathbf{p} = \mathbf{P}_2\mathbf{p}$.

A matrix satisfying the conditions $p_{ij} \geqslant 0$ and (3.1) is called a 'stochastic matrix'. If in addition

$$\sum_j p_{ij} = 1, \quad i = 1, \dots, n, \tag{3.18}$$

the matrix is said to be 'doubly stochastic' and the equation $p = \mathbf{Pp}$ always has a solution

$$\mathbf{p} = (n^{-1}, \dots, n^{-1})'$$

consisting of equal elements. $\mathbf{P}_2$ is clearly of this kind but not all doubly stochastic matrices satisfy the conditions of Theorem 3.1.

Theorem 3.1 can be extended to deal with a wider class of Markov chains in the following manner.

THEOREM 3.2. *If there exists an integer r such that $p_{ij}^{(r)} > 0$, for all i, j, i.e. after r steps every state is accessible from every other state, then there exists a vector $\mathbf{p} = (p_1, \dots, p_n)'$ for which each $p_i > 0$, and*

$$|p_{ij}^{(l)} - p_i| \leqslant (1 - n\delta_r)^s \quad (all\ t), \tag{3.19}$$

where

$$\delta_r = \min_{ij} p_{ij}^{(r)}, \tag{3.20}$$

and s is the integral part of tr^{-1}. Moreover $\mathbf{p} = \mathbf{Pp}$.

Proof. Apply Theorem 3.1 to the transition matrix $(p_{ij}^{(r)})$. Then there exists a vector $\mathbf{p} = (p_1, \ldots, p_n)'$ such that

$$|p_{ij}^{(rs)} - p_i| < (1 - n\delta_r)^s. \tag{3.21}$$

If $t = rs + u$ where $0 \leqslant u < r$, then

$$|p_{ij}^{(t)} - p_i| = \left| \sum_k p_{ik}^{(rs)} p_{kj}^{(u)} - p_i \sum_k p_{kj}^{(u)} \right|$$

because $\sum_k p_{kj}^{(u)} = 1$.

The right-hand side is therefore equal to

$$\left| \sum_k (p_{ik}^{(rs)} - p_i) p_{kj}^{(u)} \right| \leqslant \sum_k |p_{ik}^{(rs)} - p_i| p_{kj}^{(u)}$$

$$\leqslant (1 - n\delta_r)^s \sum_k p_{kj}^{(u)}$$

$$\leqslant (1 - n\delta_r)^s,$$

which is the required result.

This result settles the asymptotic behaviour of $\mathbf{P}_1^t$. $\mathbf{P}_1^2$ clearly has all its elements greater than zero and $\mathbf{P}_1^t \, \mathbf{p}_0$ must converge to a vector $\mathbf{p}$ which is the only vector satisfying the required conditions, and $\mathbf{p} = \mathbf{P}^2 \mathbf{p}$. The only such vector is again $(\frac{1}{4}, \frac{1}{4}, \frac{1}{4}, \frac{1}{4})'$. $\mathbf{P}_3$ and $\mathbf{P}_4$ do not come into the scope of these two theorems since in neither case is there an integer r such that $\mathbf{P}_3^r$ or $\mathbf{P}_4^r$ consists only of non-zero elements. For $\mathbf{P}_3$ the probability ultimately becomes entirely concentrated in the first state. To see this we calculate $\mathbf{P}_3^3$ and observe that there is then a non-zero probability of a transition from states E_2, E_3, and E_4 to state E_1. Let α be a non-zero lower bound of these three probabilities. Since for no power of $\mathbf{P}_3$ is it possible to obtain a transition out of state E_1, the combined sum of the probabilities of states E_2, E_3, and E_4 must decrease by a ratio not greater than $(1 - \alpha)$ in every set of three steps. The asymptotic state is therefore $p = (1, 0, 0, 0)'$. State E_1 is said to be an 'absorbing' state, and the other states are 'transient'.

The behaviour of $\mathbf{P}_4^t$ is different again. It is easy to verify that $\mathbf{P}_4^4$ is the unit matrix. The asymptotic behaviour is that the system passes cyclically through the states E_1, E_2, E_3, E_4, E_1, $E_2, \ldots$ and so on.

We now classify the different kinds of states in a Markov chain in such a way that we can describe, qualitatively at least, the asymptotic behaviour of their probabilities. We do this by considering which states are accessible from which.

We say that a state E_j is a 'consequent' of state E_i of order l if it is accessible from E_i in l steps (i may be equal to j or different from it), i.e. that there exists l so $p_{ji}^{(l)} > 0$. If E_j is a consequent of E_i of order l, it may also be a consequent of other orders and we shall always use the smallest such order.

We now show that if j is a consequent of i its order is not greater than $n-1$. For suppose its order was $m > n-1$. This means that there exist a series of states $E_{i_1}, E_{i_2},..., E_{i_{m-1}}$ such that

$$p_{i_1 i} p_{i_2 i_1} \cdots p_{j i_{m-1}} > 0.$$

Since there are $m+1 > n$ suffices in this product, at least two of them are equal and the terms between them can be removed. Thus the smallest order is $m \leqslant n-1$. $n-1$ is clearly a possible value of m.

We now prove an elementary lemma which is necessary for the classification of states.

LEMMA 3.1. *Suppose that we have a set, S, of positive integers whose highest common factor is d, and which is such that if a, b are integers of S, so is $a+b$. In particular a and b may be the same so that it follows that if a is a member of S so is every positive integral multiple of a. Then there exists a number N_0 so that Nd belongs to S for every integer $N > N_0$.*

Proof. By the definition of d there exists a finite set of integers $(m_1,...,m_k)$ belonging to S and such that their highest common factor is d, and by the properties of highest common factors there exists a set of integers, positive or negative, $(l_1,...,l_k)$, such that

$$l_1 m_1 + ... + l_k m_k = d. \tag{3.22}$$

Some of these l's must be negative, unless one is unity and the rest zero. If the latter is the case, all multiples of d must be in S. Suppose this is not so and remove all the negative terms in (3.22) to the right-hand side where they sum to a positive integer p. Then p belongs to S and so does $p+d$, and therefore

$$rp + q(p+d) = (r+q)p + qd$$

belongs to S for any integers r, $q > 0$. Now let N be any integer greater than $N_0 = p^2$. Then Nd can be written in the form $(Ap+B)d$ where $A \geqslant p$ and $B < p$, so that it belongs to S. This proves the lemma.

If state E_j is a consequent of state E_i and state E_i is a consequent of state E_j we say that they 'communicate' or are 'mutually accessible'. Clearly if E_i communicates with E_j, and E_j with E_k, then E_i communicates with E_k.

We can divide the states of the chain into two types, 'transient' states and 'non-transient' states. A state E_i is 'non-transient' if it is a consequent of every state which is consequent to it, i.e. we cannot find a state E_j such that it is possible to move from E_i to E_j in one or more steps but such that it is not possible to move from E_j to E_i. A state is 'transient' if there is such a state E_j.

The set of non-transient states $E_1,..., E_n$ can then be divided into 'classes' such that every state in a class communicates with every other state in that class, and no state of any other class.

If E_i is a non-transient state and communicates with a state E_j, E_j is also non-transient for if not there would be a state E_k different from E_i and E_j

such that $E_j \to E_k$ is possible and $E_k \to E_j$ impossible, from which it follows that $E_i \to E_k$ is possible and $E_k \to E_i$ impossible. Hence all the members of a class of states are either all transient or all non-transient. The states of a finite class of non-transient states all communicate with each other, and such a class is said to be 'ergodic' for reasons which will appear later.

Now arrange the n states in order so that the first n_1 states form an ergodic class C_1, the next n_2 another ergodic class C_2, and so on until there are finally left over n_k states which are transient and which form a class we denote by T. It is of course possible that there are no transient states, and also that some or all of the n_i are unity. The matrix $\mathbf{P} = (p_{ij})$ then has the form

$$\mathbf{P} = \begin{bmatrix} \mathbf{P}_1 & 0 & 0 & . & . & . & \mathbf{R}_{1,k} \\ 0 & \mathbf{P}_2 & 0 & . & . & . & \mathbf{R}_{2,k} \\ 0 & 0 & \mathbf{P}_3 & . & . & . & \mathbf{R}_{3,k} \\ . & . & . & . & . & . & . \\ . & . & . & . & . & . & . \\ . & . & . & . & . & \mathbf{P}_{k-1} & \mathbf{R}_{k-1,k} \\ . & . & . & . & . & . & \mathbf{R}_{k,k} \end{bmatrix}, \tag{3.23}$$

where the $\mathbf{P}_i$ are $n_i \times n_i$ matrices and the $R_{i,k}$ are $n_i \times n_k$ matrices. It is then obvious that $\mathbf{P}^t$ $(t = 2, 3, \ldots)$ has the form

$$\mathbf{P}^t = \begin{bmatrix} \mathbf{P}_1^t & 0 & 0 & . & . & . & \mathbf{S}_1 \\ 0 & \mathbf{P}_2^t & 0 & . & . & . & \mathbf{S}_2 \\ 0 & 0 & \mathbf{P}_3^t & . & . & . & \mathbf{S}_3 \\ . & . & . & . & . & . & . \\ . & . & . & . & . & . & . \\ . & . & . & . & . & \mathbf{P}_{k-1}^t & \mathbf{S}_{k-1} \\ . & . & . & . & . & 0 & \mathbf{S}_k \end{bmatrix}, \tag{3.24}$$

where the $\mathbf{S}_i$ depend on t. Once the system gets into any of the classes $C_1, \ldots, C_{k-1}$ it will always remain in that class. We now show that if the system starts in one of the states of class T it is certain to leave this class and enter one of the ergodic classes C_i, or put in probability language, that the probability of remaining in class T after t steps tends to zero as t increases. If this is true it will follow that to study the asymptotic behaviour of the chain it is only necessary to study the asymptotic behaviour within a single ergodic class.

THEOREM 3.3. *If the process starts in a transient state the probability that it remains in such a state converges to zero geometrically as t increases.*

Proof. Let E_i be a transient state. Then there is a state E_j belonging to one of the ergodic classes, and an integer t such that $p_{ji}^{(t)} > 0$. Choose such a state E_j and integer t for every transient state E_i. For all the states E_i in the class T let $\alpha > 0$ be a lower bound for the $p_{ji}^{(t)}$, and t_0 an upper bound for the t's. Divide the sequence of values of t into blocks of length t_0 which we can take as $(1, 2, \dots, t_0)$, $(t_0 + 1, \dots, 2t_0)$, …. Then the probability that the process is still in the class T after Nt_0 steps ($N = 1, 2, \dots$) will be bounded above by

$$(1 - \alpha)^N,$$

which converges to zero. This is equivalent to saying that if

$$\mathbf{p}(t) = \mathbf{p}^t \mathbf{p}(0),$$

then the last n_k elements of $\mathbf{p}(t)$ converge to zero, whatever the value of $\mathbf{p}(0)$.

Suppose that $\mathbf{p}(0)$ consists of zero elements except for unity in the ith place. Then $\mathbf{p}(t)$ will give the probabilities of the states at time t when the system started in state E_i. If i belongs to one of the ergodic classes the system will always remain in that class, and it follows if $\mathbf{p}(t)$ is to be independent of the initial state there can be only one ergodic class.

We now consider the behaviour inside a single ergodic class, and this can be regarded as the behaviour of a Markov chain consisting of states which all communicate with each other. Denote the transition matrix of such a chain by $\mathbf{P}$.

Since all the states communicate, there are, for each i, values of $t \geqslant 1$ such that $p_{ii}^{(t)} > 0$. The greatest common divisor of all such integers t, for a given i, is called the 'period', d_i, of the state E_i. Notice that this does not imply that $p_{ii}^{(d)} > 0$, but by Lemma 3.1 there exists an integer N_0 such that for all $N > N_0$, $p_{ii}^{(Nd_i)} > 0$, and furthermore $p_{ii}^{(t)} = 0$ if t is not a multiple of d_i.

It now follows that the periods of all the states of an ergodic class are the same, and their common value can therefore be called the period of the class. For suppose E_i is accessible from E_j in n steps, and E_j from E_i in m steps. Then

$$p_{jj}^{(m+n+Nd_i)} > p_{ji}^{(m)} p_{ii}^{(Nd_i)} p_{ij}^{(n)}, \tag{3.25}$$

and this is positive for all $N > N_0$. Thus d_j must divide d_i. Similarly d_i divides d_j, so that $d_i = d_j = d$, the period of the class.

In general $\mathbf{P}^t \mathbf{p}(0)$ does not converge unless $d = 1$, but the values for $t = Nd, Nd+1, \dots, Nd+d-1$ individually clearly converge to fixed vectors as N increases, and hence the Cesàro sum,

$$t^{-1} \sum_{s=1}^{t} \mathbf{P}^s \mathbf{p}(0) \tag{3.26}$$

always converges to a limit, and this limit is one of the solutions, with non-negative elements, of the equation

$$\mathbf{p} = \mathbf{P} \mathbf{p}.$$

That at least one such solution always exists follows from Theorem 3.2.

We also observe from the above discussion that in an ergodic class with period $d > 1$, the states can be divided into d distinct classes, $D_1, ..., D_d$, say, such that if the system is in a state of class D_i it must move at the next step to a state of class D_{i+1} if $i < d$, or to D_1 if $i = d$.

It is convenient to use the words 'regular' and 'positively regular' for the cases where the vector $\mathbf{P}^t\mathbf{p}(0)$ converges to a vector independent of $\mathbf{p}(0)$, and where in addition this vector has all its elements positive. We are now in a position to classify the asymptotic behaviour of finite Markov chains in general, and we have in fact proved the following theorem.

THEOREM 3.4. *Any finite Markov chain belongs to one of the following classes.*

I. There are no transient states and only one ergodic class for which $d = 1$.

Such a chain is positively regular and is exemplified by $\mathbf{P}_1$ and $\mathbf{P}_2$ given by (3.6) and (3.7).

II. There are no transient states and only one ergodic class for which $d > 1$.

The states can then be divided into d different sets such that the system moves through them in a definite order. Thus given t_0,

$$\mathbf{P}^{(td + t_0)}\mathbf{p}(0) \quad (t \to \infty)$$

will converge to a vector which is, in general, dependent both on $\mathbf{p}(0)$ and t_0. The chain is not regular.

III. There are no transient states and more than one ergodic class for each of which we may have $d = 1$, or $d > 1$.

The chain is not regular and what happens asymptotically depends on which class the initial state lies.

IV. There are transient states and one ergodic class for which $d = 1$ or $d > 1$.

If the system starts in a transient state it will enter the ergodic class sooner or later with probability one. What happens then depends on d. If $d = 1$ the system is regular but not positively regular. If $d > 1$ the system is not regular.

V. There are transient states and more than one ergodic class.

This is a more complicated version of IV. The chain is not regular even if the d_i are all unity.

3.3. The algebraic theory of Markov chains

The asymptotic behaviour of Markov chains can also be described in terms of the algebraic theory of the matrix $\mathbf{P}$ and depends on the roots of this matrix, which are the roots of the nth order equation

$$|\lambda \mathbf{1} - \mathbf{P}| = 0, \tag{3.27}$$

where $\mathbf{1}$ is the unit diagonal matrix. We follow the account given in Bartlett (1955).

We consider first the special case where all the roots of this equation are distinct and we write them as $\lambda_1, ..., \lambda_n$ in some order such that

$$|\lambda_1| \geqslant |\lambda_2| \geqslant ... \geqslant |\lambda_n|. \tag{3.28}$$

We first show that there is always at least one of these roots equal to unity, and that no other root has a modulus greater than unity, so that we can choose $\lambda_1 = 1$. To do this we observe that the existence of a root λ_i of (3.27) is equivalent to the existence of a non-null row vector $\mathbf{u}_i'$ such that

$$\mathbf{u}_i' \mathbf{P} = \lambda_i \mathbf{u}_i', \tag{3.29}$$

or to the existence of a non-null column vector $\mathbf{v}_i$ such that

$$\mathbf{P}\mathbf{v}_i = \lambda_i \mathbf{v}_i. \tag{3.30}$$

Now the columns of $\mathbf{P}$ sum to unity so that we can certainly take $\mathbf{u}' = (1, ..., 1)$ with $\lambda_i = 1$ in (3.29), and thus $\lambda = 1$ must be a root. Suppose λ is a root not equal to unity and

$$\mathbf{u}'(\lambda \mathbf{1} - \mathbf{P}) = 0. \tag{3.31}$$

Let the elements of $\mathbf{u}$, which cannot all be equal to zero, have absolute values with a least upper bound b. Then $|\lambda| b$ will be the least upper bound of the moduli of the elements of $\lambda \mathbf{u}'$. $\lambda \mathbf{u}'$ is, however, equal to $\mathbf{u}'\mathbf{P}$, and since the columns of $\mathbf{P}$ consist of non-negative quantities adding to unity, an upper bound of the moduli of the elements of $\mathbf{u}'\mathbf{P}$ is b so that $|\lambda| b \leqslant b$, and $|\lambda| \leqslant 1$.

Consider the n row vectors $\mathbf{u}_1', ..., \mathbf{u}_n'$, and arrange them in a matrix

$$\mathbf{U} = \begin{bmatrix} \mathbf{u}_1' \\ \vdots \\ \mathbf{u}_n' \end{bmatrix},$$

and similarly define

$$\mathbf{V} = (\mathbf{v}_1, ..., \mathbf{v}_n).$$

Then from (3.29) and (3.30) we have

$$\mathbf{UP} = \mathbf{\Lambda U}, \quad \mathbf{PV} = \mathbf{V\Lambda}, \tag{3.32}$$

where $\mathbf{\Lambda}$ is the diagonal matrix of values $\lambda_1, ..., \lambda_n$. It is well known that so long as the λ_i are all unequal, the $\mathbf{u}_i$ (and similarly the $\mathbf{v}_i$) are linearly independent. Hence $\mathbf{U}$ and $\mathbf{V}$ are non-singular matrices, and we have

$$\mathbf{P} = \mathbf{U}^{-1}\mathbf{\Lambda U} = \mathbf{V\Lambda V}^{-1}, \tag{3.33}$$

which is a 'spectral' decomposition of $\mathbf{P}$.

We also have, for $i \neq j$,

$$\mathbf{u}_i' \mathbf{P}\mathbf{v}_j = \mathbf{u}_i' \lambda_j \mathbf{v}_j = \mathbf{u}_i' \lambda_i \mathbf{v}_j, \tag{3.34}$$

and since

$$\lambda_i \neq \lambda_j,$$

we must have

$$\mathbf{u}_i' \mathbf{v}_j = 0 \quad (i \neq j), \tag{3.35}$$

and $\mathbf{UV}$ is a diagonal matrix with non-zero diagonal elements. Hence $\mathbf{u}_i' \mathbf{v}_i \neq 0$ for every i and we can multiply them by non-zero constants, if necessary, to ensure that we have

$$\mathbf{u}_i' \mathbf{v}_i = 1 \quad (\text{all } i),$$

and then

$$\mathbf{UV} = \mathbf{1}. \tag{3.36}$$

From (3.33) and (3.36) we have

$$\mathbf{P} = \mathbf{V\Lambda U} = \sum_1^n \lambda_i \mathbf{v}_i \mathbf{u}_i'$$

$$= \sum_1^n \lambda_i \mathbf{A}_i, \tag{3.37}$$

where the matrices $\mathbf{A}_i = \mathbf{v}_i \mathbf{u}_i'$ satisfy the equations:

$$\mathbf{A}_i \mathbf{A}_j = \mathbf{v}_i \mathbf{u}_i' \mathbf{v}_j \mathbf{u}_j' = 0, \quad \text{for } i \neq j,$$

$$= \mathbf{v}_i \mathbf{u}_j' = \mathbf{A}_i, \quad \text{for } i = j, \tag{3.38}$$

and

$$\sum_i \mathbf{A}_i = \sum_i \mathbf{v}_i \mathbf{u}_i' = \mathbf{VU} = \mathbf{1}. \tag{3.39}$$

From these relations it follows that since

$$\mathbf{P} = \mathbf{V\Lambda U}$$

$$= \sum_i \lambda_i \mathbf{v}_i \mathbf{u}_i' = \sum_i \lambda_i \mathbf{A}_i, \tag{3.40}$$

for $t \geqslant 1$ we have

$$\mathbf{P}^t = \sum_i \lambda_i^t \mathbf{A}_i, \tag{3.41}$$

which is the fundamental result which we require. $\mathbf{p}(0)$, the initial column vector of probabilities of states, can be expressed in terms of the linearly independent set $(\mathbf{v}_1, \ldots, \mathbf{v}_n)$ by the equation

$$\mathbf{p}(0) = \sum_{i=1}^n \alpha_i \mathbf{v}_i, \tag{3.42}$$

where the α_i can be verified to be equal to $\mathbf{u}_i' \mathbf{p}(0)$. We then have

$$\mathbf{p}(t) = \mathbf{P}^t \mathbf{p}(0) = \sum \lambda_i^t \mathbf{A}_i \mathbf{p}(0)$$

$$= \sum_i \lambda_i^t \mathbf{v}_i \{\mathbf{u}_i' \mathbf{p}(0)\}$$

$$= \sum_i \{\mathbf{u}_i' \mathbf{p}(0)\} \lambda_i^t \mathbf{v}_i. \tag{3.43}$$

This gives an explicit formula for the way in which $\mathbf{p}(t)$ changes with t. In particular if $\lambda = 1$ is the only root of modulus unity, $\mathbf{p}(t)$ will converge asymptotically to a vector $\mathbf{p}(\infty)$. If $\mathbf{v}_1$ corresponds to $\lambda_1 = 1$, $\mathbf{u}_1'$ can be taken as $(1, 1, ..., 1)$, and α_1 is $\mathbf{u}_1' \mathbf{p}(0) = 1$, since the sum of the elements of $\mathbf{p}(0)$ is unity. In this case $\mathbf{p}(\infty)$ will be independent of $\mathbf{p}(0)$ and the matrix and chain will be regular.

In the particular case where

$$\sum_j p_{ij} = 1 \quad \text{(all } i)$$

the matrix $\mathbf{P}$ is 'doubly stochastic'. Clearly for a doubly stochastic matrix which is also regular, i.e. $\mathbf{p}(\infty)$ is independent of $\mathbf{p}(0)$, all the states will be asymptotically equally probable.

Now consider stochastic matrices in general which may have multiple roots. The spectral analysis in the above manner is then rather complicated but the behaviour of $\mathbf{p}(t)$ can be demonstrated in the following manner (Bartlett 1955). Consider the nth order equation in λ,

$$f(\lambda) = |\lambda \mathbf{1} - \mathbf{P}| = 0.$$

This is a scalar equation in a scalar quantity λ. It is well known (e.g. Mirsky 1955) that if λ in the polynomial $f(\lambda)$ is replaced by the matrix $\mathbf{P}$, we have the matrix identity,

$$f(\mathbf{P}) = \mathbf{0}, \tag{3.44}$$

the matrix on the right being an $n \times n$ matrix consisting entirely of zeros. It follows that

$$f(\mathbf{P})\mathbf{p}(t) = \mathbf{0}, \tag{3.45}$$

the matrix $\mathbf{0}$ being now a column vector of zeros. Suppose that $f(\lambda)$ is written as

$$a_n \lambda^n + a_{n-1} \lambda^{n-1} + ... + a_0. \tag{3.46}$$

Then (3.45) can be written as

$$a_n \mathbf{P}^n \mathbf{p}(t) + a_{n-1} \mathbf{P}^{n-1} \mathbf{p}(t) + ... + a_0 \mathbf{p}(t)$$

$$= a_n \mathbf{p}(t+n) + a_{n-1} \mathbf{p}(t+n-1) + ... + a_0 \mathbf{p}(t)$$

$$= \mathbf{0}. \tag{3.47}$$

Hence the ith element of the vector $\mathbf{p}(t)$ also satisfies this equation. We can write this as

$$a_n p_i(t+n) + a_{n-1} p_i(t+n-1) + ... + a_0 p_i(t) = 0. \tag{3.48}$$

This is an ordinary difference equation with constant coefficients. If μ_j $(j = 1, ..., m \leqslant n)$ are the distinct non-zero roots of $f(\lambda) = 0$, i.e. the distinct non-zero values of the λ_i, then (3.48) has the general solution

$$p_i(t) = \sum_j Q_j(t)\mu_j^t, \tag{3.49}$$

where each $Q_j(t)$ is a polynomial in t of order not greater than $m_j - 1$, where m_j is the multiplicity of the λ_i corresponding to μ_j, and equation (3.49) holds for $t \geqslant n$, at least.

The $Q_j(t)$ have coefficients which can, in principle, be determined by the values of $p_i(0),\ldots, p_i(n)$ but this is usually very laborious. The values must naturally be such that $p_i(t) \leqslant 1$ for all t.

When the roots are complex they must be of the form $\lambda_i \exp(2\pi i j k^{-1})$, and hence the Cesàro limit, as defined by (3.26), must always exist.

We can now relate the matrix theory to the previous theory by considering the representation (3.23). Each $\mathbf{P}_i$ is a stochastic matrix corresponding to a chain in which all states communicate with each other. It therefore contributes one root equal to unity which is simple but may have other roots of modulus unity. $\mathbf{R}_{kk}$, on the other hand, is such that its columns add to numbers less than unity and therefore cannot have any root of modulus unity. Thus the regular case of a Markov chain in which $\mathbf{p}(\infty)$ is independent of $\mathbf{p}(0)$ can only occur when there is only one $\mathbf{P}_i$, and in this case there must be no root of unit modulus other than the simple root $\lambda = 1$. For the positively regular case we must have in addition the condition that $\mathbf{R}_{kk}$ does not occur.

3.4. Some examples

Consider the chain defined by the matrix $\mathbf{P}_1$ of (3.6) which is clearly regular and positively regular. The characteristic equation (3.27) is easily seen to be

$$(\lambda - 1)(\lambda + \tfrac{1}{3})^3 = 0,$$

so that there is only one root of modulus unity. The other roots are all equal to $-\tfrac{1}{3}$. The elements of $\mathbf{p}(t)$ therefore converge to a vector independent of $\mathbf{p}(0)$ at a rate dominated by a geometric sequence with ratio $\tfrac{1}{3}$.

The characteristic equation of $\mathbf{P}_2$ ((3.7)) is also easily found to be

$$(\lambda - 1)(\lambda - \tfrac{1}{3})^3 = 0,$$

and the behaviour of the chain is similar to the previous case.

The characteristic equation of $\mathbf{P}_3$ ((3.8)) is

$$(\lambda - 1)\lambda(\lambda^2 - 2p + p^2) = 0.$$

Here we know that the first state is absorbing and its probability tends to unity. The probabilities of the other states tend to zero at a rate dominated by a geometric sequence with ratio $\sqrt{(2p - p^2)}$.

The characteristic equation of $\mathbf{P}_4$ is obviously $\lambda^4 - 1 = 0$, whose roots are ± 1, $\pm i$, all of which lie on the unit circle $|\lambda| = 1$. This is the typical situation in a purely cyclical chain.

9

As an illustration of the algebraic theory of the previous section we now carry out the complete analysis in a very simple case. Let **P** be the matrix

$$\begin{pmatrix} p & Q \\ q & P \end{pmatrix},$$

where $0 < p = 1-q < 1$, $0 < P = 1-Q < 1$. This is clearly a simple chain which is positively regular. The characteristic equation is

$$\begin{vmatrix} \lambda-p & -Q \\ -q & \lambda-P \end{vmatrix} = \lambda^2-(p+P)\lambda+p+P-1$$

$$= (\lambda-1)(\lambda-p-P+1)$$

$$= 0. \tag{3.50}$$

The row vectors $\mathbf{u}_1'$ and $\mathbf{u}_2'$ can be taken as

$$\mathbf{u}_1' = (1,1), \qquad \lambda_1 = 1,$$

$$\mathbf{u}_2' = (q, -Q), \qquad \lambda_2 = p+P-1,$$

and similarly we can take $\mathbf{v}_1$ and $\mathbf{v}_2$ to be

$$\mathbf{v}_1 = \begin{pmatrix} Q(Q+q)^{-1} \\ q(Q+q)^{-1} \end{pmatrix}, \qquad \lambda_1 = 1,$$

$$\mathbf{v}_2 = \begin{pmatrix} (Q+q)^{-1} \\ -(Q+q)^{-1} \end{pmatrix}, \qquad \lambda_2 = p+P-1.$$

Then

$$\mathbf{U} = \begin{pmatrix} 1 & 1 \\ q & -Q \end{pmatrix}, \quad \mathbf{V} = \begin{pmatrix} Q(Q+q)^{-1}, & (Q+q)^{-1} \\ q(Q+q)^{-1}, & -(Q+q)^{-1} \end{pmatrix},$$

and we can verify (3.33), (3.35), and (3.36). The matrices $\mathbf{A}_1$ and $\mathbf{A}_2$ are

$$\mathbf{A}_1 = \begin{pmatrix} Q(Q+q)^{-1} & Q(Q+q)^{-1} \\ q(Q+q)^{-1} & q(Q+q)^{-1} \end{pmatrix}, \quad \mathbf{A}_2 = \begin{pmatrix} q(Q+q)^{-1} & -Q(Q+q)^{-1} \\ -q(Q+q)^{-1} & Q(Q+q)^{-1} \end{pmatrix},$$

which satisfy (3.38) and (3.39). If the initial vector is

$$\mathbf{p}(0) = \begin{pmatrix} p_1 \\ p_2 \end{pmatrix}, \quad p_1+p_2 = 1,$$

we have

$$\mathbf{p}(0) = \alpha_1 \mathbf{v}_1 + \alpha_2 \mathbf{v}_2,$$

where

$$\alpha_1 = 1, \quad \alpha_2 = qp_1 - Qp_2,$$

and thus

$$\mathbf{p}(t) = \begin{pmatrix} Q(Q+q)^{-1} \\ q(Q+q)^{-1} \end{pmatrix} + (qp_1 - Qp_2)(p+P-1)^t \begin{pmatrix} (Q+q)^{-1} \\ -(Q+q)^{-1} \end{pmatrix}, \tag{3.51}$$

which can be checked by evaluating $\mathbf{p}(1)$, $\mathbf{p}(2),\dots$.

3.5. The Ehrenfest model

This model was devised by P. and T. Ehrenfest (1906) to illustrate certain principles in statistical physics, and its theory shows many of the difficulties encountered in dealing with particular Markov chains. Suppose that we have two boxes containing i and $N-i$ balls where $0 \leqslant i \leqslant N$. The state of the system is determined by the integer i. At successive discrete intervals of time one of the N balls is chosen at random and moved to the other box. Thus i can change only to $i-1$ or $i+1$, and does so with probabilities iN^{-1} and $(N-i)N^{-1}$ respectively.

This is clearly a Markov chain with $N+1$ states and transition probabilities,

$$p_{i-1,i} = iN^{-1},$$

$$p_{i+1,i} = 1 - iN^{-1},$$

$$p_{ji} = 0, \quad \text{if } i = j \quad \text{or} \quad |i-j| > 1. \tag{3.52}$$

There are no absorbing states. Every state is accessible from every other state but states E_i with i even are accessible from states E_j with j even in an even number of steps only, and from states E_j with j odd in an odd number of steps only, with a similar statement when i is odd. The process therefore has a periodicity 2 and we will expect the matrix to have simple roots $\lambda = 1, -1$, and no other roots of unit modulus. Thus the matrix $\mathbf{P}^2$ transforms even states into even states and odd states into odd states.

Suppose the initial vector of probability states is $\mathbf{p}(0)$, and consider the asymptotic behaviour of $\mathbf{p}(2t)$, $\mathbf{p}(2t+1)$ $(t = 1, 2, \ldots)$. Let p_0, p_1 be the sums of the elements in $\mathbf{p}(0)$ with even and odd suffices respectively. The two classes of states, $(E_0, E_2, \ldots)$ and $(E_1, E_3, \ldots)$ can be regarded as separate Markov chains with transition probability matrices derived from $\mathbf{P}^2$ provided we consider only the instants of time $t = 0, 2, \ldots$. Both of these chains are clearly positively regular and the corresponding parts of $\mathbf{p}(t)$ must converge to sets of probabilities independent of the initial state, so long as it is known whether the latter is even or odd. Furthermore $\mathbf{P}^2$ has transition probabilities

$$p_{i-2,i} = i(i-1)N^{-2},$$

$$p_{i,i} = \{i(N-i+1) + (N-i)(i+1)\}N^{-2},$$

$$p_{i+2,i} = (N-i)(N-i-1)N^{-2},$$

$$p_{ji} = 0, \quad \text{if } j = i \pm 1 \quad \text{or} \quad |j-i| > 2. \tag{3.53}$$

Consider first the set of states with even suffices at times $t = 0, 2, \ldots$. With transition matrix $\mathbf{P}^2$ all such states are accessible from each other and none are absorbing. The probabilities of the various states therefore converge to constants which we write P_0, P_2,..., and which satisfy the equations:

$$P_0 = N^{-1}P_0 + 2N^{-2}P_2,$$

$$P_2 = (1 - N^{-1})P_0 + (5N - 8)N^{-2}P_2 + 12N^{-2}P_4,$$

$$P_{2m} = (N - 2m + 2)(N - 2m + 1)N^{-2}P_{2m-2} + \{N(4m + 1) - 8m^2\}N^{-2}P_{2m}$$
$$+ (2m + 2)(2m + 1)N^{-2}P_{2m+2}. \tag{3.54}$$

These equations are satisfied by any numbers proportional to

$$\binom{N}{2m}2^{-N},$$

and since the P_{2m} must add to p_0 we have using $(1 + 1)^N = 2^N$, $(1 - 1)^N = 0$,

$$P_{2m} = 2p_0\binom{N}{2m}2^{-N},$$

and similarly $P_{2m}(t)$ at time $t = 1, 3,...$ will converge to

$$2p_1\binom{N}{2m}2^{-N},$$

and $P_{2m+1}(t)$ will converge to

$$2p_0\binom{N}{2m+1}2^{-N}, \quad \text{or} \quad 2p_1\binom{N}{2m+1}2^{-N}$$

when t tends to infinity through odd or even values of t.

These results can be summarized by saying that given any initial vector,

$$P_i(t) - \binom{N}{i}2^{-N}[p_0\{1 + (-1)^{i+t}\} + p_1\{1 - (-1)^{i+t}\}] \tag{3.55}$$

will converge to zero as t increases.

The model was originally devised to simulate the exchange of heat between two heat containers which are in contact but are otherwise isolated.

Let the random variable X_t be equal to the suffix i of the state, i.e. to the number of balls in the first box at the stage t. Consider the conditional distribution of X_{t+1} given X_t. Then

$$E(X_{t+1} - X_t | X_t) = 1 - 2X_t N^{-1}, \tag{3.56}$$

so that, taking expectations over X_t,

$$E(X_{t+1} - \tfrac{1}{2}N) = (1 - 2N^{-1})E(X_t - \tfrac{1}{2}N), \tag{3.57}$$

and therefore

$$E(X_t) = \tfrac{1}{2}N + (1 - 2N^{-1})^t(X_0 - \tfrac{1}{2}N). \tag{3.58}$$

The mean of X_t therefore converges to $\tfrac{1}{2}N$ and (3.58) is analogous to the law of cooling in the theory of heat. In a similar way we can determine the behaviour with time of $E(X_t^2)$, but here we simply consider its asymptotic value. Since $X_{t+1} - X_t = \pm 1$ we have

$$E(X_{t+1} - X_t)^2 = 1 = E(X_{t+1}^2) + E(X_t^2) - 2E(X_{t+1} X_t).$$

Using (3.56) and (3.58) we obtain the non-homogeneous difference equation

$$E(X_{t+1}^2) = 1 + (1 - 4N^{-1})E(X_t^2) + N + 2(1 - 2N^{-1})^t(X_0 - \tfrac{1}{2}N).$$

This can be easily solved and we find that the limit of $E(X_t^2)$ exists as t increases, and

$$\lim_t E(X_t^2) = \tfrac{1}{4}N^2 + \tfrac{1}{4}N$$

so that

$$\lim_t E\{X_t - E(X_t)\}^2 = \tfrac{1}{4}N, \qquad (3.59)$$

which we could have obtained directly from (3.55). These are the same first two moments as those of a binomial distribution with probability $\tfrac{1}{2}$ and index N but, as we have seen, there is no asymptotic distribution because of the dependence on the parity of t. If, however, we average the asymptotic distributions for t even and t odd, given by (3.55), we do obtain a binomial with probability $\tfrac{1}{2}$ and index N.

3.6. The algebraic theory of the Ehrenfest model

The algebraic treatment of the above model along the lines of § 3.3 is not simple and was first achieved by Kac (1947), when N is even. To obtain the roots of (3.27) with $\mathbf{P}$ defined by (3.52) we try to find vectors $\mathbf{v}_i$ satisfying

$$\mathbf{P}\mathbf{v}_i = \lambda_i \mathbf{v}_i$$

for some value λ_i. Let $\mathbf{v}_i = (x_0, x_1, \ldots, x_N)'$ be such a vector and λ the corresponding value of λ_i. Let $N = 2M$. Then using (3.52) we have to satisfy the equations:

$$N^{-1}x_1 = \lambda x_0,$$
$$x_0 + 2N^{-1}x_2 = \lambda x_1,$$
$$(N-1)N^{-1}x_1 + 3N^{-1}x_3 = \lambda x_2,$$
$$\cdots$$
$$N^{-1}x_{N-1} = \lambda x_N. \qquad (3.60)$$

These are a set of $N+1$ equations for the ratios of $N+1$ unknowns. We solve these by first solving a more general set of an infinite number of equations in an infinite number of unknowns. These are defined by:

$$N^{-1}x_1 = \lambda x_0,$$
$$x_0 + 2N^{-1}x_2 = \lambda x_1,$$
$$N^{-1}(N+2-k)x_{k-2} + kN^{-1}x_k = \lambda x_{k-1}, \quad k = 3, 4, \ldots. \qquad (3.61)$$

If we can find a solution of these such that $x_{N+1} = 0$, the first $N+1$ x's will be a solution of (3.60). To do this we use a generating function and write

$$f(z) = \sum_0^\infty x_i z^i. \tag{3.62}$$

Multiplying the equations in (3.61) by 1, z, z^k,... and adding, we get

$$N\lambda f(z) = \sum_{i=1}^\infty i x_i z^{i-1} + \sum_{i=0}^\infty (N-i) x_i z^{i+1}$$

$$= f'(z) + Nzf(z) - z^2 f'(z),$$

which can be written as

$$f'(z) = N\frac{\lambda - z}{1 - z^2} f(z). \tag{3.63}$$

We have to find a power series solution of this equation with a constant term x_0. Straightforward integration gives

$$f(z) = x_0(1-z)^{\frac{1}{2}N(1-\lambda)}(1+z)^{\frac{1}{2}N(1+\lambda)}. \tag{3.64}$$

This will be a polynomial in z if (writing $M = \frac{1}{2}N$) $M(1-\lambda)$ and $M(1+\lambda)$ are both non-negative integers. Hence the roots of the matrix $\mathbf{P}$ are given by

$$\lambda_k = 2(k-M)N^{-1}, \quad k = 0,..., N.$$

Putting $\lambda = 2(k-M)N^{-1}$ in (3.64) we can take the x_i to be proportional to the coefficients $a_0^{(k)}$, $a_1^{(k)}$,..., $a_N^{(k)}$ in the polynomial

$$\sum_{i=0}^N a_i^{(k)} z^k = (1-z)^{N-k}(1+z)^k. \tag{3.65}$$

We can therefore take the matrix $\mathbf{V}$ defined in §3.2 to be the matrix

$$(a_i^{(k)})$$

where i is the row suffix, and k the column suffix, and both run from 0 to N.

To find the $\mathbf{u}_i'$ we must use a different method. Since the roots are all unequal we have to find $\mathbf{U}$ so (3.36) is satisfied, that is, we must find the inverse of $\mathbf{V}$. Consider the equation

$$\mathbf{VU} = \mathbf{1},$$

and denote the elements of $\mathbf{U}$ by $b_j^{(k)}$, so that $j = 0,..., N$ is the row suffix, and $k = 0,..., N$ is the column suffix. Then we must have

$$\sum_{s=0}^N a_k^{(s)} b_s^{(j)} = \delta_{jk}, \tag{3.66}$$

where

$$\delta_{jk} = 1, \quad \text{if } j = k,$$

$$= 0, \quad \text{otherwise.}$$

Multiplying by z^k and adding, we get

$$z^j = \sum_{k=0}^{N} \delta_{jk} z^k = \sum_{k=0}^{N} z^k \sum_{s=0}^{N} a_k^{(s)} b_s^{(j)}$$

$$= \sum_{s=0}^{N} b_s^{(j)} \sum_{k=0}^{N} a_k^{(s)} z^k$$

$$= \sum_{s=0}^{N} b_s^{(j)} (1-z)^{N-s} (1+z)^s;$$

and thus

$$\frac{z^j}{(1-z)^N} = \sum_{s=0}^{N} b_s^{(j)} \left(\frac{1+z}{1-z}\right)^s. \tag{3.67}$$

Putting

$$w = (1+z)(1-z)^{-1}, \quad z = -(1-w)(1+w)^{-1},$$

we get

$$\sum_{s=0}^{N} b_s^{(j)} w^s = 2^{-N}(-1)^j (1-w)^j (1+w)^{N-j},$$

so that

$$b_s^{(j)} = 2^{-N}(-1)^j a_s^{(N-j)}. \tag{3.68}$$

Using this result in (3.43) we get

$$p_{ij}^{(t)} = 2^{-N}(-1)^j \sum_{k=0}^{N} \left(\frac{k-M}{M}\right)^t a_k^{(N-j)} a_i^{(k)}. \tag{3.69}$$

A more detailed study of this model, with references, is given by Kemperman (1961), who also refers to the analogue of this model with continuous time.

3.7. The gambler's ruin problem

This is again a random walk problem but has a simpler theory than the Ehrenfest model. We suppose there are $N+1$ states $E_0, E_1, ..., E_N$, such that from any position E_i with $i \neq 0, 1$, the process can only move to $i+1$ or $i-1$, and does so with probabilities p and $q = 1-p$, where $0 \leqslant p \leqslant 1$. When the system reaches E_0 or E_N, however, it stays there.

This problem is classical and was first discussed by C. Huygens who did not, however, obtain a general solution. The early history of the solutions by Bernoulli and de Moivre is interestingly described in Thatcher (1957). It was originally proposed in the form of a problem about gambling. Consider two players, A and B, who play repeatedly a game of chance in which A's chance of winning is p, and B's chance is $q = 1-p$. At each game the stake is one unit. Let A start with i units of capital and B with $N-i$ units. If the state of the system is described by the capital A has in hand we have just the situation described in the previous paragraph.

There are two absorbing states, E_0 and E_N, and therefore two roots of the transition matrix equal to unity. Our first task is to determine the probabilities of finally ending in the states E_0 and E_N given that the initial state is E_i.

Suppose that P_i and $Q_i = 1 - P_i$ are the probabilities of ending in the states E_N and E_0 given that the initial state is E_i. Then Q_i is the probability of A being ruined. If we start from any state E_i ($i \neq 0, N$), and considering what happens at the first game, we have

$$P_i = pP_{i+1} + qP_{i-1}. \tag{3.70}$$

This is a simple difference equation and may be solved by a variety of methods. Clearly $P_0 = 0$ and $P_N = 1$, in order that (3.70) should hold for $i = 1, \ldots, N-1$. Suppose first that $p \neq q$. Then the theory of difference equations shows that the general solution of (3.70) is of the form

$$P_i = \alpha + \beta(qp^{-1})^i, \tag{3.71}$$

where α, β are arbitrary constants. Since $P_0 = 0$ and $P_N = 1$, we can find α and β, and then

$$P_i = \frac{(qp^{-1})^i - 1}{(qp^{-1})^N - 1}. \tag{3.72}$$

When $p = q$ this does not work, and the general solution of (3.70) is of the form

$$P_i = \alpha + \beta i.$$

Fitting this to the end conditions we find that

$$P_i = iN^{-1}. \tag{3.73}$$

This could also have been obtained by considering the limit of (3.72) as $p = 1 - q \rightarrow \frac{1}{2}$.

It is instructive to derive the above result in another way. The suffix i of the state E_i at time t is a random variable which we write as NX_t ($X_t = 0, N^{-1}, \ldots, 1$). Suppose that we could find a function, $f(X)$, such that the expectation of $f(X_{t+1})$, given X_t, is $f(X_t)$, i.e.

$$E\{f(X_{t+1}) \mid X_t\} = f(X_t). \tag{3.74}$$

This is known as the 'martingale' property (see §1.26). Then by applying this repeatedly,

$$E\{f(X_t)\} = f(X_0), \tag{3.75}$$

where the expectation is now taken conditional on X_0. There are exactly two absorbing states and X_t must finally be equal to 0 or to N. Given E_i as the initial state, let their final probabilities be Q_i and P_i as before. Then from (3.74) we have

$$Q_i f(0) + P_i f(1) = f(X_0) = f(iN^{-1}),$$

and

$$P_i = \frac{f(iN^{-1}) - f(0)}{f(1) - f(0)}. \qquad (3.76)$$

In the present case there exists a function such that (3.74) holds exactly, and in fact

$$f(X) = (qp^{-1})^{NX} \qquad (3.77)$$

can be verified to satisfy (3.74) from which we obtain (3.75) so long as $p \neq \frac{1}{2}$. When $p = \frac{1}{2}$, (3.74) still holds but is of no use since $f(X)$, as given by (3.77), is then a constant. However, in that case we can replace (3.77) by

$$f(X) = X, \qquad (3.78)$$

so that (3.74) holds. This type of martingale argument is often useful.

We have already seen that it is certain that sooner or later the process must end in the state E_0 or E_N. Furthermore, the probability of remaining in states other than E_0 and E_N must converge to zero as fast as some geometric sequence since all such states are transient. It follows from this that the distribution of the time, T_i say, which it will take for the process to enter E_0 or E_N starting from E_i, is such that all its moments exist. We find the mean time to enter one or the other of these states. Write this as

$$t_i = E(T_i). \qquad (3.79)$$

Considering what may happen at the first step we find

$$t_i = pt_{i+1} + qt_{i-1} + 1, \quad i = 1,..., N-1. \qquad (3.80)$$

This is again a linear difference equation of the first order but is not homogeneous. We solve it by finding a particular solution, and adding a general solution of the form (3.71). With a little experimenting it is easy to discover that $i(q-p)^{-1}$ is a particular solution when $p \neq q$. Putting

$$t_i = i(q-p)^{-1} + \alpha + \beta(qp^{-1})^i,$$

and choosing α, β, so that $t_0 = t_N = 0$, we get

$$t_i = i(q-p)^{-1} - N(q-p)^{-1}\frac{1-(qp^{-1})^i}{1-(qp^{-1})^N}, \qquad (3.81)$$

when $p \neq q$. To obtain the result for $p = q$, we let $p \to \frac{1}{2}$ in (3.81), and get

$$t_i = i(N-i). \qquad (3.82)$$

This can be verified to satisfy (3.80).

The mean time can also be found from the distribution of T_i which we now obtain, following Feller's (1957b) account.

Suppose that $q_i(t)$ is the probability of entering the state E_0 at the tth step, and $p_i(t)$ the probability of entering E_N at the tth step, where E_i is the initial state. It is sufficient to find $q_i(t)$ since $p_i(t)$ can be found by symmetry. For

$i = 1,\ldots, N-1$, we have

$$q_i(t+1) = pq_{i+1}(t)+qq_{i-1}(t),$$ (3.83)

where

$$q_i(0) = 0, \quad i > 0,$$
$$q_0(t) = 0, \quad t > 0,$$
$$q_0(0) = 1,$$
$$q_N(t) = 0, \quad t \geqslant 0.$$

Define the generating function, $Q_i(z)$ by,

$$Q_i(z) = \sum_{t=0}^{\infty} q_i(t)z^t,$$ (3.84)

and using (3.83) we must have

$$Q_i(z) = pzQ_{i+1}(z)+qzQ_{i-1}(z),$$ (3.85)
$$Q_0(z) = 1,$$
$$Q_N(z) = 0.$$ (3.86)

We look for a solution of (3.85) of the form

$$Q_i(z) = \alpha\lambda_1^i + \beta\lambda_2^i,$$

where λ_1 and λ_2 are functions of z which satisfy the equation

$$\lambda = pz\lambda^2 + qz.$$

The roots of this equation are given by

$$\lambda_1, \lambda_2 = (2pz)^{-1}\{1 \pm \sqrt{(1-4pqz^2)}\}.$$ (3.87)

Using the boundary conditions (3.86) and the fact that $\lambda_1, \lambda_2 = qp^{-1}$, we get

$$Q_i(z) = (qp^{-1})^i(\lambda_1^{N-i} - \lambda_2^{N-i})(\lambda_1^N - \lambda_2^N)^{-1}.$$ (3.88)

In this expression we have to find the coefficient of z^t. Any difference of the form

$$\lambda_1^n - \lambda_2^n,$$

where n is a positive integer, will have $\sqrt{(1-4pqz^2)}$ as a factor to an odd power. Hence the ratio of two such differences, and in particular $Q_i(z)$, will be a rational function of z. We shall show that all the roots in the denominator of $Q_i(z)$ are distinct. Define a quantity ϕ by the equation

$$\cos\phi = \tfrac{1}{2}(pq)^{-\frac{1}{2}}z^{-1}.$$

ϕ is real for z sufficiently large. Using (3.87) we get

$$\lambda_1, \lambda_2 = (qp^{-1})^{-\frac{1}{2}}(\cos\phi \pm i\sin\phi)$$
$$= (qp^{-1})^{\frac{1}{2}}\exp(\pm i\phi),$$

and

$$Q_i(z) = \frac{(qp^{-1})^{\frac{1}{2}i}\sin(N-i)\phi}{\sin N\phi}. \tag{3.89}$$

Sin $N\phi$ is equal to zero when $\phi = m\pi N^{-1}$, $m = 0, 1,..., N$. The corresponding values of z are

$$z_m = \tfrac{1}{2}(pq)^{-\frac{1}{2}}(\cos m\pi N^{-1})^{-1}. \tag{3.90}$$

These are the zeros of the polynomial in the denominator of (3.88). For $m = 0, N$, we get roots z_0, z_N which are also roots of the numerator of (3.89). Thus if N is odd the roots are given by $m = 1,..., N-1$, whilst if N is even we must omit the value $\tfrac{1}{2}N$ since there is no corresponding finite value of z. Finally, we must observe that on representing (3.88) as a ratio of polynomials in z it is possible for the order of the numerator to exceed that of the denominator by unity. It follows that in expressing (3.88) as a sum of partial fractions

$$a_0 + a_1 z + \sum_{s=1}^{N-1} \frac{b_s}{z_s - z},$$

it is necessary to include the terms a_0, $a_1 z$. Multiplying by $z_s - z$ and letting z tend to z_s, we find that

$$b_s = (pq^{-1})^{\frac{1}{2}i}\frac{\sin(\pi s i N^{-1})\sin(\pi s N^{-1})}{2N(pq)^{\frac{1}{2}}\cos^2(\pi s N^{-1})}.$$

Expanding the denominators of the partial fractions in power series, we get, for $t > 1$,

$$p_i(t) = N^{-1}2^t p^{\frac{1}{2}(t-i)}q^{\frac{1}{2}(t+i)}\sum_{s=1}^{N-1}\cos^{t-1}(\pi s N^{-1})\sin(\pi s N^{-1})\sin(\pi s i N^{-1}). \tag{3.91}$$

This is the probability that A is ruined at the tth trial. The probability that B is ruined at this trial is found by interchanging p and q, i and $N-i$, and from the sum of these probabilities we can find the distribution of T_i.

Another way of looking at this problem is illuminating. Suppose that the process enters the state E_0 at the tth step when it started from E_i. Then there has been a total of t steps, either to the right or to the left. Furthermore, the difference between the number of steps to the right and to the left must equal $-i$. Hence the number of steps to the right and to the left must be respectively $\tfrac{1}{2}(t-i)$, $\tfrac{1}{2}(t+i)$. $q_i(t)$ must equal the sum of the probabilities of all the different paths, from the state E_i at $t = 0$ to the state E_0 at time t, which do not pass through E_0 and E_N at times $1, 2,..., t-1$. In all such paths the total number of steps to the right (and similarly to the left) are the same, and hence $q_i(t)$ is equal to

$$Kp^{\frac{1}{2}(t-i)}q^{\frac{1}{2}(t+i)}, \tag{3.92}$$

where K is the number of such paths. This agrees with (3.91) so far as the factors in p and q are concerned. The problem is therefore reduced to counting the number of such paths and we do so in a manner which leads to a quite different looking expression for (3.91).

Suppose that we first consider the number of paths that lead from E_i to E_0 in t steps without ever passing through E_0 before t, whether or not such paths pass through E_N. Clearly at time $t-1$ the path will have to be at the point corresponding to E_1. The total number of paths from E_i at time 0 to E_1 at $t-1$, irrespective of whether they pass through E_0 or E_N, is clearly

$$\binom{t-1}{\frac{1}{2}(t-i)}.$$

Now consider the number of paths from E_i at time 0 to E_0 at time t which do not pass through E_0. This will be obtained by subtracting from the above the number of paths which do pass through E_0. Consider the set of all paths from E_{-i} at time 0 to E_1 at time t. All such paths pass through E_0. Given any such path reflect about the line corresponding to E_0 the part of the path joining E_{-i} to E_0 for the first time. The result will be a path from E_i at time 0 to E_1 at time t which passes through E_0 and all such paths will be given by this construction. Thus the number of such paths is

$$\binom{t-1}{\frac{1}{2}(t-i-2)},$$

and the total number of paths from E_i to E_1 which do not pass through E_0 is

$$\binom{t-1}{\frac{1}{2}(t-i)} - \binom{t-1}{\frac{1}{2}(t-i-2)} = \frac{i}{t}\binom{t}{\frac{1}{2}(t-i)}, \tag{3.93}$$

where $t-i$ must be even. (3.93) is therefore also equal to the number of paths from E_i which reach E_0 for the first time at time t.

Now suppose that the second gambler, B, has an indefinitely large capital and consider the probability that A is ruined at the nth game when his initial capital was i. In a sense this is a Markov chain with an infinite number of possible states but in any finite time only a finite number of these can be reached, and hence to discover what can happen in a finite number of steps there is no need to invoke the more elaborate theory of Markov chains with an infinite number of states which will be considered in § 3.16. Using the above result we therefore see that the probability that A is ruined at the nth step is

$$\frac{i}{n}\binom{n}{\frac{1}{2}(n-i)}p^{\frac{1}{2}(n-i)}q^{\frac{1}{2}(n+i)}. \tag{3.94}$$

n can take the values $i, i+2, i+4,\ldots$ and (3.94) is a term in a proper discrete distribution on these integers. In fact the generating function of this distribution is

$$\left\{\frac{1-(1-4pqz^2)^{\frac{1}{2}}}{2pz}\right\}^i.$$

To verify this we observe that the distribution (3.94) must be the ith convolution power of the same distribution with $i = 1$, since the path must make a first passage from i to $i-1$, from $i-1$ to $i-2$, and so on. For $i = 1$, however, the above generating function clearly generates (3.94).

Now return to the problem of finding the number of paths from E_i to E_0 which pass through neither E_0 nor E_N. The above technique is an example of the 'method of images' which is similarly used in electrostatics, and we can extend this method further to solve the problem with two boundaries. We solve the slightly more general problem of finding the number of paths from E_i to E_j $(0 < i < N, 0 \leqslant j \leqslant N)$ which do not pass through E_0 or E_N, and have n steps. Let $K_n(i,j)$ be this number. Then clearly we have

$$K_{n+1}(i,j) = K_n(i+1,j) + K_n(i-1,j),$$

where $n = 0, 1,..., i, j = 0,..., N$, and we have also the boundary conditions

$$K_n(i,j) = 0, \quad \text{if } i \leqslant 0, \quad i \geqslant N, \quad \text{all } n > 0,$$

$$K_0(i,j) = 1, \quad \text{if } i = j,$$

$$= 0, \quad \text{if } i \neq j.$$

Write

$$V(i,j) = \binom{n}{\frac{1}{2}(n-j+i)},$$

with the usual convention that this is equal to zero when $\frac{1}{2}(n-j+1)$ is not one of the integers $0, 1,..., n$. This satisfies the required difference equation for any integral value of j, but not the boundary conditions. We look for a solution which does satisfy the boundary conditions by considering sums of terms of this kind. By arguing in a manner similar to the solution of electrostatic problems with two parallel plates by using an infinite sequence of images, we see that

$$\sum_k \{V(j-i-2kN,0) - V(j+i-2kN,0)\} \tag{3.95}$$

is a solution which does satisfy the required boundary conditions. This sum is extended over all positive and negative integral values of k, only a finite number of the terms being non-zero.

Multiplying by

$$p^{\frac{1}{2}(n+j-i)}q^{\frac{1}{2}(n-j+i)},$$

we obtain the required solution. When $j = 0$ this is equivalent to (3.91) but an analytical proof of this fact is not simple.

It is possible to modify this problem so that the states E_0 and E_N are 'reflecting' instead of 'absorbing', so that once the system is in E_0 (or E_N) it must move at the next step to E_1 (or E_{N-1}). When both are reflecting the solution is given by Feller (1957), and when one is reflecting and the other absorbing by Kac (1947).

Another method of obtaining the mean time to absorption in various states is to modify the chain in such a way that when the process enters an absorbing state it returns to the initial state at the next step (Ewens 1963). If there is only one absorbing state but the probabilities to right and left are arbitrary, the mean time to absorption can be found by solving the resulting difference equation (Exercise 3.1).

3.8. Recurrence times

Formula (3.91) is a formula for a 'first passage time distribution', i.e. the distribution of the time at which the system first enters some specified state. This is very closely related to the distribution of the time at which a state first recurs given that the system is in that state at time $t = 0$.

Thus suppose the system is in state E_j at time $t = 0$. Write π_j^t for the probability that it is again in state E_j at time $t > 0$ without having been in E_j at times $1, 2,..., t-1$. π_j^t ($t = 1, 2,...$) is the 'recurrence time distribution' but is not in general a proper probability distribution unless

$$\sum_{t=1}^{\infty} \pi_j^t = 1. \tag{3.96}$$

The quantity on the left is in fact the probability that the system ever returns to E_j. If we turn back to Theorem 3.4 we see that all non-transient states in a finite Markov chain are such that the probabilities of eventual return are unity. This cannot be true for a transient state for from Theorem 3.3 we see that given $\varepsilon > 0$, there exists N so that the probability of eventual return after N steps is less than ε. If the probability of eventual return is unity the state is 'recurrent', and it must follow that the probability of eventual return after N steps is unity also, which cannot therefore be true for a transient state.

The mean recurrence time is the expected time taken for recurrence to occur and is therefore equal to

$$\sum_{t=1}^{\infty} t\pi_j^t. \tag{3.97}$$

We now show that for a finite Markov chain this is necessarily finite for a recurrent state. We have already defined $p_{jj}(t)$ to be the probability that the system is in state E_j at time t, if it was in state E_j at $t = 0$, whatever happened in between. Then by an obvious enumeration of cases, we have

$$p_{jj}(t) = \pi_j^t + \sum_{s=1}^{t-1} \pi_j^{t-s} p_{jj}(s). \tag{3.98}$$

Define the generating functions

$$P(z) = \sum_{t=1}^{\infty} p_{jj}(t)z^t,$$

$$\pi(z) = \sum_{t=1}^{\infty} \pi_j^t z^t,$$

which clearly converge for $|z| < 1$ at least. Then by multiplying (3.98) by z^t and adding we have

$$P(z) = \pi(z) + \pi(z)P(z),$$

so that

$$\pi(z) = P(z)\{1 + P(z)\}^{-1}. \tag{3.99}$$

To prove that (3.97) is finite when the state E_j is recurrent write

$$\tau_j^t = \pi_j^t + \pi_j^{t+1} + \dots \tag{3.100}$$

for $t = 1, 2, \dots$. These quantities may be infinite in general, but in any case are non-negative, and

$$\sum_{t=1}^{\infty} t\pi_j^t = \sum_{t=1}^{\infty} \tau_j^t. \tag{3.101}$$

Hence (3.97) will be finite if we can prove that τ_j^t is bounded by a decreasing geometric sequence. To prove this we observe that since E_j is recurrent, return to E_j is certain, and τ_j^t is the probability that no return occurs at times $1, 2, \dots, t-1$. On the other hand, we know that since the chain is finite E_j belongs to an ergodic class which may be periodic or aperiodic (period = 1). Suppose that it has period δ and consider the class of all states which are accessible from E_j in δ steps. Then there exists an integer δ' such that in $\delta'\delta$ steps every one of these states, including E_j, is accessible from each other with non-zero probability. Let p be the non-zero lower bound of these probabilities. Then considering the times $t = n\delta'\delta$ $(n = 1, 2, \dots)$ only, we see that the probability that E_j has not recurred by the time $n\delta'\delta$ is not greater than $(1-p)^{n\delta'\delta}$, and hence tends to zero. Thus both series in (3.101) are convergent.

We can in fact, prove the following more specific theorem.

THEOREM 3.5. *If E_j is an aperiodic recurrent state in a finite Markov chain, the mean recurrence time is p_{jj}^{-1}, the limit of the reciprocal of $p_{jj}(t)$.*

Proof. Since E_j is a recurrent state in an aperiodic chain we know that $p_{jj}(t)$ converges to a non-zero limit, and by our previously established theory (§ 3.3) we know that this convergence is dominated by a geometric sequence. We can therefore write

$$P(z) = p_{jj}(1-z)^{-1} + R(z),$$

where p_{jj} is the limit of $p_{jj}(t)$, and $R(z)$ is a power series whose coefficients tend to zero at least as fast as some geometric series. This implies that $R(1)$

is finite. From (3.99) we then have

$$\pi(z) = \frac{p_{jj}(1-z)^{-1}+R(z)}{1+p_{jj}(1-z)^{-1}+R(z)},$$

$$\pi'(z) = \frac{p_{jj}+(1-z)^2 R'(z)}{\{1-z+p_{jj}+(1-z)R(z)\}^2}.$$

Theorem 3.5 is intuitively plausible by the following argument. Let $A(N)$ be the expected number of times E_j occurs in a very long series of events of length N. Then $N^{-1}A(N)$ can be shown to be asymptotically equal to p_{jj}. On the other hand, the average distance between successive observed occurrences of E_j will be about $N(X-1)^{-1}$ where X is the number of times X occurs. As it is very plausible that XN^{-1} converges in probability to its expectation $A(N)N^{-1}$ (and this can be proved without much difficulty) the result follows.

When the chain is periodic, Theorem 3.5 requires a slight modification and we easily obtain the following

THEOREM 3.6. *If E_j is a recurrent state with period δ in a finite Markov chain, the mean recurrence time is δp_{jj}^{-1}, where p_{jj} is the limit of $p_{jj}(t\delta)$.*

This is easily proved by considering the chain defined by the matrix $\mathbf{P}^\delta$.

3.9. The modifications of a chain

To answer some types of questions about Markov chains it is useful to modify the transition probabilities in such a way that the answer to the given question remains the same whilst the transition matrix $\mathbf{P}$ is altered to a more manageable form. Thus suppose we wish to study the first passage time distribution from one state (say E_2) to another (say E_1). This is the distribution of the smallest value of t for which E_1 occurs given that the system was in state E_2 at time $t = 0$. This distribution is sometimes most easily obtained by considering another chain with the same transition probabilities except that all transitions out of E_1 are prohibited. If the original matrix was

$$\mathbf{P} = (p_{ij}),$$

the new one will have the form

$$\mathbf{P}_1 = \begin{pmatrix} 1 & \mathbf{a'} \\ \mathbf{0} & \mathbf{B} \end{pmatrix}, \tag{3.102}$$

where $\mathbf{0}$ is a column vector of $n-1$ zeros, $\mathbf{a'}$ is a row vector of $n-1$ quantities a_i equal to the p_{ij} in the corresponding positions in $\mathbf{P}$, and $\mathbf{B}$ is similarly the $(n-1)\times(n-1)$ matrix consisting of the corresponding values in $\mathbf{P}$.

Then starting from an initial state defined by a column vector $\mathbf{p}_0$, with zeros everywhere except in the second row (since we start from E_2), the value of the first element of

$$(\mathbf{P}_1)^t \mathbf{p}_0$$

will give the probability of being in the state E_1 at time t in the modified process, and this is equal to the probability, in the first process, that the first transition to E_1 has occurred at or before the time t.

Bartlett (1955) gives a useful matrix representation of this situation. Since $\mathbf{p}_0$ has a zero first element we can write it as

$$\mathbf{p}_0 = \begin{pmatrix} 0 \\ \boldsymbol{\pi}_0 \end{pmatrix},$$

where $\boldsymbol{\pi}_0$ is a column vector of $n-1$ elements. We then have

$$(\mathbf{P}_1)^t\mathbf{p}_0 = \begin{pmatrix} 1 & \mathbf{a}' \\ 0 & \mathbf{B} \end{pmatrix}^t \begin{pmatrix} 0 \\ \boldsymbol{\pi}_0 \end{pmatrix} = \begin{pmatrix} \mathbf{a}'_t & \boldsymbol{\pi}_0 \\ \mathbf{B}^t & \boldsymbol{\pi}_0 \end{pmatrix}, \tag{3.103}$$

where $\mathbf{a}'_t$ is a certain row vector depending on t. As it is simpler to work with $\mathbf{B}^t$ rather than $\mathbf{a}'_t$ we define $[\mathbf{x}]$ as a symbol representing the sum of the elements of any column vector $\mathbf{x}$, and then see that the probability of having arrived in state E_1 by time t is

$$1 - [\mathbf{B}^t\boldsymbol{\pi}_0]. \tag{3.104}$$

The probability of arriving in E_1 for the first time at time t is therefore

$$[\mathbf{B}^{t-1}\boldsymbol{\pi}_0] - [\mathbf{B}^t\boldsymbol{\pi}_0] = [(\mathbf{B}^{t-1} - \mathbf{B}^t)\boldsymbol{\pi}_0],$$

and the generating function of the time of absorption is

$$\sum_{t=1}^{\infty} [(\mathbf{B}^{t-1} - \mathbf{B}^t)\boldsymbol{\pi}_0]z^t = [z(1-\mathbf{B})(1-\mathbf{B}z)^{-1}\boldsymbol{\pi}_0], \tag{3.105}$$

where $\mathbf{B}^0$ is interpreted as $\mathbf{1}$. The formula (3.105) can sometimes be evaluated in a simple form.

Another useful modification can be used when there are several absorbing states and it is desired to find the probabilities of ultimate absorption in each, given the initial state. Thus suppose, for example, that there are n states $E_1, \ldots, E_n$ of which E_1 and E_n are absorbing. Consider the probabilities p_{ij} $(i = 1, \ldots, n)$ of moves from any other state E_j. Then starting from any initial state $E_2, \ldots, E_{n-1}$, the probabilities of ultimate absorption in E_1 and E_n will be unaltered if we replace p_{ij} by p'_{ij} $(i = 1 \ldots n, j = 2, \ldots, n-1)$

$$p'_{ij} = \lambda, \quad j \neq 0, n,$$

$$p'_{ij} = (1-\lambda)p_{ij}(1-p_{jj})^{-1}, \quad i \neq j, \tag{3.106}$$

and $0 \leqslant \lambda < 1$. In particular it is sometimes useful to take $\lambda = 0$ so that the system cannot stay in one of the states $E_2, \ldots, E_{n-1}$ for two successive values of t. By a suitable choice of λ the analytic theory of the process is thus sometimes simplified.

It is sometimes possible to lump various states together to form a single state or to break up into a number of states. Thus suppose that there are n

10

states $E_1, \ldots, E_n$, renumbered if necessary, and that

$$(n_1, \ldots, n_m)$$

is a set of integers satisfying $n_i > 0$, $\sum n_i = n$. Then we can define 'lumped states' $A_1, \ldots, A_m$ to consist of the states

$$(E_1, \ldots, E_{n_1})(E_{n_1+1}, \ldots, E_{n_1+n_2}) \ldots$$

respectively. We wish to know whether the resulting system with m states $A_1, \ldots, A_m$ can be regarded as a Markov chain. It is clear that this cannot be true in general and some further restriction is necessary.

A Markov chain is defined to be 'lumpable' (Kemeny and Snell 1960), with respect to the partition $(A_1, \ldots, A_m)$, if the probabilities in the lumped process are such that it behaves as a Markov chain (i.e. the behaviour at time $t+1$ depends on what is known to have happened at time t, but given this, is independent of what has happened earlier), and if in addition the apparent transition probabilities between the states A_i do not depend on the initial distribution of states in the original process.

Consider the sum of all values of p_{ij}, where i belongs to a set A_s, and j is fixed. This can be written $p_{A_s j}$. Then with the above definition the following theorem is immediate.

THEOREM 3.7. *A necessary and sufficient condition that a Markov chain be lumpable with a partition $(A_1, \ldots, A_m)$ is that $p_{A_j j}$ should depend on the lumped state, A_k, to which E_j belongs but, apart from this, not on E_j.*

The condition may be expressed in a different way by saying that for every pair of lumped states, A_s and A_k (including the case $s = k$), the probability of moving from a state E_j in A_k to A_s is independent of j provided k is known. Applying this to the transition from the initial state the necessity of the second part of the definition is clear.

It is also occasionally convenient to perform the reverse operation to lumping by replacing one or more states by a number of states. In particular if the stationary probabilities are all rational non-zero numbers, and the chain is positively regular, this can be done in such a way that asymptotically all states are equally probable. As in other probability problems such an introduction of distinctions which are not really relevant to the original problem can sometimes simplify the whole situation.

Kemeny and Snell, to whom the theory of lumpable chains is due, also consider a definition of 'weak lumpability' which occasionally is useful.

3.10. The Markovian property and the reversed chain

The state of a Markov chain can be described by random variables $X(t)$, where t corresponds to the time at which the state is observed $(t = 0, 1, 2, \ldots)$. Then if $X(t)$ takes different values for different states, and if we are dealing with a finite Markov chain, the distribution of $X(t+1)$ conditional on $X(t)$, $X(t-1),\ldots$ is independent of $X(t-1)$, $X(t-2),\ldots$. The system is then said to show the 'Markovian property' or to be a Markov 'process', and such systems are nearly always easier to deal with than non-Markovian ones.

We can write this as follows (Bartlett 1955). Suppose that any set of X's up to and possibly including $X(t-1)$ is denoted by U. Then the conditional distribution of $X(t+1)$ given $X(t)$ and U can be written

$$P\{X(t+1)|X(t),U\},$$

and if the Markovian property is satisfied this is always equal to

$$P\{X(t+1)|X(t)\}.$$

We now consider the joint distribution of $X(t+1),\ldots,X(t+n)$ given $X(t)$ and $X(t+n+1)$. By V we denote any set of X's corresponding to times greater than $X(t+n+1)$.

Consider the conditional distribution

$$P\{X(t+1),\ldots,X(t+n)|U,X(t),X(t+n+1),V\}$$
$$= \frac{P\{U,X(t),X(t+1),\ldots,X(t+n+1),V\}}{P\{U,X(t),X(t+n+1),V\}}$$
$$= \frac{\begin{array}{c}P\{V|U,X(t),X(t+1),\ldots,X(t+n+1)\}\\ \times P\{X(t+1),\ldots,X(t+n+1)|U,X(t)\}P\{U,X(t)\}\end{array}}{P\{V|U,X(t),X(t+n+1)\}P\{X(t+n+1)|U,X(t)\}P\{U,X(t)\}},$$

and using the Markovian property this becomes

$$\frac{P\{V|X(t+n+1)\}P\{X(t+1)\ldots X(t+n+1)|X(t)\}}{P\{V|X(t+n+1)\}P\{X(t+n+1)|X(t)\}}$$
$$= P\{X(t+1),\ldots,X(t+n)|X(t),X(t+n+1)\}.$$

Thus the joint distribution of $X(t+1),\ldots,X(t+n)$ conditional on $X(t)$, $X(t+n+1)$ is not altered by any specification of X's outside the range $t,\ldots,t+n+1$. This is known as the 'nearest neighbour' property. A large number of problems, especially those involving physical applications, show features of this kind.

Since the idea of a nearest neighbour system does not involve any preferred direction of time the above discussion suggests that we consider what happens when a Markov chain is observed as t decreases in value instead of increasing. Using the above notation we have

$$P\{X(t+1),\ldots,X(t+n)|X(t+n+1),V\}$$
$$= \frac{P\{X(t+1),\ldots,X(t+n+1),V\}}{P\{X(t+n+1),V\}}$$
$$= \frac{P\{V|X(t+1),\ldots,X(t+n+1)\}P\{X(t+1),\ldots,X(t+n+1)\}}{P\{V|X(t+n+1)\}P\{X(t+n+1)\}},$$

and using the Markovian property this becomes

$$\frac{P\{V\,|\,X(t+n+1)\}P\{X(t+1),\ldots,X(t+n+1)\}}{P\{V\,|\,X(t+n+1)\}P\{X(t+n+1)\}}$$

$$= P\{X(t),\ldots,X(t+n)\,|\,X(t+n+1)\}, \qquad (3.107)$$

which shows that the Markovian property holds when time is reversed.

Thus a Markov chain observed in the reverse direction of time will be a Markov process. However, it will not in general be a Markov chain because the observed transition probabilities will not be independent of t. To see this we have

$$p\{X(t), X(t+1)\} = p\{X(t+1)\,|\,X(t)\}p\{X(t)\}$$

$$= p\{X(t)\,|\,X(t+1)\}p\{X(t+1)\}$$

so that

$$p\{X(t)\,|\,X(t+1)\} = \frac{p\{X(t+1)\,|\,X(t)\}p\{X(t)\}}{p\{X(t+1)\}}.$$

This will be independent of t only if

$$\frac{p\{X(t)\}}{p\{X(t+1)\}}$$

is independent of t for each pair of values of $X(t)$ and $X(t+1)$. We must therefore confine ourselves to the case where the chain is in a stationary state so that the initial vector $\mathbf{p}_0$ satisfies $\mathbf{p}_0 = \mathbf{P}\mathbf{p}_0$, and all states are accessible from each other. Write π_{ij} for the apparent transition probability from E_j to E_i when the chain is observed in the reverse order. Then considering the joint probabilities at two successive instants we must have

$$p_{ji} P_i = \pi_{ij} P_j,$$

where $(P_1,\ldots,P_n)$ are the stationary probabilities. Thus

$$\pi_{ij} = p_{ji} P_i P_j^{-1}, \qquad (3.108)$$

which is always defined since $P_j > 0$ for all j if the chain is non-periodic. Furthermore, it is clear that

$$\sum_i \pi_{ij} = 1.$$

Thus the π_{ij} defined by (3.108) are the transition probabilities of a Markov chain and are, of course, not in general equal to p_{ij}.

3.11. Finite Markov processes in continuous time

So far we have considered random processes in which the state of the system can only change at discrete given instants of time. We now consider processes in which changes can occur at any instant on a continuous time scale.

Still keeping the restriction that there are only a finite number of states $E_1,..., E_N$, we suppose that the probabilities of these states at any instant t (where t is now continuously variable) are given by $p_j(t)$ $(j = 1,..., N)$, and these can be arranged, as before, in a column N vector $\mathbf{p}(t)$. We must then have, for all t,

$$p_j(t) \geqslant 0, \quad \sum_{j=1}^{N} p_j(t) = 1. \tag{3.109}$$

Considering the possible situations at two instants of time, t and s, where $t < s$, we are led to define the conditional probabilities, $p_{ij}(s, t)$, that the system is in state i at time s if it was known to have been in state j at time t. These are the elements of a matrix which we write as $\mathbf{P}(s, t)$. We assume that the process is 'Markovian', i.e. that given the state of the system at the present time, its future development is independent of the past. From this, it follows that if $u < t < s$,

$$\mathbf{P}(s, u) = \mathbf{P}(s, t)\mathbf{P}(t, u). \tag{3.110}$$

Many of the interesting cases which occur in practice have the further property of 'stationarity' which here implies that for all s, t $(t < s)$, $\mathbf{P}(s, t)$ depends only on $s - t$, so that we can write it as $\mathbf{P}(s - t)$, and (3.110) becomes

$$\mathbf{P}(s - u) = \mathbf{P}(s - t)\mathbf{P}(t - u). \tag{3.111}$$

Equation (3.110) is known as the 'Chapman–Kolmogorov' equation and some such equation must be true of every Markovian process. This is sometimes taken as characterizing Markovian processes in general but is in fact not sufficient, further restrictions being necessary when t is continuous.

We consider only the stationary case so that $\mathbf{P}(s, t)$ is a function of $s - t$ only, the following results being usually easily generalized. It is natural to suppose that the total probability of transition from any state E_j to any other of the states during any interval of time (t, s) is $O(s - t)$ when $s - t$ becomes small, and that the probability of more than one change of state occurring in that interval is $o(s - t)$. In fact we go a little further and assume that for each state E_j, and $\delta > 0$,

$$\lim_{\delta \to 0} \delta^{-1}\{1 - P_{jj}(\delta)\} = c_j, \tag{3.112}$$

a finite constant. Thus if the system is in state E_j at time t, the chance of a change of state occurring during $(t, t + \delta)$ is $c_j \delta + o(\delta)$, and taking $P_{jj}(0) = 1$, as we must, $P_{jj}(t)$ has a derivative at $t = 0$.

If a change does occur during $(t, t + \delta)$ it must be to one of the states $E_1,..., E_{j-1}, E_{j+1},..., E_N$, and we can therefore define conditional probabilities,

p_{ij}, such that for $i, j = 1, \ldots, N$

$$p_{ij} \geqslant 0,$$

$$p_{jj} = 0,$$

$$\sum_{i=1}^{N} p_{ij} = 1, \quad j = 1, \ldots, N. \tag{3.113}$$

We then have, for $\delta > 0$,

$$\lim_{\delta \to 0} \delta^{-1} P_{ij}(\delta) = c_j p_{ij}, \quad i \neq j. \tag{3.114}$$

The assumptions (3.112) and (3.113), together with the Chapman–Kolmogorov equation, are sufficient to derive the whole theory of Markov chains with a finite number of states and a continuous time parameter.

We shall continue to call such processes 'chains' in spite of the fact that this word might be better reserved for processes with t discrete.

The main tool for the study of such processes is the theory of two sets of differential equations for the transition probabilities, $P_{ij}(t)$ which are known as Kolmogorov's forward and backward equations (Kolmogorov 1931). The idea behind the derivation of these is to consider (3.111) where u, t, s are instants of time such that $u < t < s$, and carry out limiting processes.

Allowing t to tend for s (or to u) and using (3.112) and (3.113), differential equations can be found for the $P_{ij}(t)$.

Thus to derive the forward Kolmogorov equation write δ for $s-t$, put $u = 0$, and we have

$$P_{ij}(t+\delta) = \sum_{k=1}^{N} P_{ik}(\delta) P_{kj}(t). \tag{3.115}$$

Using (3.112) and (3.113) this becomes

$$P_{ij}(t+\delta) = \{1 - \delta c_i + o(\delta)\} P_{ij}(t) + \sum_{k \neq i} \{\delta c_k p_{ik} + o(\delta)\} P_{kj}(t),$$

and letting $\delta \to 0$ we get

$$\frac{d}{dt} P_{ij}(t) = -c_i P_{ij}(t) + \sum_{k \neq i} c_k p_{ik} P_{kj}(t). \tag{3.116}$$

This can be written in vector form

$$\frac{d}{dt} \mathbf{P}(t) = -\mathbf{C}\mathbf{P}(t) + \mathbf{D}\mathbf{C}\mathbf{P}(t)$$

$$= (\mathbf{D} - \mathbf{1})\mathbf{C}\mathbf{P}(t), \tag{3.117}$$

where $\mathbf{C}$ is a purely diagonal matrix with elements c_j, and $\mathbf{D}$ is the matrix (p_{ik}) which has zeros in its leading diagonal. Thus we can write the above matrix equation in the form

$$\frac{d}{dt} \mathbf{P}(t) = \mathbf{A}\mathbf{P}(t), \tag{3.118}$$

where
$$A = DC - C.$$

In all such equations the derivative of a matrix is defined to be the matrix of derivatives of its elements. It is also to be noticed that an equation similar to (3.118) holds for the vector $\mathbf{p}(t)$ giving the probability of the states at any time t, because
$$\mathbf{p}(t) = \mathbf{P}(t)\mathbf{p}(0),$$
and thus
$$\frac{d}{dt}\mathbf{p}(t) = \mathbf{A}\mathbf{p}(0). \tag{3.119}$$

Equations (3.116), (3.117), (3.118), and (3.119) are known as Kolmogorov's 'forward' equations for the process because they are derived by considering the system during an interval (t_0, t_1) (where $t = t_1 - t_0 > 0$), and examining the effect of a small change in t_1. If we take the same transition matrix from t_0 to t_1, and allow t_0 to vary we obtain another set of equations which are Kolmogorov's 'backward' equations. Thus we can write, in analogy with (3.115),
$$P_{ij}(t + \delta) = \sum_{k=1}^{N} P_{ik}(t)P_{kj}(\delta)$$
$$= P_{ij}(t)\{1 - \delta c_j + o(\delta)\} + \sum_{k \neq j} P_{ik}\{\delta c_j p_{kj} + o(\delta)\},$$

and from this we derive the system of differential equations
$$\frac{d}{dt}P_{ij}(t) = -c_j P_{ij}(t) + \sum_{k \neq j} c_j p_{kj} P_{ik}(t). \tag{3.120}$$

Using the same definitions of $\mathbf{C}$ and $\mathbf{D}$ we can write (3.120) in the matrix form
$$\frac{d}{dt}\mathbf{P}(t) = -\mathbf{P}(t)\mathbf{C} + \mathbf{P}(t)\mathbf{DC}$$
$$= \mathbf{P}(t)(\mathbf{D} - 1)\mathbf{C}$$
$$= \mathbf{P}(t)\mathbf{A}. \tag{3.121}$$

This is to be contrasted with (3.117).

Both of these sets of equations consist of n^2 equations for n^2 unknowns, but the forward equations (3.116) can be split into n independent sets of n equations for n unknowns, each set corresponding to one value of the second suffix j. Thus (3.118) can be regarded as n independent vector differential equations for the n column vectors of $\mathbf{P}(t)$, the history of each such vector being determined by its own initial value independently of the others.

On the other hand, the backward equations (3.121) can be regarded as n independent sets of n equations for n unknowns, each set corresponding to one value of the first suffix i and therefore to one of the rows of $\mathbf{P}(t)$.

In spite of this quite different structure the assumption of stationarity by which $P(s, t)$ can be replaced by a function $P(s-t)$ makes it clear that the two matrix equations (3.118) and (3.121) are identical, and that as a consequence $PA = AP$. This result is no longer true in the more general case where $P(s, t)$ is not a function of $s-t$ alone.

3.12. The solution of the Kolmogorov equations

By the theory of finite sets of linear differential equations the Kolmogorov equations for a process with a finite number of states have a unique solution, $P(t)$, which will always be a probability transition matrix. This follows since the sum of each column has a zero derivative and therefore remains identically equal to one. No element can become zero in a finite time if it starts from an initial non-zero value. This follows from (3.116) which implies that

$$\frac{d}{dt} P_{ij}(t) \geqslant - c_i P_{ij}(t).$$

By comparison with the solution of the equation

$$\frac{d}{dt} y = -c_i y,$$

it follows that

$$P_i(t) \geqslant P_i(0) e^{-c_i t} > 0.$$

Taking the equations in the backward form

$$\frac{d}{dt} P(t) = P(t) A, \qquad (3.122)$$

it is possible to write down a formal solution in matrix form.

Since the solution of

$$\frac{d}{dt} y(t) = a y(t)$$

is

$$y(t) = y(0) \exp at,$$

the solution of (3.122) may be guessed to be

$$P(t) = P(0) \exp At = \exp At, \qquad (3.123)$$

if this can be given a meaning. In fact given any matrix M we define $\exp M$ to be the matrix

$$\lim_{N \to \infty} \left\{ 1 + \sum_{s=1}^{N} (s!)^{-1} M^s \right\}, \qquad (3.124)$$

if this limit has a meaning. Suppose that m is an upper bound for the modulus of any element in M, and M is an $n \times n$ matrix. Then the elements of M^2 are bounded by nm^2, those of M^3 by $n^2 m^3$, and so on. It therefore follows that the matrix series obtained in the limit from (3.124) converges for any matrix of finite elements, and thus (3.123) always has a meaning since $\mathbf{P}(0)$ is the unit matrix. Furthermore, $\exp \mathbf{A}t$ commutes with $\mathbf{A}$, thus verifying the result $\mathbf{PA} = \mathbf{AP}$.

3.13. The imbedding of Markov chains

To any Markov process in continuous time of the above kind we can associate several Markov chains, one of which is the chain whose transition matrix is $\mathbf{P} = \mathbf{P}(1) = \exp \mathbf{A}$, the matrix of transition probabilities of the process for a unit time interval. This is known as the 'discrete skeleton' of the process (Kingman 1962, 1963c).

We have $\mathbf{P} = \exp \mathbf{A} = \exp(\mathbf{DC} - \mathbf{C}) = \exp(\mathbf{D} - \mathbf{1})\mathbf{C}$. $\mathbf{C}$ is a diagonal matrix of positive elements and $\mathbf{D} - \mathbf{1}$ is a matrix with non-negative elements off the diagonal, and -1's down the diagonal. Let λ_i be the roots of $\mathbf{A}$. Then the roots, Λ_i say, of $\mathbf{P}$ can be numbered to be equal to $\exp \lambda_i$, and the latent vector of $\mathbf{A}$ for the root λ_i will be a latent vector of $\mathbf{P}$ for the root Λ_i.

The asymptotic behaviour of $\mathbf{P}(t)$ can be studied, in part, by considering the asymptotic behaviour of $\mathbf{P}^n$. The states can clearly be split up into groups as before such that the system can get into a group but not out of it. It is thus sufficient to study one of these groups. Although the matrix $\mathbf{D}$ can be the transition matrix of a cyclic Markov chain, $\mathbf{P}$ cannot. To see this suppose that λ is a root of $\mathbf{A}$ so that

$$|\mathbf{A} - \lambda\mathbf{1}| = 0.$$

Then there is at least one non-null solution, $\mathbf{x}_1'$, of the equation

$$\mathbf{x}'(\mathbf{A} - \lambda\mathbf{1}) = 0.$$

Let x_m be a component of $\mathbf{x}_1$ which is at least as large in absolute value as any other component. Then, since $\sum_j a_{jm} = 0$,

$$|x_m||a_{mm} - \lambda| = \left|\sum_{j \neq m} a_{jm} x_j\right|$$

$$\leqslant |x_m|\left|\sum_{j \neq m} a_{jm}\right|$$

$$\leqslant |x_m||a_{mm}|,$$

so that

$$|a_{mm} - \lambda| \leqslant |a_{mm}|. \tag{3.125}$$

λ must therefore lie in or on a circle of centre a_{mm} (which is negative) whose radius is $|a_{mm}|$. It is therefore either equal to zero (which may be a multiple root) or has its real part negative. Hence the corresponding Λ may equal 1, which may be a multiple root, but if it is not equal to 1 it must lie inside the unit circle and not on it. Hence a cyclic chain is not possible.

From this it follows that

$$\lim_{t \to \infty} \mathbf{p}(t) = \lim_{t \to \infty} \mathbf{P}(t)\mathbf{p}(0)$$

always exists, in contrast to the discrete case, but of course this limit may depend on $\mathbf{p}(0)$. Lederman (1950) discusses the spectral theory for finite continuous chains, and Kingman (1962) considers the interesting problem of determining when a transition matrix, $\mathbf{P}$, can be represented as a 'skeleton', $\mathbf{P} = \exp \mathbf{A}$.

Another Markov chain closely related to the Markov process in continuous time is the chain whose transition matrix is $\mathbf{D}$. This is known as the 'imbedded' chain. This is not a general transition matrix but has the special property that $p_{ii} = 0$ for all i. Between its characteristic roots and the roots of $\mathbf{A}$ and $\mathbf{P}$ there is no simple relation yet some properties of the original process can be deduced from the study of the chain defined by $\mathbf{D}$. For example, if the process starts off in a state E_i and certain other states $E_1, \ldots, E_k$ are absorbing states, the probabilities, $\pi_1, \ldots, \pi_k$, say that the process ends in one of these absorbing states will be the same whether we consider the continuous Markov process or either of the two chains defined by $\mathbf{P}$ and $\mathbf{D}$. The distribution of the time of absorption into these states will, however, be quite different.

Thus given any continuous Markov process of the type defined by the differential equations (3.118) we can always extract from it a Markov chain in discrete time whose transition matrix $\mathbf{D}$ is defined by $\mathbf{A}$ because the diagonal elements of $\mathbf{A}$, which must all be negative, define $\mathbf{C}$, and $\mathbf{D}$ is then fixed. Conversely, given any transition matrix $\mathbf{D}$ with zeros in the diagonal, we can imbed the corresponding discrete Markov chain into a continuous process by using (3.118) with any diagonal matrix $\mathbf{C}$ whose diagonal elements are all strictly positive.

The intuitive idea of such a process is that the system, if initially in a state E_j, stays in that state for a length of time which has a negative exponential distribution with mean c_j^{-1}, and then moves to some other state k with probability p_{kj} such that

$$p_{jj} = 0,$$

$$\sum_k p_{kj} = 1.$$

Now suppose that we start from a matrix, $\mathbf{D'}$, of transition probabilities p'_{kj} for a general Markov chain so that we no longer impose the condition $p'_{jj} = 0$. Choosing any set of strictly positive numbers $c_1, \ldots, c_N$, we consider a Markovian process in which the system stays in E_j for a length of time with a negative exponential distribution of mean c_j^{-1}, and then either jumps to a different state, E_k, with probability p_{kj} or remains in E_j for a further length of time which again has a negative exponential distribution of mean c_j^{-1}, and so on.

To deal with this situation consider the distribution of the length of time, T say, between entering a state E_j and first leaving it. We shall show that this is negative exponential with mean $c_j^{-1}(1-p_{jj})^{-1}$.

To do this suppose that T consists of S successive intervals within which nothing happens and at the end of each of which the system is given the opportunity of leaving the state E_j with probability $(1-p_{jj})$. Then the number S of such intervals has a geometric distribution

$$\text{pr}(S = s) = p_{jj}^{s-1}(1-p_{jj}), \quad s = 1, 2, \ldots. \tag{3.126}$$

Each interval has a negative exponential distribution (§ 7.17) with mean c_j^{-1}. It can be proved (Exercise 7.13) that the distribution of T is then negative exponential with mean $c_j^{-1}(1-p_{jj})^{-1}$. We therefore define a diagonal matrix $\mathbf{C}$ with diagonal elements $c_j(1-p_{jj})$, and a corresponding matrix $\mathbf{D}$ of transition probabilities p'_{kj} such that

$$p'_{jj} = 0,$$
$$p'_{kj} = p_{kj}(1-p_{jj})^{-1}, \quad k \neq j. \tag{3.127}$$

The process is then of the same form as considered in the last section, and the discrete Markov chain with matrix (p_{kj}) is imbedded in the continuous Markov process.

3.14. Finite birth processes

This type of process was apparently first discussed by McKendrick (1914, 1926), a lieutenant-colonel in the Indian Medical Service (see Irwin 1963). We suppose that there are $N+1$ states which we denote by $E_0, \ldots, E_N$, where the suffix can be imagined to correspond to the number of individuals in a certain population. We suppose further that $p_{ij} = 0$ except for $p_{j+1,j} = 1$, say $(j = 0, 1, \ldots, N-1)$ and write $\lambda_i > 0$ for p_i. Then the forward Kolmogorov equations for the $P_{ij}(t)$ are

$$\frac{dP_{ij}(t)}{dt} = \lambda_{i-1} P_{i-1,j}(t) - \lambda_i P_{ij}(t), \tag{3.128}$$

where $i \geq 1$, and

$$\frac{dP_{0j}(t)}{dt} = -\lambda_0 P_{0j}(t). \tag{3.129}$$

Then if the process starts in the state E_j, it will never move into any state with suffix less than j, i.e. $P_{ij}(t) = 0$, for $i < j$ and all t.

The above equations can be solved explicitly without much difficulty. It will be seen that there is no loss in generality if we assume that the initial state is E_0 and we can therefore replace $P_{ij}(t)$ by $P_i(t)$. (3.129) is

$$\frac{dP_0(t)}{dt} = -\lambda_0 P_0(t)$$

whose solution, given the initial conditions, must be

$$P_0 = e^{-\lambda_0 t}. \tag{3.130}$$

The other $P_i(t)$ can then be obtained in succession so that we find, for example, that if $\lambda_1 \neq \lambda_0$,

$$P_1(t) = \lambda_0(\lambda_1 - \lambda_0)^{-1}(e^{-\lambda_0 t} - e^{-\lambda_1 t}),$$

and using induction and assuming all the λ_i are unequal,

$$P_i(t) = (-1)^i \lambda_0 \dots \lambda_{i-1} \sum_{j=0}^{i} \frac{e^{-\lambda_j t}}{(\lambda_j - \lambda_0)\dots(\lambda_j - \lambda_i)}, \tag{3.131}$$

the factor $(\lambda_j - \lambda_j)$ being omitted from the denominators. The case where some of the λ_i are equal can be derived from this by a limiting process. One particularly interesting case arises when all the λ_i are taken to be equal ($= \lambda$, say). Then solving in succession, as before, we find (as is otherwise obvious)

$$P_i(t) = \frac{1}{i!}(\lambda t)^i e^{-\lambda t}, \quad i = 0, 1, \dots, N,$$

and

$$P_N(t) = e^{-\lambda t}\left\{e^{\lambda t} - 1 - \lambda t - \dots - \frac{1}{N-1!}(\lambda t)^{N-1}\right\}. \tag{3.132}$$

In the imbedded discrete chain only transitions of the form $E_i \rightarrow E_{i+1}$ are possible. The system therefore moves from E_0 to E_1, E_1 to E_2,..., E_{N-1} to E_N in the first N steps, and then remains in E_N. These equations (with general λ_i) also occur in the theory of radioactive transformations and their solution has been discussed by several authors, the most elegant method being the use of Laplace transforms (see Bateman 1932, p. 46, and Sedgwick 1942).

From what we have seen before if T_i is the length of time spent in the state E_i, $\lambda_i T_i$ is distributed in a negative exponential distribution with unit mean. Anticipating for the moment the theory in §7.17 we can say that $2\lambda T_i$ 'is distributed as χ^2 with 2 degrees of freedom'. The time to reach the state E_i from E_0 is then distributed as $\frac{1}{2}\lambda^{-1}\chi^2$ where the χ^2 has $2i$ degrees of freedom. This can be seen by using (2.72) but follows more directly from the result that the sum of i independent quantities, each distributed as χ^2 with j (say) degrees of freedom, is distributed as χ^2 with ij degrees of freedom.

Another interesting special case of the pure birth process arises when we put λ_i proportional to i and thus write $\lambda_i = i\lambda$. This can be interpreted in the following way. Suppose that the state E_i means that we are concerned with a population of i individuals such that each has independently the probability $\lambda dt + o(dt)$ of producing an offspring in any interval $(t, t+dt)$, and death is not possible. Furthermore, we suppose either that E_N corresponds to the condition that the population is of size N or more, or alternatively that when the population reaches this size, reproduction is impossible.

Then the forward Kolmogorov equations become

$$\frac{dP_1(t)}{dt} = -\lambda P_1(t),$$

$$\frac{dP_i(t)}{dt} = \lambda(i-1)P_{i-1}(t) - i\lambda P_i(t), \quad i = 2,\ldots, N-1,$$

$$\frac{dP_N(t)}{dt} = \lambda(N-1)P_{N-1}(t). \tag{3.133}$$

Clearly $P_0(t)$ must remain constant and so we ignore it and suppose that the population starts with one individual. Solving these equations step by step, or substituting in (3.131), we find that $P_i(t)$ is equal to

$$e^{-\lambda t}(1 - e^{-\lambda t})^{i-1}, \tag{3.134}$$

for $i = 1,\ldots, N-1$, and to

$$(1 - e^{-\lambda t})^{N-1}, \tag{3.135}$$

for $i = N$ (a truncated geometric distribution).

This is known as the 'finite birth process'. Its generalization to an infinite number of states is the 'simple birth (or Yule–Furry) process' and will be considered in § 3.20.

3.15. Finite birth and death processes, and the random walk

Generalizing the above situation we now suppose p_{ij} to be non-zero when $i-j = \pm 1$, and zero otherwise. We can write

$$p_{j+1,j} = \lambda_j(\lambda_j + \mu_j)^{-1}, \quad j = 0, 1,\ldots, N-1,$$

$$p_{j-1,j} = \mu_j(\lambda_j + \mu_j)^{-1}, \quad j = 1, 2,\ldots, N,$$

and

$$p_{ij} = 0, \quad \text{if } i = j, \quad \text{or } |i-j| > 0.$$

Then if all $\lambda_j > 0$, $\mu_j > 0$, every state is ultimately accessible from every other state and there are no absorbing states. In one step only the nearest neighbouring states are accessible and the process is therefore sometimes called a 'general one-step random walk'.

The resulting forward Kolmogorov equations are

$$\frac{dP_i(t)}{dt} = \lambda_{i-1}P_{i-1}(t) - (\lambda_i + \mu_i)P_i(t) + \mu_{i+1}P_{i+1}(t), \tag{3.136}$$

which we can take to hold for $i = 0, 1,\ldots, N$ if we make the convention that $\lambda_{-1} = \lambda_N = \mu_0 = \mu_{N+1} = 0$. These equations are no longer recursive, and are therefore more difficult to solve in general than (3.133). Since there are only a finite number of states any solution starting from prescribed initial conditions must be unique. Using this fact we can take Laplace transforms of both sides of (3.136) and the Laplace transform of $P_i(t)$ can then be expressed as the ratio of two determinants. For most definitions of λ_i and μ_i this is not much help since the determinants are often too difficult to evaluate.

However, the stationary distribution $\{P_i\}$, to which the $P_i(t)$ tend, is easily obtainable. From (3.136) we have

$$\lambda_{i-1}P_{i-1}-(\lambda_i+\mu_i)P_i+\mu_{i+1}P_{i+1}=0, \tag{3.137}$$

and thus for $i=0, 1,$

$$-\lambda_0 P_0+\mu_1 P_1=0,$$
$$\lambda_0 P_0-(\lambda_1+\mu_1)P_1+\mu_2 P_2=0.$$

Beginning with the first of these we easily verify that the P_i must be proportional to

$$\frac{\lambda_0 \dots \lambda_{i-1}}{\mu_1 \dots \mu_i}. \tag{3.138}$$

Hence we have

$$P_i=\frac{\lambda_0 \dots \lambda_{i-1}}{\mu_1 \dots \mu_i}\left(1+\sum_{j=1}^{N}\frac{\lambda_0 \dots \lambda_{j-1}}{\mu_1 \dots \mu_j}\right)^{-1}. \tag{3.139}$$

The convergence of the $P_i(t)$ to these values is usually much more difficult to specify.

The imbedded chain is a discrete random walk of the type considered before in which it is only possible to move the states immediately to the right or to the left. If we write the probabilities of moving from i to $i+1$ respectively as $p_i=\lambda_i(\lambda_i+\mu_i)^{-1}$ and $q_i=1-p_i=\mu_i(\lambda_i+\mu_i)^{-1}$, it is sufficient to consider a discrete Markov chain with transition probabilities

$$\begin{aligned} p_{ij}&=p_j, && \text{if } i=j+1,\\ &=q_j=1-p_j, && \text{if } i=j-1,\\ &=0, && \text{otherwise.} \end{aligned}$$

This can sometimes be used to find the probabilities of absorption in various states. These will clearly be the same for the continuous Markov process, and for the chain imbedded in it. Thus if $p_j=p=1-q$, for $j=1,\dots,n-1$, and $p_0=q_N=0$, we have the familiar gambler's ruin problem considered in §3.7.

It is important to notice that it is possible to imbed other Markov chains in this continuous process. Thus we could take

$$\begin{aligned} p_{ij}&=(1-\alpha_j)p_j, && \text{if } i=j+1,\\ &=(1-\alpha_j)q_j, && \text{if } i=j-1,\\ &=\alpha_j, && \text{if } i=j,\\ &=0, && \text{otherwise,} \end{aligned}$$

where the α_i are any numbers such that $0\leqslant\alpha_j<1$.

3.16. Markov chains with an infinite number of states

Some of the theory of finite Markov chains can be extended to the case where the number of possible states is enumerably infinite. The ordinary matrix theory then only generalizes in a rather difficult way, and for the most part other methods have to be used. (For a detailed account of the non-matricial approach see Chung 1960.) We shall only consider sufficient of the theory to make possible the discussion of the most interesting special cases.

We suppose that the system can be in one of an enumerable sequence of states conventionally denoted by E_0, E_1,... (it is sometimes more convenient to denote them by ... E_{-1}, E_0, E_1,... but the appropriate modification to the theorems are then usually easy to make). As before, at discrete instants of time the system undergoes transitions from the state it is in, E_j say, to other states such as E_i, with probabilities p_{ij} where

$$p_{ij} \geqslant 0,$$

$$\sum_{i=0}^{\infty} p_{ij} = 1,$$

and the (p_{ij}) form an infinite matrix. Then the probabilities of transition from E_j to E_i in t steps will be given by a matrix $(p_{ij}^{(t)})$ which is defined by the recurrence relation

$$p_{ij}^{(t+1)} = \sum_{k=0}^{\infty} p_{ik} p_{kj}^{(t)}. \tag{3.140}$$

If $p_j(0)$, $j = 0, 1,...$ is an initial probability distribution of the states, the distribution at time t is given by

$$p_i(t) = \sum_{j=0}^{\infty} p_{ij}^{(t)} p_j(0). \tag{3.141}$$

The asymptotic behaviour of $p_i(t)$ can be rather more complex than in the case of a finite number of states, and even when $p_i(t)$ converges to values $p_i(\infty)$ it is not necessary that $\{p_i(\infty)\}$ be a probability distribution, or even that any of the $p_i(\infty)$ be non-zero. This is obviously illustrated by the example where $p_{ij} = 1$ when $i = j+1$, and is zero otherwise.

A set of numbers, p_i, $i = 0, 1,...$, such that

$$p_i \geqslant 0,$$

$$\sum_i p_i = 1,$$

$$p_i = \sum_j p_{ij} p_j,$$

will be called a 'stationary distribution' of the process. Our main aim is to find the conditions under which such stationary distributions exist, are unique, and are such that $p_i(t)$ converges to p_i.

As before we say that two states, E_i and E_j, 'communicate' if there exist integers m and n such that $p_{ij}^{(m)} > 0$, $p_{ji}^{(n)} > 0$. Such a relationship is clearly symmetric in the sense that if E_i communicates with E_j, E_j must communicate with E_i. Furthermore, it is transitive, i.e. if E_i communicates with E_j, and E_j with E_k, then E_i communicates with E_k. Thus this relation splits the set of all states into classes.

The period, d, of a state E_i is defined to the greatest common divisor of all integers, n, such that $p_{ii}^{(n)} > 0$. Then, as before, it is clear that the property of having a period d is a 'class property', i.e. if d is the period of E_i and E_i communicates with E_j, d is also the period of E_j. If $d = 1$ the class is said to be 'aperiodic'.

The existence of periodic states precludes the possibility of proving that $p_{ij}^{(n)}$ converges without imposing further conditions. We can, however, show, as in §3.2, that the Cesàro limit always exists.

THEOREM 3.8. *In any Markov chain with a finite or infinite number of states, the limit*

$$\pi_{ij} = \lim_n n^{-1} \sum_{s=1}^{n} p_{ij}^{(s)} \tag{3.142}$$

exists for all i, j, and satisfies the equations

$$\pi_{ij} = \sum_{k=1}^{\infty} \pi_{ik} p_{kj}^{(n)} = \sum_{k=1}^{\infty} p_{ik}^{(n)} \pi_{kj} = \sum_{k=1}^{\infty} \pi_{ik} \pi_{kj}, \tag{3.143}$$

and the inequality

$$\sum_{i=1}^{\infty} \pi_{ij} \leqslant 1.$$

This theorem is due to Kolmogorov (1936), and the following proof to Yosida and Kakutani (1939).

Proof. Consider the sequences of numbers

$$q_{ij}^{(n)} = n^{-1} \sum_{s=1}^{n} p_{ij}^{(s)}.$$

These are bounded, and hence by using the Cantor diagonal process we can pick out an increasing sequence of positive integers $n_1, n_2,...$ such that

$$\lim_{\nu \to \infty} q_{ij}^{(n_\nu)}$$

exists for all (i, j). We define π_{ij} to be equal to this limit. We now have, for each n,

$$\left| \sum_{k=1}^{\infty} q_{ik}^{(n)} p_{kj} - q_{ij}^{(n)} \right| = \left| \sum_{k=1}^{\infty} n^{-1} \sum_{\nu=1}^{n} p_{ik}^{(\nu)} p_{kj} - q_{ij}^{(n)} \right|$$

$$= \left| n^{-1} \sum_{\nu=2}^{n+1} p_{ij}^{(\nu)} - q_{ij}^{(n)} \right|$$

$$= n^{-1} \left| p_{ij}^{(n+1)} - p_{ij} \right|$$

$$\leqslant n^{-1}. \tag{3.144}$$

Thus the limit of

$$\sum_{k=1}^{\infty} q_{ik}^{(n\nu)} p_{kj}$$

as $\nu \to \infty$ is π_{ij} so that

$$
\begin{aligned}
\pi_{ij} &= \lim_{\nu \to \infty} \sum_{k=1}^{\infty} q_{ik}^{(n\nu)} p_{kj} \\
&= \sum_{k=1}^{\infty} \lim_{\nu \to \infty} q_{ik}^{(n\nu)} p_{kj} \\
&= \sum_{k=1}^{\infty} \pi_{ik} p_{kj},
\end{aligned}
\tag{3.145}
$$

because

$$\sum_{k} p_{kj}$$

is an absolutely convergent series. By repeated application of (3.145) we get $(n = 1, 2, \ldots)$

$$\pi_{ij} = \sum_{k=1}^{\infty} \pi_{ik} p_{kj}^{(n)} = \sum_{k=1}^{\infty} \pi_{ik} q_{kj}^{(n)}. \tag{3.146}$$

In a similar way it is easy to show that

$$\lim_{\nu \to \infty} \sum_{k=1}^{\infty} p_{ik} q_{kj}^{(n\nu)} = \pi_{ij}. \tag{3.147}$$

It is obvious from this, by considering first a finite sum, that

$$\pi_{ij} \geqslant \sum_{k=1}^{\infty} p_{ik} \pi_{kj}. \tag{3.148}$$

We wish to prove that this is an equality. If this were not so we could find i and j so that (3.148) was a strict inequality. We would then have

$$\sum_{i=1}^{\infty} \pi_{ij} > \sum_{k=1}^{\infty} \sum_{i=1}^{\infty} p_{ik} \pi_{kj} = \sum_{k=1}^{\infty} \pi_{kj}, \tag{3.149}$$

which is impossible. Thus

$$
\begin{aligned}
\pi_{ij} &= \sum_{k=1}^{\infty} p_{ik} \pi_{kj} = \sum_{k=1}^{\infty} p_{ik}^{(n)} \pi_{kj} \\
&= \sum_{k=1}^{\infty} q_{ik}^{(n)} \pi_{kj}.
\end{aligned}
\tag{3.150}
$$

Putting n_ν for n in the last expression and letting ν tend to infinity we get

$$\pi_{ij} = \sum_{k=1}^{\infty} \pi_{ik} \pi_{kj}. \tag{3.151}$$

11

We still have to prove that the right-hand side of (3.142) converges to a limit as n increases. Suppose that i and j are suffices such that this limit does not exist. Then in addition to the sequence, $\{n_\nu\}$, it will be possible to find another increasing sequence of positive integers, $m_1, m_2,...$, say, so that

$$Q_{ij} = \lim_{\nu \to \infty} q_{ij}^{(m_\nu)} \tag{3.152}$$

exists for all i, j, and for the particular values of i, j, under consideration $Q_{ij} \neq \pi_{ij}$. Replacing n by m_ν in (3.146), (3.150), and letting ν tend to infinity, we get

$$\pi_{ij} = \lim_{\nu \to \infty} \sum_{k=1}^{\infty} \pi_{ik} q_{kj}^{(m_\nu)}$$

$$\geqslant \sum_{k=1}^{\infty} \pi_{ik} Q_{kj}, \tag{3.153}$$

and

$$\pi_{ij} = \sum_{k=1}^{\infty} Q_{ik} \pi_{kj}. \tag{3.154}$$

The reason why we get an inequality in (3.153) and equality in (3.154) is that π_{ik} is not, in general, the term of a convergent series when summed over k, but is the term of a convergent series when summed over i. π_{ij} and Q_{ij} enter into the proofs of (3.153) and (3.154) in a completely symmetrical way and hence we can also deduce that

$$Q_{ij} \geqslant \sum_{k=1}^{\infty} Q_{ik} \pi_{kj}, \tag{3.155}$$

and

$$Q_{ij} = \sum_{k=1}^{\infty} \pi_{ik} Q_{kj}. \tag{3.156}$$

From (3.153) and (3.156) we have

$$\pi_{ij} \geqslant Q_{ij},$$

and similarly from (3.154) and (3.155) we have

$$Q_{ij} \geqslant \pi_{ij},$$

so that $\pi_{ij} = Q_{ij}$, and the limit in (3.142) must exist for any (i,j). The fact that

$$\sum_{i=1}^{\infty} \pi_{ij} \leqslant 1$$

follows at once from the existence of the limit. This completes the proof of Theorem 3.8.

To obtain theorems on the convergence of $p_{ij}^{(n)}$ in the ordinary sense we now have to classify the behaviour of the states in a more elaborate manner than for finite chains.

Write $K_{ij}^{(n)}$ for the probability that the system is in state E_i at $t = n$ if it was in E_j at $t = 0$, and not in E_i at $t = 1,..., n-1$. Here i may equal j or not. If $i = j$ this is the probability of a 'first return' to E_j at time $t = n$, and if $i \neq j$ the probability of a 'first passage' from E_j to E_i at time $t = n$. We put L_{ij} for the probability that the system ever returns to E_j, i.e.

$$L_{ij} = \sum_{n=1}^{\infty} K_{ij}^{(n)}, \tag{3.157}$$

which is also defined when $i = j$. We say that a state E_i is 'recurrent' or 'non-recurrent' according as $L_{ii} = 1$, $L_{ii} < 1$. Clearly the property of being recurrent is a class property.

Finally, we write M_{ij} for the probability that if the system starts in the state E_j, it is subsequently in E_i an infinite number of times. Mathematically we define this to be

$$M_{ij} = 1 - \lim_{m \to \infty} \lim_{n \to \infty} R_{ij}^{(m,n)},$$

where $R_{ij}^{(m,n)}$ is the probability that the system is in E_i at most m times during the sequence $t = 0, 1,..., n$, if it was in E_j at $t = 0$. The relationship between M_{ij} and L_{ij} is given by the following theorem.

THEOREM 3.9. $M_{ii} = 1$ if $L_{ii} = 1$, and $M_{ii} = 0$ if $L_{ii} < 1$. Thus M_{ii} can only take the values $0, 1$.

Proof. Let $S_{ii}^{(m)}$ be the probability that the system is in the state E_i at least m times during the sequence $t = 0, 1, 2,...$ when it starts in E_i at $t = 0$. Then

$$S_{ii}^{(m+1)} = L_{ii} S_{ii}^{(m)} = (L_{ii})^m, \tag{3.158}$$

so that

$$M_{ii} = \lim_n S_{ii}^{(n)} = 0, \quad \text{if } L_{ii} < 1$$

$$= 1, \quad \text{if } L_{ii} = 1.$$

LEMMA 3.2. $M_{ij} = 0$, or $M_{ij} > 0$, according as

$$\sum_{l=1}^{\infty} p_{ij}^{(l)} \tag{3.159}$$

converges or diverges. Here i may equal j or not.

Proof. Suppose this series converges to the finite sum S. Starting from the state E_j, let X_n be the number of times the state E_i occurs during the sequence $t = 1, 2,..., n$. Then

$$E(X_n) = \sum_{l=1}^{n} p_{ij}^{(l)} \leq S.$$

Using Tchebychev's inequality (§1.9) the probability that $X_n \geq x$ (x an integer) is not greater than Sx^{-1}, which tends to zero as $x \to \infty$. Hence $M_{ij} = 0$.

Now suppose the series is divergent, and consider a transition from E_j to E_i ($i = j$ included) in n steps. This can occur in n mutually exclusive ways specified by the suffix $k = 1, 2,..., n$, which gives the first time after $t = 0$ at which the system enters E_i. Then

$$p_{ij}^{(n)} = \sum_{k=1}^{n} p_{ii}^{(n-k)} K_{ij}^{(k)}, \qquad (3.160)$$

where we make the convention that $p_{ii}^{(0)} = 1$. We then have

$$\sum_{n=1}^{N} p_{ij}^{(n)} = \sum_{n=1}^{N} \sum_{k=1}^{n} p_{ii}^{(n-k)} K_{ij}^{(k)}$$

$$= \sum_{k=1}^{N} K_{ij}^{(k)} \sum_{n=k}^{N} p_{ii}^{(n-k)}$$

$$\leqslant \sum_{k=1}^{N} K_{ij}^{(k)} \left(1 + \sum_{j=1}^{N} p_{ii}^{(n)}\right). \qquad (3.161)$$

Letting N tend to infinity we have

$$\sum_{n=1}^{\infty} p_{ij}^{(n)} \leqslant L_{ij}\left(1 + \sum_{n=1}^{\infty} p_{ii}^{(n)}\right). \qquad (3.162)$$

Instead of letting $N \to \infty$ in (3.161), put $i = j$ and we get, for all j,

$$\sum_{n=1}^{N} p_{jj}^{(n)} \left(1 + \sum_{n=1}^{N} p_{jj}^{(n)}\right)^{-1} \leqslant \sum_{k=1}^{N} K_{jj}^{(k)}.$$

If the series (3.159) is divergent for some pair of values i, j, it follows from (3.162) that

$$\sum_{n=1}^{\infty} p_{ii}^{(n)}$$

must also diverge because $L_{ij} \leqslant 1$, and therefore

$$L_{ii} = \sum_{n=1}^{\infty} K_{ii}^{(n)} \geqslant 1,$$

so that $L_{ii} = 1$, and $M_{ii} = 1$ by Theorem 3.9. But $L_{ij} > 0$ and $M_{ij} = L_{ij} M_{ii}$ so that $M_{ij} > 0$ as required, which completes the proof of Lemma 3.2. When $L_{ii} = M_{ii} = 1$, the state is recurrent but the mean recurrence time,

$$T_i = \sum_{n=1}^{\infty} n K_{ii}^{(n)} \qquad (3.163)$$

may be finite, or positively infinite.

If $T_i < \infty$, the state E_i is said to be 'positive', and if $T_i = \infty$, it is said to be 'null'. A recurrent state which is positive, and not periodic, is said to be 'ergodic'. It is easy to see that the property of being positive (or null) is a class property and so, therefore, is the property of being ergodic.

A set of states is 'closed' if no transition out of it is possible. The process is 'irreducible' if it contains no closed set of states other than the set of all states. Thus in an irreducible process every state is ultimately accessible from every other state. We are now in a position to prove the main theorems on the convergence of $p_{ij}^{(n)}$.

THEOREM 3.10. *If a Markov chain with an infinite number of states is aperiodic and irreducible, and if all its states are ergodic, then for every pair of suffices i, j,*

$$\lim_{n \to \infty} p_{ij}^{(n)} = \pi_i,$$

where

$$\pi_i = T_i^{-1} > 0, \tag{3.164}$$

and the π_i satisfy the equations

$$\pi_i = \sum_{j=1}^{\infty} p_{ij} \pi_j, \tag{3.165}$$

$$\sum_{i=1}^{\infty} \pi_i = 1.$$

Moreover the π_i are uniquely defined by the two latter equations.

Proof. We shall first show that $p_{ii}^{(n)}$ converges to π_i as n increases, where $\pi_i = T_i^{-1}$. From (3.160) we have

$$p_{ii}^{(n)} = \sum_{k=1}^{n} p_{ii}^{(n-k)} K_{ii}^{(k)}. \tag{3.166}$$

Thus $p_{ii}^{(n)}$ is a weighted average of $p_{ii}^{(n-1)}, p_{ii}^{(n-2)}, \ldots$, with non-negative weights adding to unity.

Write

$$R_n = \sum_{k=n+1}^{\infty} K_{ii}^{(k)}, \quad n \geqslant 1,$$

$$R_0 = 1.$$

Then

$$T_i = \sum_{k=1}^{\infty} k R_{ii}^{(k)}$$

$$= \sum_{k=0}^{\infty} R_k. \tag{3.167}$$

Furthermore,

$$K_{ii}^{(k)} = R_{k-1} - R_k,$$

and inserting this in (3.166) we get

$$p_{ii}^{(n)} = \sum_{k=1}^{n} p_{ii}^{(n-k)} (R_{k-1} - R_k)$$

and

$$\sum_{k=0}^{n} p_{ii}^{(n-k)} R_k = \sum_{k=0}^{n-1} p_{ii}^{(n-1-k)} R_k. \tag{3.168}$$

Define the left-hand side of (3.168) to be A_n, say, then

$$A_n = A_{n-1} = \ldots = A_1 = A_0 = p_{ii}^{(0)} R_0 = 1.$$

We know that $p_{ii}^{(n)} \leqslant 1$ for all n, and hence as n tends to infinity it must have finite upper and lower limits U and L, such that $0 \leqslant L \leqslant U \leqslant 1$. To prove that $p_{ii}^{(n)}$ converges we must show that $U = L$.

Then there exists an increasing sequence of integers $n_1, n_2, \ldots$ such that, as s increases,

$$p_{ii}^{(n_s)} \to U.$$

Let j be such that $K_{ii}^{(j)} > 0$. We shall prove that

$$\lim_{s \to \infty} p_{ii}^{(n_s - j)} \to U.$$

If this is not so there will exist a number $U_1 < U$ and a subsequence $\{n_s'\}$ of $\{n_s\}$ such that

$$\lim_{s \to \infty} p_{ii}^{(n_s' - j)} \to U_1. \tag{3.169}$$

Choose ε so that

$$0 < \varepsilon < \tfrac{1}{2}(U - U_1) K_{ii}^{(j)}.$$

We know that $R_n \to 0$ so that we can choose $N > j$ so large that $R_n < \varepsilon$ for $n \geqslant N$. We then have

$$p_{ii}^{(n)} \leqslant p_{ii}^{(n-1)} K_{ii}^{(1)} + \ldots + p_{ii}^{(n-N)} K_{ii}^{(N)} + \varepsilon,$$

for any $n \geqslant N$. Taking upper limits on both sides of this inequality for increasing values of n in the subsequence $\{n_s'\}$ we have

$$U \leqslant U(1 - K_{ii}^{(j)}) + U_1 K_{ii}^{(j)} + \varepsilon,$$

because

$$U = \overline{\lim_{n}} \, p_{ii}^{(n)},$$

$$= \lim_{s} p_{ii}^{(n_s)}.$$

Then

$$U \leqslant U - (U - U_1) K_{ii}^{(j)} + \varepsilon$$

$$\leqslant U - \varepsilon, \tag{3.170}$$

which is a contradiction. Thus there is no such value U_1.

It follows that if any sequence $\{n_s\}$ is such that

$$p_{ii}^{(n_s)}$$

converges to U, so also must the sequences

$$p_{ii}^{(n_s-j)}, p_{ii}^{(n_s-2j)},\ldots \quad (s = 1, 2,\ldots).$$

Consider the two cases $K_{ii}^{(1)} > 0$, $K_{ii}^{(1)} = 0$, separately. Suppose first that $K_{ii}^{(1)} > 0$. Then $p_{ii}^{(n_s-j)}$ must converge to U for every positive integer j.

We can now carry out a similar argument using the lower bound and conclude that if $m_1, m_2,\ldots$ is any increasing sequence of integers such that

$$p_{ii}^{(m_s)} \to L, \quad s \to \infty,$$

then

$$p_{ii}^{(m_s-j)} \to L,$$

for every positive integer j.

From (3.168) we have

$$\sum_{k=0}^{n} p_{ii}^{(n-k)} R_k = 1. \tag{3.171}$$

We know that $R_k \to 0$. We can therefore choose N so large that $R_{N+1} < \varepsilon$ for some prescribed $\varepsilon > 0$. Then from (3.171)

$$p_{ii}^{(m_s)} R_0 + \ldots + p_{ii}^{(m_s-N)} R_N + \varepsilon \geqslant 1$$

for all sufficiently large m_s, where m_s belongs to the above-defined sequence. Keeping N fixed, and letting m_s tend to infinity, we have

$$L(R_0 + \ldots + R_N) + \varepsilon \geqslant 1.$$

Using the same argument with the sequence $\{n_s\}$ we get

$$U(R_0 + \ldots + R_N) \leqslant 1,$$

so that

$$U(R_0 + \ldots + R_N) \leqslant \varepsilon + L(R_0 + \ldots + R_N),$$

and therefore, since ε is arbitrary,

$$U \leqslant L.$$

But by definition $L \leqslant U$, so that

$$L = U = \left(\sum_0^\infty R_N\right)^{-1} = T_i^{-1}.$$

This settles the case $K_{ii}^{(1)} > 0$.

Now suppose $K_{ii}^{(1)} = 0$. Since the process is aperiodic, every state is aperiodic, and the set of all integers j such that $K_{ii}^{(j)} > 0$, has a greatest common divisor equal to unity. There must therefore exist a finite set of integers $l_1,\ldots, l_\nu$, which are such that

$$K_{ii}^{(l_s)} > 0, \quad s = 1,\ldots,\nu,$$

and which have a greatest common divisor equal to unity. For any fixed integers $p_1,\ldots, p_\nu$, it will follow that

$$p_{ii}^{(n_s - l_1 p_1 - \cdots - l_\nu p_\nu)} \to U.$$

Every integer greater than $l_1, l_2,\ldots, l_\nu$ can be written in the form

$$j = l_1 p_1 + \ldots + l_\nu p_\nu,$$

and therefore

$$p_{ii}^{(n_s - j)} \to U$$

for every $j > l_1 l_2,\ldots, l_\nu$. By modifying n_s the same is clearly true for every $j > 0$, and thus $U = L$, and $p_{ii}^{(n)}$ converges to M_1^{-1}.

This settles the convergence of $p_{ii}^{(n)}$ for every i, and it is now relatively simple to discuss the convergence of $p_{ij}^{(n)}$ for $i \neq j$. Write π_i for

$$T_i^{-1} = \lim p_{ii}^{(n)}.$$

We shall prove that

$$p_{ij}^{(n)} \to \pi_i.$$

The process is irreducible and hence

$$\sum_{n=1}^{\infty} K_{ij}^{(n)} = 1,$$

for each pair i, j, because otherwise there would be a non-zero probability of the system moving from i to j, and then never returning. Given any $\varepsilon > 0$ there must exist a positive integer N such that

$$\sum_{n=1}^{N} K_{ij}^{(n)} > 1 - \varepsilon.$$

Consider the equation

$$p_{ij}^{(n)} = \sum_{k=1}^{n} p_{ii}^{(n-k)} K_{ij}^{(k)}. \tag{3.172}$$

By choosing n large the first N terms on the right-hand side can be made to differ arbitrarily little from

$$\pi_i \sum_{k=1}^{N} K_{ij}^{(k)},$$

and hence from π_i, whilst the sum of the remaining terms is less than ε. Thus

$$p_{ij}^{(n)} \to \pi_i.$$

To prove (3.165) we start from the equation

$$p_{ij}^{(m+1)} > \sum_{k=1}^{N} p_{ik}^{(1)} p_{kj}^{(m)}$$

for some fixed N. Letting m tend to infinity we have

$$\pi_i \geqslant \sum_{k=1}^{N} p_{ik}^{(1)} \pi_k$$

$$\geqslant \sum_{k=1}^{\infty} p_{ik}^{(1)} \pi_k, \qquad (3.173)$$

because N is arbitrary. If this were a strict inequality for some value of i we could sum over all values of i and obtain

$$\sum_{i=1}^{\infty} \pi_i > \sum_{k=1}^{\infty} \pi_k,$$

which is impossible. Thus (3.165) must hold.

Finally, we must prove the uniqueness of the solution. Suppose that $x_1, x_2, \ldots$ is a sequence of non-negative quantities such that

$$x_i = \sum_{j=1}^{\infty} p_{ij} x_j, \qquad (3.174)$$

and

$$\sum_{i=1}^{\infty} x_i$$

is convergent. Multiply (3.174) by $p_{ki}^{(n-1)}$, and sum over i. Then by induction we have

$$x_j = \sum_{i=1}^{\infty} p_{ji}^{(n)} x_i,$$

for $n = 1, 2, \ldots$.

By the bounded convergence of the series of x_i we can let n tend to infinity and obtain

$$x_j = \pi_j \sum_{i=1}^{\infty} x_i, \qquad (3.175)$$

so $x_j \pi_j^{-1}$ is independent of j. Rescaling the x_j so that

$$\sum_{i=1}^{\infty} x_i = 1,$$

we get $x_j = \pi_j$, and the π_j must be unique. This completes the proof of Theorem 3.10.

When particular processes are aperiodic and irreducible it is usually quite easy to verify that this is so. The main problem in applying Theorem 3.10 is therefore to show that the process is ergodic. There are a number of criteria which are sufficient to establish this, and depend on properties of the p_{ij} (see, for example, Foster 1953 and the references there given). However, the simplest way of showing that the chain is ergodic is to show that there exists a stationary distribution. That this implies ergodicity follows from the following theorem.

THEOREM 3.11. *If the process is irreducible and aperiodic, and $T_i = \infty$ for some i, then $T_i = \infty$ for all i, and*

$$p_{ij}^{(n)} \to 0$$

for all i, j.

The process is then not ergodic. If a stationary distribution exists this result cannot happen and the process must therefore be ergodic, and the stationary distribution unique.

Proof. Since the chain is irreducible, if any state is null ($T_i = \infty$) all other states must be null also. We can now follow through the proof of Theorem 3.10 observing that R_n is finite and tends to zero, but $\sum_1^\infty R_i$ is infinite, so that we can conclude that

$$p_{ii}^{(n)} \to 0$$

for all i. This is incompatible with the existence of a stationary distribution, and the theorem is therefore proved.

3.17. The Borel–Tanner process

As an illustration of the above theory we consider the generation of the Borel–Tanner distribution (§ 2.19) by a random process which occurs in the theory of queues (D. G. Kendall 1951). We suppose that customers arrive at a servicing counter in a Poisson process with mean λ, i.e. the number arriving in any interval of length T has a Poisson distribution with mean λT, and the numbers arriving in different intervals are independent.

The customers are served in the order of their arrival, each service requiring a fixed time which we can take as unity. The total number of customers waiting to be served or being served is then the random variable of a random process in continuous time and would therefore be more properly discussed later in this chapter. However, as Kendall shows, we can easily imbed in the system a Markov chain with an enumerable number of states and discrete time.

To do this we define the discrete instants of time to be those instants, t, at which the service of a customer is completed, and we define the state of the system, X, to be the number of customers left behind in the queue when a customer departs. We can now define the transition probabilities,

$$p_{ij} = \mathrm{pr}(X_{t+1} = i \mid X_t = j).$$

In fact

$$p_{i0} = \lambda^i e^{-\lambda}(i!)^{-1}, \quad i = 0, 1, \ldots,$$

$$p_{ij} = 0, \quad j > 0, \quad i < j-1,$$

$$= \lambda^{i-j+1} e^{-\lambda}(i-j+1!)^{-1}, \quad i \geq j-1.$$

The process is clearly irreducible and aperiodic, so it will follow that the probabilities of the various states will converge to a stationary distribution $\{P_i\}$ if we can show that such a stationary distribution exists. In fact, however, it is simpler in this case to use a criterion of Foster (1951) which shows that a stationary distribution exists when $\lambda < 1$. Kendall has found the generating function of this distribution by an ingenious argument.

Write

$$\delta(X) = 0, \quad \text{if } X > 0,$$
$$= 1, \quad \text{if } X = 0. \tag{3.176}$$

Then

$$X_{t+1} = X_t - 1 + \delta(X_t) + Y_t, \tag{3.177}$$

where Y_t has a Poisson distribution with mean λ. When the stationary distribution has been reached, X_{t+1} and X_t will have the same distribution. Taking the expectation of (3.177) on both sides we get

$$E\{\delta(X_t)\} = 1 - \lambda. \tag{3.178}$$

Squaring both sides of (3.177) and taking expectations we get

$$E(X_{t+1}^2) = E(X_t^2) + E\{1 + \delta(X_t)^2 - 2X_t + 2\delta(X_t)X_t$$
$$+ 2X_t Y_t - 2\delta(X_t) - 2Y_t + 2\delta(X_t)Y_t\}.$$

Using the fact that the process is stationary,

$$E(X_t) = \frac{\lambda(2-\lambda)}{2(1-\lambda)}. \tag{3.179}$$

Now write

$$P(z) = \sum_{n=0}^{\infty} P_n z^n, \tag{3.180}$$

for the generating function of the distribution of the number of customers waiting or being served. We then have

$$P(z) = E(z^{X_t}) = E(z^{X_{t+1}})$$
$$= E\{z^{X_{t-1}+Y_t} \mid \delta(X_t) = 0\}\text{pr}\{\delta(X_t) = 0\}$$
$$+ E\{z^{X_t} \mid \delta(X_t) = 1\}\text{pr}\{\delta(X_t) = 1\}$$
$$= \lambda z^{-1} e^{\lambda(z-1)} E\{z^{X_t} \mid \delta(X_t) = 0\} + (1-\lambda)e^{\lambda(z-1)}. \tag{3.181}$$

We also have

$$P(z) = E(z^{X_t} \mid X_t = 0)\text{pr}(X_t = 0)$$
$$+ E(z^{X_t} \mid X_t > 0)\text{pr}(X_t > 0)$$
$$= 1 - \lambda + \lambda E\{z^{X_t} \mid \delta(X_t) = 0\}.$$

Using the last equation in (3.181) we get

$$P(z) = \frac{e^{\lambda(z-1)}(1-\lambda)(z-1)}{z - e^{\lambda(z-1)}}. \tag{3.182}$$

From this all the higher moments can be found. We can verify, with a little labour, that the probabilities given by this generating function satisfy the equations for a stationary distribution, and hence that the process is ergodic.

The Borel–Tanner distribution is defined to be the distribution of the time which will elapse until the queue is first empty, it being known that at $t = 0$ there are exactly r customers in the queue, of which the first is about to be served (§2.19). It is obvious from this definition that this distribution is the rth convolution power of the similar distribution with $r = 1$, which gives us an immediate intuitive interpretation of equation (2.131) in Chapter 2.

The time required clearly takes only the discrete values $j = r, r+1, \ldots$. We shall prove that the required distribution is given by (2.128) where α is replaced by λ. We do this by an argument which is a special case of the more general theory of the next section.

An interval during which a customer is being served is said to be a 'busy interval'. Let k be the total number of busy intervals before the queue is empty, and $z^r \pi(z)^r$ the generating function of its probability distribution which is of this form from the above argument. The servicing intervals of the r initial customers may be called the 'first generation' of intervals, the servicing intervals of all those individuals which arrived during the first generation of intervals, the 'second generation' of intervals, and so on. Then $\pi(z)$ would be the probability generating function of the total number of second, third, fourth,... generations of intervals if there is one initial interval. Write

$$\pi(z) = p_0 + p_1 z + p_2 z^2 + \ldots. \tag{3.183}$$

Consider first the case where $r = 1$ so that there is only one member of the first generation of intervals. During this interval let j be the number of customers who arrive. j is equal to the number of second generation intervals and will have a generating function

$$e^{\lambda(z-1)} = e^{-\lambda}(1 + \lambda z + \tfrac{1}{2}\lambda^2 z^2 + \ldots).$$

During each of these intervals customers causing third generation intervals may arrive, and therefore the total number of second, third,... generation intervals will have the generating function

$$e^{-\lambda}\{1 + \lambda z \pi(z) + \tfrac{1}{2}\lambda^2 z^2 \pi(z)^2 + \ldots\}$$

$$= e^{\lambda\{z\pi(z)-1\}}$$

$$= \pi(z). \tag{3.184}$$

If we equate the powers of z in the last equation we find

$$p_0 = e^{-\lambda},$$

which is otherwise obvious, and a series of equations which define each p_i in terms of $p_{i-1}, p_{i-2}, \ldots, p_0$. Hence it is sufficient to find any generating function $\pi(z)$ satisfying (3.184) since it is uniquely determined.

If we now take (2.123) we can put

$$p_0 = 0,$$
$$p_i = \alpha^{-1}i^{i-1}(\alpha e^{-\alpha})^i(i!)^{-1}, \quad i > 0.$$

That these add to unity follows from (2.126).

Replacing α by λ, we must then verify that

$$P(z) = z^{-1}\sum_{i=1}^{\infty} \lambda^{-1}i^{i-1}(z\lambda e^{-\lambda})^i(i!)^{-1} \tag{3.185}$$

satisfies (3.184).

Now from (2.126) in §2.19 we know that if two quantities, X and Y, are related by the equation (where $0 < X < 1$)

$$X = \sum_{i=1}^{\infty} Y^i i^{i-1}(i!)^{-1}, \tag{3.186}$$

then

$$Y = Xe^{-X}. \tag{3.187}$$

Put

$$Y = \lambda z e^{-\lambda}. \tag{3.188}$$

Then the right-hand side of (3.186) is equal to

$$\lambda z P(z),$$

so that

$$Y = \lambda z e^{-\lambda} = \lambda z P(z)e^{-\lambda z P(z)},$$

which is equivalent to

$$P(z) = e^{\lambda\{zP(z)-1\}},$$

as required. (3.185) is therefore the uniquely determined generating function. The result for k initial customers then follows from (2.131).

This treatment of the distribution of the time from an initial queue of k customers to the first instant at which there are no customers can be generalized and such problems are extensively discussed in books on queueing such as Syski (1960), Saaty (1961), and Takacs (1962).

3.18. The multiplicative process

This is a process in discrete time whose state is defined by a random variable taking the values 0, 1, 2,... and whose transition probabilities have a special form which arises in the following way which is a generalization of the situation described above in §3.17.

Consider a population consisting of X_1 individuals at time $t = 0$. Between $t = 0$ and $t = 1$ each of these give rise to $0, 1, 2,\ldots$ offspring with probabilities $p_0, p_1,\ldots$ where $\sum p_i = 1$. The total number of such offspring, X_2, defines the state of the system at $t = 1$. Between $t = 1$ and $t = 2$ each of the X_2 individuals in turn gives rise to offspring with the same probability distribution. Let

$$P(z) = p_0 + p_1 z + p_2 z^2 + \ldots. \qquad (3.189)$$

This is the probability generating function of X_2 conditional on $X_1 = 1$. The probability generating function of X_3 conditional on $X_1 = 1$ will then be

$$p_0 + p_1 P(z) + p_2 P(z)^2 + \ldots = P\{P(z)\}, \qquad (3.190)$$

and carrying the argument further, the probability generating function of X_n will be

$$P[P\{\ldots P(z)\ldots\}], \qquad (3.191)$$

('P' occurring $n-1$ times) provided $X_1 = 1$. For $X_1 = k$, since all the individuals are independent, the probability generating function will be

$$[P[P\{\ldots P(z)\ldots\}]]^k. \qquad (3.192)$$

Thus p_{ij}, the transition probability, will be given by

$$p_{00} = 1,$$

$$p_{i0} = 0, \quad i > 0,$$

$$p_{ij} = \text{coefficient of } z^i \text{ in } P(z)^j, \quad j > 0.$$

Such a system is known as a multiplicative process and has an extensive literature. A detailed treatment is given by Harris (1963).

Since the distribution of X_n is the kth convolution of its distribution for $k = 1$ we need consider only the latter case. The lower moments are easily obtained. Write $P_n(z)$ for (3.191). Write m_1, m_2 for the first and second moments about zero of the distribution determined by $P(z)$ so that

$$m_1 = P'(1), \quad m_2 - m_1 = P''(1).$$

Then (for $m_1 \neq 1$)

$$P'_n(1) = P'_{n-1}(1)P'(1)$$

$$= \{P'(1)\}^n = m_1^n. \qquad (3.193)$$

Similarly

$$P''_n(1) = P''_{n-1}(1)m_1^2 + m_1^{n-1}(m_2 - m_1).$$

Thus if $m_1^{(n)}$ and $m_2^{(n)}$ are the first and second moments of $P_n(z)$ we have $m_1^{(n)} = m_1^n$, and solving the recurrence relation

$$m_2^{(n)} = m_1^{2n} + (m_2 - m_1^2)m_1^{n-1}\frac{(1 - m_1^n)}{(1 - m_1)}. \qquad (3.194)$$

When $m_1 = 1$ we get $m_1^{(n)} = 1$, $m_2^{(n)} = 1 + n(m_2 - 1)$.

If $m_1 < 1$ it follows from (3.193) by using Tchebychev's inequality that X_n will converge to zero in probability. We shall see that this is the only case in which the process is ergodic.

The principal problem in the study of such processes is to determine the asymptotic behaviour of the probability that X_n is zero. This is solved by the following theorem.

THEOREM 3.12. *If $m_1 \leqslant 1$, X_n converges to zero so that the probability that the population becomes extinct is unity unless $p_0 = 0$. If $m_1 > 1$ the probability that $X_n = 0$ converges to the unique root of the equation $z = P(z)$ other than $z = 1$, and in this case the process is not ergodic.*

Proof. The result with $m_1 < 1$ has already been proved. Suppose $m_1 \geqslant 1$. Let the process start with one initial individual. The probability of extinction at the first generation is $p_0 = P(0)$, at or before the second generation $P\{P(0)\} = P_2(0)$, and at or before the nth generation $P_n(0)$. No extinction is possible unless $p_0 > 0$ which we assume. $P(x)$ is an increasing function of x and thus $P_{n+1}(0) \geqslant P_n(0)$ implies

$$P_{n+2}(0) = P\{P_{n+1}(0)\} \geqslant P\{P_n(0)\} = P_{n+1}(0). \tag{3.195}$$

Hence $P_n(0)$ is a non-decreasing function of n as is otherwise obvious, and since it is bounded above, must increase to an upper limit, z_0, which satisfies the equation

$$z_0 = P(z_0). \tag{3.196}$$

Suppose that $z_1 > 0$ is some other root of this equation. Then $P(z_1) > p_0 = P(0)$. Moreover if $P_n(0) < z_1$ we have

$$P_{n+1}(0) = P\{P_n(0)\} < P(z_1) = z_1. \tag{3.197}$$

Thus z_1 is necessarily greater than z_0, and z_0 is the smallest positive root.

Since $P''(z) > 0$ the curve $y = P(z)$ is strictly concave above and can therefore intersect the line $y = z$ in at most two points. One of these points is $y = z = 1$, and the other may be less or greater than unity. The curve is definitely above the line $y = z$ at $z = 0$ because $p_0 > 0$ and therefore intersects the line $p = z$ at a point below or above $y = z = 1$ according as $P'(1) > 0$ or $P'(1) < 0$. If $P'(1) = 1$ there is a double point at $y = z = 1$ where the line $y = z$ is tangential to the curve.

We have therefore shown that if $P'(1) \geqslant 1$, i.e. $m_1 \geqslant 1$, the probability of extinction converges to z_0 and $0 \leqslant z_0 < 1$.

If $p_0 = 0$ we have $m_1 \geqslant 1$, and the probability of extinction is zero. This completes the proof since the probabilities of the other states must be zero and the process is not ergodic.

This problem has arisen independently in a wide variety of situations and its elementary theory rediscovered by a number of different writers. The earliest appears to have been F. Galton who proposed the following problem in the *Educational Times* (1873). Suppose that each male in a population has a number of sons whose distribution has a generating function $P(z)$. It is required to find the probability that his male line of descendants ultimately dies out. This is usually known as the 'problem of family names'. A partial solution was given by H. W. Watson (1874), a paper which was later reproduced in Galton (1889). A complete solution was first given by J. F. Steffensen (1930, 1933) in reply to a reproposal of the problem by A. K. Erlang (1929). The results were applied to empirical data by Lotka (1931).

Fisher (1930) discussed the same problem applied to survival of rare mutants, and the subject has been extensively developed in a genetical connexion (see Moran 1962a for references and also Harris 1963). Woodward (1948) rediscovered the theory in connexion with cascade electron multipliers, and in § 3.17 we have used the theory to find the distribution of the first time at which a queue becomes empty when the service time is a constant. In this application what is required is the distribution of the total number of descendants. Consider this for a general offspring distribution (Otter 1949, Hawkins and Ulam 1944, Good 1949) which has a generating function $P(z)$. Suppose that the generating function of all the descendants up to and including the sth generation, and also including the original parent, is $G_s(z)$. Then clearly

$$P\{zG_s(z)\} = G_{s+1}(z). \tag{3.198}$$

If

$$\lim_{s \to \infty} G_s(z) = G_\infty(z)$$

exists, then

$$P\{zG_\infty(z)\} = G_\infty(z). \tag{3.199}$$

Otter discusses the conditions under which this equation has a solution, $G_\infty(z)$, when $P(z)$ is known. The first two moments of the distribution defined by $G_s(z)$ are easily obtained by differentiation.

Another interesting set of results arise when 'immigration' is introduced into the system. Consider a multiplicative process generated by a probability generating function $P(z)$ such that $P'(i) < 1$. Starting from a finite number of individuals the process is certain to die out but this can be obviated by introducing Y_t new individuals between t and $t+1$, where the Y_t are independent random variables having the same distribution with probability generating function $R(z)$, say. Then if the process starts at $t = 1$ with one individual we have

$$P_2(z) = P\{P(z)\}R(z),$$

$$P_3(z) = P_2\{P(z)\}R(z),$$

and so on. Except for degenerate cases it is usually easy to show that this process is ergodic, and $P_n(z)$ will tend to a generating function, $\pi(z)$, satisfying the equation

$$\pi(z) = \pi\{P(z)\}R(z). \tag{3.200}$$

Very few solutions of this equation are known. If $P(z)$ is the generating function of a binomial distribution so that $P(z) = q + pz$, and $R(z)$ is the generating function of a Poisson distribution with mean λ it is easy to verify that $\pi(z)$ must be

$$e^{\lambda q^{-1}(z-1)},$$

which corresponds to a Poisson distribution with mean λq^{-1}. This is the result of the obvious fact, mentioned in the previous chapter (§2.13), that if we have a number of objects which are distributed in a Poisson distribution with mean Λ, say, and each object is chosen or rejected independently with probability p, the total number chosen will again be a Poisson variate with mean Λp. For another case see Exercise 3.3.

Much more is known about the simple multiplicative process than is given above. In particular if $m_1 \leqslant 1$ and $m_2 < \infty$, the distribution of X_n conditional on $X_1 \neq 0$ can be shown to tend to a limiting distribution, and if $m_1 > 1$, $m_2 < \infty$, the distribution of $X_n m_1^{-n}$ also converges to a definite distribution. The theory can also be extended to cases where there is more than one kind of individual so that if there are k kinds of individual the state of the system is defined by a set of integers $(n_1, \ldots, n_k)$ where $n_i = 0, 1, 2, \ldots$.

3.19. Markov process in continuous time with an infinite number of states

We again suppose that the state of the system is defined by a variate $X(t) = 0, 1, \ldots$ where t is continuous. The probabilities of these values at time t will be defined to be quantities $p_i(t)$ where $p_i(t) \geqslant 0$. Instead of the condition that these add to unity we will see that it is necessary to impose only the wider condition

$$\sum_{0}^{\infty} p_i(t) \leqslant 1. \tag{3.201}$$

The transition probabilities from state j at time t to state i at time $s > t$ can again be written as $p_{ij}(s, t)$, but we confine our attention to the case where these are functions of $s - t$ only so that we write them as $p_{ij}(s - t)$. Introducing, as in §3.11, quantities c_j $(j = 0, 1, \ldots)$ and p_{ij} $(i, j = 0, 1, \ldots)$ we obtain by an exactly parallel argument Kolmogorov's forward differential equations

$$\frac{d}{dt} P_{ij}(t) = -c_i P_{ij}(t) + \sum_{\substack{k=0 \\ k \neq i}}^{\infty} c_k p_{ik} P_{kj}(t), \tag{3.202}$$

which parallel (3.116) but involve an infinite number of equations for an infinite number of unknown functions. Similarly we obtain Kolmogorov's backward equations

$$\frac{d}{dt} P_{ij}(t) = -c_j P_{ij}(t) + \sum_{\substack{k=0 \\ k \neq i}}^{\infty} c_j p_{kj} P_{ik}(t). \tag{3.203}$$

Notice that for each value of j, the equations (3.202) for $i = 0, 1, 2,...$ are an infinite set of equations for the functions $P_{0j}, P_{1j}, P_{2j},...$ whereas (3.203) are, for each i, an infinite set of equations for $P_{i0}, P_{i1}, P_{i2},....$ (3.202) and (3.203) can be written in matrix form but this is less useful than in the finite case since infinite matrices are more difficult to deal with.

Unlike the finite case the theory of the existence and uniqueness of the solutions of (3.202) and (3.203) is not simple (Feller 1957b, Ledermann and Reuter 1954). Here we only state the main results in the form of a theorem.

THEOREM 3.13. *Given that $c_j \geqslant 0$ for all j there exist solutions of the systems* (3.202), (3.203) *which satisfy the conditions*

$$0 < P_{ij}(t) \leqslant 1, \quad \text{all } i, j. \tag{3.204}$$

$$\sum_{i=0}^{\infty} P_{ij}(t) \leqslant 1, \quad \text{all } j, \tag{3.205}$$

$$P_{ij}(t+s) = \sum_{k=0}^{\infty} P_{ik}(t)P_{kj}(s), \quad t > 0, s > 0, \tag{3.206}$$

(the Chapman–Kolmogorov equation), and

$$\lim_{t \to 0} P_{ij}(t) = 1, \quad \text{if } i = j,$$

$$= 0, \quad \text{if } i \neq j. \tag{3.207}$$

Such solutions do not in general result in an equality in (2.205) but if they do, they are unique.

A simple proof of this theorem will be found in a paper by Feller (1957a). Before considering any further general results it is useful to consider some particular examples.

3.20. The pure birth process and the imbedded chain

This is a particularly simple process because the only transitions allowed are from each state i to $i+1$. Putting $P_i(t)$ for the probabilities of the system being in states $i = 0, 1,...$ at time t the forward equations become

$$\frac{d}{dt} P_0(t) = -\lambda_0 P_0(t),$$

$$\frac{d}{dt} P_i(t) = \lambda_{i-1} P_{i-1}(t) - \lambda_i P_i(t), \quad i > 0, \tag{3.208}$$

where λ_0, λ_1,... are a sequence of non-negative constants, and the initial condition may be taken as $P_0(0) = 1$, $P_i(0) = 0$ $(i > 0)$. We have already considered a finite model of this kind in § 3.14. Since the equations (3.208) can be solved in succession, no difficulty is found in general in finding a solution which may, however, not be proper. In particular, if all the λ_i are equal to λ, say, the solution, as in § 3.14, is

$$P_i(t) = e^{-\lambda t}(\lambda t)^i (i!)^{-1}, \quad i \geqslant 0. \tag{3.209}$$

This is known as the Poisson process. The $P_i(t)$ are uniquely determined and

$$\sum_{i=0}^{\infty} P_i(t) = 1.$$

In the next simplest process we put $\lambda_i = \lambda i$ and obtain, as in § 3.13,

$$P_i(t) = e^{-\lambda t}(1 - e^{-\lambda t})^{i-1},$$

$$P(z) = e^{-\lambda t}z(1 - z + ze^{-\lambda t})^{-1}, \tag{3.210}$$

where the initial conditions have now been necessarily modified to

$$P_0(0) = 0,$$

$$P_1(0) = 1,$$

$$P_i(0) = 0, \quad i > 1. \tag{3.211}$$

Here again the $P_i(t)$ are uniquely defined and

$$\sum_{i=1}^{\infty} P_i(t) = 1.$$

This is the Yule–Furry process (Yule 1924, Furry 1937). We see from (3.210) that the distribution at any time is geometric and the moments are easily found. In particular the first moment is

$$m_1(t) = e^{\lambda t}, \tag{3.212}$$

and the second moment

$$m_2(t) = 2e^{2\lambda t} - e^{\lambda t}. \tag{3.213}$$

From these results it is easy to obtain the transition probabilities, $P_{ij}(t)$ by convoluting the distribution $\{P_i(t)\}$ and thus obtaining negative binomial distributions.

Just as in the case of continuous processes with a finite number of states we can imbed a discrete Markov chain in each of the above two processes and in both cases the chain is defined by the simple condition that at each instant of time at which a jump is possible, the state changes from j to $j+1$ so that

$$p_{ij} = 1, \quad \text{if } i = j+1$$

$$= 0, \quad \text{otherwise.}$$

Now consider the general case in which the λ_i have any finite positive values. We then encounter the remarkable phenomenon that with suitable λ_i an infinite number of transitions may occur in a finite time, and in this case (3.201) is not an equality (Feller 1957*b*, p. 405). We may state the condition for this in the form of a theorem.

THEOREM 3.14. *In a simple birth process the necessary and sufficient condition that we always have*

$$\sum_0^\infty P_i(t) = 1,$$

is that

$$\sum_0^\infty \lambda_i^{-1} \tag{3.214}$$

is divergent, it being assumed that all $\lambda_i > 0$.

   Proof. Write

$$S_n(t) = P_0(t) + \ldots + P_n(t).$$

Then

$$\frac{d}{dt} S_n(t) = -\lambda_n P_n(t).$$

Let i be the initial state so that $P_0(t) = \ldots = P_{i-1}(t) = 0$, $P_i(0) = 1$. Taking $n \geqslant i$, we have

$$1 - S_n(t) = -\int_0^t \frac{d}{du} S_n(u)\, du$$

$$= \lambda_n \int_0^t P_n(u)\, du. \tag{3.215}$$

For fixed t, this is necessarily a non-increasing function of n, and therefore has a limit, $\mu(t)$. We then have

$$\mu(t) \leqslant \lambda_n \int_0^t P_n(u)\, du,$$

for all $n \geqslant i$. Hence

$$\int_0^t S_n(u)\, du = \sum_{j=1}^n \int_0^t P_j(u)\, du \geqslant \mu(t)(\lambda_i^{-1} + \ldots + \lambda_n^{-1}).$$

But from (3.215) $S_n(t) \leqslant 1$, and therefore

$$\int_0^t S_n(u)\, du \leqslant t$$

so that

$$t \geqslant \mu(t)(\lambda_i^{-1} + \ldots + \lambda_n^{-1}).$$

Thus if (3.214) diverges we must have $\mu(t) = 0$ for all t. As n increases, (3.215) must tend to zero for each fixed t and

$$\sum_{0}^{\infty} P_i(t) = 1, \quad \text{all } t.$$

Conversely suppose (3.214) is convergent. Then

$$\int_0^t S_n(u) \, du = \sum_{j=1}^n \int_0^t P_j(u) \, du$$

$$\leqslant \sum_{j=1}^n \lambda_j^{-1} \{1 - S_j(t)\}$$

$$\leqslant \sum_{j=1}^n \lambda_j^{-1},$$

and hence

$$\int_0^t S_n(u) \, du$$

is bounded for all t and all n, which is impossible if $S_n(t) \to 1$ for all t.

A simple example in which (3.214) is convergent occurs if we put $\lambda_i = \lambda i^2$. What happens then is that there is a non-zero probability that an infinite number of jumps occurs in a finite time. This may be looked at in a different way if we anticipate slightly the theory of continuous random variables discussed in Chapters 5, 6 and 7. Suppose the system starts in the state E_1. It is easy to see that the time which will elapse before the jump to E_2 is a random variable which has a negative exponential distribution with mean λ_1^{-1}. Similarly the time which elapses between the jump $E_1 \to E_2$ and the jump $E_2 \to E_3$ will have mean λ_2^{-1}. Thus the mean time to the Nth jump is

$$\lambda_1^{-1} + \ldots + \lambda_N^{-1}. \tag{3.216}$$

If the series (3.214) converges, (3.216) is bounded above and thus by Tchebychev's inequality there must exist a number T such that the probability that the Nth jump occurs before T is greater than $\frac{1}{2}$, independently of N. Hence the probability of an infinite number of jumps before T is also $\geqslant \frac{1}{2}$. In this process an explicit solution for the $P_i(t)$ can in fact be obtained (John 1961).

3.21. The Pólya process

This is a non-stationary pure birth process in which λ_n is a function of t which we assume to be

$$\lambda_n(t) = \frac{1 + an}{1 + at}. \tag{3.217}$$

Notice that the constant a occurs in both numerator and denominator. This is done for algebraic convenience and can easily be changed by rescaling t. Direct solution of the forward Kolmogorov equations gives

$$P_0 = (1+at)^{-a^{-1}},$$

$$P_j = t^j(1+at)^{-a^{-1}-j}\left\{\frac{(1+a)...\{1+(j-1)a\}}{j!}\right\}, j = 1, 2,..., \qquad (3.218)$$

where the initial state is E_0. (3.218) is a negative binomial distribution and

$$P_t(z) = (1+at)^{-a^{-1}}\left(1-\frac{atz}{1+at}\right)^{-a^{-1}}. \qquad (3.219)$$

The first two moments are easily found to be

$$m_1(t) = t,$$

$$m_2(t) = t+(1+a)t^2.$$

The similarity of (3.219) and (3.210) is understood if we take as a new time variable $\tau = \log(1+at)$ and put $a = 1$. We have already seen in §2.14 that the negative binomial can arise by compounding the parameter in a Poisson distribution whereas here it has arisen as the distribution in a random process. Bartlett (1955, p. 55) has shown that this result can be generalized. The Pólya process can also be obtained as a limit from the Pólya urn scheme considered in §2.16.

As pointed out in §2.14, the distribution (3.218) also has implications for the theory of accident proneness. The number of accidents occurring to workers in factories is usually better fitted by a negative binomial than by a Poisson distribution. As shown in §2.14, this can be explained if we regard 'accident-proneness' as a characteristic varying from worker to worker with something like a gamma distribution. However, the above theory shows that if all workers begin with an equal proneness but are more liable to have an accident if they have had one or more already, in the manner defined by (3.217), the same observed negative binomial distribution will be found. Thus it is not possible to distinguish between the two theories. For a further interesting discussion see Irwin (1964) and Greenwood (1950).

3.22. The birth and death process

Here we assume that transitions both of the form $E_i \to E_{i+1}$ and of the form $E_i \to E_{i-1}$ are possible, the former with coefficient λ_i and the latter, corresponding to a death, with coefficient μ_i. The forward Kolmogorov equations are then

$$\frac{d}{dt}P_0(t) = -\lambda_0 P_0(t)+\mu_1 P_1(t), \qquad (3.220)$$

and

$$\frac{d}{dt}P_i(t) = \lambda_{i-1}P_{i-1}(t) - (\lambda_i + \mu_i)P_i(t) + \mu_{i+1}P_{i+1}(t). \tag{3.221}$$

Systems of this kind are more difficult to solve than those of simple birth processes because they cannot be solved recursively beginning with $P_0(t)$. Here we can only describe a few of the simpler facts about such processes and refer the reader to Kendall (1949), Bharucha–Reid (1960), Syski (1960), and a series of papers by Karlin and McGregor (1957a, b, 1958 and later papers) for further details.

Conditions for the uniqueness of the solution of the forward Kolmogorov equations similar to those of Theorem 3.14 are rather more complicated (see Bharucha–Reid 1960 for details and references) and here it is sufficient to state that if $\lambda_n > 0$ for $n \geqslant i$, and

$$\sum_{n=i}^{\infty} \frac{\mu_i \cdots \mu_n}{\lambda_i \cdots \lambda_n} \tag{3.222}$$

is divergent, such a solution is unique and the probabilities add to unity. Other useful sufficient conditions are also known.

The process which has been studied in the most detail is the simple linear birth and death process defined by taking $\lambda_i = i\lambda$, $\mu_i = i\mu$, where $\lambda, \mu > 0$ (the still simpler case where $\lambda_i = \lambda > 0$, $\mu_i = \mu > 0$, is a random walk and will be considered in § 3.23 and Chapter 10). The equations can then be solved in various ways. One method (Kendall 1949) is to set up a partial differential equation for the generating function

$$P(z, t) = \sum_{n=0}^{\infty} P_n(t)z^n, \tag{3.223}$$

and solve this for suitable initial conditions, whilst another method is to use Laplace transforms.

The solution with $P_1(0) = 1$ can be verified to be given by the generating function

$$P(z, t) = \{\mu(1 - e^{(\lambda-\mu)t}) - (\lambda - \mu e^{(\lambda-\mu)t})z\}\{\mu - \lambda e^{(\lambda-\mu)t} - \lambda(1 - e^{(\lambda-\mu)t})z\}^{-1}$$

$$\tag{3.224}$$

and was first obtained by C. Palm. From this it follows that if $\lambda \neq \mu$,

$$P_0(t) = \xi(t),$$

$$P_n(t) = \{1 - \xi(t)\}\{1 - \eta(t)\}\eta(t)^{n-1}, \quad n = 1, 2, \ldots, \tag{3.225}$$

where

$$\xi(t) = \mu(e^{(\lambda-\mu)t} - 1)(\lambda e^{(\lambda-\mu)t} - \mu)^{-1},$$

$$\eta(t) = \lambda\mu^{-1}\xi(t). \tag{3.226}$$

From these we obtain the first two moments

$$m_1 = e^{(\lambda-\mu)t},$$

$$m_2 = m_1^2 + (\lambda+\mu)(\lambda-\mu)^{-1}e^{(\lambda-\mu)t}(e^{(\lambda-\mu)t}-1). \tag{3.227}$$

From (3.225) we see that the probability of ultimate extinction is unity if $\lambda < \mu$ and $\mu\lambda^{-1}$ if $\lambda > \mu$. The case $\lambda = \mu$ can be considered by taking the limit, for t fixed, as $\lambda \to \mu$ and we then get

$$P_0(t) = \lambda t (1 + \lambda t)^{-1}, \tag{3.228}$$

so that extinction is also certain if $\lambda = \mu$, similarly to the situation in the simple multiplicative process (§ 3.18). Interesting results are also obtained if we put $\lambda < \mu$ and prevent extinction by introducing some degree of immigration.

Thus we could assume, in addition to the transitions resulting from birth and death, that in each interval $(t, t+dt)$ there is a probability $\kappa \, dt$ that a new individual migrates into the population. The forward equations are then easy to write down and their solution (Kendall 1948, 1949) corresponds to the generating function, given no initial individuals, and $\lambda \neq \mu$,

$$P(z, t) = \left(\frac{\lambda-\mu}{\lambda e^{(\lambda-\mu)t}-\mu}\right)^{\kappa\lambda^{-1}} \left\{1 - z \frac{\lambda(e^{(\lambda-\mu)t}-1)}{\lambda e^{(\lambda-\mu)t}-\mu}\right\}^{-\kappa\lambda^{-1}}. \tag{3.229}$$

When $\lambda = \mu$ the solution is

$$P(z, t) = (1 - \lambda t)^{-\kappa\lambda^{-1}} \left(1 - \frac{\lambda t z}{1 + \lambda t}\right)^{-\kappa\lambda^{-1}}. \tag{3.230}$$

The most interesting feature about these results is their limit when $t \to \infty$. (3.230) does not result in a limiting distribution but (3.229) becomes, if $\lambda < \mu$,

$$P(z, \infty) = (1 - \lambda\mu^{-1})^{\kappa\lambda^{-1}}(1 - \lambda z\mu^{-1})^{-\kappa\lambda^{-1}}, \tag{3.231}$$

which, like (3.229), is a negative binomial. This result was first given by McKendrick and has been rediscovered several times. It again illustrates the fact that the negative binomial can arise in ways quite different from that described in § 2.14. Similarly, when $\lambda = 0$, the distribution has the generating function of a Poisson distribution

$$P(z, t) = \exp\{\kappa\mu^{-1}(1 - e^{-\mu t})(z-1)\} \tag{3.232}$$

which can be found from (3.229) by a limiting process. It is also easy to see that if we take (3.231) and let $\kappa \to 0$ but consider the distribution conditional on the population size being greater than zero, we get a logarithmic distribution (Kendall 1948).

It is instructive to look at these results in a slightly different way. Consider the stationary distribution (3.231) which arises when $\lambda < \mu$ and the process has started up an infinite time previously. The descendants of each immigrant are distributed independently, and the population at time t can therefore be regarded as the sum of an infinite number of random variables typified by the descendants of the immigrant, if any, which entered the population during the small interval $(t-\tau, t-\tau+d\tau)$. These descendants have the generating function

$$(1-\kappa\,d\tau)+P(z,\tau)\kappa\,d\tau$$

where $P(z,\tau)$ is given by (3.229).

The product of all such generating functions can be written as

$$\prod[1-\kappa\{1-P(z,\tau)\}d\tau] = \exp\left[-\kappa\int_0^\infty\{1-P(z,\tau)\}d\tau\right]. \qquad (3.233)$$

This is known as a 'product integral', and the formal calculations easily justified rigorously. On inserting the value of $P(z,t)$ given by (3.224), and taking $\lambda < \mu$, (3.231) follows at once.

3.23. The simple queue

The subject of queues has a very large literature and we here consider only the simplest case as an example of the above theory. (See Syski 1960, Cox and Smith 1961, Takacs 1962, and Saaty 1961 for a full discussion.)

We suppose that customers arrive at a single service point at instants of time which form a simple Poisson process so that the number arriving in any interval of length T is a Poisson variate with mean λT. These are served in the order of their arrival, and the service times of each are independently distributed in negative exponential distributions with means μ^{-1}.

The number of customers being served at any time t is then a Markovian variate and contains all the information on which the future behaviour of the system depends. In fact given that there are j customers being served, the probability of a new customer arriving in the interval $(t, t+dt)$ is $\lambda\,dt+o(dt)$, and the probability of a customer completing his service in this interval is $\mu\,dt+o(dt)$ if there is any customer being served, and zero otherwise. Hence we can use Kolmogorov's forward equations. If there are initially j_0 customers in the queue (including the one being served), and $P_j(t)$ $(j = 0, 1, 2, ...)$ is the probability of j customers at time t, we have

$$\frac{d}{dt}P_0(t) = -\lambda P_0(t)+\mu P_1(t),$$

$$\frac{d}{dt}P_i(t) = \lambda P_{i-1}(t)-(\lambda+\mu)P_i(t)+\mu P_{i+1}(t). \qquad (3.234)$$

These are identical with (3.220) and (3.221) on putting $\lambda_i = \lambda$, $\mu_i = \mu$ in the latter. It will be found that there is a stationary state if and only if $\lambda\mu^{-1} < 1$. That this is a necessary and sufficient condition can also be made plausible by the following argument which is often useful in similar problems and can be made rigorous in most cases without much difficulty.

Suppose that at time t there are j customers awaiting service and consider the distribution of the number of customers awaiting service at $t+T$, where T is to be large. The number of customers which have arrived in this interval is a Poisson variate with mean and variance equal to λT. The number of customers which are served in this time will be not greater than a random variable with mean and variance μT (the number which would be served if so many were waiting that the server was never idle). Hence the difference has a mean at least $(\lambda - \mu)T$ if $\lambda > \mu$ and a standard deviation which is $O(T^{\frac{1}{2}})$. Using Tchebychev's inequality the number waiting must therefore become arbitrarily large with probability arbitrarily near unity as T increases. A slightly more complicated argument shows that this does not happen if $\lambda < \mu$, but the case $\lambda = \mu$ requires further discussion.

The solution of the equations (3.234) for equilibrium was first given by Erlang (Brockmeyer *et al.* 1948). Putting the left-hand side of (3.234) equal to zero and solving in terms of P_0, where P_i is the equilibrium probability, we get

$$P_i = (\lambda \mu^{-1})^i P_0,$$

and hence, since the sum must equal unity,

$$P_i = (1 - \lambda \mu^{-1})(\lambda \mu^{-1})^i, \tag{3.235}$$

which again illustrates another way in which the geometric distribution arises in the theory of random processes. The time-dependent solution of (3.234) is much more complicated and will not be given here (see, for example, Syski 1960).

Another quantity of interest is the distribution of the time a new customer will have to wait if he arrives at a queue in a stationary state. This is the 'waiting time distribution'. If the queue is empty the waiting time of a newly arrived customer will be zero. The probability of this occurring is $(1 - \lambda \mu^{-1})$ and hence the waiting time distribution has a concentration of this amount at zero.

On the other hand, when a new customer arrives there may be $n > 0$ customers in the queue. The probability of this is

$$(1 - \lambda \mu^{-1})(\lambda \mu^{-1})^n,$$

and each of these will have a service time with a negative exponential distribution (§7.17) with mean μ^{-1}. Hence their sum will have a gamma distribution of the form

$$\mu^n (n-1!)^{-1} e^{-\mu x} x^{n-1} \, dx.$$

Multiplying these and summing, the probability that the waiting time lies in the interval $(x, x+dx)$ $(x > 0)$ become

$$\sum_{n=1}^{\infty} \mu^n(1-\lambda\mu^{-1})(\lambda\mu^{-1})^n(n-1!)^{-1}e^{-\mu x}x^{n-1}\,dx = \lambda(1-\lambda\mu^{-1})e^{(\lambda-\mu)x}\,dx.$$

(3.236)

In the references given above much more complicated queueing systems with general inter-arrival and service time distributions are considered.

Another very simple queueing system (due to D. G. Kendall) leads to a probability distribution considered in the last chapter. Suppose that at a taxi-rank customers and taxis arrive at random, i.e. their times of arrival are distributed in Poisson processes of the types considered in § 3.20. Furthermore, suppose that at time zero there are no taxis or customers waiting and that in any interval of time $(t, t+dt)$ the probabilities of arrival of taxis and customers are $\lambda\,dt+o\,(dt)$, and $\mu\,dt+o\,(dt)$ respectively. Let X be a random variable equal to the number of taxis waiting minus the number of customers waiting. We suppose that customers are picked up immediately a taxi arrives so that taxis and customers cannot be waiting at the same time. Then the distribution of X at time t is the distribution of the difference between two Poisson variables of parameters λt and μt and is given by an expression of the form (2.76). If $\lambda > \mu$ $(\lambda < \mu)$ the queue of taxis (customers) will grow indefinitely whilst if $\lambda = \mu$ we will be equally likely to have a queue of either, although the expected size of this queue will grow indefinitely large as t increases.

Bibliographical notes

The subject of random processes has an enormous literature. For general surveys see Bailey (1964), Bartlett (1955), Bharucha–Reid (1960), Chung (1960), Cox and Miller (1965), Doob (1953), Dynkin (1960), Rosenblatt (1962), and Takacs (1960). For process in continuous time see also Hannan (1960), and Lévy (1948a), for multiplicative processes Harris (1963), and for queues Cox and Smith (1961), Saaty (1961), Syski (1960), and Takacs (1963).

Much more work has been done on finite Markov chains than is described in this chapter, and in addition to some of the above books the reader should also consult Feller (1957b), Fréchet (1937, Vol. II) and Kemeny and Snell (1960). Non-homogeneous chains in which the transition probabilities are not independent of the time have been considered by Hajnal (1956, 1958), and (in a special case) by Moran (1962b). The distribution of the numbers of transitions of particular kinds has been studied by Whittle (1955). It is also possible to represent the nth order transition probabilities, $p_{ij}^{(n)}$, by Fourier–Stieltjes integrals.

For recent Russian work on random processes, and in the analytic theory of probability generally, see La Salle (1962).

Random walks of various types are further considered in Chapter 10. An original approach to first passage and recurrence distributions in discrete Markov chains is given by Skellam and Shenton (1957). A numerical example of an artificial realization of a birth and death process is given by Kendall (1950a).

Exercises

Exercise 3.1. Consider a Markov chain with $n+1$ states $E_0,..., E_n$, such that

$$p_{i-1,i} = q_i,$$

$$p_{i,i} = r_i,$$

$$p_{i+1,i} = p_i,$$

for $i = 0,..., N$, where $p_i + r_i + q_i = 1$, $p_0 > 0$, $q_0 = 0$, and $q_n = p_n = 0$. Then E_n is the sole absorbing state. If T_i is the mean time to absorption, starting from a state E_i, prove that

$$T_i = a_i + a_{i+1} + ... + a_{n-1},$$

where

$$a_i = \frac{1}{p_i} + \frac{q_i}{p_i p_{i-1}} + \ ... \ + \frac{q_1 ... q_i}{p_0 ... p_i}.$$

Exercise 3.2. In the immigration model described by equation (3.200) suppose that m_1, m_2 are the first and second moments, about the origin, of the distribution generated by $P(z)$, and M_1, M_2 correspond similarly to $R(z)$. Prove that the first and second moments of the distribution generated by $\pi(z)$ are

$$M_1(1-m_1)^{-1}$$

and

$$\frac{M_1(m_2 - m_1 + 2M_1 \, m_1)}{(1-m_1)(1-m_1^2)} + \frac{M_2 - M_1}{1-m_1^2} + \frac{M_1}{1-m_1}.$$

Exercise 3.3. If $0 < a < 1, 0 < b < 1, k > 0$, prove that a solution of (3.200) is given by the generating functions,

$$\pi(z) = (1-a)^k(1-az)^{-k},$$

$$P(z) = \{1 - b - (a-b)z\}\{1 - ab - a(1-b)z\}^{-1},$$

$$R(z) = \left(\frac{1-a}{1-ab}\right)^k \left\{1 - z\frac{a(1-b)}{1-ab}\right\}^{-k}.$$

Exercise 3.4. Consider a Markov process at discrete instants of time whose state is defined in the following way. There are N particles each of which can exist in one of n states $E_1,..., E_n$. At discrete intervals of time two of the N particles are chosen at random. If their states are E_k and E_l they move to states E_i and E_j respectively with probability $\pi(i, j \,|\, k, l) = \pi(j, i \,|\, k, l) = \pi(i, j \,|\, l, k)$. Prove that the resulting process is a Markov chain with N^n states and call this the M_3-process. Consider similarly the process in which the state of the system is defined by the vector $(s_1,..., s_n)$ where s_i is the number of particles in state E_i ($\sum s_i = N$). Show that this is a Markov chain with

$$\binom{N+n-1}{n-1}$$

states and call it the M_2-process. Finally, consider the Markov chain with n^2 states (i, j) whose transition matrix is $\pi(i, j \mid k, l)$ and call it the M_1-process. Show that if the M_1-process is completely regular so also are the M_2 and M_3 processes but that there exist systems for which M_2 and M_3 are completely regular but M_1 is not.

Exercise 3.5. Following the definitions of Exercise 3.4 suppose further that

$$\pi(i, j \mid k, l) = 0, \quad \text{unless} \quad i+j = k+l,$$

and

$$\pi(i, j \mid k, l) = \pi(k, l \mid i, j),$$

for all i, j, k, l. Suppose further that in the chain M_1 each of the matrices corresponding to $i+j = k+l =$ constant is completely regular. Then starting from any initial state in which

$$\sum i s_i = E,$$

prove that the joint distribution of the s_i in the M_2-process converges to a stationary distribution with probabilities

$$\frac{N!}{s_1! \ldots s_n!}$$

where κ is the coefficient of z^E in

$$(z^n - 1)^N (1 - z^{-1})^{-N}.$$

Exercise 3.6. A particle, starting from $x = 0$, performs a random walk on the integral points $x = 0, \pm 1, \ldots$. At discrete instants of time it moves from x to $x+1$ or $x-1$ with probabilities p and $q = 1-p$. Let L_n and R_n be the total number of steps to the left and right after n moves $(L_n + R_n = n)$. A 'stop rule' is defined so that the process stops when L_n, R_n reach certain values. Prove that for any stop rule which is such that the process must stop before $L_n = N$ or $R_n = N$, N some positive integer,

$$E(pL_n - qR_n) = 0,$$

where L_n and R_n are the total number of moves to the left and right. This result shows that decisions by parents on the number of children they will have, based on the sex of the children they do have, has no effect on the sex ratio of their contribution to the population.

Exercise 3.7. Consider a Markov chain with three states of 0, 1, 2. Let $\bar{p}_{00}^{(n)}, \bar{p}_{10}^{(n)}, \bar{p}_{01}^{(n)}, \bar{p}_{11}^{(n)}$ be the probabilities of transitions $0 \to 0, 0 \to 1, 1 \to 0, 1 \to 1$, conditional on no transition occurring to state 2 (e.g. p_{00} is $p_{00}(1-p_{20})^{-1}$, $p_{00}^{(2)}$ is $p_{00}^{(2)}(1-p_{20}^{(2)})^{-1}$, etc., in the usual notation). Prove that in general

$$p_{00}^{(2)} \neq (\bar{p}_{00}^{(1)})^2 + \bar{p}_{01}^{(1)} \bar{p}_{10}^{(1)}.$$

4. Probability and Measure Theory

4.1. In this chapter we outline as much of the theory of measure as is necessary to provide a foundation for the parts of probability theory dealt with in this book. In these we are not concerned with the general theory of random processes and hence do not need a theory of measure in function spaces. In fact those places in the rest of the book where measure theory is unavoidable deal with cases, such as the theory of infinite sequences of random events, in which the required results are intuitively plausible, and this chapter is therefore not essential to the understanding of the other chapters.

We need to construct a theory of measure in both finite and (enumerably) infinite-dimensional Euclidean space, and we begin with the former. This theory is easy to extend to more general spaces but as we are here concerned only with an introduction to the subject, it is simpler to confine the discussion to Euclidean space.

4.2. An n-dimensional Euclidean space, R_n, is defined to consist of all sets of n real numbers $(x_1,...,x_n)$, which we describe as the coordinates of a point X. The distance, $d(X, Y)$, between two points, X and $Y = (y_1,...,y_n)$, is defined as the non-negative quantity, $d(X, Y)$, satisfying

$$d(X, Y)^2 = \sum_1^n (x_i - y_i)^2. \qquad (4.1)$$

In this space we consider sets of points defined in various ways. We denote the set of points consisting of all the points of the space by Ω, and other sets by capital letters such as $A, B,...$. We can now set up a 'field' of sets in the same way as in §1.2, the 'space' in that case consisting of all points $E_1, E_2,...$ of an enumerable set and the sets $A, B,...$ of selections out of these. For classes we use the notation $\{A\}$ to mean the 'class of sets such as A'.

Given two sets A and B we can define their 'product', AB, as the set consisting of those points which belong to both, and their 'sum', $A+B$, as the set consisting of those points which belong to A, to B, or to both. As before these operations are associative,

$$(AB)C = A(BC), \quad (A+B)+C = A+(B+C),$$

commutative,

$$AB = BA, \quad A+B = B+A,$$

and distributive,

$$A(B+C) = AB+AC.$$

We can also extend the definitions of product and sum to enumerable sequences $(A_1, A_2,...)$ so that

$$\prod_1^\infty A_i, \quad \sum_1^\infty A_i \qquad (4.2)$$

are sets consisting respectively of those points in the space which belong to all A_i, and to one of the A_i at least.

If B is a set contained in A (i.e. all points of B are points of A) we write $B \subset A$, or $A \supset B$, and we then define $A - B$ as the set of all points of A not in B. Then the set $\Omega - A$ is always defined although it may be empty (the empty set will be denoted by 0). We write $\bar{A}$ for $\Omega - A$, and call it the 'complement of A'.

Given an infinite sequence of sets, A_1, A_2,... we have defined their infinite product and infinite sum, but there are two other sets depending on the class (A_i) which are also important.

The 'upper limit' of the sequence is written $\limsup A_i$, and is defined to be the set consisting of all points which belong to an infinite number of the A_i. We can in fact write

$$\limsup A_i = \prod_{n=1}^\infty \sum_{i=n}^\infty A_i, \qquad (4.3)$$

since every such point must belong to $\sum_{i=n}^\infty A_i$ for every n, and if a point does not belong to any A_i, or to only a finite number of them, there will be a value of n such that it does not belong to $\sum_{i=n}^\infty A_i$.

The 'lower limit' of the sequence A_i is defined to consist of all those points which belong to all the A_i except possibly a finite number, and thus we can write

$$\liminf A_i = \sum_{n=1}^\infty \prod_{i=n}^\infty A_i, \qquad (4.4)$$

and then

$$\liminf A_i \subset \limsup A_i.$$

Just as in § 1.4 we can introduce 'indicator functions' corresponding to each set. These are functions of points X in the space which are equal to 0 or 1. Then the indicator of the set A is a function, $I_A(X)$, which is equal to unity when X belongs to A, and to zero otherwise. Using this notation, the indicator functions of $\limsup A_i$ and of $\liminf A_i$ can be easily seen to be

$$\limsup_i I_{A_i}(x),$$

and

$$\liminf_i I_{A_i}(x),$$

where these definitions are taken in the usual sense of limits.

4.3. Fields of sets

A field of sets, F, is a class of sets which is closed under all finite operations of the above type, i.e. if A and B are any sets belonging to the field so also are AB, $A+B$, and $\Omega-A$, $\Omega-B$. This is often not a narrow enough definition, and so we define a σ-field as a class of sets which is closed under all enumerable set operations, i.e. given any enumerable sequence of sets A_1, A_2,... belonging to the field, $\prod_1^\infty A_i$ and $\sum_1^\infty A_i$ also belong to the field.

The terms 'field' and 'σ-field' are not those used in all books. We follow here the terminology of Loève (1963) and not, for example, that of Halmos (1950) who calls a field an algebra.

Given any class of sets we can construct σ-fields which contain it, for example, the class of all sets (if that be regarded as a logically satisfactory notion). Given any two σ-fields the class of sets which belong to both obviously form a σ-field, and similarly the sets belonging to all σ-fields of some class of σ-fields form a σ-field. Thus given any class of sets, there is a uniquely defined σ-field which is contained in all σ-fields which contain the given class. This unique σ-field is said to be the minimal σ-field over the class of sets. It certainly contains all the sets which can be generated by taking all enumerable sequences of finite or enumerable operations on the sets of the given class. The minimal σ-field over a class of sets is sometimes called the Borel field over the given class but we will restrict this term to the field considered in the following section.

To construct a useful theory of measure we must construct σ-fields of this type from classes of sets defined initially in some particular way.

4.4. The Borel field in Euclidean space

A point $X(x_1,...,x_n)$ is said to be a limit point of a set A if there is an infinite sequence X_1, X_2,... of distinct points belonging to A such that

$$\lim_{n \to \infty} d(X, X_n) = 0. \tag{4.5}$$

A set is said to be closed if it contains all its limit points, if any. Thus in one-dimensional Euclidean space R_1 an interval $(a \leqslant x \leqslant b)$, which contains its end points, is closed, and in R_n the sphere $\sum x_i^2 \leqslant d$ is closed.

An open set A is the complement of a closed set. An open set can also be defined as a set A which is such that given any point P belonging to it, it is possible to find a non-zero quantity d such that all the points distant d or less from P also belong to the set. This definition is equivalent to the previous one since if such a distance d can be found, P cannot be a limit point of $\Omega-A$, and so all the limit points of $\Omega-A$ must belong to $\Omega-A$. Conversely if P is not a limit point of $\Omega-A$ there must exist $d > 0$ so that any point distant less than d from P, and different from P, must belong to A.

The sum of a finite or infinite number of open sets is obviously open, and taking complements we see that the common part of a finite or infinite number of closed sets is closed.

An open interval in R_n is defined to be a set of points whose coordinates satisfy the inequalities

$$a_i < x_i < b_i \quad (i = 1,\dots,n),$$

where the a_i may be $-\infty$ and the b_i, $+\infty$. An open interval is clearly an open set. Every open set can be represented as a finite or enumerable sum of open intervals. To prove this consider all the points in the open set with rational coordinates and construct around each the largest open interval which contains this point at its centre. The number of such intervals is enumerable and all points of the open set must belong to at least one of them.

In one dimension it is convenient to denote the open, closed, and half-open intervals,

$$a < x < b,$$
$$a \leqslant x \leqslant b,$$
$$a \leqslant x < b,$$
$$a < x \leqslant b$$

by the symbols (a, b), $[a, b]$, $[a, b)$, and $(a, b]$. The advantage of using half-open intervals is that if two of them abut, their sum is again a half-open interval.

Given the set of all open sets in R_n consider the smallest σ-field containing them. This also contains all closed sets, and is known as the Borel field in R_n.

4.5. Measure functions

A set function is a numerically valued function which is defined for every set in some class of sets, i.e. it associates with each such set a real number. Such functions are usually (but not always) defined for all sets in some σ-field. A set function, $\phi(A)$, is said to be σ-additive on a σ-field, if

$$\phi(\Sigma A_i) = \Sigma \phi(A_i) \tag{4.6}$$

for every finite or enumerable set of non-overlapping sets A_i such that $\phi(\Sigma A_i)$ is bounded. It is σ-additive on a field if this is true whenever ΣA_i also belongs to the field (even if the field is not a σ-field).

If in addition $\phi(A)$ is always non-negative it is said to be a 'measure'. It may be regarded intuitively as the amount of 'mass' contained in the set A. If a set B is contained in A then $\phi(A) = \phi(B) + \phi(A - B) \geqslant \phi(B)$. Hence if $\phi(\Omega)$ is finite every set in the σ-field has a finite measure. If $\phi(\Omega) \neq 1$ we can divide every value by $\phi(\Omega)$ and obtain a new measure

$$P(A) = \frac{\phi(A)}{\phi(\Omega)}, \tag{4.7}$$

13

for which $P(\Omega) = 1$, and $0 \leqslant P(A) \leqslant 1$ for every A in the σ-field. $P(A)$ is then said to be a probability measure and is a generalization of the definition of probability in enumerable sets of events given in §1.3. We shall always assume that it is defined on a σ-field, unless otherwise explicitly stated. However it is defined, on a field or σ-field, ϕ is a probability measure if it is non-negative, σ-additive whenever ΣA_i also belongs to the field, and $\phi(\Omega) = 1$. The σ-field and its corresponding measure is known as a 'probability space', and the sets, $\{A\}$, are 'events'.

Given a sequence of sets A_1, A_2, A_3,... we have already defined $\lim \sup A_i$ and $\lim \inf A_i$. When these are equal, the set they represent is called the limit of the sequence A_i and we write

$$A = \lim A_i.$$

We then say the sequence is 'convergent'.

We say that a sequence, A_i, is monotone if we always have

$$A_1 \subset A_2 \subset A_3 \subset ..., \quad \text{or} \quad A_1 \supset A_2 \supset A_3 \supset .... \tag{4.8}$$

Then by the definition of $\lim \sup A_i$, and $\lim \inf A_i$, we see that every monotone sequence is convergent.

A class of sets such that every monotone sequence of such sets converges to a set belonging to the class is said to be a monotone class.

THEOREM 4.1. *A field which is a monotone class is a σ-field, and conversely.*

Proof. Clearly every σ-field contains the limit of every monotone sequence and is therefore a monotone class. Conversely every enumerable sum or product

$$\Sigma A_i, \quad \Pi A_i$$

is a monotonic limit of finite sums or products, and therefore belongs to the class, so that the result of any finite or enumerable sequences of set operations is again a member of the class.

A measure, $P(A)$, is said to be continuous if

$$P(\lim A_i) = \lim P(A_i) \tag{4.9}$$

for every increasing monotone sequence $\{A_i\}$, and for every decreasing monotone sequence such that $P(A_i)$ is finite from some i onwards.

THEOREM 4.2. *A probability measure on a σ-field is continuous; conversely if a non-negative set function is finitely additive, continuous, and such that $P(\Omega) = 1$, it is a probability measure, i.e. it is σ-additive.*

Proof. Suppose $P(A)$ is a probability measure, and that A_1, A_2,... is an increasing sequence of sets. Then $\lim A_n$ exists and

$$\lim A_n = A_1 + (A_2 - A_1) + (A_3 - A_2) + ...$$

Since the terms on the right are disjoint we have

$$P(\lim A_n) = P(A_1) + P(A_2 - A_1) + \cdots$$
$$= \lim P(A_n).$$

By taking complements the same result is true for a decreasing sequence.

Conversely suppose that $P(A)$ is a non-negative finitely additive continuous set function such that $P(\Omega) = 1$. Then for any sequence $A_1, A_2, \ldots$ of disjoint sets,

$$P\left(\sum_1^\infty A_n\right) = P\left(\lim \sum_1^n A_i\right) = \lim P\left(\sum_1^n A_i\right)$$
$$= \lim \sum_1^n P(A_i) = \sum_1^\infty P(A_i),$$

so that $P(A)$ is σ-additive and therefore a probability measure.

4.6. The extensions of probability measures

Suppose that a probability measure $P(A)$ is defined for all sets of some class $\{A\}$. If another class $\{B\}$ contains all the sets of $\{A\}$ a probability measure on the class $\{B\}$ will be said to be an extension of the probability measure $P(A)$ if it is equal to $P(A)$ for all sets A of $\{A\}$. We then have the following theorem.

THEOREM 4.3. *A probability measure on a field $\{A\}$ can be extended to the minimal σ-field containing $\{A\}$ and this extension is unique.*

To carry out the proof we introduce the idea of an 'outer measure' which is due to Carathéodory.

A non-negative function, $\phi(A)$, defined for all sets in some space, is said to be an 'outer measure' if

$$\phi(A) \leqslant \phi(B) \tag{4.10}$$

whenever $A \subset B$,

$$\phi(\Sigma A_i) \leqslant \Sigma \phi(A_i) \tag{4.11}$$

for any finite or enumerable set of sets A_i, and

$$\phi(A) = 0 \tag{4.12}$$

when A is the empty set. Notice that such a measure is defined on the σ-field consisting of all subsets of the space.

We say that a set A is measurable (ϕ) if, for any set B,

$$\phi(B) \geqslant \phi(AB) + \phi(\bar{A}B). \tag{4.13}$$

From (4.11) we always have

$$\phi(B) \leqslant \phi(AB) + \phi(\bar{A}B)$$

and hence the condition (4.13) is equivalent to insisting that

$$\phi(B) = \phi(AB) + \phi(\bar{A}B) \qquad (4.14)$$

for every set B of the space.

Now suppose that $\phi(A)$ is a measure defined on some field F. We define a set function, $\phi_1(B)$, for all sets B in the space by writing $\phi_1(B)$ for the greatest lower bound of

$$\Sigma\phi(A_i)$$

where ΣA_i is the sum of any finite or enumerable class of sets of F which contains B, i.e. $B \subset \Sigma A_i$. ΣA_i is said to be a 'cover' of B. $\phi_1(B)$ is called the 'outer extension' of the measure ϕ. We shall prove that $\phi_1(B)$ is an outer measure which is equal to $\phi(A)$ for sets A of the given field, and is therefore the extension of $\phi(A)$ to an outer measure.

If A is a set of the field F, it is also a set of the space and hence can be taken as a cover, so that

$$\phi_1(A) \leqslant \phi(A). \qquad (4.15)$$

On the other hand, if ΣA_i is a covering of A, where the A_i belong to F, we have

$$A \subset \Sigma A_i,$$

and therefore

$$\phi(A) \leqslant \Sigma\phi(A_i),$$

so that

$$\phi(A) \leqslant \phi_1(A).$$

With (4.15) this means that

$$\phi(A) = \phi_1(A). \qquad (4.16)$$

We now have to show that $\phi_1(A)$ is an outer measure. Since the empty set belongs to the field F we have

$$\phi_1(A) = 0,$$

when A is the empty set, and if $A \subset B$,

$$\phi_1(A) \leqslant \phi_1(B),$$

because every cover of B is a cover of A. We now prove property (4.11). Suppose

$$A = \Sigma A_i.$$

Let $\varepsilon > 0$. For each A_i there is a covering of sets B_{ij} $(j = 1, 2, ...)$ belonging to F such that

$$\sum_j \phi(B_{ij}) \leqslant \phi_1(A_i) + \varepsilon 2^{-i}.$$

The sets B_{ij} (all i, j) are then a covering of A such that

$$\sum_{i,j} \phi(B_{ij}) \geqslant \phi_1(A)$$

and

$$\phi_1(A) \leqslant \sum_i \phi_1(A_i) + \varepsilon,$$

$$\leqslant \sum_i \phi_1(A_i),$$

since ε is arbitrary. Thus $\phi_1(A)$ is an outer measure.

We now show that the class of all sets measurable (ϕ_1) is a σ-field and that when restricted to this field, ϕ_1 is a measure.

In the definition (4.13) of the measurability of a set A, the sets A and $\bar{A}$ occurred symmetrically so that if A is measurable so is $\bar{A}$.

If A and B are measurable so is AB because for given any set C of the space

$$\phi_1(C) = \phi_1(AC) + \phi_1(\bar{A}C)$$

$$= \phi_1(ABC) + \phi_1(A\bar{B}C) + \phi_1(\bar{A}BC) + \phi_1(\bar{A}\bar{B}C)$$

since A and B are measurable. Therefore

$$\phi_1(C) \geqslant \phi_1(ABC) + \phi_1(A\bar{B}C + \bar{A}BC + \bar{A}\bar{B}C),$$

and the right-hand side is equal to

$$\phi_1(ABC) + \phi_1\{(\overline{AB})C\}$$

so that AB is measurable. Thus the class of measurable sets is a field. On this field ϕ_1 is additive because if A and B are disjoint and measurable,

$$\phi_1(A+B) = \phi_1\{(A+B)A\} + \phi_1\{(A+B)\bar{A}\}$$

$$= \phi_1(A) + \phi_1(B).$$

Now consider an enumerable sequence $A_1, A_2, \ldots$ of disjoint measurable sets. We want to prove that

$$\phi_1(\textstyle\sum A_i) = \sum_1^\infty \phi_1(A_i), \tag{4.17}$$

if the right-hand side is finite, and that $\sum A_i$ belongs to the field which is therefore a σ-field. Then for any $\varepsilon > 0$ and n sufficiently large,

$$\phi_1\left(\sum_1^\infty A_i B - \sum_1^n A_i B\right) \leqslant \sum_{n+1}^\infty \phi_1(A_i B) \leqslant \varepsilon.$$

It therefore follows that

$$\phi_1(B) + \varepsilon \geqslant \phi_1\left(\sum_1^n A_i B\right) + \phi_1\left\{\left(\Omega - \sum_1^n A_i\right)B\right\} + \phi_1\left(\sum_{n+1}^\infty A_i B\right)$$

$$\geqslant \phi_1\left\{\left(\Omega - \sum_1^n A_i\right)B\right\} + \phi_1\left(\sum_1^\infty A_i B\right)$$

$$\geqslant \phi_1\left\{\left(\Omega - \sum_1^\infty A_i\right)B\right\} + \phi_1\left(\sum_1^\infty A_i B\right).$$

Since this is true for all ε, $\sum_1^\infty A_i$ belongs to the field which is therefore a σ-field. Then

$$\phi_1\left(\sum_1^\infty A_i\right) \leqslant \sum_1^\infty \phi_1(A_i),$$

and

$$\phi_1\left(\sum_1^\infty A_i\right) \geqslant \lim_n \phi_1\left(\sum_1^n A_i\right)$$

$$\geqslant \lim_n \sum_1^n \phi_1(A_i)$$

$$\geqslant \sum_1^\infty \phi_1(A_i),$$

so that

$$\phi_1\left(\sum_1^\infty A_i\right) = \sum_1^\infty \phi_1(A_i), \tag{4.18}$$

and ϕ_1 is σ-additive, and therefore a probability measure.

We now prove that every set in the field $\{A\}$ is measurable $\{\phi_1\}$. If A is a set of the field, $\{A\}$, and B is any set, then for any $\varepsilon > 0$ there is a covering of B consisting of sets $A_1, A_2, \ldots$ of $\{A\}$ such that

$$\phi_1(B) + \varepsilon \geqslant \sum \phi(A_i)$$

$$\geqslant \sum \phi(AA_i) + \sum \phi(\bar{A}A_i),$$

because A belongs to $\{A\}$. Thus

$$\phi_1(B) + \varepsilon \geqslant \phi_1(AB) + \phi_1(\bar{A}B),$$

so that A is measurable (ϕ_1). Hence the field $\{A\}$ is contained in the σ-field of sets measurable (ϕ_1). The minimal σ-field over $\{A\}$ is therefore contained in this σ-field, and ϕ_1 is a measure of this σ-field.

To prove it is a unique extension suppose ϕ_1 and ϕ_2 are two measures on the σ-field of sets measurable ϕ which are extensions of ϕ on the field $\{A\}$. Consider the subset of this σ-field consisting of all sets for which ϕ_1 and ϕ_2 have the same value. Call this subset M. M clearly contains $\{A\}$ and for any monotonic sequence $A_1, A_2, \ldots$ in M we have

$$\phi_1(\lim A_n) = \lim \phi_1(A_n) = \lim \phi_2(A_n) = \phi_2(\lim A_n),$$

so that the set M is a monotone class and therefore a σ-field by Theorem 4.1. Hence M contains the minimal σ-field over $\{A\}$, and is therefore identical with it. This completes the proof of Theorem 4.3. Notice that we have nowhere had to use the fact that the space is finite dimensional.

Given a measure, ϕ_1, on a σ-field $\{A\}$ it is often possible to extend it to a larger field in the following way. There may well exist sets C whose outer measure, $\phi_1(C)$, is zero but which do not belong to the σ-field $\{A\}$. For every set which can be written as $A + C$ where A belongs to the σ-field and $\phi_1(C)$ is zero, write

$$\phi(A+C) = \phi(A).$$

Then it can be easily verified that all such sets form a σ-field which includes $\{A\}$, and ϕ is a σ-additive measure on this field. Such a measure is said to be 'complete'. This is, in fact, the way ordinary Lebesgue measure is constructed from Borel measure on the Borel field.

4.7. Examples of probability measures

(1) Consider sets of points in the half-open interval $(0, 1]$. Let the class F_1 consist of all finite sets of non-overlapping half-open intervals $(a, b]$, i.e. points x such that $a < x \leqslant b$, where $0 \leqslant a < b \leqslant 1$. Such a class is clearly a field. On this field define $\phi(A)$ to be the sum of the lengths of the intervals comprising A.

Let F_2 be the smallest σ-field over the field F_1. This is known as the Borel field on the half-open interval $0 < x \leqslant 1$. It clearly includes all open sets, for any open set is a finite or enumerable sum of open intervals, $a < x < b$, say, and any such open interval can be represented as an enumerable sum of the half-open intervals

$$a+n(n+1)^{-1}(b-a) < x \leqslant a+(n+1)(n+2)^{-1}(b-a)$$

for $n = 0, 1, 2, \ldots$. Since F_2 contains all open sets it contains all closed sets, and it is the smallest σ-field over the class of all open and closed sets in the interval $0 < x \leqslant 1$.

For any set A in F_1 we have defined $\phi(A)$ to be the sum of all the lengths of the intervals comprising A. Then ϕ can be extended as above to a unique measure ϕ_1 on the Borel field F_2, and ϕ_1 is said to be 'Borel measure'. If this measure is now completed by adding to the field F_2 all sets in $0 < x \leqslant 1$ which have outer measure zero, we obtain a field F_3 and a complete measure ϕ_2. F_3 is said to be the field of all 'Lebesgue measurable' sets in $0 < x \leqslant 1$ and ϕ_2 is 'Lebesgue measure'.

The same construction can be used to construct the Lebesgue measure of all Lebesgue measurable sets in a half-open rectangle

$$a_i < x_i \leqslant b_i \quad (i = 1, \ldots, n)$$

in n-dimensional Euclidean space.

(2) Consider the one-dimensional Euclidean space $-\infty < x < \infty$. Let $F(x)$ be a non-decreasing function, continuous to the right, and such that

$$\lim_{x \to \infty} F(x) = 1, \tag{4.19}$$

$$\lim_{x \to -\infty} F(x) = 0. \tag{4.20}$$

We have already described (§1.1) such a function as a 'distribution function'. To every half-open interval, $a < x \leqslant b$, we ascribe the measure

$$\phi(a < x \leqslant b) = F(b) - F(a),$$

where we make the convention that $F(\infty) = 1$, $F(-\infty) = 0$. F_1 is defined as above to be the class of all finite sets of non-overlapping intervals, half-open to the left. This is clearly a field. The smallest σ-field F_2 containing this field is the Borel field, and since ϕ is an additive measure on the field F_1, it can be extended to a σ-additive probability measure ϕ_1 on the Borel field F_2. By adding all sets whose outer measure with respect to ϕ_1 is zero we obtain a σ-field which is known as the Lebesgue–Stieltjes field with respect to the function $F(x)$, and the corresponding measure is a Lebesgue–Stieltjes probability measure. This is the type of measure we shall use for random variables in one dimension, for $F(x)$ can be interpreted as the probability that a given random variable X is less than or equal to the value x, and thus $F(x)$ can be defined as the measure of $(-\infty, x]$.

THEOREM 4.4. *A probability measure on the real line* $(-\infty < x < \infty)$ *determines a probability distribution function, $F(x)$, uniquely, and vice versa.*

Proof. The proof follows from the remarks above except for the fact that in order to show that ϕ on F_1 can be extended to ϕ_1 on F_2, we have to show that ϕ is an additive measure on F_1. This is obvious for finite additivity. We therefore have to show that $\phi(\sum A_i) = \sum \phi(A_i)$ when A_i are sets of the field F_1, and it is known that $\sum A_i$ belongs to F_1. It is clearly sufficient to do this when $\sum A_i$ is a half-open interval $(a, b]$ which is the sum of an enumerable sequence of half-open intervals $(a_n, b_n]$. The latter are non-overlapping and hence

$$\phi\{(a, b]\} = F(b) - F(a)$$

$$\geqslant \sum_1^n \phi\{(a_n, b_n]\} \quad \text{for all } n,$$

and therefore

$$\phi\{(a, b]\} \geqslant \sum_1^\infty \phi\{(a_n, b_n]\}.$$

We now have to prove the reverse inequality. Let ε be small and choose $a' > 0$ so $F(a') - F(a) < \varepsilon$. To each interval $(a_n, b_n]$ associate an open interval (a_n, b_n') such that $b_n' > b_n$ and

$$F(b_n') - F(a_n) < (1 + \varepsilon)\{F(b_n) - F(a_n)\}.$$

This is possible because $F(x)$ is continuous to the right. Then the closed set $[a', b]$ is contained inside the sum of the intervals (a_n, b_n'). By the Heine–Borel theorem it can be covered by a finite set of these, which we write as $I_1, \ldots, I_N$. For each I_i write $\phi(I_i)$ for the difference of the values of $F(x)$ at its two ends.

We then have

$$\phi\{(a,b]\} \leqslant F(b) - F(a') + \varepsilon$$

$$\leqslant \sum_{1}^{N} \phi(I_i) + \varepsilon$$

$$\leqslant (1+\varepsilon) \sum_{1}^{N} \phi(I'_n) + \varepsilon,$$

where I'_n is the interval of type $(a_j, b_j]$ corresponding to (a_j, b'_j). Proceeding to the limit we have

$$\{(a,b]\} \leqslant \sum_{1}^{\infty} \phi\{(a_n, b_n]\},$$

and the proof is complete.

It should be noted that the field of sets over which the Lebesgue–Stieltjes measure is defined depends on the function $F(x)$ because the completion process adds to the Borel field all sets which are of measure zero, and these depend on $F(x)$. On the other hand, the Borel field is defined independently of $F(x)$ and for this reason it is sometimes useful to restrict attention to Borel measures since all such measures are defined on the same field.

The particular case where $F(x)$ is a continuous function which is linear inside a given interval, and constant outside it, results in a probability measure which is proportional to the Lebesgue measure of the parts of the sets included in the interval.

(3) Now consider n-dimensional Euclidean space R_n. The considerations of (2) can be applied in a similar manner. The space consists of all points $(x_1, \ldots, x_n)$ where the x_i are finite. Since the distance between two points can be defined by (4.1), the definition of open and closed sets and their properties are obvious.

An interval in this space half-open to the left is defined as the set of finite values $(x_1, \ldots, x_n)$ satisfying inequalities of the form $a_i < x_i \leqslant b_i$ $(i = 1, \ldots, n)$, where the a_i can be $-\infty$, and the b_i $+\infty$. We write this interval as $(a_i, b_i]$. F_1 is defined to be the class of all finite sums of such intervals and is clearly a field.

Let $F(x_1, \ldots, x_n)$ be a function of $(x_1, \ldots, x_n)$ such that F is non-decreasing and continuous to the right in each variable, and

$$0 \leqslant F(x_1, \ldots, x_n) \leqslant 1,$$

$$\lim F(x_1, \ldots, x_n) = 1$$

as all $x_i \to \infty$, and

$$\lim F(x_1, \ldots, x_n) = 0,$$

if one or more of the $x_i \to -\infty$. These conditions are in fact not quite sufficient to define a distribution function. Let $(a_i, b_i]$ $(i = 1, \ldots, n)$ define an n-dimensional interval and write Δ_i for the operator which turns $F(a_1, \ldots, a_n)$

into $F(a_1,..., a_{i-1}, b_i a_{i+1},..., a_n) - F(a_1,..., a_n)$. Then we must also have

$$\Delta_1 \Delta_2 ... \Delta_n F(a_1,..., a_n) \geqslant 0 \qquad (4.21)$$

for every choice of (a_i, b_i) where $a_i \leqslant b_i$. This ensures that the measure ascribed to every interval is non-negative.

Then to every interval $(a_i, b_i]$ we ascribe the probability measure given by the left-hand side of (4.21), and this defines an additive measure over the field F_1. By a similar argument to that used above we can show that this measure is σ-additive for enumerable sums of disjoint intervals which are also finite sums of intervals.

The smallest σ-field over F_1 is the Borel field, F_2, in n-dimensional space and the measure can, as before, be extended to a probability measure on F_2, and finally, for any particular function $F(x_1,..., x_n)$, completed to make a Lebesgue–Stieltjes probability measure. $F(x_1,..., x_n)$ can be interpreted as the probability that an n-dimensional random variable $(X_1,..., X_n)$ is such that $X_i \leqslant x_i$ for all i.

4.8. Product measure

A particularly important example of a probability measure in n-dimensional Euclidean space is given by product measures. Suppose the n-dimensional space is regarded as the 'product' of n one-dimensional spaces on each of which is defined a probability measure $\phi_i(A)$ ($i = 1,..., n$). To each of these corresponds a non-decreasing distribution function $F_i(x)$. Then the function

$$F(x_1,..., x_n) = F_1(x_1)... F_n(x_n)$$

satisfies all the conditions in example (3) of §4.7, and therefore defines a measure, ϕ, on the Borel field in n-dimensional space, and this measure can be completed if desired. ϕ is the 'product measure' generated by the separate measures on the n one-dimensional spaces.

We have therefore proved that given probability measures in n one-dimensional spaces it is possible to define a measure in the n-dimensional space which is their product. This has been done by using the cumulative distribution function $F(x_1,..., x_n)$, and is possible because we are concerned with Euclidean spaces. In fact, however, the product measure theorem applies to the product of any set of spaces, and we therefore give an alternative proof which can be easily generalized to such spaces and is along the lines of the proof of the more difficult result for infinite product spaces which we consider in the next section.

It is sufficient to carry out the proof for the product of two spaces each of which is a σ-field of sets on a real line from $-\infty$ to ∞. Let the sets on these lines be denoted by A_i and B_i, and the two lines by R_1 and R_2. A set in the product space, $R_1 \times R_2$, is a product set, $A \times B$, if its points can be represented as (x, y) where x belongs to A, and y to B. Let the probability measures on R_1 and R_2 be denoted by ϕ_1 and ϕ_2. If A and B are sets belonging to the corresponding fields, we define the measure of $A \times B$ to be

$$\phi(A \times B) = \phi_1(A)\phi_2(B).$$

This defines an additive set function on the field in $R_1 \times R_2$ which consists of all finite sums of sets which are product sets of the form $C = A \times B$. Let F be the smallest σ-field on the class of all such product sets. Let ϕ_P be the finitely additive set function defined on the field of all finite sums of product sets. We wish to show that this can be extended to a probability measure on the σ-field F. ϕ_P is clearly non-negative, and to apply Theorem 4.3 all that is necessary is to show that it is σ-additive on this field. This means that if C is a finite sum of product sets which is also an enumerable sum of sets, C_i, which are themselves finite sums of product sets, then

$$\phi_P(C) = \Sigma\phi_P(C_i).$$

By Theorem 4.2 this will follow if we show that ϕ_P is a continuous measure.

Suppose now that C_n is a decreasing sequence of sets, each a finite sum of product sets, and such that

$$\phi_P(C_n) > \varepsilon > 0.$$

We prove that $C = \lim C_n$ is not a null set.

If y is a point on the line R_2, let $C_n(y)$ be the intersection of C_n with the line through y parallel to R_1. Let B_n be the set of values of y for which

$$\phi_1\{C_n(y)\} > \tfrac{1}{2}\varepsilon.$$

B_n is clearly measurable (ϕ_2). Then

$$\phi_2(B_n) + \tfrac{1}{2}\varepsilon\{1 - \phi_2(B_n)\} \geqslant \phi_1(C_n) > \varepsilon,$$

and therefore

$$\phi_2(B_n) > \tfrac{1}{2}\varepsilon.$$

Since this is true for all n, $\lim B_n$ cannot be empty and so $\lim C_n$ cannot be empty either. This proves the result. For the product of more than two one-dimensional spaces the proof is exactly similar.

4.9. Probability measure in infinite dimensional Euclidean space

We define the infinite dimensional Euclidean space R_∞, to consist of 'points' defined by infinite sequences $(x_1, x_2, \ldots)$ of real finite numbers. This can be regarded as the 'product' of an enumerably infinite set of one-dimensional spaces. We do not define a distance in R_∞ and hence do not consider open and closed sets, yet we still wish to construct in it a σ-additive field of sets on which we can define a probability measure.

We suppose that x_i lies on a line which we denote by R_i. On this line we can define as before the class of all Borel sets which form a σ-field which we denote as B_i. Given sets A_1, A_2,... in the spaces R_1, R_2,... we define the product set

$$A_1 \times A_2 \times A_3 \times ...$$

to be the set of all points $(x_1, x_2,...)$ in R_∞ such that x_1 belongs to A_1, x_2 to A_2, and so on.

Suppose that for a finite set of indices $i_1,..., i_n$ we have a set $A_{i_1...i_n}$ in the space $R_{i_1} \times ... \times R_{i_n}$. The 'product' of this set with all the spaces R_j ($j \neq i_1,...,i_n$) is said to be a 'cylinder'.

In particular suppose that for a finite set of indices $i_1,..., i_n$ we have prescribed sets $A_{i_1},..., A_{i_n}$, and that for all other indices the A's are taken as the whole line ($-\infty < x_i < \infty$). Then we say that the product set is a 'product cylinder' in R_∞. Let the sets A_i be Borel sets in their corresponding one-dimensional spaces R_i. Then the class of all sets in R_∞ which are finite sums of such product cylinders is clearly a field since the sum or common part of two such sets is clearly again a member of the class, as is also the complement of such a set.

The minimal σ-field constructed over this field is said to be the Borel product field, B, and it is on this field that we construct probability measures. To do this we suppose that a probability measure, $\phi_i(A_i)$, is defined in each space R_i for every Borel measurable set A_i in R_i. For any product cylinder defined by $A_{i_1},..., A_{i_n}$ in $R_{i_1},..., R_{i_n}$ we define its measure to be

$$\phi_{i_1}(A_{i_1})\phi_{i_2}(A_{i_2})...\phi_{i_n}(A_{i_n})$$

and for any set in the field consisting of finite sums of disjoint cylinders we can define the measure to be the sum of the measures of the individual cylinders. This defines a finitely additive set function. Then by using the method of §4.8 we can extend this measure to provide a measure for all sets of the form $A_{i_1...i_n} \times \prod R_j$, where j is taken over all suffices not equal to $i_1,...,$ or i_n, and $A_{i_1...i_n}$ is a set in $R_{i_1} \times ... \times R_{i_n}$.

THEOREM 4.5. *Denote the field formed by all finite sums of product cylinders by C, and the finitely additive set function defined on it by the above method by ϕ_C. Then ϕ_C can be extended to be a σ-additive probability measure, ϕ_B, on the Borel product field B.*

From Theorem 4.3 it is sufficient to show that ϕ_C is σ-additive on the field C, i.e. that for any *finite* sum, A, of measurable product cylinders which is an enumerable sum of disjoint *finite* sums of measurable product cylinders, A_i, then

$$\phi_C(A) = \Sigma\phi_C(A_i).$$

From the continuity Theorem 4.2 it is sufficient to show that ϕ_C is continuous since we already know it is finitely additive. Suppose that A_i ($i = 1, 2,...$) is a decreasing sequence of sets each of which is a finite sum of measurable product cylinders. We want to show that if $\lim A_i$ is empty, $\phi_C(A_i)$ converges to zero. Suppose that $\phi_C(A_i) > \varepsilon > 0$ for every i. We show that $A = \lim A_i$ cannot be empty. Each set A_i is a finite sum of product cylinders. Remembering the suffices i if necessary we can write each A_n as a cylinder

$$A_n = A_{nm} \times \Omega_m,$$

where Ω_m is the space

$$R_{m+1} \times R_{m+2} \times \ldots$$

and A_{nm} is a set in the space $R_1 \times \ldots \times R_m$.

We have defined ϕ_C as a finitely additive set function on the field of all cylinders in Ω_0. Define $\phi^{(i)}$ as the similar finitely additive set function on the field of all cylinders in Ω_i. If $\omega_1, \ldots, \omega_k$, are points in $R_1, \ldots, R_k$, respectively, we define $A(\omega_1, \ldots, \omega_k)$ to be the section of A_n at the point $(\omega_1, \ldots, \omega_k)$.

Let B_1^n be the set of all ω_1 such that

$$\phi^{(1)}\{A_n(\omega_1)\} > \tfrac{1}{2}\varepsilon.$$

Then adding over the sets B_1^n and $R_1 - B_1^n$ we get

$$\phi(B_1^n) + \tfrac{1}{2}\varepsilon\{1 - \phi(B_1^n)\} > \phi_C(A_n) > \varepsilon.$$

Write

$$B_1 = \lim_n B_1^n.$$

Then

$$\phi(B_1) \geqslant \tfrac{1}{2}\varepsilon,$$

and there must be a point $\overline{\omega_1}$ in B_1 such that

$$\phi^{(1)}\{A_n(\overline{\omega_1})\} > \tfrac{1}{2}\varepsilon$$

for every n.

We now apply the same argument to $A^n(\omega_1)$ in the space $\Omega_1 = R_2 \times R_3 \times \ldots$ and obtain a point $\overline{\omega_2}$ such that

$$\phi^{(2)}\{A_n(\overline{\omega_1}, \overline{\omega_2})\} > \tfrac{1}{4}\varepsilon.$$

Continuing this process we obtain a point

$$(\overline{\omega_1}, \overline{\omega_2}, \ldots)$$

which belongs to all A_n, so that A is not empty. This completes the proof.

ϕ_C is then a continuous set function and therefore σ-additive, and Theorem 4.5 is proved. We can therefore always define a σ-additive product probability measure in a space which is a finite or enumerable product of one-dimensional Euclidean spaces, and this measure can be completed if required.

4.10. Consistent probabilities in product spaces

We now return to finite dimensional spaces R_n. Many probability measures on such spaces are not product measures but nevertheless the relationship between them and the measures which they 'induce' in the spaces obtained by selecting some only of the coordinates of R_n needs to be examined. Thus by selecting a particular m $(m < n)$ out of the n coordinates $(x_1,...,x_n)$ we obtain a space R_m, and the remaining $n-m$ coordinates define a space R_{n-m} such that R_n can be regarded as the Cartesian product $R_m \times R_{n-m}$.

Given a probability measure, $P_n(A)$, defined for all Borel sets in R_n we can define the marginal probability measure, $P_m(A')$, for all measurable sets A' in the chosen space R_m, by putting $P_m(A')$ equal to $P_m(A)$ where A is the cylinder set in the space R_n whose base consists of A' in R_m. If $P_n(A)$ is known to be a probability measure on the Borel field in R_n then $P_m(A')$ is defined for every Borel set in every R_m, and is itself a σ-additive probability measure known as the 'marginal probability measure' in the space R_m.

Given $P_n(A)$ the marginal probability measures induced in this way are consistent in the sense that if R_k is a subspace of R_l and R_m, two subspaces of R_n, the probability measures induced in R_k by the marginal probability measures in R_l and R_m coincide.

4.11. Measurable functions

Having constructed a theory of measure in Euclidean spaces we can now apply it to the definition of an integral and to do this we first consider some properties of real functions.

We denote points in R_n or R_∞ by such symbols as x, it being understood that this stands for the whole set of coordinates $(x_1, x_2,...)$ which may be finite or infinite in number. A function $f(x)$ is defined when given any point x, the real numerical value $f(x)$ is determined.

It is convenient to write $f^+(x)$ for the function equal to $f(x)$ when $f(x) > 0$, and equal to zero otherwise. Similarly we put $f^-(x)$ for a function equal to $f(x)$ when $f(x) < 0$, and equal to zero otherwise. Then

$$f(x) = f^+(x) + f^-(x),$$

$$|f(x)| = f^+(x) - f^-(x). \tag{4.22}$$

In the function $f(x)$, x is a point in an n-dimensional or infinite dimensional space R, whilst $f(x)$ is a real number and can be regarded as a point in a one-dimensional space R_1. Hence the function $f(x)$ can be regarded as a mapping of the set of points, x, on which $f(x)$ is defined (which is called the 'domain' of $f(x)$ and need not be the whole of R) on to a part or the whole of R_1. The latter is called the 'range' of $f(x)$.

If we have a set E in R, the values of $f(x)$ for x in E are denoted by $f(E)$ and are called the 'image' in R_1 of E. Similarly the set consisting of all points x such that $f(x)$ belongs to a set F in R_1 is known as the 'inverse image' of F

and is written $f^{-1}(F)$. For A and B in R_1 we then have

$$f^{-1}(A+B) = f^{-1}(A) + f^{-1}(B),$$
$$f^{-1}(A-B) = f^{-1}(A) - f^{-1}(B), \qquad (4.23)$$

with similar results for infinite sums and products. (The corresponding result for $f(A-B)$ is not true although that for $f(A+B)$ is correct.) It then follows that the inverse image of a σ-field, i.e. the class of all inverses of the sets of such a field, is itself a σ-field.

From now on we choose a definite σ-field of sets, B, in the space R and since we shall later associate with it a σ-additive measure, we call a set 'measurable' if and only if it belongs to the σ-field B. Thus 'measurability' is defined in terms of the σ-field and not in terms of the measure.

A measurable function can be defined in several equivalent ways. Perhaps the simplest is to insist that the set of points x for which $f(x) < a$ is always a measurable set (i.e. a member of the σ-field B) for every real value of a. We then have:

THEOREM 4.6. *If $f(x)$ is measurable, the inverse images of all Borel sets are measurable, and vice versa.*

Proof. From the above definition the inverse images of all intervals half-open to the right are members of B, and therefore the same is true for all sets obtained from such intervals by finite or enumerable set operations, and therefore for all Borel sets. The converse is obvious.

Another nearly equivalent definition which is useful in defining integrals is based on the idea of a 'simple function' which is defined to be a function which takes only an enumerable set of values $f_1, f_2, \ldots$ on an enumerable class of disjoint measurable sets $A_1, A_2, \ldots$ which we usually suppose exhaust the whole set on which $f(x)$ is defined. We then have:

THEOREM 4.7. *A measurable function is the limit of a convergent sequence of simple functions, and vice versa.*

Proof. Every simple function is obviously measurable and if $f(x)$ is the limit of a sequence of simple functions $f_1(x), f_2(x), \ldots$ the set of values x such that $f(x) \geqslant a$ is clearly the limit as $\varepsilon \to 0$ of the limit of the sets for which $f_n(x) > a - \varepsilon$, where $\varepsilon > 0$. But the latter sets are measurable and hence the set of x at which $f(x) \geqslant a$ is measurable. Hence the set where $f(x) < a$ is measurable also. Similarly if $f(x)$ is measurable the sets for which

$$m2^{-n} \leqslant f(x) < (m+1)2^{-n},$$

for $m = 0, \pm 1, \ldots, n = 1, 2, \ldots$ are also measurable and we can take $f_n(x) = m2^{-n}$ where m is the integer for which this inequality holds.

It is clear that if $f(x)$ is measurable, so also are $f^+(x)$ and $f^-(x)$, and conversely.

THEOREM 4.8. *If $f(x), g(x)$ are measurable functions so also is $f(x) \pm g(x)$, $f(x)g(x)$, $f(x)g(x)^{-1}$ if $g(x) \neq 0$, and $\lim_n f_n(x)$ if the latter exists everywhere.*

This follows at once from Theorem 4.7.

THEOREM 4.9. *If $f(x)$ is measurable so also is $F\{f(x)\}$ when $F(y)$ is a continuous function of y, or the limit of such functions (and in fact for any Baire function).*

Proof. It is sufficient to prove the result when $F(y)$ is continuous. If $f_n(x)$, $n = 1, 2,...$, is a sequence of simple functions converging to $f(x)$ so also is $F\{f_n(x)\}$ a sequence of simple functions converging to $F\{f(x)\}$ and the result follows from Theorem 4.8.

4.12. Measure theory and convergence

Consider a sequence of functions, $\{f_n(x)\}$, $n = 1, 2,...$, defined on some space and measurable with respect to some σ-field on which there is defined a measure, $P(E)$, which for the moment we do not assume to be a probability measure. Then in the usual way we can define what is meant by the convergence of $\{f_n(x)\}$ to a function $f(x)$ at some point x, or for all x in the field of definition. In probability theory and many other parts of analysis it is more useful to use definitions of convergence more general than this, and, as we shall see in the next chapter, these correspond to important definitions of 'convergence' for random variables (for a valuable discussion of convergence of real functions see Littlewood 1944).

We say that a sequence of functions, $f_n(x)$, converges 'almost everywhere' if it converges at every point in the field of definition outside a set of measure zero. Inside the latter it may or may not converge and it is not implied in the definition that the set of points where convergence does not occur is measurable, although this can in fact be proved (Exercise 4.4). When $f_n(x)$ converges almost everywhere to $f(x)$ we write

$$f_n(x) \rightarrow f(x) \quad \text{(a.e.).} \tag{4.24}$$

It then easily follows that the Cauchy criterion holds so that

$$\lim_{\substack{m \rightarrow \infty \\ n \rightarrow \infty}} |f_m(x) - f_n(x)| = 0, \tag{4.25}$$

except possibly on a set of measure zero, and conversely if the latter holds, there exists a function, $f(x)$, such that (4.24) holds.

It may happen that we have two sequences $f_n(x)$ and $g_n(x)$ such that

$$f_n(x) \rightarrow f(x) \quad \text{(a.e.),}$$

$$g_n(x) \rightarrow f(x) \quad \text{(a.e.),}$$

and the sets of zero measure in the two definitions are not the same. We could then take the sum of these sets to define the convergence almost everywhere. However, it is often more convenient to define the function $f(x)$ only 'up to an equivalence' in the sense that we regard all functions as being the same if they differ in pairs only in some set of outer measure zero.

A sequence of functions, $f_n(x)$, is said to converge to $f(x)$ 'in measure' if for each $\varepsilon > 0$, the measure of the set where $|f_n(x) - f(x)|$ is greater than ε, tends to zero. We write this as

$$f_n(x) \rightarrow f(x), \ (P),$$

or

$$P\{|f_n(x) - f(x)| > \varepsilon\} \rightarrow 0, \tag{4.26}$$

where the inequality inside the brackets defines the set whose P-measure is to be found. Although we write '(P)' here, this definition holds whether or not the measure is a probability measure. The Cauchy property of convergence remains true for this definition but we delay proof of this for the moment (it is a consequence of Theorem 4.11) until we consider the relations between the three types of convergence so far defined.

Firstly, it is clear that if the sequence $\{f_n(x)\}$ converges everywhere then it converges almost everywhere. However, it need not converge in measure as we see by defining $f_n(x) = 1$ in $(n, n+1)$ and zero elsewhere, and use ordinary Lebesgue measure. This converges everywhere to zero in the ordinary sense but not in measure. Here we have taken a general measure. For a probability measure, however, the implication is correct and in fact we have the following:

THEOREM 4.10. *If the sequence $f_n(x)$ converges everywhere, or almost everywhere, and the measure is bounded (in particular a probability measure) then $f_n(x)$ also converges in measure.*

Proof. Let $\varepsilon > 0$, and $M(E)$ be a finite measure such that $m(\Omega) < \infty$. Write E_n for the set of points where $|f_n(x) - f(x)| > \varepsilon$. Then by definition

$$\lim_n M(E_n) = 0,$$

because $\lim_n E_n$ has measure zero by Theorem 4.2 so that the result follows.

The reverse implication is not correct even for bounded measures. To show this suppose $f_n(x)$ is defined on the closed interval $\langle 0, 1 \rangle$ to be equal to zero outside a closed interval I_n inside which it equals unity. If I_n is taken to be the nth interval in the sequence

$$\langle 0, \tfrac{1}{2} \rangle, \ \langle \tfrac{1}{2}, 1 \rangle, \ \langle 0, \tfrac{1}{4} \rangle, \ \langle \tfrac{1}{4}, \tfrac{1}{2} \rangle, \ldots$$

it is clear that $f_n(x)$ converges to zero in measure but not at any point in the ordinary sense. However, it is possible to obtain a partial reverse implication which turns out to be important.

THEOREM 4.11. *Suppose that $f_n(x)$ converges in measure to a measurable function $f(x)$. Then it is possible to pick out a subsequence, n_k $(k = 1, 2, \ldots)$, such that $f_{n_k}(x)$ converges to $f(x)$ almost everywhere. Similarly if $f_n(x)$ is a sequence which is mutually convergent in measure, i.e. the analogous Cauchy*

14

condition holds so that

$$\lim_{n} M\{|f_{n+m}(x)-f_n(x)| > \varepsilon\} = 0, \tag{4.27}$$

uniformly in $m > 0$, then there exists a subsequence $f_{n_k}(x)$ which converges mutually almost everywhere and therefore converges almost everywhere to a certain function $f(x)$. Furthermore $f_{n_k}(x)$ converges in measure to $f(x)$.

Proof. If $f_n(x)$ converges to a function $f(x)$ in measure, then (4.27) holds, so we assume the latter. For any integer k there exists an integer $n(k)$ such that for all $n \geqslant n(k)$,

$$M\{|f_{n+\nu}(x)-f_n(x)| \geqslant 2^{-k}\} > 2^{-k}, \tag{4.28}$$

where ν is an arbitrary positive integer. We define n_1 to be $n(1)$, n_2 to be the greater of n_1+1 and $n(2)$, and so on. Put $g_k(x) = f_{n_k}(x)$ for convenience. Then

$$M\{|g_{k+1}(x)-g_k(x)| \geqslant 2^{-k}\} \leqslant 2^{-k}, \tag{4.29}$$

and the set, E say, where at least one of the inequalities

$$|g_{k+1}(x)-g_k(x)| \geqslant 2^{-k}, \quad k = n, n+1,\ldots$$

holds therefore has measure not greater than

$$\sum_{k=n}^{\infty} 2^{-k} = 2^{1-n}. \tag{4.30}$$

Given any $\varepsilon > 0$ we choose n so $2^{1-n} < \varepsilon$, and so that outside the set E,

$$|g_{n+\nu}(x)-g_n(x)| \leqslant \sum_{k=n}^{\infty} |g_{k+1}(x)-g_k(x)| \leqslant 2^{1-n}. \tag{4.31}$$

The measure of the common part of all such sets for all $\varepsilon > 0$ is therefore zero, and $g_{n+\nu}(x)-g_n(x)$ must tend to zero almost everywhere as n increases and ν has any positive integral values. This means that for the sequence, $g_n(x)$, the Cauchy criterion holds almost everywhere and hence that there exists a function, which we write as $f(x)$, such that $g_n(x)$ tends to $f(x)$ almost everywhere.

Outside the set E we have

$$|g_{n+\nu}(x)-g_n(x)| \leqslant 2^{1-n} < \varepsilon,$$

so that letting $\nu \to \infty$ we get, for all x not in E,

$$|f(x)-g_n(x)| \leqslant 2^{1-n} < \varepsilon. \tag{4.32}$$

Then

$$M\{|f(x)-g_n(x)| \geqslant \varepsilon\} < \varepsilon \tag{4.33}$$

from (4.30). As $n \to \infty$, $\varepsilon \to 0$ and therefore $g_n(x)$ converges in measure to $f(x)$.

It is obvious that convergence in measure of a sequence $f_n(x)$ implies mutual convergence in measure, and by the above proof the reverse is also true. Thus convergence in measure also possesses the Cauchy property.

In §5.11 and in Chapter 8, we will show how functions measurable on every space with a probability measure can be identified with 'random variables', and how the above types of convergence can be made to correspond to various modes of convergence of random variables.

4.13. Integration

Consider measurable functions, $f(x)$, where x belongs to a space in which there are defined a σ-field, F, and a measure, $M(E)$, for sets E belonging to the σ-field. In nearly all applications in probability theory the measure, $M(E)$, will be non-negative and such that $M(\Omega) = 1$ where Ω is the whole space, i.e. a probability measure. In any case we assume $M(E)$ is 'σ-finite', i.e. Ω is the sum of an enumerable sequence of sets on which $M(E)$ is finite.

The integral of $f(x)$ over a set E in F can be defined in various ways but the simplest is to define it first for 'simple' functions and then, analogously with Theorem 4.7, for functions which are the limits of simple functions and are therefore measurable. This method of procedure can be contrasted with the definition of Riemann integrals in elementary analysis which is simplest when applied to continuous functions in contrast to 'simple' functions which are, in general, everywhere discontinuous.

If $f(x)$ takes the non-negative values $f_1, f_2,\ldots$ on measurable sets $E_1, E_2,\ldots$ where ΣE_i is the whole space, we define the integral of $f(x)$ on a measurable set A to be

$$\int_A f_n(x)\,dM(x) = \Sigma f_i\, m(AE_i), \tag{4.34}$$

where the notation '$dM(x)$' indicates that we are using the measure $M(E)$. The expression (4.34) may be infinite. The integral of a non-negative measurable function $f(x)$ is defined to be the limit

$$\int_A f(x)\,dM(x) = \lim_{n \to \infty} \int_A f_n(x)\,dM(x) \tag{4.35}$$

where $f_n(x)$, $n = 1, 2,\ldots$, is a non-decreasing sequence of simple functions which converge to $f(x)$ everywhere and are never greater than it. We have to prove that such a sequence exists, and that the limit exists and is unique. It is clearly sufficient to suppose that A is the whole space.

Such a sequence exists for we can always define the sets E_i to be the measurable sets where

$$i2^{-n} < f(x) \leqslant (i+1)2^{-n}, \tag{4.36}$$

and n is any positive integer. On these sets we take $f_n(x)$ to be equal to $i2^{-n}$, and the convergence of $f_n(x)$ to $f(x)$ follows. $f_n(x)$ is clearly non-decreasing.

The limit in (4.35) exists for any such sequence. Let U be the least upper bound of

$$\int f_n(x) \, dM(x)$$

for all such sequences of functions and assume first that U is finite. There exists some sequence of simple functions $g_n(x)$ such that

$$\lim_{n \to \infty} \int g_n(x) \, dM(x) = U,$$

and

$$g_n(x) \leqslant f(x) \quad \text{everywhere.}$$

Since the measure, $M(E)$, is σ-finite it is sufficient to prove the uniqueness where the functions are defined on a subset of the whole space which has finite measure. Suppose $h_n(x)$ is any other non-decreasing sequence of simple functions such that $h_n(x) \leqslant f(x)$ and such that each $h_n(x)$ tends to $f(x)$ at every point x. If E_i is one of the sets on which $g_n(x)$ is defined, the set at which $h_n(x) < g_N(x) - \varepsilon$, for any given N and $\varepsilon > 0$, must be one of a decreasing sequence of sets tending to zero. We can split the sum for $h_n(x)$ up into sums of smaller sets each contained in one of the E_i and then in each of these

$$\lim_{n \to \infty} \int h_n(x) \, dM(x) \geqslant \int g_N(x) \, dM(x).$$

From this the result follows. When U is infinite a similar argument holds and the integral is regarded as infinite.

For a function which takes both signs we define the integral to be

$$\int f(x) \, dM(x) = \int f^+(x) \, dM(x) + \int f^-(x) \, dM(x),$$

with the restriction that the integrals on the right-hand side are not both infinite. When both are finite $f(x)$ is said to be 'integrable'.

Using this definition the ordinary expected properties of an integral are easily established. In particular we have the following results:

(1) If

$$\int_A f(x) \, dM(x)$$

exists, the integral over any measurable set contained in A also exists.

(2) If the integrals of $f(x)$ and $g(x)$ both exist and are not both infinite with opposite signs, their sum is equal to the integral of $f(x) + g(x)$, which exists.

(3) If $f(x) \geqslant g(x)$ then $\int f(x) \, dM(x) \geqslant \int g(x) \, dM(x)$, if both integrals exist.

(4) If $f(x)$ is integrable with a finite integral so is $|f(x)|$, and any measurable $g(x)$ such that

$$|g(x)| \leqslant |f(x)|$$

also has a finite integral.

The integral we have defined in this way is known as a 'Lebesgue–Stieltjes' integral. For many purposes in probability theory it is not absolutely necessary to use an integral of such generality. If $M(x)$ is a distribution function and $f(x)$ is a function which is reasonably well behaved (e.g. continuous) it is possible to define a 'Riemann–Stieltjes' integral,

$$\int_a^b f(x)\, dM(x),$$

which is extended over the interval (a, b), and does not require the theory of measure. This approach will be briefly described in the next chapter.

4.14. Limiting properties of integrals

THEOREM 4.12. *If $f_n(x)$ is a non-decreasing sequence of non-negative functions which converge to a function $f(x)$ then the integral of $f(x)$ exists and*

$$\lim_{n \to \infty} \int f_n(x)\, dM(x) = \int f(x)\, dM(x). \tag{4.37}$$

Proof. This follows at once by approximating to each $f_n(x)$ by an increasing sequence of simple function $f_{ns}(x)$ $(s = 1, 2, \ldots)$. We can then replace the sequence $f_n(x)$ by a new sequence of simple functions $g_n(x) = \max_{1 \leqslant l \leqslant n} f_{ln}(x)$. These are clearly a set of simple functions which are non-decreasing and converge to $f(x)$.

THEOREM 4.13. *If $f(x)$ is integrable with a finite integral, $M(E) \to 0$ implies*

$$\int_E f(x)\, dM(x) \to 0. \tag{4.38}$$

Proof. It is sufficient to suppose $f(x) \geqslant 0$. Since $f(x)$ is finitely integrable, given $\varepsilon > 0$ we can choose a number N such that the integral over the set E_1, where $f(x) > N$, is less than ε. Then

$$\int_E f(x)\, dM(x) = \int_{EE_1} f(x)\, dM(x) + \int_{E - EE_1} f(x)\, dM(x)$$

$$< \varepsilon + NM(E),$$

from which the result follows.

A more subtle theorem is the following.

THEOREM 4.14 (the Fatou–Lebesgue theorem). *Suppose $f_n(x)$ is a sequence of integrable functions such that*

$$g(x) \leqslant f_n(x) \leqslant h(x), \tag{4.39}$$

where $g(x)$ and $h(x)$ are finitely integrable. Then

$$\int \lim \inf f_n(x)\, dM(x) \leqslant \lim \inf \int f_n(x)\, dM(x), \tag{4.40}$$

$$\int \lim \sup f_n(x)\, dM(x) \geqslant \lim \sup \int f_n(x)\, dM(x). \tag{4.41}$$

Proof. It is clearly sufficient to prove the first inequality. Suppose first that the $f_n(x)$ are non-negative. Write

$$k_n(x) = \underset{m \geqslant n}{\text{lower bdd}}\, f_m(x).$$

Then $k_n(x)$ is a non-decreasing function of n which converges to $\lim \inf f_n(x)$, and hence

$$\int \lim \inf f_n(x)\, dM(x) = \lim_n \int k_n(x)\, dM(x)$$

$$\leqslant \lim_n \inf \int f_n(x)\, dM(x). \tag{4.42}$$

This, and the similar argument with an upper limit, proves the result when the $f_n(x)$ are non-negative or non-positive, and the general case is obtained by splitting $f_n(x)$ into $f_n^+(x)$ and $f_n^-(x)$.

THEOREM 4.15 (the Dominated Convergence Theorem). *If $f_n(x)$ is a sequence of integrable functions such that $|f_n(x)| \leqslant g(x)$ where $g(x)$ is finitely integrable, and if $f_n(x)$ converges to an integrable function $f(x)$ almost everywhere, or in measure, then*

$$\int f_n(x)\, dM(x) \to \int f(x)\, dM(x), \tag{4.43}$$

and

$$\int |f_n(x) - f(x)|\, dM(x) \to 0. \tag{4.44}$$

Proof. By splitting $f_n(x) - f(x)$ into its positive and negative parts, (4.44) implies (4.43). Now $|f_n(x) - f(x)| \leqslant 2g(x)$ almost everywhere. It is therefore sufficient to prove the result when the $f_n(x)$ are non-negative and $f(x) = 0$. If $f_n(x)$ converges to zero almost everywhere the result follows from (4.41). If $f_n(x)$ converges to zero in measure, write U for the upper limit of

$$\int f_n(x)\, dM(x).$$

This is finite and non-negative since $0 \leqslant f_n(x) \leqslant g(x)$. Let n_k be a subsequence of the n such that

$$\lim_k \int f_{n_k}(x)\, dM(x) = U.$$

By Theorem 4.11 we can select a subsequence, n_l ($l = k_1, k_2, \ldots$) out of the sequence n_k in such a way that $f_{n_l}(x)$ converges to zero almost everywhere, and for this subsequence

$$U = \lim_l \int f_{n_l}(x)\, dM(x) = 0.$$

Hence $U = 0$ and the theorem is proved.

This result makes it possible to prove the standard theorems on differentiating an integral with respect to a parameter under the integral sign.

4.15. The distribution of geometric elements

So far we have been concerned with the probability distribution of 'random variables' or sets of random variables. For some applied problems it is important to extend this concept so that we can talk about the distribution of geometric objects. Such objects may be points, lines, planes, solid bodies, or other geometrical figures. To discuss the distribution of such objects we must first define a set of coordinates or parameters which will define their position uniquely. Thus points in Euclidean space can be defined by their coordinates, lines in three dimensions by their intersections with a fixed plane and their directions or by their Plücker coordinates, planes in three dimensions whose equations are of the form

$$ux + vy + wz + 1 = 0, \tag{4.45}$$

by (u, v, w), and so on.

A distribution of such objects is then determined when we prescribe a probability distribution for the set of parameters. We can denote such a set of parameters by $(\theta_1, \ldots, \theta_k)$, taking values in all or part of a Euclidean parameter space of dimension k.

The properties of geometric objects in Euclidean spaces which interest the geometer are characterized by the fact that they are invariant under certain groups of transformations. For the most part the groups of transformations of Euclidean space which is here relevant is the group consisting of all translations, reflections, and rotations. If we apply this group to the set of all points (or all lines, or all planes) in space we see that there is no probability measure which is invariant under such transformations. For the purposes of illustration we may consider the simple case of all points in n-dimensional space, R_n. It is then obvious that it is impossible to define a distribution of a point P in R_n which is invariant in this way, for such a distribution would have to ascribe the same measure to every unit cube and could not therefore be normalized to give unit measure to the whole space.

Two ways out of this difficulty exist. In the first, which we discuss in §4.17, we consider not absolute probabilities of lying in some set E but the conditional probability of lying in E when it is known that the point lies in a set

$F \supset E$. In the second approach we no longer consider individual points but random 'fields' of points ('field' being used now in a different sense from § 4.3), such that the number of points occurring in any region of a certain finite measure is a random variable having a Poisson distribution.

Let E be a set in the parameter space and suppose that it belongs to a field of sets on which is defined a measure, $M(E)$. Then we assume that the number of geometric elements which occur whose parameters lie in E is a Poisson variate with mean $\lambda M(E)$ ($\lambda > 0$). This distribution enables one to assume that for all disjoint sets $E_1, E_2, ..., E_k$, the numbers of elements with parameters in each set are independently distributed.

The distribution in any set E has generating function

$$\exp \lambda(z-1) \int_E dM, \tag{4.46}$$

and conditional on there being exactly N such elements, the joint distribution of their parameters $\theta_1, ..., \theta_N$, is

$$\frac{dM(\theta_1)...dM(\theta_N)}{\left\{\int_E dM(\theta)\right\}^N}. \tag{4.47}$$

In this formula the parameters, θ_i, may represent vectors if more than one quantity is needed to parametrize each geometric element, as is usually the case.

$M(E)$ is a measure but need not be such that $M(\Omega) < \infty$, where Ω is the whole of the parameter space. We are therefore now free to choose M in such a way that it is invariant under the group of transformations in the parameter space which is induced by all translations, reflections and rotations in the space of the geometric elements. If we consider points in R_3, their parameter space can be taken as R_3 itself, and the natural invariant measure is ordinary Lebesgue measure. The determination of suitable invariant measures for other classes of geometric elements can involve a considerable amount of analysis.

4.16. Examples of geometric distributions

Consider lines in a plane and suppose they are defined by equations of the form

$$ux+vy+1 = 0. \tag{4.48}$$

Translations, reflections, and rotations in the plane correspond to linear transformations of the coordinates of the form

$$X = a+x\cos\alpha-y\sin\alpha,$$
$$Y = b+x\sin\alpha+y\cos\alpha, \tag{4.49}$$

where $-\infty < a, b < \infty, 0 \leqslant \alpha < 2\pi$. The equation of the line after transformation will be of the form

$$UX + VY + 1 = 0, \tag{4.50}$$

where

$$u = (U \cos \alpha + V \sin \alpha)(aU + bV + 1)^{-1},$$

$$v = (-U \sin \alpha + V \cos \alpha)(aU + bV + 1)^{-1}. \tag{4.51}$$

If we have a measure function of the form

$$\int_E \int f(u, v) \, du \, dv,$$

it must be always equal to

$$\int_E \int f(U, V) \, dU \, dV = \int_E \int f(u, v) \frac{\partial(u, v)}{\partial(U, V)} \, dU \, dV.$$

Evaluating the Jacobian it is easy to verify that $f(u, v)$ must be proportional to

$$(u^2 + v^2)^{-\frac{3}{2}}, \tag{4.52}$$

so that we can take

$$M(E) = \lambda \int_E \int (u^2 + v^2)^{-\frac{3}{2}} \, du \, dv. \tag{4.53}$$

We have ignored lines which pass through the origin but these must have total measure zero. In a similar way it is possible to consider lines and planes in R_3, and rotations in R_3. For the last mentioned three parameters are required and the measure is chosen to be invariant under the group generated by the group of all rotations in R_3. A detailed discussion with references and many applications will be found in Kendall and Moran (1963), and Deltheil (1926).

4.17. Conditional probability

Suppose that E and F are measurable sets in a probability space with probability measure $P(E)$. The conditional probability of the event represented by the set E, given that F has occurred, is defined to be

$$P(E|F) = \frac{P(EF)}{P(F)}, \tag{4.54}$$

whenever $P(F) > 0$.

Note that if $P(F)$ is zero this definition does not work. What is important in many cases is to be able to deal with problems of this kind in the special case where we are considering a distribution function, $F(x, y)$, in two dimensions and find it convenient to be able to talk about the distribution of X for a fixed value of Y, where X and Y are random variables for which

$$\text{pr}(X \leqslant x, \ Y \leqslant y) = F(x, y).$$

A full solution of this problem requires very elaborate analysis for which the reader is referred to Doob (1953), Chapter I. For most problems in probability it is sufficient to restrict attention to the case where $F(x, y)$ is of the form

$$\int_{-\infty}^{x} \int_{-\infty}^{y} f(x, y) \, dx \, dy$$

where $f(x, y)$ is a Lebesgue measurable function. A theorem which enables this to be done is given in the next chapter.

For a fixed F and $P(F) > 0$, (4.54) is a probability measure function on the σ-field of sets E, as is easily verified. However, for this to be true it is not necessary that $P(E)$ be a probability measure and in fact any σ-finite measure function could be used. Thus we can use for $M(E)$ any of the invariant measures considered above. In this way we can make conditional probability statements about, say, a random line in space. For example, we may ask for the probability that a random line in space, which is known to intersect a bounded convex body K, will also intersect another bounded convex body K_1 contained in K.

Renyi (1955) and Csaszar (1955) have constructed an axiomatic theory of probability which begins from the definition of functions of two sets which have the properties of conditional probabilities. Suppose that $\{A\}$ is a σ-field of sets A, and $\{B\}$ is a subset of this field. A numerical function, $P(A|B)$, of two such sets is defined to be a conditional probability measure if:

(1) $P(A|B) \geqslant 0$, $P(B|B) = 1$, for B in $\{B\}$;
(2) for any fixed B, $P(A|B)$ is a probability measure;
(3) if A and B belong to $\{A\}$, and B and C to $\{B\}$, then

$$P(A|BC)P(B|C) = P(AB|C).$$

Under a further weak restriction it can then be shown that $P(A|B)$ can be written in the form

$$P(A|B) = \frac{Q(AB)}{Q(B)}, \tag{4.55}$$

where $Q(A)$ is a non-negative measure on the σ-field $\{A\}$, and $Q(B) > 0$ when B is in $\{B\}$.

A very detailed discussion of the theory of measures of this kind is given in the two papers referred to above, and Renyi, in particular, points out the usefulness of the idea in the statistical theory of inference and in number theory. (See also the remarks on 'density' in Chapter V of Kac 1959a.)

For a further consideration of conditional distributions see §5.4.

Bibliographical notes

The first treatment of probability as a measure is the fundamental work of Kolmogorov (1933). Detailed and instructive treatments will also be found in Loève (1963), Kingman and Taylor (1966), and Halmos (1950). See also Kamke (1932), Richter (1956), Tornier (1936), and Pitt (1963).

Just as in the case of finite fields (Chapter 1) we can ask whether it is possible to represent a given abstract σ-field, $\{A_i\}$, as a field of sets. This type of 'representation theorem' has been discussed by several writers beginning with Stone (1936) and an outline of the results with further references is given in Halmos (1950).

The idea of using measures which are invariant under the operations of transformation groups has led to the beautiful measure theory of Haar to which an introduction is given in Chapter XI of Halmos (1950). The theory of probability fields on more general algebraic structures than Euclidean space is treated in detail in a highly illuminating manner by Grenander (1963) who considers groups, semigroups, Banach and Hilbert spaces, and algebras. Many, but not all, of the general theorems in the remaining chapters of this book can be thus generalized but these generalizations are often not trivial.

Exercises

Exercise 4.1. For sets $\{A\}$ in any space define the 'symmetric difference' of A and B to be

$$A \bigtriangleup B = (A - AB) + (B - AB)$$

$$= A + B - AB.$$

Prove

$$A \bigtriangleup B = B \bigtriangleup A,$$

$$A \bigtriangleup (B \bigtriangleup C) = (A \bigtriangleup B) \bigtriangleup C,$$

$$A(B \bigtriangleup C) = (AB) \bigtriangleup (AC).$$

If $\{A_n\}$ is a sequence of sets, and $D_1 = A_1$, $D_{n+1} = D_n \bigtriangleup A_n$, prove that $\lim D_n$ exists if and only if $\lim A_n = 0$.

Exercise 4.2. A space of points $\{x\}$ is a 'quasi'-metric space if there exists a function, $d(x, y)$, of all pairs of points such that

(1) $d(x, y) \geqslant 0$,
(2) $d(x, y) = d(y, x)$
(3) $d(x, y) \leqslant d(x, z) + d(z, y)$.

If $\{A\}$ is a field of sets on which there is defined a measure $M(A)$ prove that it can be made into a quasi-metric space by defining a distance between sets, $d(A, B)$, by

$$d(A, B) = M(A \bigtriangleup B)$$

where $\bigtriangleup$ is the symmetric difference operator used in Exercise 4.1.

Exercise 4.3. Using the definitions (4.3) and (4.4) prove formally that

$$\lim \inf A_i \subset \lim \sup A_i,$$

and that both limits are independent of the order of the A_i.

Exercise 4.4. If $f_n(x)$ converges to $f(x)$ almost everywhere, where x is a point in a set in which there is defined a σ-field and a measure, prove that the set of points where $f_n(x)$ does not converge to $f(x)$ is measurable.

Exercise 4.5. If $f_n(x)$ converges to $f(x)$ almost everywhere with respect to a bounded measure $P(E)$, and ε is any positive number, there exists a measurable set E such that $P(E) < \varepsilon$, and $f_n(x)$ converges to $f(x)$ uniformly outside E (Egorov's theorem).

5. Random Variables and Continuous Distributions

5.1. Distribution functions

IN THE previous chapter we have defined probability in terms of measure theory and thus shown how a random variable can be defined by setting up a probability measure on its range of variation. A great deal of probability theory can, however, be developed without using the general theories in the last chapter and we therefore start again here by defining distribution functions and Riemann–Stieltjes integration. We shall refer to the results of the last chapter only at a few particular points and most of the rest of the book can therefore be read independently.

A distribution function, $F(x)$, is defined to be a non-decreasing function on the range $(-\infty, \infty)$ such that $0 \leqslant F(x) \leqslant 1$. $F(x)$ is to be interpreted as the probability that a random variable, X, is less than or equal to the value x, so that $F(x)$ is continuous to the right. We could have defined it as continuous to the left but this is a slightly less convenient convention. We describe such a distribution as being 'continuous' because in the general definition we do not necessarily impose any restriction of the random variable to a discrete set of values, and thus the possible values of the variable are, in general, a continuous set. However, the definition also includes all discrete random variables such as those considered in previous chapters.

The distribution is said to be 'proper' if

$$\lim_{x \to \infty} F(x) = 1, \tag{5.1}$$

and

$$\lim_{x \to -\infty} F(x) = 0. \tag{5.2}$$

This implies that there is no concentration of probability at $\pm \infty$. If either of these conditions is not satisfied the distribution is said to be 'improper'. We usually take distributions to be proper unless otherwise stated.

We define the 'graph' of $F(x)$ to be the continuous curve defined by the set of points which consists of all points $\{x, F(x)\}$ together with all the vertical segments joining the pairs of points $\{x, F(x-0)\}$, $\{x, F(x)\}$, where x is a point of discontinuity.

In a similar manner we can define a distribution function for a finite set of random variables $(X_1,..., X_k)$ by a non-decreasing function $F(x_1,..., x_k)$ which we interpret as the probability that $(X_1 \leqslant x_1,..., X_k \leqslant x_k)$. We then have

$$0 \leqslant F(x_1,..., x_k) \leqslant 1, \tag{5.3}$$

but this, combined with the fact that F is non-decreasing in each x_i is, as we have seen in Chapter 4, not now sufficient. Denote by $\triangle_i$ the operation of differencing F with respect to x_i over an interval $(x_i, x_i + \delta_i)$ where $\delta_i > 0$. Thus

$$\triangle_i F(x_1,..., x_k) = F(x_1,..., x_i + \delta_i,..., x_k) - F(x_1,..., x_k). \tag{5.4}$$

Then

$$\triangle_1 \triangle_2 ... \triangle_k F(x_1,..., x_k) \geqslant 0 \tag{5.5}$$

will be the probability that

$$(x_1 < X \leqslant x_1 + \delta_1,..., x_k < X_k \leqslant x_k + \delta_k),$$

and this must be non-negative. Then any function satisfying (5.3) and (5.5) which is non-decreasing in each x_i will serve as a distribution function for k random variables. Such a distribution is said to be proper if

$$\lim_{\text{all } x_i \to \infty} F(x_1,..., x_k) = 1, \tag{5.6}$$

and

$$\lim_{\text{any } x_i \to -\infty} F(x_1,..., x_k) = 0. \tag{5.7}$$

Letting some of the x_i tend to infinity we obtain the marginal distribution function for the random variables corresponding to the remaining suffices. We say that the random variables $X_1,..., X_k$ are jointly 'independent' if

$$F(x_1,..., x_k) = F_1(x_1)...F_k(x_k), \tag{5.8}$$

where

$$F_i(x_i) = \lim F(x_1,..., x_k), \tag{5.9}$$

as all the x's except x_i tend to infinity. Similarly any set $X_i,..., X_s$, $(s < k)$ are independent if their joint distribution, obtained by letting $x_{s+1},..., x_k$ tend to infinity, satisfies a similar relation. As before, if $k > 2$, it is easy to see that independence in pairs does not imply independence.

In Chapter 4 we have seen that the definition of a distribution function is equivalent to the setting up of a probability measure on the σ-ring of all Borel sets on the line (or plane), since the increment of the distribution function over an interval gives us the measure to be ascribed to this set when considered as open to the left and closed to the right. From this we can obtain the measure of all open and all closed sets, and hence of the smallest σ-ring containing them. If the distribution is proper, the measure is then 'normed'.

We say that a point x is a point of increase of $F(x)$ if we cannot find any open interval containing x in its interior in which $F(x)$ is constant. This is equivalent to saying that for every $\varepsilon > 0$, $F(x+\varepsilon) - F(x-\varepsilon) > 0$. The set of all points of increase of a distribution is known as its spectrum and is clearly a closed set.

Distribution functions are used for other purposes than the description of probability distribution and may be useful for the specification of the manner in which any finite quantity or mass is distributed over a spatial range. Thus in the theory of stationary stochastic processes the 'energy spectrum' is determined by a distribution function. They can also be used to specify the whole set of values taken by a set of quantities. If these quantities are $x_1, ..., x_n$ we may write

$$F_n(x) = n^{-1} \{\text{number of } x\text{'s } \leqslant x\}. \tag{5.10}$$

In statistics this is a convenient way of describing a sample of n values produced in some probabilistic manner. If in fact the n quantities are n actual values taken by n independent random variables having the same distribution, $F(x)$, we say that $F_n(x)$ is a sample distribution. The central problem of statistical theory is to devise means of drawing inferences about the nature of $F(x)$ when $F_n(x)$ alone is known. We return to this problem later (§ 5.6). Distributions of the form $F_n(x)$ are also useful in providing a description of all the real roots of an algebraic equation. Note that if we want to describe all the roots, whether real or complex, we have to use a corresponding distribution in two dimensions, i.e. in the complex plane.

Since $F(x)$ is a non-decreasing function the limits

$$F(x+0) = \lim_{\delta \to 0} F(x+\delta), \tag{5.11}$$

$$F(x-0) = \lim_{\delta \to 0} F(x-\delta) \tag{5.12}$$

always exist and if they differ, there is a positive concentration of probability at the value x. The number of such simple discontinuities can be at most enumerable, for consider those discontinuities where the jump is not less than N^{-1}, when N is a positive integer. The number of such jumps can be at most N and is therefore finite. Considering a sequence of values of N tending to infinity we see that all such discontinuities can be enumerated. Define $F_1(x)$ to be the sum of all jumps for values of the variable less than or equal to x. Then $F_1(x)$ is a non-decreasing function with at most an enumerable number of discontinuities and $F(x) - F_1(x)$ is a non-decreasing function which is continuous. We can therefore write

$$F(x) = F_1(x) + F_2(x). \tag{5.13}$$

By a more elaborate analysis (Loève 1963, p. 130) it can be shown that we can write

$$F_2(x) = F_{2c}(x) + F_{2s}(x), \tag{5.14}$$

where $F_{2c}(x)$ and $F_{2s}(x)$ have derivatives almost everywhere (i.e. except possibly in a set of Lebesgue measure zero) but $F_{2c}(x)$ is the integral of its derivative,

$$F_{2c}(x) = \int_{-\infty}^{x} F_{2c}'(t)\, dt, \qquad (5.15)$$

whereas $F_{2s}'(t) = 0$ almost everywhere, although $F_{2s}(x)$ is not constant. The representation

$$F(x) = F_1(x) + F_{2c}(x) + F_{2s}(x) \qquad (5.16)$$

is known as the 'Lebesgue decomposition'.

It is also possible to prove a Lebesgue decomposition theorem for distributions in n-dimensional Euclidean space but the structure of the singular component is then very complex.

The continuous component is defined in such a way that it can be represented as the Lebesgue integral of an integrable non-negative density function. We can thus write it as

$$F_{2c}(x_1,\ldots,x_n) = \int_{-\infty}^{x_1} \cdots \int_{-\infty}^{x_n} f(t_1,\ldots,t_n)\, dt_1 \ldots dt_n.$$

In nearly all particular problems of probability we have to deal either with discrete distributions for which only $F_1(x)$ is non-zero, or distributions with a probability density function $F_{2c}'(x)$ for which $F_1(x)$ and $F_{2s}(x)$ are zero, or occasionally with a sum of these two. Distributions of the form $F_{2s}(x)$ usually occur only in constructing counter-examples or in other purely theoretical problems.

5.2. The Stieltjes integral

We have already defined and used Lebesgue–Stieltjes integrals in the previous chapter. For the sake of those who wish to skip the latter it is desirable to provide a short summary of the main properties of Stieltjes integrals and in doing this we consider only the Riemann–Stieltjes integral which is sufficient for most purposes. A more detailed account will be found in Graves (1946) and in Widder (1946).

Consider the integration of a function $f(x)$ with respect to a function $g(x)$ on the finite closed interval $[a, b]$. Let $x_1,\ldots,x_n$ be numbers in this interval such that

$$a = x_1 < x_2 < \ldots < x_{n-1} < x_n = b,$$

and

$$d = \max(x_{i+1} - x_i)$$

Write

$$S = \sum_{i=1}^{n-1} f(z_i)\{g(x_{i+1}) - g(x_i)\}. \qquad (5.17)$$

where each z_i lies in $[x_i, x_{i+1}]$ but is otherwise arbitrary. Then if the limit of

S exists as $d \to 0$, independently of the choice of the x_i and z_i subject to the restriction that $(x_{i+1} - x_i) \leqslant d$, $f(x)$ is said to be Stieltjes integrable with respect to $g(x)$, and the limit is written as

$$\int_a^b f(x) \, dg(x). \tag{5.18}$$

It follows from the linearity of (5.17) used in the above definition that (5.18) is linear in $f(x)$ for given $g(x)$, and linear in $g(x)$ for given $f(x)$. This implies that, provided the right-hand sides exist, the equations

$$\int_a^b \{f_1(x) + f_2(x)\} \, dg(x) = \int_a^b f_1(x) \, dg(x) + \int_a^b f_2(x) \, dg(x), \tag{5.19}$$

$$\int_a^b f(x) \, d\{g_1(x) + g_2(x)\} = \int_a^b f(x) \, dg_1(x) + \int_a^b f(x) \, dg_2(x), \tag{5.20}$$

and

$$\int_a^b \alpha f(x) \, d\{\beta g(x)\} = \alpha\beta \int_a^b f(x) \, dg(x) \tag{5.21}$$

hold, and the left-hand sides must exist.

From the definition as the limit of (5.17) we see that we get into trouble if $f(x)$ and $g(x)$ have discontinuities at the same points. For this and other reasons quite severe restrictions have to be imposed on $f(x)$ and $g(x)$ in order to make the integrals exist.

We also have, if $a < b < c$

$$\int_a^b f(x) \, dg(x) + \int_b^c f(x) \, dg(x) = \int_a^c f(x) \, dg(x) \tag{5.22}$$

provided both sides exist.

THEOREM 5.1. *If*

$$\int_a^b f(x) \, dg(x)$$

exists, then so does

$$\int_a^b g(x) \, df(x) \tag{5.23}$$

and we have the formula for integrating by parts,

$$\int_a^b f(x) \, dg(x) + \int_a^b g(x) \, df(x) = f(b) \, g(b) - f(a) \, g(a). \tag{5.24}$$

Proof. Consider the sum (5.17) and, with the same subdivision, define

$$S_1 = \sum_{i=1}^{n-1} \{f(z_i) - f(z_{i+1})\} g(x_{i+1})$$

$$= S + f(z_1) g(x_1) - f(z_n) g(x_n).$$

15

Proceeding to the limit and using the fact that the limit of S exists for any choice of the z_i inside the closed intervals x_i, x_{i+1} (and therefore for $z_1 = x_1$, $z_n = x_n$) it follows that (5.23) exists and (5.24) holds.

To establish convenient conditions for the existence of a Riemann–Stieltjes integral we must introduce the idea of a function of bounded variation. Suppose $f(x)$ is defined on the closed interval $[a, b]$. We partition this by points x_i $(i = 1, \ldots, n)$ so

$$a = x_1 < x_2 < \ldots < x_n = b.$$

Write

$$V_1 = \sum_{i=1}^{n-1} |f(x_{i+1}) - f(x_i)|, \tag{5.25}$$

$$V_2 = \sum{}'\{f(x_{i+1}) - f(x_i)\}, \tag{5.26}$$

$$V_3 = -\sum{}''\{f(x_{i+1}) - f(x_i)\}, \tag{5.27}$$

where $\sum'$, $\sum''$ denotes summations over all values of i for which the individual differences are positive and negative respectively. Let W_1, W_2, W_3 be the least upper bounds of V_1, V_2, V_3, for all such partitions. Then

$$W_1 = W_2 + W_3. \tag{5.28}$$

W_1 is said to be the total variation of $f(x)$ over $[a, b]$. When W_1 is finite, $f(x)$ is said to be of bounded variation on $[a, b]$. W_1, W_2, and W_3 are functions of the interval $[a, b]$, and are clearly additive functions of such intervals in the sense that if $a < b < c$ and W_1, W_1', and W_1'' correspond to the closed intervals $[a, c]$, $[a, b]$, and $[b, c]$, then $W_1 = W_1' + W_1''$, and similarly for W_2, W_3. Since

$$f(b) - f(a) = W_2 - W_3, \tag{5.29}$$

any function of bounded variation can be written as the difference of two non-decreasing functions. It therefore has at most a denumerable number of discontinuities. These results clearly hold also when the interval is infinite.

THEOREM 5.2. *If $f(x)$ is continuous and $g(x)$ is of bounded variation both the integrals in (5.24) exist for any finite closed interval for which these conditions hold.*

Proof. Let $f(x)$ be continuous in the closed interval $[a, b]$, and $g(x)$ of bounded variation over it. Since a function of bounded variation can be represented as the difference of two non-decreasing functions it is sufficient to prove the result when $g(x)$ is non-decreasing. By adding and subtracting x we can suppose $g(x)$ is strictly increasing.

First suppose $g(x)$ is continuous. If we change the variable to $y = g(x)$, y increases from $g(a)$ to $g(b)$ when x increases from a to b. The inverse function, $x = G(y)$, is continuous and increasing, and $F(y) = f\{G(y)\}$ is a continuous function. The sum, S, is then of the form

$$\sum_{i=1}^{n-1} F(\eta_i)(y_{i+1} - y_i), \quad y_i \leqslant \eta_i \leqslant y_{i+1}.$$

Since $g(x)$ is continuous it is uniformly continuous, and $\max(y_{i+1}-y_i)$ will tend to zero with $d = \max(x_{i+1}-x_i)$. Since $F(y)$ is continuous the integral exists by the ordinary theory of the Riemann integral.

Now suppose $g(x)$ is not necessarily continuous except at a and b, and consider the sum

$$S_1 = \sum_{i=1}^{n-1} f(z_i)\{g(x_{i+1}+0)-g(x_i+0)\} \quad (g(x_n) = g(x_n+0)),$$

where $g(x+0)$ is the limit of $g(x)$ on the right. The difference $S-S_1$, where S is given by (5.17), can be written

$$f(z_1)\{g(a+0)-g(a)\}+ \sum_{i=1}^{n-1} \{f(z_{i+1})-f(z_i)\}\{g(x_{i+1}+0)-g(x_{i+1})\}. \quad (5.30)$$

$f(z)$ is continuous and therefore uniformly continuous on $[a, b]$. The sum in (5.30) can therefore be made less than $\varepsilon > 0$ by making d small. It is therefore sufficient to show that S_1 has a limit when $d \to 0$, independent of the mode of division. Let $G(x)$ be the sum of all the jumps of $g(x)$ in the closed interval $[a, x]$. Then $g(x)-G(x)$ is continuous and non-decreasing. Write

$$k(x) = g(x)-G(x).$$

We can then put

$$g(x+0) = G(x)+(k(x)+x)-x.$$

$k(x)+x$ and x are increasing continuous functions and the integrals with respect to them must exist. It is therefore sufficient to consider

$$S_2 = \sum_{i=0}^{n-1} f(z_i)\{G(x_{i+1})-G(x_i)\}. \quad (5.31)$$

Let $x_1', x_2',...$ be the points of discontinuity at which the jumps are $s_1, s_2,...$, these being arranged so that $s_1 \geqslant s_2 \geqslant s_3 \geqslant ....$. We can then write S_2 as

$$\sum_{j=1}^{\infty} s_j f(z_i),$$

where i depends on j, and z_i is such that $|z_i-x_j'| \leqslant d$. Since $f(x)$ is uniformly continuous, S_2 has the limit

$$\sum_{j=1}^{\infty} s_j f(x_j')$$

and hence the integral exists.

It is in fact also possible to prove that the integral exists when $f(x)$ and $g(x)$ are of bounded variation but not necessarily continuous provided their discontinuities do not occur at the same points. This result, which we need later, is proved in Widder (1946), p. 25.

5.3. Examples of distributions

The value of a general theory of distribution functions is illustrated by the fact that we can now deal with discrete distributions such as occur in Chapters 1 and 2, and distributions with a probability density (i.e. for which $F'(x)$ exists almost everywhere and $F(x)$ is its integral) at one and the same time. For example, the binomial distribution whose generating function is $(q+pz)^n$ has a distribution function whose points of increase are $0, 1,..., n$, at which the jumps in $F(x)$ are equal to

$$q^n, \binom{n}{1}q^{n-1}p,..., \binom{n}{n-1}qp^{n-1}, p^n, \quad \text{respectively.}$$

As an example of a continuous distribution consider the normal distribution whose probability density is

$$\frac{1}{\sqrt{(2\pi)}}\exp\left(-\tfrac{1}{2}x^2\right), \tag{5.32}$$

and for which the cumulative distribution is

$$F(x) = \frac{1}{\sqrt{(2\pi)}}\int_{-\infty}^{x}\exp\left(-\tfrac{1}{2}t^2\right)dt. \tag{5.33}$$

That $F(x)$ tends to unity as x tends to infinity is most easily seen by considering

$$\left\{\lim_{x\to\infty}F(x)\right\}^2 = \frac{1}{2\pi}\left\{\int_{-\infty}^{\infty}\exp\left(-\tfrac{1}{2}t^2\right)dt\right\}^2,$$

$$= \frac{1}{2\pi}\int_{-\infty}^{\infty}\int_{-\infty}^{\infty}\exp\left\{-\tfrac{1}{2}(t^2+u^2)\right\}dt\,du,$$

$$= \frac{1}{2\pi}\int_{0}^{2\pi}d\theta\int_{0}^{\infty}\exp\left(-\tfrac{1}{2}r^2\right)r\,dr = 1.$$

As an example of a multivariate distribution we now consider the multivariate normal distribution which is of frequent occurrence. We look for a multivariate distribution with a probability density proportional to $\exp(-Q)$ where Q is a quadratic form in the n variables $x_1,..., x_n$. If Q is not non-negative definite there will be a region of infinite volume in which $\exp -Q \geqslant 1$ and it will be impossible to define a normed probability measure. We must therefore insist that Q is non-negative definite and we consider first the case where it is positive definite.

It is convenient to use vector notations. Let $\mathbf{A}$ be a symmetric $n \times n$ matrix and $\mathbf{x}$ a column matrix whose transpose is $\mathbf{x}' = (x_1,..., x_n)$. Then the quadratic form Q can be written $-\tfrac{1}{2}\mathbf{x}'\mathbf{A}\mathbf{x}$, the factor $\tfrac{1}{2}$ being introduced for simplicity. If $\mathbf{A}$ is positive definite there exists an orthogonal transformation $\mathbf{T}$ such that if we write $\mathbf{x} = \mathbf{T}\mathbf{y}$ (and therefore $\mathbf{x}' = \mathbf{y}'\mathbf{T}'$) the quadratic form becomes $\mathbf{y}'\mathbf{T}'\mathbf{A}\mathbf{T}\mathbf{y}$, where $\mathbf{T}'\mathbf{A}\mathbf{T}$ is a diagonal matrix whose diagonal consists of positive terms which we write as $\sigma_1^{-2},\dots,\sigma_n^{-2}$. The Jacobian of the transformation is the determinant of $\mathbf{T}$ and has modulus unity since $\mathbf{T}$ is orthogonal. It can therefore be taken as unity and we have

$$\int_{-\infty}^{\infty}\dots\int_{-\infty}^{\infty}\exp\{-(\tfrac{1}{2}\mathbf{x}'\mathbf{A}\mathbf{x})\,dx_1\dots\}\,dx_n = \int_{-\infty}^{\infty}\dots\int_{-\infty}^{\infty}\exp\left\{-\left(\frac{1}{2}\sum_1^n\frac{y_i^2}{\sigma_i^2}\right)\right\}\,dy_1\dots dy_n$$

$$= (2\pi)^{n/2}\sigma_1\dots\sigma_n = (2\pi)^{n/2}|\mathbf{A}|^{-\frac{1}{2}}. \quad (5.34)$$

Thus the normalized distribution has a density function given by

$$f(x_1,\dots,x_n) = \frac{|\mathbf{A}|^{\frac{1}{2}}}{(2\pi)^{n/2}}\exp\left(-\tfrac{1}{2}\mathbf{x}'\mathbf{A}\mathbf{x}\right). \quad (5.35)$$

Now suppose that $\mathbf{A}$ is non-negative definite but not positive definite. This implies that there is an orthogonal matrix, $\mathbf{T}$, such that the transformation $\mathbf{x} = \mathbf{T}\mathbf{y}$ reduces $\mathbf{x}'\mathbf{A}\mathbf{x}$ to a sum of k squares, where $k < n$, so that

$$\mathbf{x}'\mathbf{A}\mathbf{x} = \mathbf{y}'\mathbf{T}'\mathbf{A}\mathbf{T}\mathbf{y} = \sum_1^k y_i^2\sigma_i^{-2}. \quad (5.36)$$

Since an orthogonal transformation has a unit Jacobian we would have to have a probability density proportional to $\exp(-\tfrac{1}{2}\sum_1^k y_i^2\sigma_i^{-2})$ and this is impossible since integration over the whole space would result in an infinite value for the integral. To interpret the distribution we must therefore suppose that all the probability is concentrated in the subspace defined by $y_{k+1} = y_{k+2} = \dots = y_n = 0$, so that we have a singular distribution in the whole space.

Whether singular or not, the distribution is symmetrical (but not in general spherically symmetrical) about the point $(0,\dots,0)$ which corresponds to the set of means of the quantities $x_1,\dots,x_n$ and can be shifted to any other point $\mathbf{a} = (a_1,\dots,a_n)$ by taking the probability density proportional to

$$\exp\{-\tfrac{1}{2}(\mathbf{x}-\mathbf{a})'\mathbf{A}(\mathbf{x}-\mathbf{a})\}. \quad (5.37)$$

The second-order moments are evaluated in §5.12.

5.4. Conditional distributions

If we have a multivariate distribution which is purely discrete so that the probability density is concentrated at an enumerable number of points, it is easy to define the conditional distribution of some of the random variables when all the others are given specified values, by the methods of Chapter 1. When the distribution is not purely discrete the problem becomes much more complicated and in the completely general case requires some very elaborate measure theory. As such a situation is not often met with, we confine ourselves here to consideration of the case where the probability distribution is given by a probability density and refer the reader for the more general case to Doob (1953), Kolmogorov (1933), and Loève (1963).

Consider for simplicity a two-variate case in which

$$\mathrm{pr}(X_1 \leqslant x_1,\ X_2 \leqslant x_2) = F(x_1, x_2)$$

$$= \int_{-\infty}^{x_1} \int_{-\infty}^{x_2} f(t_1, t_2)\, dt_1\, dt_2. \tag{5.38}$$

This implies that $f(t_1, t_2)$ is integrable in the sense of Lebesgue in the whole two-dimensional plane. To obtain a definition of conditional probability we use the following theorem which is due to Fubini (Saks 1937, p. 76).

THEOREM 5.3. *If $f(x, y)$ is non-negative and Lebesgue integrable in the whole plane $-\infty < x < \infty$, $-\infty < y < \infty$, then*

(1) *$f(x, y)$ is Lebesgue measurable in x for almost all y, i.e. except possibly for a set of values of y of measure zero, and is Lebesgue measurable in y for almost all x.*

(2) *The integrals*

$$\int_{-\infty}^{\infty} f(x, y)\, dy, \quad \int_{-\infty}^{\infty} f(x, y)\, dx$$

exist, and are finite for almost all x and y respectively, and their values, considered as functions of x and y, are Lebesgue integrable.

(3)
$$\int_{-\infty}^{\infty} \int_{-\infty}^{\infty} f(x, y)\, dx\, dy = \int_{-\infty}^{\infty} \left\{ \int_{-\infty}^{\infty} f(x, y)\, dx \right\} dy$$

$$= \int_{-\infty}^{\infty} \left\{ \int_{-\infty}^{\infty} f(x, y)\, dy \right\} dx. \tag{5.39}$$

A proof of this theorem will be found in the above reference. It is important to notice that the assumption of the theorem is that $f(x, y)$ is jointly integrable with respect to x and y, for it can be shown that the assumption of the existence of the repeated integrals does not, of itself, imply the existence of the double integral. Similar theorems can be established for integration in more than two dimensions.

Suppose now that we have a probability distribution given by (5.38). We define the conditional probability distribution of X_1, given X_2, by

$$\mathrm{pr}(X_1 < x_1 \,|\, X_2 = x_2) = \frac{\displaystyle \int_{-\infty}^{x_1} f(t_1, x_2)\, dt_1}{\displaystyle \int_{-\infty}^{\infty} f(t_1, x_2)\, dt_1}$$

$$= G(x_1 \,|\, x_2), \quad \text{say}, \tag{5.40}$$

for all values of x_2 such that the denominator is finite and non-zero. When the denominator is zero we define $G(x_1 \,|\, x_2)$ to be zero. Then since the integral with respect to x_2 exists and is finite for all x_2 with the possible exception of a set of measure zero, the conditional probability is defined almost everywhere and

$$\text{pr}(X_1 < x_1 | X_2 < x_2) = F(x_1, x_2)/F(\infty, x_2) \tag{5.41}$$

whenever the denominator is non-zero, and is zero otherwise. In a similar way we can define integrals of suitably regular functions with respect to conditional distributions.

Returning now to the example of the multivariate normal distribution in n variates $X_1, \ldots, X_n$ as given by (5.35) we see that the conditional distribution of k of these, $X_1, \ldots, X_k$, say, given $X_{k+1}, \ldots, X_n$, is again a multivariate normal distribution whose means are linear functions of $X_{k+1}, \ldots, X_n$.

If $f(X, Y, \ldots)$ is a function of several random variables which have a joint probability distribution, $F(x, y, \ldots)$, we define its expectation by

$$E\{f(X, Y, \ldots)\} = \int \int \ldots \int f(x, y, \ldots) \, dF(x, y, \ldots), \tag{5.42}$$

whenever this integral exists absolutely, i.e. whenever the corresponding integral of $|f(x, y, \ldots)|$ is finite. It is then easy to prove the following important theorems.

THEOREM 5.4. *Whatever the joint distribution*

$$E\{f(X) + g(Y)\} = E\{f(X)\} + E\{g(Y)\}, \tag{5.43}$$

whenever these expectations exist.

THEOREM 5.5. *If X and Y have a joint distribution which is such that X and Y are independent,*

$$E\{f(X)g(Y)\} = E\{f(X)\}E\{g(Y)\}, \tag{5.44}$$

whenever these expectations exist. If X and Y are not independent, this may or may not be true.

5.5. Sets of distribution functions

It is convenient to regard the set of all distribution functions as a topological space, i.e. a set of objects amongst which we can define some concept of 'nearness' or convergence. One way of doing this is to introduce a metric, i.e. a distance, between pairs of the elements, or 'points', of the space. We say that a set is a metric space if there exists a numerical function, $d(x, y) = d(y, x)$, of pairs of points x, y such that for any points x, y, z we have

(1) $d(x, y) = 0$ implies $x = y$, and vice versa;

(2) $d(x, y) + d(y, z) \geqslant d(x, z)$. \hfill (5.45)

The second of these conditions is the 'triangle inequality' and implies that $d(x, y) \geqslant 0$, for all x, y.

Given the set of all distribution functions we wish to introduce such a function in a useful way. We might, for example, take

$$d(F(x), G(x)) = \max_{x} |F(x) - G(x)|.$$

This would not be a natural definition in probability theory for consider the case where $F(x)$ has a unit jump at $x = 0$. Then the distance between $F(x)$ and $F(x+\delta)$ $(\delta > 0)$ is always unity but as δ tends to zero $F(x+\delta)$ must tend to $F(x)$ in any sense that would be useful in probability theory.

A satisfactory definition of the distance was introduced by Lévy. Let $F(x)$ and $G(x)$ be distribution functions. Since they are non-decreasing their graphs, $y = F(x)$, $y = G(x)$, intersect any line $x+y = a$ $(-\infty < a < \infty)$ in unique points which always exist because the graphs include the vertical segments at points of discontinuity x_0 running from $F(x_0-0)$ to $F(x_0+0)$. Let $d(F, G)$ be the least upper bound of the distance between these two points for all values of a. Clearly (1) is satisfied. If $F(x)$, $G(x)$, and $H(x)$ are distributions define $d(F, G, a)$ to be the distance between the intersections of the graphs of $F(x)$ and $G(x)$ on the line $x+y = a$. Then

$$d(F, G, a) + d(G, H, a) \geqslant d(F, H, a),$$

and taking least upper bounds we get

$$d(F, G) + d(G, H) \geqslant d(F, H),$$

so that (2) is satisfied and $d(F, G)$ is a distance.

Using this definition we can define the convergence of a sequence of distributions, $F_n(x)$ $(n = 1, 2, ...)$ to a distribution $F(x)$, by the condition that $d(F_n, F)$ tends to zero. This definition satisfies the Cauchy property that mutual convergence of a sequence implies the existence of a distribution function to which the sequence converges and vice versa. The latter being obvious suppose the sequence $\{F_n\}$ is mutually convergent, i.e. given $\varepsilon > 0$ there exists n_0 such that for all $m, n > n_0$, $d(F_m, F_n) < \varepsilon$. Consider the intersections of the sequence $\{F_n(x)\}$ with a line $x+y = a$. These are mutually convergent and hence converge to a unique point. The set of such points obtained by varying a must then, by an obvious argument, define a non-decreasing function $F(x)$. To prove this is a proper distribution we have to show that $F(\infty) = 1$, $F(-\infty) = 0$. If this were not so suppose that $F(\infty) = 1 - \varepsilon < 1$. Then given any n we could, since $F_n(\infty) = 1$, choose a and $N > n$ both large so that $d(F_n, F_N)$ is greater than $\tfrac{1}{2}\varepsilon$ and this contradicts the assumption that $d(F_n, F_N)$ converges to zero. Thus the Cauchy property is true, and a metric space for which this is so is said to be 'complete'.

It is also possible to define the convergence of a sequence of distribution functions without introducing a metric into the space. We shall say that such a sequence converges 'weakly' to a function $F(x)$ if $F_n(x)$ converges to $F(x)$ in the ordinary sense at all continuity points of $F(x)$. This definition does not assume that $F(x)$ is itself a distribution function, and that this is not necessarily true can be seen from the example of a sequence of normal distributions with zero means and standard deviations equal to n. Then for all fixed x, $F_n(x)$ tends to $\frac{1}{2}$ so that there is no proper distribution to which $F_n(x)$ converges.

We say that a sequence $\{F_n(x)\}$ converges 'completely' to a function $F(x)$ if it converges weakly to $F(x)$, and if $F(x)$ is a proper distribution, i.e. $F(-\infty) = 0$, $F(\infty) = 1$. Notice that in both these definitions the convergence is defined in terms of a function to which the sequence converges and the Cauchy property does not arise.

THEOREM 5.6. *Complete convergence is equivalent to convergence in the Lévy metric.*

Proof. Suppose first that $d(F_n, F)$ tends to zero. We wish to prove $F_n(x) \to F(x)$ whenever x is a continuity point of $F(x)$. For any point x we have

$$F_n(x) \leqslant 2^{-\frac{1}{2}}d(F_n, F) + F\{x + 2^{-\frac{1}{2}}d(F_n, F)\},$$

and if x is a continuity point x_0,

$$F_n(x_0) - F(x_0) \leqslant 2^{-\frac{1}{2}}d(F_n, F) + F\{x_0 + 2^{-\frac{1}{2}}d(F_n, F)\} - F(x_0)$$

implies that as $d(F_n, F) \to 0$,

$$\overline{\lim_{n}}\{F_n(x_0) - F(x_0)\} \leqslant 0.$$

Similarly

$$\underline{\lim_{n}}\{F_n(x_0) - F(x_0)\} \geqslant 0,$$

and the result is proved.

Now suppose $F_n(x)$ tends to $F(x)$ at every continuity point of $F(x)$. We wish to show that $d(F_n, F)$ tends to zero. Given $\eta > 0$, choose continuity points a, b such that $F_k(a) < \eta$, $F(b) > 1 - \eta$, and then a set of values $x_1, ..., x_n$ such that

$$a = x_1 < x_2 < ... < x_n = b,$$

and so that $x_2 - x_1$, $x_3 - x_2, ..., x_n - x_{n-1}$, are all less than η, and all the x_i are continuity points. Since there are only a finite number of these we can find n_0 such that $|F_n(x_i) - F(x_i)| < \eta$ for all $n > n_0$, and every x_i. Then

$$d(F_n, F) < 2^{\frac{1}{2}}\eta.$$

Since η is arbitrary, $d(F_n, F)$ must tend to zero.

We can therefore make the set of all distribution functions into a metric space (and thus into a topological space) by using the Lévy metric, or by using the above definition of convergence into a topological space without a metric but with the same topological structure. Further results on the convergence of distributions are given in the next chapter.

5.6. The estimation of a distribution

Particular subspaces of the space of all distributions are those determined by distributions which contain one or more parameters. Thus we might have a distribution $F(x)$ depending on k parameters $\theta_1,\ldots,\theta_k$. The central problem of statistics, with which we are not concerned in this book, is to draw inferences about the set of quantities $(\theta_1,\ldots,\theta_k)$ from a 'sample' of values $(X_1,\ldots,X_n)$, where the X_i are independently distributed in this distribution. In particular in order to be able to estimate all the θ_i it is necessary that they should be 'identifiable', a simple concept which has given rise to a good deal of confusion. Suppose that the values $(\theta_1,\ldots,\theta_k)$ are confined to a set θ in the k dimensional parameter space. Then we say that $\theta_1,\ldots,\theta_k$ are individually identifiable if there are no two distinct points in θ to which correspond the same distribution. Thus in a normal distribution with mean m_1+m_2 and variance σ^2, the three parameters m_1, m_2, σ^2 are not all identifiable, but the two parameters m_1+m_2 and σ^2 are identifiable.

In this book we shall not be concerned with the problem of estimating parameters, but we shall consider the simplest aspects of the related problem of estimating a complete distribution of unknown form from a sample $(X_1,\ldots,X_n)$ of independent variates.

From this sample we construct the 'sample distribution' $F_n(x)$, which we define to be equal to n^{-1} (number of $X_i \leqslant x$). This is a step function with n steps and as n increases we may expect it to become closer and closer to $F(x)$. If we measure the difference by $d(F_n, F)$ we see that $d(F_n, F)$ is itself a random variable. Its distribution is complicated and will not be considered in this book. A survey of the very elaborate literature dealing with this problem is given by Darling (1957).

Suppose, however, that we merely wish to estimate $F(x)$ at a particular point x so that the natural estimator to use is $F_n(x)$. It is clear that $nF_n(x)$ is distributed in a binomial distribution with probability $F(x)$ and index n. Since $E\{F_n(x)\} = F(x)$ it is said to be unbiased and its variance is $n^{-1}F(x)\{1-F(x)\}$, which tends to zero as n tends to infinity. Similarly we may wish to estimate the probability $F(b)-F(a)$ of X lying in the half-open interval $a < x \leqslant b$ and the natural estimator is $F_n(b)-F_n(a)$. This is again unbiased and has variance

$$n^{-1}\{F(b)-F(a)\}\{1-F(b)+F(a)\}. \tag{5.46}$$

If we let the length, $b-a$, of the interval tend to zero, this variance tends to zero but so also does the quantity estimated and the relative error increases. It is illuminating to consider the particular case where $F(x)$ is the integral of its derivative so that we can write

$$F(x) = \int_{-\infty}^{x} f(t)\,dt.$$

Suppose that we wish to estimate $f(t)$ at a particular point $t = x$ where its first three derivatives exist. We choose a small number $a > 0$ and consider

$$G_n(x) = a^{-1}\{F_n(x + \tfrac{1}{2}a) - F_n(x - \tfrac{1}{2}a)\}. \tag{5.47}$$

Then $aG_n(x)$ is again a binomial variate and

$$E\{G_n(x)\} = a^{-1} \int_{x - \frac{1}{2}a}^{x + \frac{1}{2}a} f(t)\, dt, \tag{5.48}$$

$$\text{var}\{G_n(x)\} = \frac{1}{na^2} \int_{x - \frac{1}{2}a}^{x + \frac{1}{2}a} f(t)\, dt \left\{ 1 - \int_{x - \frac{1}{2}a}^{x + \frac{1}{2}a} f(u)\, du \right\}. \tag{5.49}$$

As a becomes small the variance is asymptotically equal to

$$(na)^{-1} f(x)$$

so that the smaller we make a the larger n will have to be to get the same accuracy.

On the other hand, if a is not small the estimator will be biased. In fact, expanding $f(u)$ in a Taylor series around $u = x$ we get

$$E(G_n) = a^{-1} \int_{x - \frac{1}{2}a}^{x + \frac{1}{2}a} f(u)\, du = f(x) + \tfrac{1}{24}a^2 f''(x) + 0(a^3), \tag{5.50}$$

so that the bias is of the order $\tfrac{1}{24}a^2 f''(x)$ which is not in general zero.

To obtain an accurate estimate we must therefore have a small and n large but also na large. In fact the best estimator could well be taken as that for which the sum of the variance and the square of the bias is minimized. This means using the value

$$a = \left[\frac{144 f(x)}{n\{f''(x)\}^2} \right]^{1/5}. \tag{5.51}$$

Letting n tend to infinity with a having this value we obtain a consistent estimator, i.e. one which converges in probability to the true value. In practice $f(x)/\{f''(x)\}^2$ is not known and would have to be guessed.

The above problem is similar to that occurring in estimating the spectral density of a stochastic process for which an uncertainty relation of this kind was given by Grenander (1951). (See also Whittle 1957, 1958; Rosenblatt 1956; Bartlett 1963; and Parzen 1962.) Although intriguing, the analogy with Heisenberg's uncertainty relation in physics is somewhat remote, but might be worth further exploration.

5.7. Distribution of sums

Suppose that X and Y are independent random variables and we wish to determine the probability distribution of their sum. By saying that they are independent we are implying that they have a joint distribution which can be written as

$$F(x, y) = F_1(x)F_2(y),$$

where $F_1(x)$ and $F_2(y)$ are the distributions of X and Y. Finding the distribution of $Z = X + Y$ means finding the integral

$$\int\int dF_1(x)\, dF_2(y)$$

over the part of the real plane where $x + y \leqslant z$.

THEOREM 5.7. *If $H(z)$ is the distribution function of Z we have, for almost all z,*

$$H(z) = \int\int_{x+y\leqslant z} dF_1(x)\, dF_2(y)$$

$$= \int_{-\infty}^{\infty} F_1(z-x)\, dF_2(x) = \int_{-\infty}^{\infty} F_2(z-x)\, dF_1(x), \qquad (5.52)$$

these integrals existing for all values of z with the possible exception of an enumerable set.

Proof. Since $F_1(x)$ and $F_2(x)$ are distribution functions they have only an enumerable set of points of discontinuity, and the set of values ξ which can be written $\xi = \zeta + \eta$, where $x = \zeta$ is a point of discontinuity of $F_1(x)$ and η a point of discontinuity of $F_2(x)$, is also enumerable. It follows that when z is not one of these points, $F_1(z-x)$ and $F_1(x)$, considered as functions of x, never have discontinuities at the same point. The above Riemann–Stieltjes integrals therefore exist. We have, therefore, only to prove that the double integral is equal to each of the single integrals.

We first observe that (5.52) is true when $F_1(x)$ and $F_2(x)$ are the integrals of their derivatives for we can then write

$$F_1(x) = \int_{-\infty}^{x} f_1(u)\, du, \quad F_2(y) = \int_{-\infty}^{y} f_2(v)\, dv,$$

and

$$\int\int_{x+y\leqslant z} dF_1(x)\, dF_2(y) = \int\int_{u+v\leqslant z} f_1(u)f_2(v)\, du\, dv$$

$$= \int_{-\infty}^{\infty} \left\{ \int_{-\infty}^{z-u} f_1(u)f_2(v)\, dv \right\} du$$

$$= \int_{-\infty}^{\infty} F_2(z-u)\, dF_1(u),$$

in the above notation.

For general $F_1(x)$, $F_2(x)$, write

$$G_1(x) = h^{-1} \int_0^h F_1(x+u)\,du,$$

$$G_2(y) = h^{-1} \int_0^h F_2(y+v)\,dv,$$

where h is small and positive. Then $G_1(x) \geqslant F_1(x)$, $G_2(x) \geqslant F_2(x)$, and $G_1(x)$, $G_2(x)$ are absolutely continuous distribution functions. Thus

$$\iint_{x+y \leqslant z} dF_1(x)\,dF_2(y) < \iint_{x+y \leqslant z} dG_1(x)\,dG_2(y), \tag{5.53}$$

and at any point where the left-hand side is a continuous function of z, the right-hand side can be made as near the left as we like by choosing h small. On the other hand, $G_1(x)$ and $G_2(x)$ are absolutely continuous functions so that the right-hand side of (5.53) is equal to

$$\int_{-\infty}^{\infty} G_1(z-u)\,dG_2(u) \geqslant \int_{-\infty}^{\infty} F_1(z-u)\,dG_2(u).$$

Since $G_2(u)$ is continuous we can integrate the right-hand side by parts to obtain

$$\int_{-\infty}^{\infty} G_2(z-u)\,dF_1(u) \geqslant \int_{-\infty}^{\infty} F_2(z-u)\,dF_1(u).$$

Thus

$$\int_{-\infty}^{\infty} F_2(z-u)\,dF_1(u) \leqslant \iint_{x+y < z} dF_1(x)\,dF_2(y) + \varepsilon, \tag{5.54}$$

where ε can be chosen arbitrarily small. We can similarly prove the reverse inequality by defining $G_1(x)$ to be

$$\frac{1}{h} \int_0^h F_1(x-u)\,du$$

and similarly for $G_2(x)$.

The function $H(z)$ is said to be the convolution of the distributions $F_1(x)$ and $F_2(x)$ and we write $H(z) = F_1(x) * F_2(x)$. The operation denoted by $*$ is associative and commutative, and thus has some of the properties of multiplication, but, as we shall see later, not all. It is also useful to define the nth convolution of a distribution $F(x) * F(x) \ldots * F(x)$, which we write as $F(x)^{n*}$. This is the distribution of the sum of n independent random variables all having the same distribution $F(x)$.

The device used above of proving general theorems about distributions by first smoothing them and then proceeding to the limit is a very useful one and has a simple probability interpretation. $G_1(x)$ is clearly the distribution of a random variable which is the sum of a random variable having the distribution $F_1(x)$, and an independent random variable having a distribution with a uniform probability density on the range $(0, h)$ or $(-h, 0)$.

5.8. Properties of convolutions

We expect that distributions which are convolutions will be at least as smooth and regular as their components. We shall see, however, that this is not always so if by 'smoothness' is meant 'analyticity'.

THEOREM 5.8. *If $F_1(x)$ (or $F_2(x)$) is continuous, so is $H(y)$.*

Proof. If $F_1(x)$ is continuous it is uniformly continuous for all x, for choosing any small $\varepsilon > 0$ we can find a, b such that $-\infty < a < b < \infty$ and $F_1(b) > 1-\varepsilon$, $F_1(a) < \varepsilon$. On the closed interval $[a, b]$, $F_1(x)$ is continuous and therefore there exists $\delta < b-a$ such that $|F_1(y)-F_1(x)| < \varepsilon$ whenever x lies in $[a, b]$ and $|y-x| < \delta$. Hence $|F(y)-F(x)| < 2\varepsilon$ for any values x, y on $(-\infty, \infty)$ such that $|y-x| < \delta$. Then

$$|H(y+\eta)-H(y)| \leqslant \int_{-\infty}^{\infty} |F_1(y+\eta-x)-F_1(y-x)| \, dF_2(x)$$

$$\leqslant 2\varepsilon, \tag{5.55}$$

whenever $|\eta| < \delta$.

THEOREM 5.9. *If $F_1(x)$ is absolutely continuous (and therefore the integral of its derivative, which exists almost everywhere) so is $H(y)$.*

Proof. That $F_1(x)$ is absolutely continuous means that given $\varepsilon > 0$ there exists $\delta > 0$ such that the sum of the increments of $F_1(x)$ over any set of non-overlapping intervals of total length less than δ is less than ε. The sum of any such set of increments of $H(z)$ is, however, the Stieltjes integral with respect to $F_2(x)$ of a corresponding sum for $F_1(x)$, and thus if $F_1(x)$ is absolutely continuous, so is $H(z)$.

THEOREM 5.10. *If $F_1(x)$ is such that its first n derivatives exist everywhere and*

$$|F_1^{(\nu)}(x)| < a_\nu, \quad \nu = 1,\dots, n \quad (-\infty < x < \infty)$$

then $H^{(\nu)}(z)$ exists and satisfies

$$|H^{(\nu)}(z)| < a_\nu, \tag{5.56}$$

for $\nu = 1,\dots, n-1$, and wherever $H^{(n)}(z)$ exists, it satisfies

$$|H^{(n)}(z)| < a_n.$$

Proof. Denote by $\triangle_\eta$ the operator which changes any function such as $H(z)$ into $H(z+\eta)-H(z)$. Then for $\nu = 1,\dots, n$,

$$\triangle_\eta^\nu H(z) = \int_{-\infty}^{\infty} \triangle_\eta^\nu F_1(z-x) \, dF_2(x)$$

$$= \int_{-\infty}^{\infty} \eta^\nu F_1^{(\nu)}\{z-x+\theta(z-x)\} \, dF_2(x),$$

by Taylor's theorem, where $0 \leqslant \theta(x) \leqslant \eta\nu$. Hence

$$|\triangle_\eta^\nu H(z)| \leqslant \eta^\nu a_\nu,$$

and wherever $H^{(n)}(z)$ exists it is not greater than a_n. If $\nu < n$, we also have

$$\triangle_\eta^\nu H(z) = \eta^\nu \int_{-\infty}^\infty F_1^{(\nu)}(z-x)\,dF_2(x)$$

$$+\eta^\nu \int_{-\infty}^\infty \theta(z-x)F_1^{(\nu+1)}\{z-x+\theta_1(z-x)\}\,dF_2(x),$$

where $0 \leqslant \theta_1(x) \leqslant \theta(x) \leqslant \eta\nu$, and the second integral is not greater than $\nu\eta^{\nu+1}a_{\nu+1}$, so that $H^{(\nu)}(z)$ exists and is bounded in modulus by a_ν.

Given these results it is natural to conjecture (and has in fact been stated in some works) that if $F_1(x)$ is analytic so also must be $H(z)$. This has been shown to be false by Raikov (1939) who proves that if

$$F(x) = \exp{(1-e^{e^{-x}})}, \tag{5.57}$$

so that $F(x)$ is the real part of an integral function, $F(x) * \{1 - F(-x)\}$ is not analytic at $x = 0$.

5.9. Moments

If $g(x)$ is any function of x we can define the expectation of the random variable $g(X)$ by

$$E\{g(X)\} = \int_{-\infty}^\infty g(x)\,dF(x), \tag{5.58}$$

whenever this integral exists (i.e. whenever the corresponding integral of $|g(x)|$ is finite). In particular if we take $g(x) = x^n$ $(n = 0, 1, ...)$, we get the moments

$$m_n = E(X^n) = \int_{-\infty}^\infty x^n\,dF(x), \tag{5.59}$$

which agrees with the definition of moments for discrete distributions, given in Chapter 2. We shall only say that m_n exists when the integral converges absolutely so that

$$\nu_n = \int_{-\infty}^\infty |x|^n\,dF(x), \tag{5.60}$$

which is known as the nth absolute moment, exists. Clearly $m_n = \nu_n$ when n is even.

We can also consider the moments, $\mu_n = E\{(X-a)^n\}$ around any point a and in particular about the mean m_1. By expanding $(x-m_1)^n$ we can easily write down formulae for m_n in terms of μ_n and vice versa. Thus

$$\mu_1 = 0,$$

$$\mu_2 = m_2 - m_1^2,$$

$$\mu_3 = m_3 - 3m_1 m_2 + 2m_1^3,$$

$$\mu_4 = m_4 - 4m_1 m_3 + 6m_1^2 m_2 - 3m_1^4, \tag{5.61}$$

and in general

$$\mu_n = m_n - \binom{n}{1}m_{n-1}m_1 + \binom{n}{2}m_{n-2}m_1^2 - \ldots - (-)^n\left\{1 - \binom{n}{n-1}\right\}m_1^n.$$

Similarly

$$m_2 = \mu_2 + m_1^2,$$
$$m_3 = \mu_3 + 3m_1\mu_2 + m_1^3,$$
$$m_4 = \mu_4 + 4m_1\mu_3 + 6m_1^2\mu_2 + m_1^4, \tag{5.62}$$

and in general

$$m_n = \mu_n + \binom{n}{1}\mu_{n-1}m_1 + \ldots + \binom{n}{n-2}m_1^{n-2}\mu_2 + m_1^n.$$

It will be noticed that m_1, when it exists, can be regarded as a measure of the position m_1 of the distribution on the real line since the distribution $F(x+a)$ has its first moment equal to $m_1 - a$.

Similarly $m_2^{\frac{1}{2}}$ can be regarded as a measure of the spread of a distribution and is proportional to the scale in which X is measured.

In a similar way we can define moments for distributions in two or more dimensions. Thus if $X_1,\ldots,X_n$ have a joint distribution we define $m_{rs\ldots t}$ to be

$$E(X_1^r X_2^s \ldots X_n^t), \tag{5.63}$$

when the integral defining

$$v_{rs\ldots t} = E(|X_1^r X_2^s \ldots X_n^t|), \tag{5.64}$$

converges absolutely. In particular cases moments are often most simply found from the characteristic functions of distributions and examples of this will be given in Chapter 7.

5.10. Inequalities for moments

A number of inequalities hold between moments of different orders and these are most easily discussed in terms of the absolute moments which may be easily defined in an unambiguous manner for all real non-negative values of n although non-integral moments do not often occur in applications.

THEOREM 5.11. *If* $0 < p < p'$ *and* $v_{p'} = E(|X|^{p'})$ *is finite, so also is* $v_p = E(|X|^p)$.

Proof. When $|X| \geqslant 1$ we have $|X|^p \leqslant |X|^{p'}$ and when $|X| < 1$ we have $|X|^p < 1$. Therefore $|X|^p \leqslant 1 + |X|^{p'}$ for all X and integrating with respect to $F(x)$ the result follows. This shows that the set of values p for which v_p is finite consists of an interval (which may degenerate into a point) stretching from $p = 0$ to some non-negative value which may be infinity.

THEOREM 5.12. *If X and Y are any random variables, dependent or not,*

$$E(|X+Y|^p) \leqslant C_p E(|X|^p) + C_p E(|Y|^p), \tag{5.65}$$

where

$$C_p = 1 \quad \text{if} \quad 0 < p \leqslant 1, \quad \text{and} \quad C_p = 2^{p-1} \quad \text{if} \quad p > 1.$$

Proof. This follows from the inequality

$$|x+y|^p \leqslant C_p(|x|^p + |y|^p).$$

To prove this it is sufficient to take $x = 1$ and $y \geqslant 1$. For $p > 1$ the result follows from the fact that y^p is concave above, and for $p \leqslant 1$ from the inequality $(1+y)^p < 1+y^p$, which is easily verified by differentiation.

This theorem shows that the pth absolute moment of $X + Y$ is finite if the pth absolute moments of X and Y exist. For $p \geqslant 1$ another upper bound for $E(|X+Y|^p)$ is given by the following theorem.

THEOREM 5.13 (Minkowski's inequality). *If* $p \geqslant 1$,

$$\{E(|X+Y|^p)\}^{1/p} \leqslant \{E(|X|^p)\}^{1/p} + \{E(|Y|^p)\}^{1/p}. \tag{5.66}$$

Before proving this we consider inequalities for the product of two random variables.

THEOREM 5.14 (Hölder's inequality). *If* $p > 1$, $p^{-1} + q^{-1} = 1$ *(and thus* $q > 1$*)*,

$$E(|XY|) \leqslant (E|X|^p)^{1/p}(E|Y|^q)^{1/q}. \tag{5.67}$$

Proof. For any real numbers x and y we have

$$|xy| \leqslant \frac{|x|^p}{p} + \frac{|y|^q}{q},$$

as may be easily shown by putting $y = Kx^{-1}$, where K is some constant, and minimizing the right-hand side for variations in x. Replacing x and y by $X/\{E|X|^p\}^{1/p}$ and $Y/\{E|Y|^q\}^{1/q}$, we get

$$|XY| \leqslant p^{-1}|X|^p\{E(|X|^p)\}^{(1/p)-1}\{E(|Y|^q)\}^{1/q}$$

$$+q^{-1}|Y|^q\{E(|X|^p)\}^{1/p}\{E(|Y|^q)\}^{(1/q)-1}$$

and taking expectations of both sides we get (5.67). The case $p = q = 2$ is particularly important and may be written

$$\{E(|XY|)\}^2 \leqslant E(|X|^2)E(|Y|^2). \tag{5.68}$$

This is known as Schwartz's inequality.

Proof of Theorem 5.13. In Hölder's inequality replace Y by $X+Z$ and apply to the simple inequality

$$E(|X+Z|^p) \leqslant E(|X||X+Z|^{p-1}) + E(|Z||X+Z|^{p-1})$$

$$\leqslant [\{E(|X|^p)\}^{1/p} + \{E(|Z|^p)\}^{1/p}]\{E(|X+Z|)^{(p-1)q}\}^{1/q}.$$

Dividing by $E(|X+Z|^p)^{1/q}$, and observing that $(p-1)q = p$ we get (5.66) with Y replaced by Z. So long as $p > 1$ this provides another upper bound to $E(|X+Z|)$.

16

Now let $0 < p < p'$, and replace X and Y in (5.68) by $|X|^{(p'-p)/2}$ and $|X|^{(p'+p)/2}$, so that we obtain

$$\{E(|X|^{p'})\}^2 \leqslant \{E(|X|^{p'-p})\}\{E(|X|^{p'+p})\}.$$

On taking logarithms

$$\log E(|X|^{p'}) \leqslant \tfrac{1}{2}\log E(|X|^{p'-p}) + \tfrac{1}{2}\log E(|X|^{p'+p}). \qquad (5.69)$$

From this we deduce:

THEOREM 5.15. *The function of p, $\log E(|X|^p)$, is convex from below. Suppose this function exists for some $p > 0$. Then it exists for all smaller non-zero values of p, and as $p \to 0$ it tends to 0. Hence the graph of this function passes through $(0, 0)$ and*

$$\frac{1}{r}\log E(|X|^r)$$

must be non-decreasing. Therefore:

THEOREM 5.16.

$$\{E(|X|^r)\}^{1/r}$$

is a non-decreasing function of r for $r > 0$.

THEOREM 5.17 (Liapounov's inequality). *Suppose that $0 < a < b < c$. Then*

$$v_b^{c-a} \leqslant v_a^{c-b} v_c^{b-a}. \qquad (5.70)$$

Proof. This follows at once by taking logarithms and using Theorem 5.15. Another and interesting proof is given by Good (1950b).

5.11. Random variables and the spaces L^p

In this chapter we have considered a random variable as being defined when we know its probability distribution function $F(x)$. In this definition the real line, which is, or contains, the set of values which the random variable can take, is also the space in which the probability measure is defined by means of $F(x)$. It is sometimes convenient and illuminating to separate these two spaces.

Thus suppose we have some general space, S, whose nature need not be defined, and in which there is a σ-field F of sets A and a probability measure, $P(A)$, defined on this field. In the space S we define a function $f(x)$ for each point x such that $f(x)$ takes values on the real line $(-\infty, \infty)$. If $f(x)$ is measurable the set of points in S such that $f(x) \leqslant x_0$, for any x_0, will be a set of the field F and will have a measure which we write as $P([f(x) \leqslant x_0])$ where the symbol $[f(x) \leqslant x_0]$ means the set of points x satisfying this condition. Then $P([f(x) \leqslant x_0])$ will be a distribution function as defined in § 5.1. In fact from Theorem 4.6 the set of all points x in S which correspond to any Borel set on the real line will be measurable. In this way a measure, $P(A)$, in the space S has been used to induce a measure on the real line.

We keep the space S and its measure fixed and regard all measurable functions from S to $(-\infty, \infty)$ as 'random variables'. Then 'convergence in probability' and 'almost certain convergence', concepts which we will discuss in Chapter 8, are identified with convergence in measure and convergence almost everywhere, respectively.

The expectation, $E(X)$, of a random variable X is then the integral

$$\int f(x)\, dM(x), \tag{5.71}$$

if this exists, and similarly the kth moment is

$$\int \{f(x)\}^k\, dM(x). \tag{5.72}$$

Theorems 5.11 to 5.17 can then be deduced immediately from the standard theorems on inequalities for integrable functions, and the class of all random variables with finite pth moments can be made into a linear vector space by identifying it with the class of all functions which are such that their moduli raised to the pth power are integrable. This is known as the class L^p.

In this definition two random variables, X and Y, are independent if the functions which define them, $f_X(x)$ and $f_Y(x)$ say, are such that

$$P([f_X(x) \leqslant x_0, f_Y(x) \leqslant y_0]) = P([f_X(x) \leqslant x_0])P([f_Y(x) \leqslant y_0]) \tag{5.73}$$

for all x_0, y_0. Thus a pair of such functions, $f_X(x)$ and $f_Y(x)$, independent or not, induce a probability measure into the product space of their range spaces, and furthermore if (5.73) holds, this probability measure is a product measure. Independent events can be defined as events corresponding to measurable sets in the domain of x which are such that their indicator functions are independent.

5.12. Second-order moments in the Euclidean space R_n

Consider a joint distribution of k random variables $X_1, \ldots, X_k$ defined by a distribution function $F(x_1, \ldots, x_k)$. Write m_{ij} for the expectation

$$E(X_i X_j) = \int \ldots \int x_i x_j\, dF(x_1, \ldots, x_n), \tag{5.74}$$

where i may equal j or not. For simplicity of notation we suppose that the means of all the X_i are zero. Then the symmetric matrix (m_{ij}) is made up of all the second-order moments of the distribution. If $u_1, \ldots, u_k$ are any real numbers we have

$$Q = \sum_{i,j} m_{ij} u_i u_j = E(X_1 u_1 + \ldots + X_k u_k)^2 \geqslant 0, \tag{5.75}$$

and hence the quadratic form Q is non-negative definite. We show that this is not only a necessary condition but given that the quadratic form Q is non-negative definite, there exists a k-variable distribution with m_{ij} as moments. In fact we have the following theorem.

THEOREM 5.18. *The necessary and sufficient condition that there exists a distribution of k random variables with second-order moments $m_{ij} = m_{ji}$ is that $Q \geqslant 0$.*

Proof. The necessity has already been shown and the sufficiency follows if we construct a suitable distribution. To do this we evaluate the second moments of the multinormal distribution. For a normal distribution with one random variable and zero mean we have

$$E(X^2) = (2\pi\sigma^2)^{-\frac{1}{2}} \int_{-\infty}^{\infty} x^2 \exp\left(-\frac{1}{2}\frac{x^2}{\sigma^2}\right) dx$$

$$= \sigma^2, \tag{5.76}$$

on integrating by parts. Consider the distribution (5.35). The matrix, $\{E(X_i X_j)\}$, is

$$\mathbf{T}\begin{pmatrix} \sigma_1^2 & & & 0 \\ & \cdot & & \\ & & \cdot & \\ & & & \cdot \\ 0 & & & \sigma_k^2 \end{pmatrix}\mathbf{T}', \tag{5.77}$$

on using the transformation to the random variables Y_i which are independent with zero means and second moments σ_i^2. We already know that

$$\mathbf{T}'\mathbf{AT} = \begin{pmatrix} \sigma_1^{-2} & & & 0 \\ & \cdot & & \\ & & \cdot & \\ 0 & & & \sigma_k^{-2} \end{pmatrix}$$

so that

$$\begin{pmatrix} \sigma_1^2 & & & 0 \\ & \cdot & & \\ & & \cdot & \\ 0 & & & \sigma_k^2 \end{pmatrix} = (\mathbf{T}'\mathbf{AT})^{-1}$$

$$= \mathbf{T}^{-1}\mathbf{A}^{-1}(\mathbf{T}')^{-1}. \tag{5.78}$$

Hence $\{E(X_i X_j)\} = \mathbf{A}^{-1}$. Thus if $M = (m_{ij})$ is any symmetric positive definite matrix we can use $\mathbf{A} = \mathbf{M}^{-1}$ in (5.36), and we obtain a multivariate normal distribution of the required type. If (m_{ij}) is non-negative definite, but not positive definite, there must exist u_i which are not all zero and are such that

$$\sum_{ij} m_{ij} u_i u_j = E(X_1 u_1 + \dots + X_k u_k)^2 = 0, \tag{5.79}$$

and the X_i then satisfy one or more linear equations. In this case the required distribution is singular and can be obtained by using as many of the X_i as are linearly independent and expressing the rest in terms of them.

Theorem 5.18 is very simple but has some curious geometrical consequences. Suppose that, as above, we have a joint non-singular distribution of k random variables $X_1, \dots, X_k$ with zero means. In terms of these we define a new set of random variables $Y_1, \dots, Y_k$, which also have zero means, but for which the second moments $M_{ii} = E(Y_i^2) = 1$, and the cross-moments $M_{ij} = E(Y_i Y_j) = 0$ when $i = j$. To do this we take

$$Y_1 = X_1(m_{11})^{-\frac{1}{2}}, \tag{5.80}$$

$$Y_2 = (X_2 - a_{21} X_1) b_2, \tag{5.81}$$

where a_{21} is chosen to make $E(Y_1 Y_2) = 0$, and b_2 is chosen to make $E(Y_2^2) = 1$, and so on. Thus

$$Y_i = (X_i - a_{i1} X_1 - a_{i2} X_2 - \dots - a_{ii-1} X_{i-1}) b_i \tag{5.82}$$

for $i = 3, \dots, k$, the constants being chosen to make

$$E(Y_i^2) = 1, \quad E(Y_i Y_1) = E(Y_i Y_2) = \dots = E(Y_i Y_{i-1}) = 0.$$

This is a non-singular transformation, since the distribution is non-singular, and therefore has an inverse which we write as

$$X_1 = C_{11} Y_1,$$

$$X_2 = C_{21} Y_1 + C_{22} Y_2,$$

$$\cdot \quad \cdot \quad \cdot \quad \cdot \quad \cdot \quad \cdot$$

$$X_k = C_{k1} Y_1 + \dots + C_{kk} Y_k. \tag{5.83}$$

Now consider k orthogonal unit vectors $e_1, \dots, e_k$ in a space of k or more dimensions. We can then define k vectors $v_1, \dots, v_k$ by using the coefficients in (5.83) so that

$$v_1 = C_{11} e_1,$$

$$\cdot \quad \cdot \quad \cdot \quad \cdot \quad \cdot \quad \cdot$$

$$v_k = C_{k1} e_1 + \dots + C_{kk} e_k. \tag{5.84}$$

It is then obvious that the length of the vectors v_i is $m_{ii}^{\frac{1}{2}}$, and the scalar product, $v_i . v_j$, is m_{ij}. This proves the following theorem.

THEOREM 5.19. *Given a Euclidean space of k or more dimensions with origin O, the necessary and sufficient condition that there exist k points $A_1, \dots, A_k$ such that the vectors OA_i have lengths $m_{ii}^{\frac{1}{2}}$, and the scalar products $OA_i . OA_j$*

have values m_{ij}, *is that the quadratic form*

$$Q = \sum_{ij} m_{ij} u_i u_j \qquad (5.85)$$

should be positive definite. This can be equivalently stated in the form that the necessary and sufficient condition that such a set of k points should exist is that

$$\sum_i r_i^2 u_i^2 + \frac{1}{2} \sum_{i \neq j} (r_i^2 + r_j^2 - r_{ij}^2) u_i u_j \qquad (5.86)$$

is positive definite, where r_i *is the distance of* A_i *from* O, *and* r_{ij} *is the distance from* A_i *to* A_j (P. Lévy 1948a).

This gives a necessary and sufficient condition for the existence of a simplex with $k+1$ vertices in a space of k or more dimensions, in terms of the length of the edges, but it is to be noticed that it is not expressed symmetrically in terms of these $\frac{1}{2}k(k+1)$ lengths. As there are $k+1$ vertices any of which could have been chosen as the point O, there are $k+1$ quadratic forms whose positive definiteness must be equivalent. (See also Exercise 5.5.)

5.13. Inequalities of Tchebychev type

In §1.9 we have already established a simple inequality which showed that for a discrete random variable X,

$$\text{pr}(|X - m_1'| \geq x) \leq \mu_2 x^{-2}.$$

This result clearly extends to any random variable when m_1' and μ_2 are the moments. In fact by using the same kind of argument we can immediately deduce the following theorem.

THEOREM 5.20. *If* $g(x)$ *is a non-decreasing non-negative function on the interval* $(0, \infty)$

$$\text{pr}(|X| \geq x) \leq E\{g(|X|)\}g(x)^{-1} \qquad (5.87)$$

when the right-hand side is finite and $x \geq 0$.

Proof.
$$E\{g|X|)\} = \int_{-\infty}^{\infty} g(|X|) \, dF(X)$$

$$\geq \int_{x-0}^{\infty} + \int_{-\infty}^{-x} g(x) \, dF(X)$$

$$\geq g(x)\text{pr}(|X| \geq x).$$

In a sense this is the best possible result, for given x we can choose the distribution to be concentrated at the points $\pm x$, having jumps of $\frac{1}{2}$ at each of these. (5.87) is particularly useful when $g(X) = |X|^r$ $(r > 0)$. For $r = 2$ the bound is given in terms of the second moment and the inequality is the Bienaymé–Tchebychev inequality.

Very much stronger results are possible if further assumptions are made and there is a large literature dealing with such inequalities. Here we consider only a few of the more important extensions. Detailed surveys with bibliographies are given by Savage (1961) and Godwin (1955, 1964). (See also Fréchet 1937.) General theorems on inequalities of Tchebychev type are given by Kingman (1963b).

In order to obtain results stronger than (5.87) it is necessary to have more information either about the moments or about the distribution. For example, if the distribution has a single mode we have the following theorem originally due to Gauss (1821).

THEOREM 5.21. *If the distribution, $F(x)$, has a continuous density function with a single maximum (a 'mode') at $x = 0$, then*

$$\mathrm{pr}(|X| > xm_2^{\frac{1}{2}}) \leqslant 1 - x/\sqrt{3} \quad (0 \leqslant x \leqslant 2/\sqrt{3}),$$

$$\leqslant \tfrac{4}{9}x^{-2} \quad (2/\sqrt{3} \leqslant x). \tag{5.88}$$

Proof (C. B. Winsten 1946). Here $m_2^{\frac{1}{2}}$ is a scale factor so we put $xm_2^{\frac{1}{2}} = L$ and keep L fixed whilst trying to find the smallest value of m_2 subject to the probability that $|X| > L$ being fixed also.

Write $f(x) = F'(x)$ and $f_1(x) = f(x) + f(-x)$. Then $f_1(x)$ is decreasing. Put

$$p = 1 - q = \int_{-L}^{L} dF = 1 - \int_{L}^{\infty} f_1(x)\, dx. \tag{5.89}$$

We keep p fixed and minimize

$$m_2 = \int_{0}^{L} x^2 f_1(x)\, dx + \int_{L}^{\infty} x^2 f_1(x)\, dx \tag{5.90}$$

by minimizing these two integrals separately. We do this subject to the additional condition that $f_1(x)$ is prescribed at $x = L$. Let its value be h. Then the second integral in (5.90) will be a minimum if $f_1(x) = h$ over an interval to the right of $x = L$ and zero beyond it. In fact we must have

$$f_1(x) = h \quad (L \leqslant x \leqslant L + qh^{-1}),$$

$$= 0 \quad (x > L + qh^{-1}).$$

Inside the interval $(0, L)$, $f_1(x)$ must decrease as x increases. A minimum m_2 will therefore be obtained if $f_1(x)$ is constant from 0 to L, and a concentration of probability of amount

$$p - hL$$

occurs at $x = 0$. This is not a continuous distribution of the type required but is the limit of such distributions as the first integral in (5.90) converges to its minimum.

We then have

$$m_2 = \int_0^{L+qh^{-1}} hx^2 \, dx$$

$$= \tfrac{1}{3}h(L+qh^{-1})^3. \tag{5.91}$$

Letting h vary the minimum of this expression is easily seen to occur when $h = 2qL^{-1}$. However, we also know that $h \leqslant pL^{-1}$. Suppose first that $p \leqslant 2q$, so that $p \leqslant \tfrac{2}{3}$. The minimum is obtained when $h = pL^{-1}$ because (5.91) is a decreasing function of h at this point. We then have $m_2 = L^2(3p^2)^{-1}$. If $p > 2q$ the minimum is obtained for $h = 2qL^{-1}$ and $m_2 = \tfrac{9}{4}qL^2$. We therefore have

$$\min m_2 = L^2(3p^2)^{-1} \quad (p \leqslant \tfrac{2}{3}),$$

$$= \tfrac{9}{4}qL^2 \quad (p > \tfrac{2}{3})$$

and these inequalities are equivalent to (5.88).

Although this is a real strengthening of the Bienaymé–Tchebychev inequality it usually gives very crude limits, as for example with the normal distribution. Better results can be often obtained by using higher moments, and Winsten (1946) has extended the above results for unimodal distributions to these.

A further strengthening is obtained by considering several moments. This problem was first studied by Tchebychev (1874) who gave upper and lower limits for a distribution whose first n moments are known, but he did not supply a proof. Such a proof was later given by Markov (1884), and Stieltjes (1884, 1894). (See Shohat and Tamarkin 1943.) The problem of obtaining such bounds when the first n absolute moments are given was solved by Wald (1939). (See also Royden 1953.) These results all provide limits which are rather too wide for statistical use. Mallows (1956, 1963) has strengthened Tchebychev's results by imposing further conditions on the smoothness of the distributions and in the second of these papers gives references to the extensive body of research on the more purely mathematical aspect of this class of problems.

Hoeffding and Shrikhande (1955) have shown how Tchebychev's inequality, using the second moment, can be improved if it is known that the distribution is the convolution of n identical distributions.

Multivariate generalizations have been given by many writers amongst whom we may mention Birnbaum, Raymond, and Zuckermann (1947), Camp (1948), Olkin and Pratt (1958), Marshall and Olkin (1960), and Birnbaum and Marshall (1961).

Another type of generalization occurs in Kolmogorov's inequalities (§ 8.10).

5.14. Some other measures of distributions

A number of other quantities can be used to provide information about the position and shape of a distribution function. Thus we have already used the idea of a mode above (§ 5.13). When there exists a probability density function, $f(x)$, everywhere, x is said to be a mode if $f(x)$ is a local maximum. This quantity is of little use unless there is only one mode and even then its properties do not lead to useful theorems, and are not helpful in dealing with convolutions. In fact, contrary to expectation, it is possible to construct a distribution function, $F(x)$, which is unimodal but is such that $F(x) \star F(x)$ is bimodal (see Exercise 5.3).

A median, M, of a distribution $F(x)$ is defined to be a number such that

$$F(M-0) \leqslant \tfrac{1}{2} \leqslant F(M). \tag{5.92}$$

There may be more than one median in which case they fill up all the values of a closed interval. Given the medians of two distributions little can be said in general about the median of their convolution. However, the median is a useful quantity in some more theoretical investigations because there always exists at least one such quantity, and it is therefore useful as a measure of position of an arbitrary one-dimensional distribution. In more than one dimension difficulties arise (Haldane 1948).

Measures of spread which always exist are also easy to define. The most interesting of these is a function introduced by P. Lévy which, although little used by other writers, has some curious and useful properties.

Given any proper distribution function, $F(x)$, the 'concentration function', $Q(l)$ $(l \geqslant 0)$, is defined to be the least upper bound of

$$F(x+l) - F(x-0) \tag{5.93}$$

for all x in $(-\infty, \infty)$. Conventionally we also take $Q(l) = 0$ for $l < 0$ so that $Q(l)$ is itself a proper distribution function of a random variable taking non-negative values. $Q(l)$ is thus a non-decreasing function. We take its inverse function to be the 'dispersion function', $w(\alpha)$, defining this to be constant at values where $Q(l)$ has jumps and (say) to be undefined at points where it has jumps itself (corresponding to constant values of $Q(l)$ in intervals). Then $Q(l)$ and $w(\alpha)$ are connected by the relationship $Q(w(\alpha)) = \alpha$, at points where $w(\alpha)$ is continuous and $Q(w(\alpha))$ a continuous function of $w(\alpha)$. Both describe the dispersion of $F(x)$ not by a single number but by a whole function.

Writing m_2 for the second moment about any particular point, which we can take to be zero, it follows from Tchebychev's inequality that if $x > 0$,

$$\mathrm{pr}(|X| \geqslant x) \leqslant m_2 x^{-2},$$

and therefore

$$1 - Q(2x) \leqslant m_2 x^{-2}.$$

Thus

$$Q(l) \geqslant 1 - 4m_2 l^{-2},$$ (5.94)

and

$$w(\alpha) \leqslant 2\left(\frac{m_2}{1-\alpha}\right)^{\frac{1}{2}}.$$ (5.95)

Knowledge of the second moment does not, however, provide inequalities in the opposite direction and the advantage of $Q(l)$, $w(\alpha)$, as measures of dispersion is that they can be applied to any proper distribution whether it has moments or not.

Notice also that if a sequence of distribution functions, $\{F_n(x)\}$, $n = 1, 2,...$, converges to a distribution function, $F(x)$, the dispersion functions of $F_n(x)$ converge to those of $F(x)$, but the converse is obviously not true.

Suppose that X and Y are independent random variables with second moments about their means which we write as σ_1^2 and σ_2^2. Then $X + Y$ will have a second moment about its mean equal to $\sigma_1^2 + \sigma_2^2$. Thus if we use the second moment as a measure of dispersion, the dispersion of the sum of two independent random variables will be greater than either. Moreover, the sum of n independent random variables with the same finite second moment, σ^2, will have a second moment $n\sigma^2$. If the distribution is, in addition, normal the sum will have a normal distribution of the same kind as the individual terms but with the scale increased in the ratio $n^{\frac{1}{2}}$. This is the typical kind of behaviour for distributions with finite second moment, but if the latter is not finite the effective scale factor may increase quite differently. The study of this increase in dispersion is of great interest (see § 5.15) and $Q(l)$ can be used to throw some light on it. A basic result is given by the following theorem.

THEOREM 5.22. *If X and Y are independent random variables with functions of concentration $Q_1(l)$ and $Q_2(l)$, $X + Y$ will have a function of concentration, $Q(l)$, such that*

$$Q(l) \leqslant Q_1(l),$$
$$Q(l) \leqslant Q_2(l).$$ (5.96)

Proof. Let the distribution functions of X, Y, and $X + Y$ be $F_1(x)$, $F_2(x)$, and $G(x)$. Then from (5.52) we have

$$G(x+l) - G(x) = \int \{F_1(x+l-y) - F_1(x-y)\} \, dF_2(y),$$

$$\leqslant \int Q_1(l) \, dF_2(y),$$

$$\leqslant Q_1(l).$$

Another measure which can be applied to some distributions is 'entropy'. In Chapter 1 we defined the entropy of a discrete distribution $\{p_n\}$ by

$$E = -\sum_n p_n \log p_n.$$ (5.97)

If we have a distribution on $(-\infty, \infty)$ which has a continuous probability density we might try to define its entropy by splitting the range up into a sequence of small intervals $(n\delta, (n+1)\delta)$ $(n = ..., -1, 0, 1, 2, ...)$ and applying the definition (5.97) to the probabilities

$$F((n+1)\delta) - F(n\delta),$$

finally proceeding to the limit by letting $\delta \to 0$. If we do this, however, we find that the 'entropy' becomes infinite and we therefore need some other kind of definition. In fact if the distribution has a continuous probability density, $f(x)$, we define its entropy to be

$$E = -\int f(x) \log f(x) \, dx, \tag{5.98}$$

when this exists (Shannon 1948). This may be regarded as the expectation of $-\log f(X)$, where X is a random variable having a distribution with probability density $f(x)$. Thus we do not have a univocal definition which can be applied usefully to all distributions.

This concept of entropy has mainly found application in the theory of communication, a subject too vast to be considered in this book. We shall therefore only consider the following.

THEOREM 5.23. *Of all the continuous distributions for which the probability density $f(x)$ exists, and for which the second moment is a finite prescribed constant, the normal distribution has the largest entropy.*

Proof. We have to maximize the integral (5.98) for all integrable functions, $f(x)$, subject to the side-conditions·

$$\int x^2 f(x) \, dx = \sigma^2, \tag{5.99}$$

and

$$\int f(x) \, dx = 1. \tag{5.100}$$

This is equivalent to maximizing the integral

$$\int \{-f(x)\log f(x) + \lambda f(x)x^2 + \mu f(x)\} \, dx,$$

and by applying the calculus of variations we get

$$-1 - \log f(x) + \lambda x^2 + \mu = 0 \tag{5.101}$$

as a necessary condition from which it follows that

$$f(x) = (2\pi\sigma^2)^{-\frac{1}{2}} \exp -\frac{x^2}{2\sigma^2}.$$

For this distribution E is easily seen to be

$$\tfrac{1}{2} \log 2\pi e \sigma^2,$$

which is the lower bound for the entropy of any distribution with second moment σ^2.

For a further discussion of entropy in relation to statistical inference see Kullback (1959).

5.15. The dispersion of sums

One of the most important problems in the theory of probabilities is the study of the asymptotic behaviour of the distribution of the sums of independent random variables, particularly in the case when the latter all have the same distribution. Thus suppose X_1, X_2,... is a sequence of independent random variables having the same distribution $F(x)$. Write $S_n = X_1 + ... + X_n$ and consider the asymptotic behaviour of the distribution of S_n.

Two elementary particular cases are instructive. Suppose first that the X_i have the normal distribution with density function

$$(2\pi\sigma^2)^{-\frac{1}{2}} \exp -\frac{1}{2}\frac{x^2}{\sigma^2}. \tag{5.102}$$

Then S_n will have the normal distribution with density

$$(2n\pi\sigma^2)^{-\frac{1}{2}} \exp -\frac{1}{2}\frac{x^2}{n\sigma^2}. \tag{5.103}$$

This follows at once from the following theorem.

THEOREM 5.24. *If X and Y have normal distributions with means m_1 and m_2, and standard deviations σ_1 and σ_2, $Z = X + Y$ will have a normal distribution with mean $m_1 + m_2$ and standard deviation $(\sigma_1^2 + \sigma_2^2)^{\frac{1}{2}}$.*

Proof. We can clearly take $m_1 = m_2 = 0$. Z will have a distribution with a continuous density function, $f(z)$, equal to

$$(2\pi\sigma_1\sigma_2)^{-1} \int_{-\infty}^{\infty} \exp -\frac{1}{2}\left\{\frac{(z-y)^2}{\sigma_1^2} + \frac{y^2}{\sigma_2^2}\right\} dy$$

$$= (2\pi\sigma_1\sigma_2)^{-1} \int_{-\infty}^{\infty} \exp\left\{-\frac{1}{2}\left\{\left(\frac{1}{\sigma_1^2}+\frac{1}{\sigma_2^2}\right)\left(y-\frac{z\sigma_2^2}{\sigma_1^2+\sigma_2^2}\right)^2+\frac{z^2}{\sigma_1^2}-\frac{z^2\sigma_2^2}{\sigma_1^2(\sigma_1^2+\sigma_2^2)}\right\}\right\} dy$$

$$= \{2\pi(\sigma_1^2+\sigma_2^2)\}^{-\frac{1}{2}} \exp\left\{-\frac{z^2}{2(\sigma_1^2+\sigma_2^2)}\right\}.$$

Applying this repeatedly to the sum S_n the result (5.103) follows. Thus S_n has the same distribution as the X_i but increased by the scale factor $n^{\frac{1}{2}}$, and $w(\alpha)$ for S_n is similarly equal to the $w(\alpha)$ for the X_i increased by the factor $n^{\frac{1}{2}}$.

This is roughly the sort of behaviour we expect for any distribution of X which has a finite second moment μ_2, for S_n must have a second moment equal to $n\mu_2$ and if we regard $\mu_2^{\frac{1}{2}}$ as a measure of dispersion, the latter will increase proportionally to $n^{\frac{1}{2}}$.

The situation is quite different when the second moment does not exist and we illustrate this by considering what happens if the X_n have the 'Cauchy' distribution defined by

$$f(x) = \frac{1}{\pi(1+x^2)}. \tag{5.104}$$

Consider the integral

$$\pi^{-2} \int_{-\infty}^{\infty} \frac{dy}{(1+y^2)\{n^2+(x-y)^2\}}$$

$$= \frac{\pi^{-2}}{(x^2+(n+1)^2)(x^2+(n-1)^2)}$$

$$\times \int_{-\infty}^{\infty} \left\{ \frac{2xy}{1+y^2} + \frac{x^2+n^2-1}{1+y^2} + \frac{2x^2-2xy}{n^2+(x-y)^2} + \frac{x^2-n^2+1}{n^2+(x-y)^2} \right\} dy$$

$$= \pi^{-1}\left(\frac{n+1}{n}\right)\{x^2+(n+1)^2\}^{-1}.$$

Thus X_1+X_2 is distributed with probability density

$$\frac{2}{\pi(4+x^2)}$$

and proceeding by induction, S_n is distributed with density

$$\frac{n}{\pi(n^2+x^2)}.$$

$n^{-1}S_n$ therefore has the same distribution as each of the X_i and the dispersion therefore increases as n instead of $n^{\frac{1}{2}}$.

We shall show in Chapter 9 that given any α satisfying $\frac{1}{2} \leqslant \alpha < \infty$ there exists a distribution whose nth convolution power is of the same form except for an increase in the scale by a factor n^α. Thus $\alpha = \frac{1}{2}$ corresponds to the normal distribution which is therefore one extreme possibility whilst $\alpha = 1$ corresponds to the Cauchy distribution. More generally Lévy (1937a, p. 155) has proved the following result which applies to distributions of any kind.

THEOREM 5.25. *Let S_n be the sum of n independent random variables with the same distribution $F(x)$. If $0 < \alpha < 1$, and the concentration of $F(x)$ satisfies*

$$Q(2l-0) \leqslant \alpha, \tag{5.105}$$

or equivalently

$$w(\alpha) \geqslant 2l, \tag{5.106}$$

then for any β such that $0 < \beta < 1$ there exist $k > 0$, $N > 0$, such that for $n > N$ the dispersion of S_n corresponding to the probability β, is at least $kln^{\frac{1}{2}}$.

This result bounds the rate of increase of dispersion from below and shows that it must be at least as fast as the rate of increase of dispersion, as measured by the second moment, for the case where the X_i have a finite second moment.

Bibliographical notes

For the properties of distribution functions in general Lévy (1937*a*) provides the pleasantest introduction, and Loève (1963) and Renyi (1962) should also be consulted.

Exercises

Exercise 5.1. Construct a function, $F(x_1,..., x_n)$, which increases as each x_i increases, and which satisfies (5.3), (5.6), and (5.7) but not (5.5).

Exercise 5.2. X, Y, Z are three independent random variables having distributions $F_1(x)$, $F_2(x)$, $F_3(x)$. Show that it is possible to choose these in such a way that $\text{pr}(X < Y) > \frac{1}{2}$, $\text{pr}(Y < Z) > \frac{1}{2}$, and $\text{pr}(Z < X) > \frac{1}{2}$ (Steinhaus and Trybula 1959).

Exercise 5.3. Consider a distribution whose probability density is equal to 5 when $-(30)^{-1} \leqslant x \leqslant 0$, to 1 when $0 < x \leqslant \frac{5}{6}$, and to zero elsewhere. Show that the sum of two independent random variables having this distribution has a distribution with two modes. Verify that the result holds if the distribution is modified slightly to have a single mode at (say) $-(30)^{-1}$. This result is due to K. L. Chung (Gnedenko and Kolmogorov 1954, p. 254).

Exercise 5.4. X_i ($i = 1,..., n$) and Y_i ($i = 1,..., m$) are independent random variables having normal distributions with means m_1, m_2, and variances σ_1^2, $\sigma_1^2 + \sigma_2^2$. As an estimator of σ_2^2 we take

$$T = (m-1)^{-1} \sum_1^m (Y_i - \overline{Y})^2 - (n-1)^{-1} \sum_1^n (X_i - \overline{X})^2$$

where

$$\overline{X} = n^{-1} \sum_1^n X_i, \quad \overline{Y} = m^{-1} \sum_1^m Y_i.$$

Prove that

$$E(T) = \sigma_2^2,$$

$$\text{var}(T) = \frac{2\sigma_1^4}{n-1} + \frac{2(\sigma_1^2 + \sigma_2^2)^2}{m-1}.$$

Exercise 5.5. Prove that a necessary and sufficient condition that there exist k points $A_1,..., A_k$ in n-dimensional space such that the distance of A_i from A_k is r_{ik} is that

$$\sum_{i,j=1}^k r_{ij}^2 u_i u_j \leqslant 0$$

for every set of real numbers u_i such that

$$\sum_{i=1}^k u_i = 0.$$

Exercise 5.6. Prove that $X - \max(Y_1,..., Y_n)$ is distributed in a negative exponential distribution where X, $Y_1,..., Y_n$, n are independent random variables such that

$$\text{pr}(X = m) = (e-1)e^{-m} \quad (m = 1, 2,...),$$

$$\text{pr}(n = j) = (e-1)^{-1}(n!)^{-1} \quad (n = 1, 2,...)$$

and the Y_i are uniformly distributed on $(0, 1)$ (Hammersley and Handscomb 1964).

Exercise 5.7. To estimate the integral

$$I = \int_0^1 f(x)\, dx$$

n independent random variables $X_1, \ldots, X_n$, uniformly distributed on $(0, 1)$, are obtained and

$$T_1 = n^{-1} \sum_1^n f(X_i)$$

is used as an estimator. Prove $E(T) = I$ and the variance of T is

$$n^{-1} \int_0^1 \{f(x) - I\}^2\, dx$$

(Hammersley and Handscomb 1964).

Exercise 5.8. In the above example instead of T_1 use

$$T_2 = \tfrac{1}{2} n^{-1} \sum_1^n \{f(X_i) + f(1 - X_i)\}.$$

Prove that

$$\text{var } T_2 = (2n)^{-1} \left[\int_0^1 f(x)^2\, dx + \int_0^1 f(x) f(1 - x)\, dx - 2 \left\{ \int_0^1 f(x)\, dx \right\}^2 \right]$$

(Hammersley and Handscomb 1964).

Exercise 5.9. Random variables X_i $(i = 1, \ldots, m)$ take the values x_{ij} $(i = 1, \ldots, m, \ j = 1, \ldots, n)$ when exclusive events E_j $(j = 1, \ldots, n)$ occur. The events have probabilities $p_1, \ldots, p_n$ $(\sum p_i = 1)$, and the mean values of X_i are $E(X_i)$. Let $k_1, \ldots, k_m$ be non-negative quantities such that $\sum k_i^{-1} \leqslant 1$. Then there is at least one event, E_s, such that the corresponding X's all satisfy the conditions $X_i \leqslant k_i E(X_i)$ (Shannon 1961).

Exercise 5.10. X and Y are independent random variables with distributions $F(x)$, $G(y)$. X is said to be 'stochastically larger' than Y if $F(x) \leqslant G(x)$ for all x. Prove that if this is so,

$$\text{pr}(X \geqslant Y) \geqslant \tfrac{1}{2}.$$

6. The Characteristic Function

6.1. The characteristic function and its inversion

THE most effective tool in the study of probability distributions is the use of their functional transforms. If $F(x)$ is a distribution function and $K(t, x)$ is some function of the two variables t and x, we consider the Stieltjes integral

$$\int_{-\infty}^{\infty} K(t, x)\, dF(x).$$

The usefulness of such a transform depends on the choice of the function $K(t, x)$, and by far the most useful such functions are e^{tx} and e^{itx}. We therefore define the moment generating function, $M(x)$, by

$$M(t) = \int_{-\infty}^{\infty} e^{tx}\, dF(x), \tag{6.1}$$

and the characteristic function, $\phi(t)$, by the Fourier–Stieltjes transform

$$\phi(t) = \int_{-\infty}^{\infty} e^{itx}\, dF(x). \tag{6.2}$$

For real values of t, $M(t)$ may not always exist, but for such values the characteristic function always exists since $|e^{itx}|$ is a bounded and continuous function for all finite real t and x. Similarly $M(t)$ always exist when t is purely imaginary, and if either exists, we have $\phi(t) = M(it)$. These functions are generalizations of the probability generating functions used in Chapter 2, for if a random variable takes the values 0, 1, 2,... with probabilities $p_0, p_1, p_2,...$ its probability generating function is $P(x) = p_0 + p_1 x + ...$, whilst its distribution function is constant except for jumps at 0, 1, 2,... of size $p_0, p_1, p_2,...$, so that its characteristic function is

$$\phi(t) = \sum_{n=0}^{\infty} e^{int} p_n = P(e^{it}),$$

and

$$M(t) = P(e^t).$$

In a similar way a characteristic function may be defined for a multivariate distribution function $F(x_1,..., x_n)$ by the expression

$$\phi(t_1,..., t_n) = \int_{-\infty}^{\infty} ... \int_{-\infty}^{\infty} e^{i(t_1 x_1 + ... + t_n x_n)}\, dF(x_1,..., x_n), \tag{6.3}$$

which again always exists for any set of real numbers $(t_1,...,t_n)$. Both (6.2) and (6.3) become ordinary Lebesgue integrals when the distribution function is the integral of a probability density function. Thus if

$$F(x) = \int_{-\infty}^{x} f(u)\,du,$$

(6.2) becomes

$$\phi(t) = \int_{-\infty}^{\infty} e^{itu}f(u)\,du.$$

It is also sometimes convenient to transform (6.2) into an ordinary Riemann integral involving the function $F(x)$. Thus integrating (6.2) by parts separately for the two intervals $(-\infty, 0)$, $(0, \infty)$, we get

$$\phi(t) = 1 - it\int_{-\infty}^{0} e^{itx}F(x)\,dx + it\int_{0}^{\infty} e^{itx}\{1 - F(x)\}\,dx. \qquad (6.4)$$

From this it follows that if $1 - F(x) + F(-x)$ tends to zero faster than any power of e^{-x} as $x \to \infty$, then $\phi(t)$ is an integral function. This is quite a severe condition on $F(x)$. If we only suppose that $1 - F(x) + F(-x)$ tends to zero faster than some power of e^{-x}, e^{-rx} say, with $r > 0$, then $\phi(t)$ is clearly analytic in the neighbourhood of $t = 0$. If $F(x)$ is only known to be a distribution function, however, $\phi(t)$ is at least continuous for real t, and in fact uniformly continuous on the whole real line.

To prove this let ε be any small positive quantity and a such that $1 - F(a) < \varepsilon$, $F(-a) < \varepsilon$. Then

$$|\phi(t_1) - \phi(t)| = \left| \int_{-\infty}^{\infty} (e^{it_1 x} - e^{itx})\,dF(x) \right|$$

$$\leqslant 2\int_{a}^{\infty} dF(x) + 2\int_{-\infty}^{-a} dF(x)$$

$$+ \int_{-a}^{a} \left| e^{it_1 x} - e^{itx} \right| dF(x).$$

We can make the last integral as small as we like by making $|t_1 - t|$ small and since, given ε, x must lie in the range $(-a, a)$, the resulting continuity is uniform with respect to t. Thus every characteristic function is continuous. Stronger results are not true in general as there exist characteristic functions which are not differentiable anywhere (Dugué 1957).

It is also clear from the definition (6.2) that

$$|\phi(t)| \leqslant \int_{-\infty}^{\infty} \left| e^{itx} \right| dF(x) = 1,$$

that

$$\phi(-t) = \overline{\phi(t)},$$

and that

$$\phi(0) = 1.$$

17

As we shall see later, although these conditions are necessary for a function $\phi(t)$ to be a characteristic function they are not sufficient. Notice also that if $F(x) = 1 - F(-x)$ at continuity points, $\phi(t)$ is real and even.

$\phi(t)$ is uniquely determined by $F(x)$. We shall show that the reverse is true by obtaining an explicit expression, the inversion formula, for $F(x)$ in terms of $\phi(t)$.

Defining $F(x+0)$ and $F(x-0)$ to be the limits of $F(x)$ on the right and left, this inversion formula is given by Theorem 6.1.

THEOREM 6.1

$$\tfrac{1}{2}\{F(a+h+0)+F(a+h-0)\} - \tfrac{1}{2}\{F(a+0)+F(a-0)\}$$

$$= \lim_{T \to \infty} \frac{1}{2\pi} \int_{-T}^{T} (it)^{-1}(1-e^{-ith})e^{-ita}\phi(t)\,dt. \qquad (6.5)$$

Proof. We shall show that this limit exists for all finite a and h. Consider the integral

$$\frac{1}{2\pi} \int_{-T}^{T} (it)^{-1}(1-e^{-ith})e^{-ita}\phi(t)\,dt$$

$$= \frac{1}{2\pi} \int_{-T}^{T} (it)^{-1}(1-e^{-ith})e^{-ita} \int_{-\infty}^{\infty} e^{itu}\,dF(u)\,dt.$$

The integrand is bounded and we can therefore interchange the order of integration to obtain

$$\frac{1}{2\pi} \int_{-\infty}^{\infty} \left\{ \int_{-T}^{T} (it)^{-1}(e^{it(u-a)} - e^{it(u-a-h)})\,dt \right\} dF(u)$$

$$= \frac{1}{\pi} \int_{-\infty}^{\infty} \int_{0}^{T} \left\{ \frac{\sin t(u-a) - \sin t(u-a-h)}{t} \right\} dt\,dF(u). \qquad (6.6)$$

Consider the integral

$$J(a,b) = \frac{1}{\pi} \int_{a}^{b} \frac{\sin t}{t}\,dt.$$

By splitting the range of integration into intervals $(m\pi, (m+1)\pi)$ where $m = 0, \pm 1, \pm 2$, we see that J is a bounded function of a and b, and it is well known (see, for example, Titchmarsh 1932, p. 43) that

$$\lim_{b \to \infty} \frac{1}{\pi} \int_{0}^{b} \frac{\sin t}{t}\,dt = \lim_{a \to -\infty} \frac{1}{\pi} \int_{a}^{0} \frac{\sin t}{t} = \frac{1}{2}.$$

It follows that

$$\int_{0}^{T} \frac{\sin t(u-a) - \sin t(u-a-h)}{t}\,dt = \int_{T(u-a-h)}^{T(u-a)} \frac{\sin t}{t}\,dt,$$

converges, as $T \to \infty$, to 1 when $a < u < a+h$, to $\tfrac{1}{2}$ if $u = a$ or $u = a+h$, and to zero when $u < a$ or $u > a+h$. Since this integral is also uniformly bounded, the limit of (6.6) can be obtained by taking the limiting operation inside the integral in (6.6) with respect to u, and since moreover the limits of $F(u)$ to the right and to the left exist at $u = a$, $u = a+h$, the limit of (6.6) as T increases exists and is equal to

$$\tfrac{1}{2}\{F(a+h+0)-F(a+h-0)\}+F(a+h-0)-F(a+0)+\tfrac{1}{2}\{F(a+0)-F(a-0)\}$$
$$= \tfrac{1}{2}\{F(a+h+0)+F(a+h-0)\}-\tfrac{1}{2}\{F(a+0)+F(a-0)\},$$

which is the required result.

It follows that just as $\phi(t)$ is uniquely determined by $F(x)$, so is $F(x)$ uniquely determined at all its continuity points by the values of $\phi(t)$ on the real axis of t. The above inversion theorem holds for any distribution $F(x)$ and in fact for any function $F(x)$ which is of bounded variation. If $F(x)$ has a derivative, $f(x)$, everywhere, so that a probability density exists, the formula for $\phi(t)$ simplifies, and in the reverse direction we can say that if $\phi(t)$ is absolutely integrable (i.e. if the integral of $|\phi(t)|$ over the whole real axis is finite) it follows that the derivative $F'(x)$ of $F(x)$ exists everywhere, is bounded and continuous, and

$$F'(x) = \frac{1}{2\pi} \int_{-\infty}^{\infty} e^{-itx}\phi(t)\,dt. \tag{6.7}$$

To show this we observe first that $F(x)$ is continuous from (6.5) in which the integrand is absolutely integrable on the range $(-\infty, \infty)$, and that

$$\frac{F(x+h)-F(x)}{h} = \frac{1}{2\pi} \lim_{T \to \infty} \int_{-T}^{T} \left(\frac{1-e^{-ith}}{ith}\right) e^{-itx}\phi(t)\,dt,$$

in which the integrand is uniformly bounded by $|\phi(t)|$ for all non-zero h, so that we can proceed to the limit as $h \to 0$ inside the right-hand integral to obtain (6.7). The boundedness and continuity then follow easily from (6.7). However, it can be shown (e.g. Lukacs 1960, p. 72) that not all absolutely continuous distributions have $\phi(t)$'s which are absolutely integrable. We can also always find the jump at any point for

$$F(x+0)-F(x-0) = \lim_{T \to \infty} \frac{1}{2T} \int_{-T}^{T} e^{-itx}\phi(t)\,dt.$$

To prove this we observe that for any given T the expression on the right-hand side can be written

$$\frac{1}{2T} \int_{-T}^{T} e^{it(u-x)}\,dF(u)\,dt,$$

in which the order of integration can always be interchanged to obtain

$$\int_{-\infty}^{\infty} \frac{\sin T(u-x)}{T(u-x)}\,dF(x),$$

and since $F(x)$ has limits $F(x+0)$, $F(x-0)$ to the right and left whilst the integrand is bounded, the limit of the integral as $T \to \infty$ is seen to be $\frac{1}{2}\{F(x+0) - F(x-0)\}$.

In a similar way it is possible to set up an inversion formula for a multi-dimensional characteristic function. It is most convenient to express this for values for which the cumulative distribution function $F(x_1,...,x_n)$ does not have jumps. Thus if $(a_i, a_i + h_i)$ are values for which the marginal distribution $F(\infty,...,\infty, x_i, \infty,...,\infty)$ does not have jumps, we find

$$\mathrm{pr}(a_1 \leqslant X_1 \leqslant a_1 + h_1,..., a_n \leqslant X_n \leqslant a_n + h_n)$$

$$= \frac{1}{(2\pi)^n} \lim_{T \to \infty} \int_{-T}^{T} \cdots \int_{-T}^{T} \prod_{i=1}^{n} \left(\frac{1 - e^{-it_i h_i}}{it_i}\right) e^{-i\Sigma t_i a_i} \phi(t_1,...,t_n)\, dt_1 \cdots dt_n$$

$$(6.8)$$

and this is clearly sufficient to specify $F(x_1,...,x_n)$ except at an enumerable set of points. If $\phi(t_1,...,t_n)$ is absolutely integrable on the whole space, (6.8) can be similarly modified to obtain the probability density which then exists everywhere.

6.2. The convergence of distributions

The unique correspondence between a distribution function and its characteristic function suggests that when either converges to a limit, in some sense, the other should also, and that the limits should correspond. To obtain useful theorems of this type we have to specify carefully the mode of convergence.

We have already defined the weak convergence of a sequence, $F_n(x)$, of distribution functions in Chapter 5 by the condition that there exists a (possibly improper) distribution function $F(x)$ such that $F_n(x)$ converges to $F(x)$ in the ordinary sense at all continuity points of $F(x)$. However, we are now principally interested in convergence to proper distribution functions.

THEOREM 6.2. *Let $F_n(x)$ $(n = 1, 2,...)$ be a sequence of distribution functions with characteristic functions $\phi_n(t)$. Then a necessary and sufficient condition that the sequence $F_n(x)$ converges weakly to a proper distribution function, $F(x)$, is that the sequence, $\phi_n(t)$, converges for each real t to a function $\phi(t)$ which is continuous at $t = 0$. $\phi(t)$ is then the characteristic function of $F(x)$.*

A closely related theorem imposes uniform convergence of the $\phi_n(t)$ in some interval instead of continuity of $\phi(t)$ at the origin, and may be stated as follows.

THEOREM 6.3. *A necessary and sufficient condition that the sequence $F_n(x)$ converges weakly to a proper distribution function, $F(x)$, is that the sequence $\phi_n(t)$ converges to a function $\phi(t)$ for all real t, and this convergence is uniform in some finite interval containing $t = 0$ in its interior.*

Theorems 6.2 and 6.3 are generalizations of Theorem 2.1 in § 2.5. To prove these theorems we need two analytical results in the theory of Stieltjes integrals known as the first and second theorems of Helly.

THEOREM 6.4 (Helly's first theorem). *Every sequence, $\{F_n(x)\}$, of uniformly bounded non-decreasing functions contains a subsequence $F_n(x)$ which converges weakly to a non-decreasing bounded function $F(x)$.*

Proof. Let $x_1, x_2,...$ be an enumerable set of values of x dense on the whole real line $(-\infty, \infty)$. Let $F_1^{(1)}(x), F_2^{(1)}(x),...$ be a subsequence of $\{F_n(x)\}$ which converges to some particular value when $x = x_1$. Similarly let $F_1^{(2)}(x), F_2^{(2)}(x),...$ be a subsequence of $\{F_n^{(1)}(x)\}$ which converges for $x = x_2$, and similarly define the sequences $\{F_n^{(3)}(x)\},....$ Then the sequence $F_1^{(1)}(x), F_2^{(2)}(x),...$ converges at all the points x_i to a function $F(x)$. By the way in which it is defined the function $F(x)$ is bounded and non-decreasing, and because of their monotonicity, the sequence of functions $F_n^{(n)}(x)$ must converge to $F(x)$ at all the latter's continuity points. This completes the proof of the theorem which in fact implies that in the sense of weak convergence, every sequence of distribution functions is weakly compact. This theorem starts from a given sequence, $\{F_n(x)\}$, of proper distribution functions and concludes to a limiting function, $F(x)$, which may be improper, i.e. may have all the corresponding probability concentrated at $\pm\infty$. It is clear that an arbitrary set of proper distributions need not be compact in the sense of strong convergence. In order to obtain a compactness theorem in the sense of strong convergence it is necessary to impose some further condition. It is, for example, easy to show that a sequence of distribution functions, $F_n(x)$, is compact in the sense of strong convergence if there exists a function, $g(x)$, which tends to zero as $x \to \infty$ and is such that

$$F_n(-x)+1-F_n(x) < g(x) \quad (x > 0),$$

for all n. In particular, a sequence of distribution whose dispersions and medians are bounded is strongly compact.

THEOREM 6.5 (Helly's second theorem). *Suppose that $g(x)$ is a continuous function, and that $\{F_n(x)\}$ is a sequence of distribution functions converging weakly to a function $F(x)$, for all values x in a finite closed interval $[a, b]$ for which $F(x)$ is continuous at $x = a$ and $x = b$. Then*

$$\int_a^b g(x)\,dF_n(x) \to \int_a^b g(x)\,dF(x),$$

in the sense of ordinary convergence.

Proof. It is clearly sufficient to prove this when $g(x) > 0$. Let $x_1,..., x_k$, be continuity points of $F(x)$ such that $a = x_1 < ... < x_k = b$, and such that $|g(x)-g(x_i)| \leqslant \varepsilon$ for x in (x_i, x_{i+1}) where $\varepsilon > 0$. Write $g_m(x) = \min g(x)$ for x in (x_i, x_{i+1}). Then the left-hand side is not less than

$$S = \sum_{i=1}^{k-1} \int_{x_i}^{x_{i+1}} g_m(x)\,dF_n(x)$$

where we make the convention that jumps in $F_n(x)$ at x_i and x_{i+1} make no contribution to the integrals. Then S converges to

$$\int_a^b g_m(x)\,dF(x) \geqslant \int_a^b g(x)\,dF(x) - \varepsilon(F(b) - F(a)),$$

and

$$\underline{\lim} \int_a^b g(x)\,dF_n(x) \geqslant \int_a^b g(x)\,dF(x) - \varepsilon(F(b) - F(a)).$$

Since ε is arbitrary,

$$\underline{\lim} \int_a^b g(x)\,dF_n(x) \geqslant \int_a^b g(x)\,dF(x).$$

Similarly

$$\overline{\lim} \int_a^b g(x)\,dF_n(x) \leqslant \int_a^b g(x)\,dF(x),$$

and the theorem is proved.

It is desirable to be able to use a theorem of this kind also for an infinite interval so we prove the following extension of theorem 6.5.

THEOREM 6.6. *If $g(x)$ is bounded and continuous on the infinite interval $(-\infty, \infty)$, and $\{F_n(x)\}$ is a sequence of distribution functions converging to a distribution function $F(x)$ such that $F(-\infty) = 0$, $F(+\infty) = 1$, then*

$$\int_{-\infty}^{\infty} g(x)\,dF_n(x) \to \int_{-\infty}^{\infty} g(x)\,dF(x)$$

in the sense of ordinary convergence.

Proof. We can write the difference between the above integrals as the sum of differences of integrals over $(-\infty, a)$, (a, b), (b, ∞), where a and b are continuity points of $F(x)$. Since $g(x)$ is bounded it follows from the above conditions that we can choose a and b so that the first and third of these are less than ε, where ε is any positive number, independently of n. We can then choose n so large that the second difference of integrals is less than ε by theorem 6.5. This proves the result.

Proof of Theorem 6.2. First suppose that the sequence $\{F_n(x)\}$ converges to a distribution function $F(x)$. Then putting $g(x) = e^{itx} = \cos tx + i \sin tx$ in Theorem 6.6, we see that $\phi_n(t)$ converges to $\phi(t)$ for every t, and since $\phi(t)$ is the characteristic function of $F(x)$, it is continuous.

Next suppose that $\phi_n(t)$ converges for all t to a function $\phi(t)$ which is continuous at $t = 0$. From Helly's first theorem the sequence $\{F_n(x)\}$ contains a subsequence, $\{F_{n_k}(x)\}$ $(k = 1, 2, \ldots)$, which is weakly convergent to a bounded non-decreasing function $F(x)$, which must therefore satisfy the inequalities $0 \leqslant F(x) \leqslant 1$. We must now prove that $F(+\infty) - F(-\infty) = 1$. Suppose this was not true, and that $F(+\infty) - F(-\infty) = 1 - 3\varepsilon$ where $\varepsilon > 0$. $\phi(t)$ is continuous and $\phi(0) = 1$, so that there exists a positive number η such that

$$\frac{1}{\eta}\left|\int_0^\eta \phi(t)\,dt\right| > 1 - \varepsilon. \qquad (6.9)$$

Now take any $X > 2(\varepsilon\eta)^{-1}$, and then k large enough so that

$$\int_{-X}^X dF_{n_k}(x) < 1 - 2\varepsilon.$$

Let ϕ_{n_k} be the characteristic function of $F_{n_k}(x)$. We have

$$\int_0^\eta \phi_{n_k}(t)\,dt = \int_{-\infty}^\infty \int_0^\eta e^{ixt}\,dt\,dF_{n_k}(x).$$

When $|x| \leqslant X$ the inner integral can be majorized by η, and when $|x| > X$ we have

$$\left|\int_0^\eta e^{ixt}\,dt\right| = \left|\frac{e^{ix\eta} - 1}{ix}\right| < \frac{2}{x}.$$

Thus for all sufficiently large k,

$$\frac{1}{\eta}\left|\int_0^\eta \phi_{n_k}(t)\,dt\right| < \int_{-X}^X dF_{n_k}(x) + \int_{x \geqslant |X|} \frac{2}{\eta X}\,dF_{n_k}(x)$$

$$< 1 - 2\varepsilon + \varepsilon = 1 - \varepsilon.$$

However, the sequence $\phi_{n_k}(t)$ converges uniformly to $\phi(t)$ which contradicts (6.9), which is a strict inequality.

Thus every convergent subsequence converges to a distribution function whose characteristic function is $\phi(t)$, and by Theorem 6.1 this distribution is unique.

Proof of Theorem 6.3. We first prove that if $F_n(x)$ converges to $F(x)$, the convergence of $\phi_n(t)$ to $\phi(t)$ must be uniform in every finite interval. For any $\varepsilon > 0$ choose X and N large so that for all $n > N$

$$F(-X) + 1 - F(X) < \varepsilon \quad \text{and} \quad F_n(-X) + 1 - F_n(X) < \varepsilon.$$

We then have

$$|\phi_n(t) - \phi(t)| \leqslant 2\varepsilon + \left|\int_{-X}^X e^{itx}\,dF(x) - \int_{-X}^X e^{itx}\,dF_n(x)\right|$$

and we can assume that X and $-X$ are continuity points of $F(x)$.

Let $T > 0$. We shall prove the convergence is uniform in the interval $-T \leqslant t \leqslant T$. Subdivide the interval $(-X, X)$ into N intervals by points at which $F(x)$ is continuous and denote these by

$$-X = x_0 < x_1 < \ldots < x_N = X.$$

Let the largest of these intervals be less than εT^{-1}. Then

$$\left| \int_{-X}^{X} e^{itx}\, dF(x) - \int_{-X}^{X} e^{itx}\, dF_n(x) \right|$$

$$\leqslant \left| \sum_{i=0}^{N-1} \left\{ \int_{x_i}^{x_{i+1}} e^{itx_i}\, dF(x) - \int_{x_i}^{x_{i+1}} e^{itx_i}\, dF_n(x) \right\} \right|$$

$$+ \left| \sum_{i=0}^{N-1} \left\{ \int_{x_i}^{x_{i+1}} (e^{itx} - e^{itx_i})\, dF(x) + \int_{x_i}^{x_{i+1}} (e^{itx} - e^{itx_i})\, dF_n(x) \right\} \right|.$$

Keeping t fixed we can choose n so large that the modulus of the first sum is less than ε. In each of the intervals (x_i, x_{i+1}), the derivative of e^{itx} is not greater than T in modulus, and the second sum is therefore less than 2ε, and thus $|\phi_n(t) - \phi(t)| < 4\varepsilon$ for n sufficiently large.

Conversely suppose that the convergence of $\phi_n(t)$ to $\phi(t)$ is uniform in some interval $(-T, T)$ containing the origin. Since the uniform limit of a sequence of continuous functions is continuous, $\phi(t)$ must be continuous in this interval and the theorem is proved.

It will be noticed that the above theorems relate the convergence of the sequence $\{F_n(x)\}$ to the convergence of $\phi_n(t)$ on the real axis and this is the most useful general theorem since we know that $\phi_n(t)$ exists there for any distribution function $F_n(x)$. When x is purely imaginary the characteristic function becomes the moment generating function (6.1) which is the **Laplace** transform of the distribution and does not always exist. It is, however, sometimes useful to have a corresponding theorem for moment generating functions and this is provided by Theorem 6.7 (Kac 1959).

THEOREM 6.7. *Let $\{F_n(x)\}$ be a sequence of distribution functions with moment generating functions*

$$M_n(t) = \int_{-\infty}^{\infty} e^{tx}\, dF_n(x).$$

If the $M_n(t)$ converge for all real t to a function $M(t)$ which is given by

$$M(t) = \int_{-\infty}^{\infty} e^{tx}\, dF(x),$$

where $F(x)$ is a proper distribution function, and if there exists an integral function, $N(z)$, which reduces to $M(z)$ on the real line, then $F_n(x)$ tends to $F(x)$ at every continuity point of the latter.

Proof. For any complex number $w = t + iu$ write

$$M_n(w) = \int_{-\infty}^{\infty} e^{wx}\, dF_n(x).$$

Then

$$|M_n(w)| \leqslant \int_{-\infty}^{\infty} |e^{wx}|\, dF_n(x) = M_n(t),$$

and hence $M_n(w)$ exists for all w for sufficiently large n. Taking R as any positive number we can let N be large enough for $M_n(w)$ to be finite for all $n \geqslant N$ and all w such that $|w| \leqslant R'$ where R' satisfies $0 < R < R'$. Consider the circle $|w| < R$. For every point w inside this circle we have

$$(\delta w)^{-1}\{M_n(w + \delta w) - M_n(w)\} = \int_{-\infty}^{\infty} (\delta w)^{-1}\{e^{ix(w + \delta w)} - e^{ixw}\}\, dF_n(x)$$

$$= \int_{-\infty}^{\infty} e^{ixw} ix\, dF_n(x) + \delta w \int_{-\infty}^{\infty} K(x, w)\, dF_n(x),$$

where $K(x, w)$ is bounded by $|x^2 e^{Rx}|$ for $-\infty < x < \infty$, $|w| < R$. Since $R < R'$, the integral of $K(x, w)$ is finite, and the first integral exists for $|w| < R$ since the integrand is majorized by $|x e^{Rx}|$. Thus the derivative of $M_n(w)$ must exist for $n \geqslant N$ and $|w| \leqslant R$. For these values $M_n(w)$ is analytic and bounded uniformly in n.

We can therefore choose a subsequence $M_{n_k}(w)$ $(k = 1, 2, ...)$ which converges uniformly to an analytic function. This function equals $M(x)$ on the interval $-R < x < R$, and hence equals $M(w)$ by analytic continuation for $|w| < R$. Thus in every finite interval $-R < \xi < R$ the integral

$$\int_{-\infty}^{\infty} e^{i\xi x}\, dF_{n_k}(x)$$

converges uniformly to $M(i\xi)$ and hence $F_{n_k}(x)$ converges weakly to $F(x)$. Since the same argument holds for every subsequence it follows that $F_n(x)$ converges weakly to $F(x)$.

In this theorem we have shown that convergence of distributions is implied by convergence of their characteristic functions on the imaginary axis, but we have had to impose the further conditions that the limit on the imaginary axis is a moment generating function of a distribution and that it is the value there of an integral function in the whole plane. These conditions are severe but are nevertheless often fulfilled in practice. On the relationship between uniform convergence of $\phi_n(t)$ and $F_n(x)$ see Dyson (1953).

6.3. Characteristic functions and moments

We now consider the relationship between moments of a distribution, if they exist, and the characteristic function.

THEOREM 6.8. *If the characteristic function, $\phi(t)$, has a derivative of order k at $t = 0$, all the moments exist up to m_k if k is even, and up to m_{k-1} if k is odd.*

Proof. It is clearly sufficient to suppose k is even for if not, $k-1$ is even. The existence of the kth derivative at the origin implies that the kth symmetric derivative exists there. This is defined to be

$$\lim_{t \to 0} (2t)^{-k} \sum_{s=0}^{k} (-1)^s \binom{k}{s} \phi((k-2s)t),$$

$$= \lim_{t \to 0} \int_{-\infty}^{\infty} \left\{ \frac{e^{itx} - e^{itx}}{2t} \right\}^k dF(x),$$

$$= (i)^k \lim_{t \to 0} \int_{-\infty}^{\infty} \left\{ \frac{\sin tx}{tx} \right\}^k x^k \, dF(x).$$

Since k is even, this is not less in modulus than

$$\lim_{t \to 0} \int_a^b \left\{ \frac{\sin tx}{tx} \right\}^k x^k \, dF(x) = \int_a^b x^k \, dF(x),$$

where a and b are any finite numbers. Since the latter integral is thus bounded for all a, b, the kth moment must exist and therefore all moments of order less than k. Zygmund has given an example to show that this is a best possible result (Lukacs 1960).

THEOREM 6.9. *If the kth absolute moment of a distribution exists, the characteristic function is differentiable k times, and*

$$\phi^{(k)}(t) = \int (ix)^k e^{itx} \, dF(x).$$

Proof. Since the kth absolute moment exists, the integral above is absolutely convergent and the order of integration and differentiation can be interchanged giving the above equality. For a further result see Pitman (1956).

THEOREM 6.10. *If the absolute moment exists we can write*

$$\phi(t) = \sum_{s=0}^{k-1} m_s \frac{(it)^s}{s!} + t^k \int_0^1 \frac{(1-u)^{k-1}}{(k-1)!} \phi^{(k)}(tu) \, du.$$

Proof. We know that

$$e^{itx} = \sum_{s=0}^{k-1} \frac{(itx)^s}{s!} + (itx)^k \int_0^1 \frac{(1-u)^{k-1}}{k-1!} e^{iutx} \, du,$$

and integrating this with respect to $F(x)$ we obtain the result by using Theorem 6.9, and observing that we can invert the order of integration. When k is even it is useful to strengthen this theorem as follows.

THEOREM 6.11. *If k is even, and m_k is finite,*

$$\phi(t) = \sum_{s=0}^{k-1} m_s \frac{(it)^s}{s!} + \rho m_k \frac{(it)^k}{k!}, \qquad (6.10)$$

for real t, where $|\rho| \leqslant 1$, and as $t \to 0$, $\rho \to 1$.

Proof. In the formula

$$e^{itx} = \sum_{s=0}^{k-1} \frac{(itx)^s}{s!} + (itx)^k \int_0^1 \frac{(1-u)^{k-1}}{k-1!} e^{iutx}\, du,$$

the integral is less in modulus than

$$\int_0^1 \frac{(1-u)^{k-1}}{k-1!} |e^{iutx}|\, du = \frac{1}{k!},$$

and

$$e^{itx} = \sum_{s=0}^{k-1} \frac{(itx)^s}{s!} + \rho' \frac{(itx)^k}{k!}$$

where $|\rho'| \leqslant 1$, and t and x are real. Integrating this we obtain (6.10). Now for any finite x interval, $e^{iutx} \to 1$ as $t \to 0$ uniformly in x (since u is bounded), and thus $\rho' \to 1$ as $t \to 0$ uniformly in any finite x-interval. By definition

$$\rho m_k = \int_{-\infty}^\infty \rho' x^k\, dF(x),$$

so that

$$|\rho - 1| m_k \leqslant \int_{-\infty}^\infty |\rho' - 1| x^k\, dF(x),$$

$$\leqslant \varepsilon + \int_{-A}^A |\rho' - 1| x^k\, dF(x),$$

by choosing A large enough and dependent only on $F(x)$. We can then choose t small enough and dependent on A for $|\rho' - 1|$ to be so small uniformly in $(-A, A)$ that

$$|1 - \rho| m_k \leqslant 2\varepsilon.$$

Hence as $t \to 0$, $\rho \to 1$.

If all the moments exist we can write

$$\phi(t) = \sum_{s=0}^\infty m_s \frac{(it)^s}{s!}$$

in the interval of convergence of this series if this interval exists.

6.4. Unique determination by moments

We are now in a position to return to a question left open in the last chapter, namely how far moments determine a distribution, and the related problem of the relationship between convergence of moments and convergence of distributions. The complete discussion of this problem is analytically too complicated to be considered here (see Shohat and Tamarkin 1943) but some of the results are often useful in particular applications of probability theory.

It must first be pointed out that a set of moments, even if all finite, do not necessarily determine a distribution uniquely. This is connected with the fact that knowledge of $\phi(t)$ in the neighbourhood of the origin does not necessarily determine it over the whole range of real values of t.

There are a number of well-known examples which illustrate this of which the simplest is the log-normal distribution (see § 7.16).

It will be easily suspected that the non-uniqueness is associated with the fact that the characteristic function is not analytic at the origin. The following theorem assuming such analyticity provides a sufficient condition for the uniqueness of determination of a· distribution by its moments which is adequate in most particular cases.

THEOREM 6.12. *If the moments, m_r, of a distribution are finite, and are such that the series*

$$\sum_{0}^{\infty} m_r z^r$$

is convergent in a circle with a radius $\rho > 0$, then the distribution is uniquely determined by its moments.

To prove this we use the following result which is of independent interest.

THEOREM 6.13. *If the characteristic function, $\phi(z)$, is analytic inside a circle of radius $\rho > 0$, it is analytic inside the strip $|\mathscr{I}z| < \rho$.*

Proof. Since $\phi(z)$ is regular inside the circle $|z| < \rho$, it follows that

$$\sum_{n=1}^{\infty} m_{2n}(2n!)^{-1}z^{2n} = \sum_{1}^{\infty} v_{2n}(2n!)^{-1}z^{2n}$$

is convergent for z real and $|z| < \rho$. Now from Theorem 5.16,

$$(v_{2n-1})^{1/(2n-1)} \leqslant (v_{2n})^{1/2n} = (m_{2n})^{1/2n},$$

so that

$$v_{2n-1} \leqslant \max(v_{2n}, 1),$$

and therefore

$$\sum_{n=1}^{\infty} v_{2n-1}(2n-1!)^{-1}z^{2n-1}$$

is convergent for $|z| < \rho$ since each term is certainly bounded by the sum of the corresponding terms in

$$\sum_{n=1}^{\infty} v_{2n}(2n-1!)^{-1}z^{2n-1},$$

and

$$\sum_{n=1}^{\infty} (2n-1!)^{-1}z^{2n-1},$$

both of which are convergent for $|z| < \rho$. It follows that

$$\int_{-\infty}^{\infty} e^{z|x|}\,dF(x) = \sum_{0}^{\infty} v_n(n!)^{-1}z^n$$

is convergent for z real and $|z| < \rho$. Then if $z = u + iv$,

$$\phi(z) = \int_{-\infty}^{\infty} e^{iux - vx} \, dF(x)$$

exists for all u real and $-\rho < v < \rho$, and can be expanded about any point on the real axis in a power series which is convergent in a circle of radius ρ.

Thus the characteristic function is analytic everywhere on the real axis and is determined uniquely by analytic continuation from its values near the origin. Since the values of the characteristic function on the real axis determine the distribution uniquely, Theorem 6.12 is proved.

Various particular cases of this result are important. Thus if

$$\overline{\lim_{n \to \infty}} \, n^{-1} m_{2n}^{1/2n} < \infty, \tag{6.11}$$

the distribution is uniquely determined since in that case there exists a number $K > 0$ such that

$$n^{-1} \nu_{2n}^{1/2n} < K < \infty,$$

and

$$\sum_{0}^{\infty} \frac{\nu_{2n}}{(2n)!} z^{2n} \leqslant \sum_{0}^{\infty} \frac{(Knz)^{2n}}{(2n)!},$$

which is convergent for $|z| < (Ke)^{-1}$ since $n!$ is asymptotically equal to $(2\pi)^{\frac{1}{2}} n^{n+\frac{1}{2}} e^{-n}$. In particular if a distribution has a finite range it is uniquely determined by its moments, for if the random variable X is such that $|X| < A > 0$, then $\nu_n < A^n$, and the above condition applies. A more important particular application is to the normal distribution for which (for zero mean and unit standard deviation),

$$\mu_{2n} = \frac{(2n)!}{2^n n!},$$

and

$$n^{-1}(\mu_{2n})^{1/2n}$$

is asymptotically equal to

$$n^{-\frac{1}{2}} e^{-\frac{1}{2}},$$

which tends to zero.

A more general sufficient condition for the uniqueness of determination of a distribution has been given by Carleman (1925) who has proved that the distribution is unique if the series

$$\sum_{1}^{\infty} (\mu_{2n})^{-1/2n} \tag{6.12}$$

diverges, and this result has been extended to multivariate distributions by Cramer and Wold (1936).

The more fundamental problem as to whether there exists a distribution whose moments are a prescribed sequence of constants has been completely solved by Hamburger, but leads to very complicated mathematics which is described in Shohat and Tamarkin (1943), who give a much more complete account of the analytical theory of moments than would be useful here.

6.5. Convergence of moments

We now study how far convergence of a sequence of distributions is related to convergence of the corresponding moments. From the above discussion of uniqueness we see that mere convergence of the moments to a set of finite constants will not ensure convergence of the corresponding distribution. Conversely convergence of a sequence of distribution functions will not by itself ensure convergence of the moments since we can alter each of the tails of the distributions in such a way that the convergence of the distributions is unaltered but so that the moments of even order, say, get larger and larger. It is therefore necessary to impose additional conditions. The following account is based on Rao and Kendall (1950).

THEOREM 6.14. *Suppose that we have a sequence of distribution functions,* $F_n(x)$, $n = 1, 2,...$, *such that for each j the jth moment,* $m_j^{(n)}$, *of* $F_n(x)$ *exists for all sufficiently large n, e.g. for* $n \geqslant N_j$, *and suppose that*

$$\lim_{n \to \infty} m_j^{(n)} = \lambda_j < \infty$$

for all j. Then if the sequence $\{F_n(x)\}$ *converges to a limit* $G(x)$ *(which must be bounded and non-decreasing) at all the continuity points of* $G(x)$, *we can conclude that*

(1) $G(x)$ *is a proper distribution having finite moments of all orders;*
(2) λ_j *is the jth moment of* $G(x)$.

Proof. $G(x)$ must be a proper distribution for λ_2 is finite, and the use of Tchebychev's inequality shows that $G(-\infty) = 0$, $G(\infty) = 1$. Hence by Theorem 6.2 the sequence of characteristic functions of F_n, which we write $\{\phi_n(t)\}$, converges to $\phi(t)$, the characteristic function of $G(x)$.

We now need a converse to Theorem 6.11 which is of some interest in itself.

LEMMA 6.1. *Suppose that* $\phi(t)$ *is the characteristic function of a distribution* $G(x)$, *and that for some positive integer k*

$$\phi(t) = \sum_{s=0}^{2k} \frac{\lambda_s}{s!} (it)^s + o(t^{2k}),$$

when $t \to 0$ *through real values. Then the first 2m moments of* $G(x)$ *exist and are equal to* λ_s.

Proof. Let δ^2 be the operator corresponding to the symmetrical second difference with interval $2h$ so that

$$\delta^2 f(t) = f(t+2h) - 2f(t) + f(t-2h).$$

Then

$$\lim_{h \to 0} \left(\frac{\delta^2}{4h^2}\right)^k \left\{\sum_{s=0}^{2k} \frac{\lambda_s}{s!} (it)^s\right\} = (-1)^k \lambda_{2k}.$$

The $o(t^{2k})$ term can be written as $t^{2k} g(t)$ where $g(t) \to 0$ as $t \to 0$. Use of standard difference operator formulae then enables one to see that

$$\left|\left(\frac{\delta^2}{4h^2}\right)^k \{t^{2k} g(t)\}\right| = \left|\sum_{s=0}^{2k} \binom{2k}{s}(-1)^s(k-s)^{2k} g(2kh - 2sh)\right|$$

$$\leqslant \sum_{s=0}^{2k} \binom{2k}{s}(k-s)^{2k} \varepsilon,$$

if $|g(t)| < \varepsilon$ for $|t| \leqslant \eta$, where $\eta \geqslant 2kh$. Thus by choosing ε small, and then t small, we see that

$$\lim_{h \to 0} \left(\frac{\delta^2}{4h^2}\right)^k \{t^{2k} g(t)\} \leqslant \sum_{s=0}^{2k} \binom{2k}{s} |k-s|^{2k} \varepsilon,$$

where ε is arbitrarily small. Hence the left-hand side is zero at $t = 0$. Therefore

$$\lim_{h \to 0} \left(\frac{\delta^2}{4h^2}\right)^k \phi(t) = (-1)^k \lambda_{2k}, \quad \text{at} \quad t = 0.$$

We also have

$$\delta^2 e^{ixt} = -(2 \sin xh)^2 e^{ixt},$$

so that by taking the operator δ^{2k} inside the integral for $\phi(t)$, we have

$$\lim_{h \to 0} \int_{-\infty}^{\infty} \left(\frac{\sin xh}{xh}\right)^{2k} x^{2k} \, dF(x) = \lambda_{2k}.$$

The left-hand side must tend to m_{2k}, finite or infinite, as $h \to 0$. Hence $m_{2k} = \lambda_{2k}$ and is finite. By Theorem 6.11 we have

$$\phi(t) = \sum_{s=0}^{2k} \frac{m_s}{s!} (it)^s + \frac{m_{2k}}{(2k)!} (it)^{2k} \varepsilon,$$

where $\varepsilon \to 0$ as $t \to 0$. By comparing the behaviour of this and the previous expansion near $t = 0$ we also see that $m_s = \lambda_s$ for odd s and the lemma is proved.

Now for $n > N_{2k}$ in Theorem 6.14 we can write, from Theorem 6.11,

$$\phi_n(t) = \sum_{s=0}^{2k-1} \frac{(it)^s}{s!} m_s^{(n)} + \rho \frac{(it)^{2k}}{(2k)!} m_{2k}^{(n)},$$

for every fixed t, and since $\phi_n(t)$, and $m_s^{(n)}$ $(s = 0, \ldots, 2k-1)$ converge to limits so also does the last term, and for each k we can write

$$\phi(t) = \sum_{s=0}^{2k-1} \frac{(it)^s}{s!} \lambda_s + R \frac{(it)^{2k}}{(2k)!} \lambda_{2k}.$$

Furthermore since $|\rho| \leqslant 1$ for each n, we have $|R| \leqslant 1$. R being bounded, the conditions of the lemma hold for every k and the moments of $G(x)$ exist and are equal to λ_s. Notice that it is not assumed that all the moments of each distribution $F_n(x)$ exist, but only that the moments of any given order exist from some point onwards.

We can now prove the two main theorems relating the convergence of moments to the convergence of distributions and vice versa, in each case imposing an additional condition.

THEOREM 6.15. *Suppose the sequence of distribution functions* $\{F_n(x)\}$ *converges to the distribution* $G(x)$ *at all the latter's points of continuity and that the moments* $m_s^{(n)}$ *exist for* $n > N_s$, *so that* $|m_j^{(n)}| < A_j < \infty$. *Then the moments,* λ_j, *of* $G(x)$ *exist, and* $m_j^{(n)} \to \lambda_j$, *for each* j.

Proof. The sequence, $\{\Phi_n(t)\}$, of characteristic functions of $\{F_n(x)\}$ converge for every real t to the characteristic function, $\phi(t)$, of $G(x)$. Furthermore when $n > N_2$, we see from Theorem 6.11 that

$$\phi_n(t) = 1 + itm_1^{(n)} + K_n(it)^2,$$

where $|K_n| < K$, a constant independent of t and n. Taking t fixed we have

$$|\overline{\lim}\, m_1^{(n)} - \underline{\lim}\, m_1^{(n)}| \leqslant t^{-1}\{\overline{\lim}|\phi_n(t)| - \underline{\lim}|\phi_n(t)|\} + 2tK$$
$$\leqslant 2tK,$$

and since this is true for arbitrarily small t we have

$$\overline{\lim}\, m_1^{(n)} = \underline{\lim}\, m_1^{(n)},$$

so that $m_1^{(n)}$ converges to a finite limit which we write as λ_1.

Next for sufficiently large n, the fourth moment exists and hence we can write

$$\phi_n(t) = 1 + itm_1^{(n)} + \tfrac{1}{2}(it)^2 m_2^{(n)} + K_n^{(2)}(it)^3$$

where $|K_n^{(2)}| < K$, independent of n and t. A similar argument then shows that the limit of $m_2^{(n)}$ exists and is finite. In this way the limits of sequences of all moments can be shown to exist and the fact that these are the moments of $G(x)$ follows from Theorem 6.14.

THEOREM 6.16. *Suppose the moments,* $m_j^{(n)}$, *of* $F_n(x)$ *exist when* $n \geqslant N_j$ *and have limits* λ_j *which are known to be the moments of a unique distribution* $G(x)$. *Then the sequence,* $\{F_n(x)\}$, *converges to* $G(x)$ *at all the latter's points of continuity.*

Since the sequence of second moments is bounded above for $n > N_2$, it is clearly possible to pick out a subsequence of $F_n(x)$ which converges to a proper distribution $H(x)$ at all its points of continuity. Then by Theorem 6.14 the moments of $H(x)$ must exist and must equal λ_j, and by uniqueness we must have $H(x) = G(x)$ at the points of continuity of either. If now the sequence $\{F_n(x)\}$ does not converge to $G(x)$ it would be possible to choose a value x_0 such that $G(x)$ is continuous at x_0 and so that $\overline{\lim}\, F_n(x_0) \neq G(x_0)$, or else $\underline{\lim} F_n(x_0) \neq G(x_0)$. Either of these possibilities would imply that it is possible to choose a subsequence of $F_n(x)$ which converges to a distribution $H(x)$ not equal to $G(x)$ at the latter's points of continuity. This is impossible, and the theorem is therefore proved.

Theorem 6.16 is important in particular probability problems, especially when used to prove the convergence of a distribution, suitably standardized to have zero mean and fixed variance, to normality, for the normal distribution is uniquely determined by its moments as we have seen above. This type of argument frequently occurs in combinatorial problems.

6.6. The characteristic function of a convolution

Instead of moments it is often useful to consider another sequence of constants describing a distribution which are functions of moments and are known as cumulants. However, before introducing these it is useful to prove the fundamental theorem giving the characteristic function of the distribution of the sum of two independent random variables.

THEOREM 6.17. *If X and Y are independent random variables with distribution functions $F_1(x)$ and $F_2(y)$, and characteristic functions $\phi_1(t)$ and $\phi_2(t)$, the characteristic function of $X + Y$ is $\phi_1(t)\phi_1(t)$.*

Proof. Suppose first that the distributions of X and Y are discrete so that X takes values $x_0, x_{\pm 1},\ldots$ with probabilities $p_0, p_{\pm 1},\ldots$ ($\Sigma p_i = 1$), and Y takes values $y_0, y_{\pm 1},\ldots$ with probabilities $p'_0, p'_{\pm 1},\ldots$ ($\Sigma p'_i = 1$). Then

$$\phi_1(t) = \sum_{-\infty}^{\infty} p_i e^{ix_i t}, \quad \phi_2(t) = \sum_{-\infty}^{\infty} p'_i e^{iy_i t},$$

and the characteristic function of $X + Y$ is

$$\sum_{i,j} p_i p'_j e^{i(x_i + y_j)t} = \phi_1(t)\phi_2(t),$$

so that the result is true for discrete distributions. On the other hand, any distribution can be approximated to as closely as we like, in the sense of weak convergence, by a discrete distribution and by Theorem 6.2 the corresponding characteristic functions will also be as close together, in any finite t-interval, as we desire. The result then follows by proceeding to the limit.

This theorem shows that the operation of composition on two distributions corresponds to multiplication of their characteristic functions, and furthermore that the product of two characteristic functions is again a characteristic function. This helps to explain why the operation of convolution, which we have denoted before by the notation $F_1 * F_2$, has many of the properties of multiplication but there is not a complete analogy, as we shall see later, since 'division' is not always possible, and when possible, not always unique. We can express this by saying that the class of all distributions is made into a semi-group (but not a group) by the operation of convolution.

18

For the characteristic function of the ratio of two independent random variables see Cramér (1962, p. 46). This was extended to non-independent random variables by Geary (1944).

Just as the product of characteristic functions is the characteristic function of the sum of independent random variables, so too is the product of the Mellin transforms of distributions on $(0, \infty)$, the Mellin transform of the distribution of the product of two independent random variables. For a systematic study of such transforms see Epstein (1948), Kabe (1958), Zolotarev (1957), and Fox (1957). For the use of Hankel transforms see Lord (1954a, b).

6.7. Cumulants

It is sometimes more convenient to deal with functions related to distributions whose addition corresponds to convolution. Since $\phi(t) = 1$ when $t = 0$, and is continuous, there exists a neighbourhood of $t = 0$ for which $\phi(t) \neq 0$, and in this neighbourhood we define

$$K(t) = \log \phi(t)$$

as the cumulant generating function of the distribution $F(x)$ even when the cumulants do not exist. We have already considered a special case of this in §2.2. $K(t)$ is also sometimes (and preferably) known as the second characteristic function. The definition of $K(t)$ is unique provided we take $\log \phi(t)$ to be that branch of the logarithm which vanishes when $\phi(t) = 1$.

Suppose that the moments $m_1, \ldots, m_{2k}$ of the distribution $F(x)$ exist. Then by Theorem 6.11 we can write

$$\phi(t) = \sum_{s=0}^{2k} m_s \frac{(it)^s}{s!} + (it)^{2k} o(t),$$

and using the expansion of $\log(1 + x)$ we obtain

$$K(t) = \sum_{s=0}^{2k} \kappa_s \frac{(it)^s}{s!} + o(t^{2k}),$$

where the numbers $\kappa_0, \ldots, \kappa_{2k}$ are the 'cumulants' of the distribution $F(x)$. These can be found in terms of the μ_s by simple expansion and we have in particular,

$$\kappa_1 = m_1,$$

$$\kappa_2 = m_2 - m_1^2,$$

$$\kappa_3 = m_3 - 3m_2 m_1 + 2m_1^2,$$

$$\kappa_4 = m_4 - 4m_3 m_1 - 3m_2^2 + 12m_2 m_1^2 - 6m_1^4, \qquad (6.13)$$

and using the inverse relation,

$$m_1 = \kappa_1,$$

$$m_2 = \kappa_2 + \kappa_1^2,$$

$$m_3 = \kappa_3 + 3\kappa_2 \kappa_1 + \kappa_1^3,$$

$$m_4 = \kappa_4 + 4\kappa_3 \kappa_1 + 3\kappa_2^2 + 6\kappa_2 \kappa_1^2 + \kappa_1^4. \tag{6.14}$$

Tables of m_n in terms of the κ_s, and κ_n in terms of the m_s up to $n = 10$ are given in M. G. Kendall and A. Stuart (1958) who also give general formulae. (See also Frisch 1926.) It will be seen that κ_n is always a polynomial in $m_1,..., m_n$, and vice versa, and thus κ_n and m_n are always finite, infinite, or undefined together.

Furthermore, it is easy to see from the definition of the cumulant generating function that the κ_n ($n \geq 2$) are dependent only on $\mu_2, \mu_3,...$ and are the same for all distributions of the form $F(x+d)$ $(-\infty < d < \infty)$. They are therefore sometimes known as 'semi-invariants' since they are invariant under translation. It will also be seen that they are often analytically simpler than the moments.

If X and Y are independent random variables with moments m_{1i}, m_{2i}, and cumulants κ_{1i} and κ_{2i} ($i = 1,...$), it is then obvious that the moments of $X + Y$ are

$$E(X+Y)^n = \sum_{s=0}^{n} \binom{n}{s} m_{1s} m_{2(n-s)},$$

whilst the nth cumulant of $X+Y$ is $\kappa_{1n} + \kappa_{2n}$ from Theorem 6.17. The simplicity of the latter result is often useful.

The same idea can be extended to bivariate and higher distributions to define multivariate cumulants and a multivariate cumulant generating function but the former are rarely used.

6.8. The Cramér–Wold theorem

Consider n random variables with a joint distribution function $F(x_1,...,x_n)$. Then the characteristic function is

$$\phi(t_1,...,t_n) = \int \cdots \int \exp\{i(t_1 x_1 + ... + t_n x_n)\} dF(x_1,...,x_n).$$

The characteristic function of the random variable $Z = a_1 X_1 + ... + a_n X_n$ is thus $\phi(a_1 t,..., a_n t)$. If therefore the distribution of Z is known for all values of $a_1,..., a_n$, the function $\phi(t_1,...,t_n)$ will be known for all real values of the t_i, and $F(x_1,...,x_n)$ is determined. Thus we have the following theorem due to Cramér and Wold (1936).

THEOREM 6.18. *If the integral of* $dF(x_1, ..., x_n)$ *is known over all 'half-spaces',*

$$a_1 x_1 + ... + a_n x_n \leqslant d,$$

the function $F(x_1, ..., x_n)$ *is determined.*

Real variable proofs of this theorem have been given by Hacaturov (1954) and by Kostelyanec and Resetnyak (1954). The above technique of reducing multivariate to single variate characteristic functions can be used to obtain other useful results.

6.9. Examples of characteristic functions

(1) *Discrete distributions.* If the random variable X takes the values 0, 1, 2,... with probabilities $p_0, p_1, p_2, ...$ the probability generating function is

$$P(z) = p_0 + p_1 z + p_2 z^2 + ...$$

Since the distribution function has jumps of magnitude $p_0, p_1, ...$ at $x = 0, 1, ...$ the characteristic function is

$$\phi(t) = p_0 + p_1 e^{it} + p_2 e^{2it} + ... = P(e^{it}).$$

$P(e^{it})$ is often more convenient to deal with than $P(z)$ especially when we wish to study the convergence of the cumulative discrete distribution (suitably rescaled) to some continuous distribution.

(2) *The rectangular distribution.* Suppose there is a uniform probability density on the range $-A \leqslant x \leqslant A$. Then the characteristic function is

$$\frac{1}{2A} \int_{-A}^{A} e^{itx} dx = \frac{\sin tA}{tA}. \tag{6.15}$$

(3) *The normal distribution.* We have already defined this in § 5.3. If $\phi(t)$ is the characteristic function of the distribution of a random variable X, the characteristic function of $aX + b$ is $\phi(at)e^{bt}$. Consequently we needed only to consider the normal distribution with mean zero and unit standard deviation whose characteristic function is

$$\frac{1}{\sqrt{(2\pi)}} \int_{-\infty}^{\infty} e^{itx - \frac{1}{2}x^2} dx = \frac{e^{-\frac{1}{2}t^2}}{\sqrt{(2\pi)}} \int_{-\infty}^{\infty} e^{-\frac{1}{2}(x - it)^2} dx,$$

$$= \frac{e^{-\frac{1}{2}t^2}}{\sqrt{(2\pi)}} \int e^{-\frac{1}{2}w^2} dw,$$

where the integral is taken along the straight line from $-\infty - it$ to $+\infty - it$. The integrand tends to zero at both ends uniformly for bounded t and rapidly. We integrate around a rectangle with corners $\pm R$, $\pm R - it$, and since there are no singular points and the integrals over the vertical sides of the rectangle tend to zero, the path of integration in the above formula for $\phi(t)$ can be shifted to the real axis and hence

$$\phi(t) = e^{-\frac{1}{2}t^2}. \tag{6.16}$$

Notice that for the normal distribution the analytical forms of $f(x)$ and $\phi(t)$ are similar.

Expansion of $\phi(t)$ provides a convenient way of obtaining the moments of the normal distribution and we get

$$m_n = 0, \qquad n \text{ odd,}$$

$$= \frac{(2m)!}{2^m m!}, \qquad n = 2m. \tag{6.17}$$

(4) *The Cauchy distribution.* This is defined to be the distribution with a probability density

$$f(x) = \frac{1}{\pi(1+x^2)}, \qquad (-\infty < x < \infty),$$

so that

$$\phi(t) = \int_{-\infty}^{\infty} \frac{e^{itx}}{\pi(1+x^2)} dx.$$

Suppose $t \geqslant 0$. Integrate the integrand along the real axis from $-R$ to $+R$ and around a semicircle of radius $R > 1$ in the upper half-plane. There is one pole inside this contour at $x = i$. Evaluating the residue at this pole, letting R tend to infinity, and observing that the integral around the semi-circle tends to zero, we get

$$\phi(t) = e^{-t}, \quad \text{if } t \geqslant 0,$$

and similarly

$$\phi(t) = e^{t}, \quad \text{if } t \leqslant 0.$$

Hence generally

$$\phi(t) = e^{-|t|}.$$

The fact that this is not analytic at $t = 0$ corresponds to the non-existence of the moments in this case.

The form of this characteristic function gives an easy way of verifying the fact, proved in § 5.15, that the mean of n independent random variables having the same Cauchy distribution also has this distribution.

From (3) and (4) we see that $\exp -|t|^\alpha$ is a characteristic function for $\alpha = 1, 2$. In Chapter 9 (§ 9.16) we shall see that this is also true for all α in the range $0 < \alpha \leqslant 2$.

(5) *The gamma or type III distribution.* This is a distribution with a continuous probability density on the positive real axis given by

$$f(x) = \frac{1}{\Gamma(k)} x^{k-1} e^{-x}, \tag{6.18}$$

where $k > 0$. Then

$$\phi(t) = \frac{1}{\Gamma(k)} \int_0^\infty x^{k-1} e^{-x(1-it)} dx.$$

This can be written

$$\frac{1}{\Gamma(k)} \int \frac{z^{k-1}}{(1-it)^k} e^{-z} dz,$$

where the integral is taken along the straight line in the complex plane from $z = 0$ to $z = +\infty(1-it)$. Using Cauchy's theorem we see that this line of integration can be rotated around the origin into the real axis, and so

$$\phi(t) = (1-it)^{-k}. \tag{6.19}$$

This can be expanded into a convergent series when t is small and so all moments are finite. From (6.19) it is clear that if X and Y are independent random variables having type III distributions with parameters k and k', $X + Y$ will also have this distribution with parameter $k + k'$.

(6) *The first passage-time distribution for the Brownian movement.* This is a distribution on the positive real axis, and has a continuous probability density defined by

$$f(x) = \frac{1}{\sqrt{(2\pi)}} e^{-1/2x} x^{-\frac{3}{2}}. \tag{6.20}$$

Putting $x = y^{-2}$, and integrating we verify that

$$\int_0^\infty f(x)\,dx = \frac{2}{\sqrt{(2\pi)}} \int_0^\infty e^{-\frac{1}{2}v^2}\,dy = 1.$$

The characteristic function is

$$\phi(t) = \frac{1}{\sqrt{(2\pi)}} \int_0^\infty \exp\left(-\frac{1}{2x} + itx\right) x^{-\frac{3}{2}}\,dx.$$

It is clear that this integral exists for $\mathscr{I}(t) \geqslant 0$ (i.e. the upper half-plane), and is an analytic function inside this domain which is continuous at finite points of the boundary. We shall find $\phi(t)$ easiest to evaluate when t is purely imaginary, and we have, putting $t = iv$,

$$\phi(t) = M(-v) = \frac{1}{\sqrt{(2\pi)}} \int_0^\infty \exp\left(-\frac{1}{2x} - xv\right) x^{-\frac{3}{2}}\,dx.$$

Since $v > 0$, put $v = u^2$ and we obtain

$$\phi(iu^2) = \frac{e^{-u\sqrt{2}}}{\sqrt{(2\pi)}} \int_0^\infty \exp\left\{-\left(u\sqrt{x} - \frac{1}{\sqrt{(2x)}}\right)^2\right\} x^{-\frac{3}{2}}\,dx$$

$$= \frac{2e^{-u\sqrt{2}}}{\sqrt{(2\pi)}} \int_0^\infty \exp\left\{-\left(\frac{u}{y} - \frac{y}{2}\right)^2\right\} dy$$

on putting $y = x^{-\frac{1}{2}}$. If we now put $u\sqrt{2}/y = z$ we get

$$\frac{2^{\frac{1}{2}} e^{-u\sqrt{2}}}{\sqrt{(2\pi)}} \int_0^\infty \exp\left\{-\left(\frac{z}{\sqrt{2}} - \frac{u}{z}\right)^2\right\} uz^{-2}\, dz.$$

Taking the average of this and the previous integral,

$$\phi(iu^2) = \frac{e^{-u\sqrt{2}}}{\sqrt{(2\pi)}} \int_0^\infty \left(1 + \frac{u\sqrt{2}}{z^2}\right) \exp\left\{-\left(\frac{z}{\sqrt{2}} - \frac{u}{z}\right)^2\right\} dz$$

$$= e^{-u\sqrt{2}}.$$

It therefore follows that for $t > 0$

$$\phi(t) = \exp - \sqrt{(-2it)}. \tag{6.21}$$

Using analytic continuation in the upper half-plane we get

$$\phi(t) = \exp - |t|^{\frac{1}{2}}(1 - i\,\mathrm{sgn}\,t)$$

for t real.

The moments of this distribution are all infinite. It shares with the normal, Cauchy, and a number of other distributions the property that if X and Y are independent random variables having the same distribution, $X + Y$ also has the same distribution when suitably scaled. From (6.16), (6.17), and (6.21) we see that if $C(X + Y)$ is to have the same distribution as X and Y, C must take the values $2^{-\frac{1}{2}}$, 2^{-1}, and 2^{-2} respectively, in these cases.

(7) *The beta distribution.* This distribution has a continuous probability density inside the interval $(0, 1)$ given by

$$f(x) = \frac{\Gamma(l+m)}{\Gamma(l)\Gamma(m)} x^{l-1}(1-x)^{m-1}, \tag{6.22}$$

where $l > 0$, $m > 0$. The characteristic function is

$$\phi(t) = \frac{\Gamma(l+m)}{\Gamma(l)\Gamma(m)} \int_0^1 e^{ixt} x^{l-1}(1-x)^{m-1}\, dx$$

$$= \frac{\Gamma(l+m)}{\Gamma(l)} \sum_{s=0}^\infty \frac{\Gamma(l+s)}{\Gamma(l+m+s)\Gamma(s+1)} (it)^s,$$

since we can expand the exponential term inside the integral and integrate term by term. We can also write this in the form (Whittaker and Watson 1935, p. 352, Example 1)

$$\phi(t) = e^{-\frac{1}{2}il}(it)^{-\frac{1}{2}(l+m)} M_{a,b}(it), \tag{6.23}$$

where $M_{a,b}(it)$ is Whittaker's confluent hypergeometric function, and $a = \frac{1}{2}(m-l)$, $b = \frac{1}{2}(l+m-1)$. (I owe this remark to Dr C. C. Spicer.)

(8) *The multivariate normal distribution.* n random variables $X_1, \ldots, X_n$ have a joint normal distribution when they have a probability density given by (5.35),

$$f(x_1, \ldots, x_n) = \frac{|\mathbf{A}|^{\frac{1}{2}}}{(2\pi)^{n/2}} \exp -\tfrac{1}{2}\mathbf{x}'\mathbf{A}\mathbf{x},$$

where $\mathbf{x}$ is the row vector $(x_1, \ldots, x_n)$, and $\mathbf{A}$ is the $n \times n$ matrix of a positive definite form. Let $\mathbf{t}'$ be the row vector $(t_1, \ldots, t_n)$. Then the characteristic function is

$$\phi(t_1, \ldots, t_n) = \frac{|\mathbf{A}|^{\frac{1}{2}}}{(2\pi)^{n/2}} \int \cdots \int \exp(i\mathbf{t}'\mathbf{x} - \tfrac{1}{2}\mathbf{x}'\mathbf{A}\mathbf{x})\, dx_1 \ldots dx_n,$$

where the integral is taken over the whole n-dimensional space. Then, as before, there exists an orthogonal transformation represented by a matrix $\mathbf{T}$ such that the transformation $\mathbf{x} = \mathbf{T}\mathbf{y}$ reduces $\mathbf{x}'\mathbf{A}\mathbf{x}$ to a sum of squares so that we can write

$$\mathbf{x}'\mathbf{A}\mathbf{x} = \mathbf{y}'\mathbf{T}'\mathbf{A}\mathbf{T}\mathbf{y} = \sum_{s=1}^{n} \sigma_s^{-2} y_s^2,$$

where $\sigma_i^{-2} > 0$. Hence the above integral becomes

$$\frac{|\mathbf{A}|^{\frac{1}{2}}}{(2\pi)^{n/2}} \int \cdots \int \exp\left(i \sum_{s=1}^{n} u_s y_s - \frac{1}{2} \sum_{s=1}^{n} \sigma_s^{-2} y_s^2\right) dy_1 \ldots dy_n \qquad (6.24)$$

where $(u_1, \ldots, u_n)$ is a row vector $\mathbf{u}'$ equal to $\mathbf{t}'\mathbf{T}$.

The integral in (6.24) is then the product of n integrals each of which is the characteristic function of a univariate normal distribution and is therefore equal to

$$|\mathbf{A}|^{\frac{1}{2}} \sigma_1 \ldots \sigma_n \exp\left\{-\frac{1}{2} \sum_{s=1}^{n} u_s^2 \sigma_s^2\right\}.$$

If we write $\mathbf{S}$ for the diagonal matrix whose diagonal elements are σ_s^2 we have $\mathbf{S}^{-1} = \mathbf{T}'\mathbf{A}\mathbf{T}$, and since

$$|\mathbf{A}| = \sigma_1^{-2} \ldots \sigma_n^{-2},$$

the characteristic function is equal to

$$\exp -\tfrac{1}{2}\mathbf{u}'\mathbf{S}\mathbf{u} = \exp\{-\tfrac{1}{2}\mathbf{t}'\mathbf{A}^{-1}\mathbf{t}\}. \qquad (6.25)$$

If we write V_{ij} for the elements of the matrix $\mathbf{A}^{-1}$ we have

$$E(X_i X_j) = V_{ij},$$

and the matrix A^{-1} is therefore known as the 'variance–covariance' matrix of the distribution.

6.10. The characterization of a characteristic function

We have already seen that any characteristic function $\phi(t)$ must be continuous and satisfy the conditions

$$\phi(0) = 1, \quad \phi(-t) = \overline{\phi(t)}, \quad |\phi(t)| \leqslant 1.$$

These are necessary but not sufficient and it is useful to obtain conditions which are both necessary and sufficient although, as we shall see, such conditions turn out to be somewhat awkward.

THEOREM 6.19 (Bochner 1932). *A function, $\phi(t)$, of a real variable t on the interval $(-\infty, \infty)$ is a characteristic function of a distribution if and only if*
   (1) $\phi(0) = 1$,
   (2) $\phi(t)$ *is continuous, and*

$$(3) \quad \sum_{i,j=1}^{n} \phi(u_i - u_j) v_i \bar{v}_j \geqslant 0$$

for any n, any set $u_1, \ldots, u_n$ of real numbers, and any set $v_1, \ldots, v_n$ of complex numbers. $\phi(t)$ is then said to be non-negative definite.

Proof of necessity. If $\phi(t)$ is a characteristic function, conditions (1) and (2) have already been proved, and the quadratic form in (3) can be written

$$\sum_{j,k=1}^{n} v_j \bar{v}_k \int e^{i(u_j - u_k)x} \, dF(x) = \int \left| \sum_{j=1}^{n} v_j e^{iu_j x} \right|^2 dF(x) \geqslant 0.$$

Proof of sufficiency. For x real and $T > 0$, consider the integral

$$I(x) = \frac{1}{T} \int_0^T \int_0^T \phi(u-v) e^{-i(u-v)x} \, du \, dv.$$

If we approximate to this integral by Riemann sums, the latter are of the form of the quadratic form in the condition of the theorem and hence, proceeding to the limit, $I(x) \geqslant 0$. Put $u = v + z$, and integrate first with respect to v for z fixed. Then

$$I(x) = \int_{-T}^{T} \left(1 - \frac{|z|}{T}\right) e^{-izx} \phi(z) \, dz \geqslant 0.$$

Write

$$\phi_T(z) = \left(1 - \frac{T}{|z|}\right) \phi(z), \quad \text{for} \quad |z| \leqslant T,$$

$$= 0, \qquad \qquad \text{for} \quad |z| > T.$$

Then

$$I(x) = \int_{-\infty}^{\infty} \phi_T(z) e^{-izx} \, dz.$$

Multiply both sides by $\{(X - |x|)/2\pi X\} e^{iux}$, and integrate over the range

$(-X, X)$, where $X > 0$. Then

$$\frac{1}{2\pi} \int_{-X}^{X} \left(1 - \frac{|x|}{X}\right) I(x) e^{iux} dx = \frac{1}{2\pi} \int_{-\infty}^{\infty} \phi_T(z) \int_{-X}^{X} \left(1 - \frac{|x|}{X}\right) e^{ix(u-z)} dx dz$$

$$= \frac{1}{2\pi} \int_{-\infty}^{\infty} \phi_T(z) \int_{0}^{X} 2\left(1 - \frac{x}{X}\right) \cos x(u-z) dx dz$$

$$= \frac{2}{\pi} \int_{-\infty}^{\infty} \phi_T(z) \frac{\sin^2 \frac{1}{2} X(u-z)}{X(u-z)^2} dz.$$

Since $\{1 - (|x|/X)\} \geq 0$ in $(-X, X)$, and $I(x) \geq 0$, the left-hand side of the above equation has the form of a characteristic function, multiplied if necessary by a positive constant. The limit of the right-hand side, as $X \to \infty$, is $\phi_T(z)$ by the same kind of argument as used in Theorem 6.1. Therefore $\phi_T(z)$ is the limit of a sequence of characteristic functions, and since it is continuous at $z = 0$, it is a characteristic function. Now let $T \to \infty$. Then since $\phi_T(z) \to \phi(z)$ the continuity theorem again applies and $\phi(z)$ is a characteristic function.

Other criteria for characteristic functions of types closely related to the above theorem have been given by Cramér (1939) and Khintchine. (See Lukacs 1960 for a good account of these.) However, the criterion given in Bochner's theorem is awkward and sometimes difficult to verify for particular cases. It is therefore useful to have easily verified criteria which are sufficient but not necessary, and a useful one is provided by the following theorem which is due to Pólya (1949b).

THEOREM 6.20. $\phi(z)$ *is a characteristic function if*
(1) $\phi(0) = 1$,
(2) $\phi(t) = \phi(-t)$,
(3) $\phi(t)$ *is continuous and convex for* $t > 0$, *i.e. for* t_1, $t_2 > 0$,
    $2\phi\{\frac{1}{2}(t_1 + t_2)\} \leq \phi(t_1) + \phi(t_2)$,
(4) $\lim_{t \to \infty} \phi(t) = 0$.

Proof. The fact that $\phi(t)$ is convex implies that it has a right-hand derivative almost everywhere which we denote by $\phi'_+(t)$ and which must be non-positive and non-decreasing, so that we must have

$$\lim_{t \to \infty} \phi'_+(t) = 0.$$

Then for all $x \neq 0$, the integral

$$p(x) = \frac{1}{2\pi} \int_{-\infty}^{\infty} e^{itx} \phi(t) dt$$

$$= \frac{1}{\pi} \int_{0}^{\infty} \phi(t) \cos tx \, dt$$

exists, and we can invert it by the inversion theorem for Fourier transforms to obtain

$$\phi(t) = \int_{-\infty}^{\infty} e^{itx} p(x)\, dx$$

$$= 2 \int_{0}^{\infty} p(x) \cos tx\, dx.$$

Hence if we can show that $p(x) \geqslant 0$ the result will follow. Integrating the previous equation by parts we get

$$p(x) = -\frac{1}{\pi x} \int_{0}^{\infty} \phi'_+(t) \sin tx\, dt$$

and since $-\phi'_+(t)$ is non-increasing and non-negative this integral is necessarily non-negative as we see by splitting it into a series of integrals over the ranges $(0, \pi)$, $(\pi, 2\pi)$,.... Thus $p(x) \geqslant 0$.

From this theorem it follows that such functions as $e^{-|t|}$, and $(1+|t|)^{-1}$ are characteristic functions.

Now let $\phi_1(t)$, $\phi_2(t)$,... be a series of characteristic functions and $\phi(t, \theta)$ a characteristic function, with variable t, which is an integrable function of θ over the set of points of increase of some distribution function $F_1(\theta)$. We can now conclude from Theorem 6.19 all the results of the following theorem.

THEOREM 6.21. *Given that the* ϕ_i *are characteristic functions,*

(1) $a\phi_1 + b\phi_2$ *is a characteristic function if*

$$a \geqslant 0, \quad b \geqslant 0, \quad a+b = 1,$$

(2) *any finite product of* $\phi_i's$ *is a characteristic function,*
(3) *any finite polynomials in the* $\phi_i's$ *with non-negative coefficients adding to unity is a characteristic function,*
(4) $\overline{\phi_1(t)}$, $|\phi_1(t)|^2$, *and* $R\phi_1(t)$ *are characteristic functions,*
(5) $\displaystyle\int_{-\infty}^{\infty} \phi(t, \theta)\, dF_1(\theta)$ *is a characteristic function,*
(6) *if* $a_i \geqslant 0$ *and* $\sum_1^\infty a_i = 1$ *then*

$$\sum_1^\infty a_i \phi_i(t)$$

is a characteristic function and so, in particular, are

$$e^{\phi_1(t)-1}, \quad \cosh \phi_1(t)/\phi_1(t)(\cosh 1)^{-1},$$

$$\sinh \phi(t)(\sinh 1)^{-1}, \quad and \quad (1-a)\phi_1(t)\{1 - a\phi_1(t)\}^{-1},$$

if $0 \leqslant a < 1$.

(7) *if* $P(z)$ *is a probability generating function,* $P(\phi_1(t))$ *is a characteristic function,*

(8) *if* $0 \leqslant a < 1$ *and* $b > 0$

$$\left(\frac{\phi_1(t)(1-a)}{1-a\phi_1(t)} \right)^b$$

is a characteristic function,

(9) $\dfrac{1}{t} \displaystyle\int_0^t \phi_1(u) \, du$

is a characteristic function.

Proof. Statements (1)–(8) above, which enable us to find the characteristic functions of mixtures and compounds of distributions, all follow from a direct application of Bochner's theorem (Theorem 6.19) on observing that the property of being non-negative definite holds for a sum of two functions if it holds for each separately. Statement (9) follows on observing that

$$\frac{1}{t} \int_0^t \phi_1(u) \, du = \int_{-\infty}^{\infty} t^{-1} \int_0^t e^{ixu} \, du \, dF(x)$$

$$= \int_{-\infty}^{\infty} x^{-1} \int_0^x e^{itu} \, du \, dF(x)$$

$$= \int_{-\infty}^{\infty} \phi_2(t, x) \, dF(x),$$

where $\phi_2(t, x)$ is the characteristic function of a uniform distribution on the interval $(0, x)$, and x is now a parameter so that the result follows from statement (5).

It is also convenient to have easily applied criteria for determining when a function is not a characteristic function.

THEOREM 6.22. *If* $\phi(t)$ *is not constant, and for* t *small we have*

$$\phi(t) = 1 + a(t) + o(t^2)$$

where $a(t) = -a(-t)$, *and* $a(t)$ *is* $o(t)$, *then* $\phi(t)$ *is not a characteristic function.*

Proof. In Theorem 6.25 we shall prove that for any characteristic function, and t, h real,

$$|\phi(t) - \phi(t+h)|^2 \leqslant 2\phi(0)\{\phi(0) - R\phi(h)\}.$$

If $\phi(t)$ above is a characteristic function so is

$$\Phi(t) = \phi(t)\phi(-t)$$
$$= 1 + o(t^2),$$

and applying the above inequality to $\Phi(t)$ we get

$$|\Phi(t) - \Phi(t+h)|^2 = o(t^2),$$

so that $\Phi'(t) = 0$ everywhere, which results in a contradiction.

Thus

$$e^{-u^4}, \quad e^{-|u|^r} (r > 2), \quad (1+u^4)^{-1},$$

are not characteristic functions. Related to these results is the following important theorem due to Marcinkiewicz (1938) which we state but do not prove. For other proofs see Dugué (1951a) and Lukacs (1958).

THEOREM 6.23. *If $P(t)$ is a polynomial of degree greater than 2, $e^{-P(t)}$ is not a characteristic function.*

6.11. $\phi(t)$ and the discontinuities of $F(x)$

We now obtain criteria for determining when a distribution has a continuous probability density everywhere and when it has jumps. The simplest of these is clearly that if $\phi(t)$ has a finite integral on the range $(-\infty, \infty)$ then $f(x)$, the probability density of the distribution, exists everywhere. We have already proved this in the course of proving Theorem 6.1.

THEOREM 6.24. *Write*

$$M_T(\phi) = \frac{2}{\pi T} \int_0^\infty |\phi(t)|^2 \left(\frac{\sin Tt}{t}\right)^2 dt \quad (T > 0),$$

$$M(\phi) = \lim_{T \to \infty} \frac{1}{2T} \int_{-T}^T |\phi(t)|^2 dt.$$

Then

(1)
$$M_T(\phi) = 2 \int_0^{2T} \left(1 - \frac{x}{2T}\right) dF^s(x),$$

where $F^s(x)$ is the convolution of the distribution of X with that of $-X$,

(2) *$M_T(\phi)$ is non-decreasing as T increases and tends to unity as $T \to \infty$, and to $M(\phi)$ as $T \to 0$,*

(3)
$$M(\phi) = \sum_{k=1}^\infty p_k^2,$$

where $p_1, p_2, \ldots$ are the jumps of the distribution $F(x)$ in some order,

(4)
$$M(\phi_1 \phi_2) \geqslant M(\phi_1) M(\phi_2),$$

(5) *if $\phi_n \to \phi$ everywhere, where ϕ is a characteristic function, and $M(\phi) = 0$, then $M(\phi_n) \to 0$, but the converse is not true.*

This theorem provides a criterion for the existence of jumps, and in particular no jumps exist if $|\phi(t)|^2$ has a finite integral since $M(\phi)$ is then necessarily zero.

Proof of Theorem 6.24. Since $F^s(x)$ is the convolution $F(x) * \{1 - F(-x)\}$, it has $|\phi(t)|^2$ as characteristic function. Integrating by parts we have

$$2 \int_0^{2T} \left(1 - \frac{x}{2T}\right) dF^s(x) = \frac{1}{2T} \int_0^{2T} \{F^s(x) - F^s(-x)\} dx$$

since $F^s(x)$ is symmetric.

Using the inversion theorem we now have

$$2 \int_0^{2T} \left(1 - \frac{x}{2T}\right) dF^s(x) = \frac{1}{2T} \int_0^{2T} dx \lim_{U \to \infty} \frac{1}{2\pi} \int_{-U}^{U} \frac{e^{itx} - e^{-itx}}{it} |\phi(t)|^2 dt,$$

and since the integrand in the inner integral is uniformly bounded, the convergence to the limit is also uniformly bounded for x in $(0, 2T)$, and the first integral and limit can be interchanged. In this way we get

$$\frac{1}{2T} \lim_{U \to \infty} \frac{1}{2\pi} \int_{-U}^{U} \frac{e^{2itT} + e^{-2itT} - 2}{(-t^2)} |\phi(t)|^2 dt$$

$$= \frac{-1}{4\pi T} \lim_{U \to \infty} \int_{-U}^{U} \frac{(e^{itT} - e^{-itT})^2}{t^2} |\phi(t)|^2 dt$$

$$= \frac{2}{\pi T} \lim_{U \to \infty} \int_0^{U} \frac{\sin^2 tT}{t^2} |\phi(t)|^2 dt$$

$$= M_T(\phi).$$

It is obvious that

$$M_T(\phi) = 2 \int_0^{2T} \left(1 - \frac{x}{2T}\right) dF^s(x)$$

is non-decreasing as T increases and tends to unity as $T \to \infty$. Furthermore, as T tends to zero

$$\frac{1}{2T} \int_0^{2T} \{F^s(x) - F^s(-x)\} dx$$

converges to the jump in $F^s(x)$ at $x = 0$. From what was said in the proof of Theorem 6.1 this jump is equal to

$$M(\phi) = \lim_{U \to \infty} \frac{1}{2U} \int_{-U}^{U} |\phi(t)|^2 dt,$$

since $|\phi(t)|^2$ is the characteristic function of $F^s(x)$. Hence

$$\lim_{T \to 0} M_T(\phi) = M(\phi).$$

Suppose that a distribution function $G(x)$, has jumps of size $a_1, a_2, \ldots$ at $x_1, x_2, \ldots$, and another distribution, $H(x)$, has jumps $b_1, b_2, \ldots$ at $y_1, y_2, \ldots$. At any point $z = x_i + y_j$, the jump of the convolution $G(x) * H(x)$ will be $\sum a_i b_j$, where the sum is taken over all i, j such that $x_i + y_j = z$. Applying this result to the distribution $F(x) * 1 - F(-x)$ we see that its jump at $x = 0$ is $\sum_1^{\infty} p_k^2$, and hence

$$M(\phi) = \sum_{k=1}^{\infty} p_k^2.$$

Now suppose $\phi_1(t)$ and $\phi_2(t)$ are the characteristic function of $G(x)$ and $H(x)$. Then

$$M(\phi_1)M(\phi_2) = \left(\sum_i a_i^2\right)\left(\sum_j b_j^2\right). \tag{6.26}$$

On the other hand, $G(x) * H(x)$ has characteristic function $\phi_1(t)\phi_2(t)$ and

$$M(\phi_1\phi_2) = \sum_k (\sum a_i b_j)^2, \tag{6.27}$$

where the inner sum is taken over all values of i and j such that $x_i + x_j = z_k$, where z_k is the position of a jump of $F(x) * G(x)$. Every term in the product (6.26) occurs in (6.27) and both are non-negative so that

$$M(\phi_1)M(\phi_2) \leqslant M(\phi_1\phi_2).$$

Now let $\phi_n(t)$ be a sequence of characteristic functions such that $\phi_n(t) \to \phi(t)$ everywhere, where $M(\phi(t)) = 0$. Since $\phi(t)$ is a characteristic function this convergence is uniform in any finite interval and $F_n(x)$, the distribution corresponding to $\phi_n(t)$, must converge everywhere to $F(x)$. Furthermore, $M(\phi(t)) = 0$ implies that $F(x)$ is continuous and hence this convergence is uniform. Thus the size of the largest jump in $F_n(x)$ must tend to zero. Let this jump be P. Then if $p_1, p_2, \ldots$ are the jumps of $F_n(x)$ we have

$$M(\phi_n(t)) = \sum_1^\infty p_k^2 \leqslant P \sum_1^\infty p_k \leqslant P,$$

and $M(\phi_n(t))$ tends to zero. That the converse is false can be seen by observing that a discrete distribution, for example the distribution with all probability concentrated at $x = 0$, can be approximated to by a contiuuous distribution such as a normal distribution with zero mean and vanishing small variance.

6.12. Some useful inequalities

We now prove a series of inequalities which are occasionally useful in more theoretical problems.

THEOREM 6.25. *If $\phi(t)$ is a characteristic function*

$$1 - R\phi(2t) \leqslant 4\{1 - R\phi(t)\}, \tag{6.28}$$

$$1 - |\phi(2t)|^2 \leqslant 4\{1 - |\phi(t)|^2\}, \tag{6.29}$$

$$|\phi(t) - \phi(t+h)|^2 \leqslant 2\{1 - R\phi(h)\}, \tag{6.30}$$

for all real t and h.

Proof. We have

$$1 - R\phi(2t) = \int (1 - \cos 2tx)\, dF(x)$$

$$= 2 \int \sin^2 xt\, dF(x)$$

$$= 2 \int (1 - \cos xt)(1 + \cos xt)\, dF(x)$$

$$\leqslant 4 \int (1 - \cos xt)\, dF(x)$$

$$\leqslant 4(1 - R\phi(t)).$$

Applying this to the characteristic function $|\phi(t)|^2$ we get (6.29). The left-hand side of (6.30) is

$$\left| \int e^{ixt}(1 - e^{ixh})\, dF(x) \right|^2,$$

and using Schwartz's inequality this is not greater than

$$\int dF(y) \int |1 - e^{ixh}|^2\, dF(x) = 2 \int (1 - \cos xh)\, dF(x) \leqslant 2\{1 - R\phi(h)\}.$$

(6.30) shows that $\phi(t)$ is uniformly continuous over the whole range of t. Moreover, if we have a sequence of characteristic functions, $\phi_n(t)$, which converge for all t to a function $\Phi(t)$ which is continuous at $t = 0$, then $\Phi(t)$ is continuous for all t. Furthermore, if $\phi_n(t)$ converges to unity in a set of values of t which has positive Lebesgue measure, then it converges to unity everywhere. To prove this suppose E is the set of positive measure. Then it is known that the set of all values of $t - t'$, where t, t' belong to E, fills up a non-vanishing interval surrounding the origin. If $\phi_n(t)$ converges to unity, so must $\phi_n(-t) = \overline{\phi_n(t)}$. Hence E can be taken as symmetric about the origin. Let t and t' be points of E. Then from (6.30)

$$|\phi_n(t) - \phi_n(t - t')|^2 < 2\{1 - R\phi_n(-t')\}.$$

$\phi_n(t)$ and $\phi_n(-t')$ both converge to unity and so therefore must $\phi_n(t - t')$. Hence $\phi_n(t)$ converges to unity for all values of t in an interval surrounding the origin and since, using (6.30) again, we have

$$|\phi_n(t) - \phi_n(2t)|^2 < 2\{1 - R\phi_n(t)\},$$

$\phi_n(t)$ must converge to unity on the whole real line.

Theorem 6.25 concerns inequalities between the differences of values of the characteristic function. We now prove two inequalities which provide bounds on $F(x)$ in terms of values of $\phi(t)$.

THEOREM 6.26. *If $u > 0$,*

$$\int_{|x| \leqslant u^{-1}} x^2 \, dF(x) \leqslant 3u^{-2}\{1 - R\phi(u)\}. \tag{6.31}$$

$$\int_{|x| \geqslant u^{-1}} dF(x) \leqslant 7u^{-1} \int_0^u \{1 - R\phi(t)\} \, dt. \tag{6.32}$$

(6.31) provides a bound for the second moment of a truncated random variable in terms of the behaviour of $\phi(t)$ in the neighbourhood of $t = 0$, whilst (6.32) gives a bound for the tails of the distribution in the same way. Thus (6.32) reflects a property of characteristic functions which we have already noted (Theorem 6.8), namely that the smoothness of $\phi(t)$ near the origin is related to the speed with which $F(-x)$ and $1 - F(x)$ tend to zero as x increases

Proof of Theorem 6.26. We have

$$1 - R\phi(u) = 1 - \int_{-\infty}^{\infty} \cos ux \, dF(x)$$

$$= \int_{-\infty}^{\infty} \{1 - \cos ux\} \, dF(x).$$

It is easily verified by repeated differentiation that

$$1 - \cos x \geqslant \tfrac{1}{2} x^2 (1 - \tfrac{1}{12} x^2),$$

for $x \leqslant 1$ at least, and hence

$$1 - R\phi(u) \geqslant \int_{|x| \leqslant u^{-1}} \tfrac{1}{2} u^2 x^2 (1 - \tfrac{1}{12} u^2 x^2) \, dF(x)$$

$$\geqslant \frac{11u^2}{24} \int_{|x| \leqslant u^{-1}} x^2 \, dF$$

$$\geqslant \frac{u^2}{3} \int_{|x| \leqslant u^{-1}} x^2 \, dF,$$

which proves (6.31). To obtain (6.32) we have

$$u^{-1} \int_0^u \{1 - R\phi(t)\} \, dt = u^{-1} \int_0^u dt \int_{-\infty}^{\infty} (1 - \cos xt) \, dF(x)$$

$$= \int_{-\infty}^{\infty} \left(1 - \frac{\sin xu}{xu}\right) dF(x).$$

The integrand is non-negative, and thus restricting the integration to the intervals where $|ux| \geqslant 1$, the left-hand side is not less than

$$(1 - \sin 1) \int_{|x| \geqslant u^{-1}} dF(x) \geqslant \frac{1}{7} \int_{|x| \geqslant u^{-1}} dF(x),$$

19

since $\sin 1 < \frac{8}{7}$. Clearly in both the above inequalities the constants are not the best possible but this is immaterial in most applications.

6.13. The extension of a characteristic function outside an interval

We may now inquire under what conditions the knowledge of a characteristic function in a finite interval will determine the distribution function, and hence the characteristic function, completely. It is clear from Theorem 6.13 that if $\phi(t)$ is analytic in any region surrounding the origin it is analytic inside a strip containing the real axis in its interior, so that $\phi(t)$ is determined uniquely for all real t by analytic continuation. Thus analyticity of $\phi(t)$ in a region around $t = 0$ implies that $\phi(t)$ and $F(x)$ are uniquely determined by the values of $\phi(t)$ in a small interval containing $t = 0$ in its interior.

An instructive example where this does not happen may be constructed as follows. The function

$$\phi_1(t) = 1 - |t|, \quad |t| < 1,$$
$$= 0, \qquad |t| \geqslant 1,$$

is a characteristic function since using (6.7) we get

$$f(x) = \frac{1}{\pi x^2}(1 - \cos x) \geqslant 0,$$

as the density in the range $(-\infty, \infty)$. This characteristic function is clearly not analytic at $t = 0$.

On the other hand, consider the discrete distribution on the points $x = 0$, $\pm \pi k$ $(k = 1, 2, ...)$, given by

$$p_0 = \mathrm{pr}(X = 0) = \tfrac{1}{2},$$
$$p_k = \mathrm{pr}(X = \pi k) = \mathrm{pr}(X = -\pi k)$$
$$= \frac{2}{k^2 \pi^2}, \quad \text{where } k = 2m+1, \ m = 0, 1, 2,...$$
$$= 0, \quad \text{where } k = 2m.$$

Then the characteristic function is given by

$$\phi_2(t) = \frac{1}{2} + \sum_{m=0}^{\infty} \frac{\cos \pi(2m+1)t}{\pi^2(2m+1)^2}.$$

This can be verified to be the Fourier series of a function which is periodic of period 2, and defined on the interval $(-1, 1)$ to be equal to $\phi_1(t)$. The fact that

$$p_0 + 2\sum_{1}^{\infty} p_k = 1$$

can be verified by observing that $\phi_2(0) = \phi_1(0) = 1$. $\phi_1(t)$ and $\phi_2(t)$ are equal for $|t| \leqslant 1$, but not in general for $|t| > 1$.

This example shows further that it is possible to have three different distributions $F_1(x)$, $F_2(x)$, and $F_3(x)$ such that

$$F_1 * F_2 = F_1 * F_3$$

so that even when 'division' is possible with the convolution operation, '*', it is not necessarily unique. We shall study this type of problem in more detail in Chapter 9.

The example of $\phi_1(t)$ shows that a characteristic function can vanish in an interval. Smith (1962) has shown, however, that this cannot happen for the characteristic function of a non-negative random variable.

6.14. The conditional characteristic function

Suppose, for simplicity, that X and Y are two random variables with a joint distribution which has a continuous probability density $f(x,y)$. The joint characteristic function is

$$\phi_1(t_1, t_2) = \int_{-\infty}^{\infty} \int_{-\infty}^{\infty} e^{it_1 x + it_2 y} f(x, y) \, dx \, dy,$$

which we shall assume to be absolutely integrable with respect to (t_1, t_2), and with respect to t_2 for each t_1. We can write

$$f(x, y) = g(x|y)h(y),$$

where $h(y)$ is the probability density of the marginal distribution of Y, and $g(x|y)$ is the conditional distribution of X given $Y = y$. The latter has a characteristic function

$$\phi_1(t_1|y) = \int_{-\infty}^{\infty} e^{it_1 x} g(x|y) \, dx,$$

and therefore

$$\phi(t_1, t_2) = \int_{-\infty}^{\infty} e^{it_2 y} h(y)\phi_1(t_1|y) \, dy.$$

Using the Fourier inversion formula we have

$$h(y)\phi_1(t_1|y) = \frac{1}{2\pi} \int_{-\infty}^{\infty} e^{-it_2 y} \phi(t_1, t_2) \, dt_2,$$

and

$$h(y) = h(y)\phi_1(0|y) = \frac{1}{2\pi} \int_{-\infty}^{\infty} e^{-it_2 y} \phi(0, t_2) \, dt_2,$$

so that

$$\phi_1(t_1|y) = \frac{\int_{-\infty}^{\infty} e^{it_2 y} \phi(t_1, t_2)\, dt_2}{\int_{-\infty}^{\infty} e^{it_2 y} \phi(0, t_2)\, dt_2}, \tag{6.33}$$

whenever $h(y) \neq 0$. This formula, which is due to Bartlett (1938), gives the conditional characteristic function for a fixed Y. By multiplying by the marginal distribution of Y and integrating over some set E (e.g. the set $y > a$) we can obtain the characteristic function of X conditional on Y lying in E.

As an example consider the bivariate normal distribution with zero means, unit standard deviations, and correlation ρ. The characteristic function is

$$\phi(t_1, t_2) = \exp\{-\tfrac{1}{2}(t_1^2 + 2\rho t_1 t_2 + t_2^2)\},$$

so that the characteristic function of the conditional distribution of X given $Y = y$ is

$$\frac{I(t_1|y)}{I(0|y)},$$

where

$$I(t_1|y) = \frac{1}{2\pi} \int_{-\infty}^{\infty} \exp\{-it_2 y - \tfrac{1}{2}(t_1^2 + 2\rho t_1 t_2 + t_2^2)\}\, dt_2$$

$$= \frac{1}{\sqrt{(2\pi)}} \exp\{-\tfrac{1}{2}\{(1-\rho^2)t_1^2 - 2\rho y i t_1\}\},$$

so that

$$\phi_1(t_1|y) = \exp\{-\tfrac{1}{2}\{(1-\rho^2)t_1^2 - \rho y i t_1\}\}.$$

This is the characteristic function of a normally distributed random variable with mean ρy and variance $(1-\rho^2)$. Now consider the distribution of X conditional on $Y > y_0$. This will be obtained by multiplying $\phi_1(t_1|y)$ by the conditional probability distribution of Y and integrating over the range (y_0, ∞), thus obtaining

$$\phi(t_1|y > y_0) = \{\exp - \tfrac{1}{2}(1-\rho^2)t_1^2\}$$

$$\times \int_{y_0}^{\infty} \frac{1}{\sqrt{(2\pi)}} \exp\{-\tfrac{1}{2}(y^2 - 2\rho y i t_1)\}\, dy \left\{\int_{y_0}^{\infty} \frac{1}{\sqrt{(2\pi)}} \exp\{-\tfrac{1}{2}y^2\}\, dy\right\}^{-1}.$$

From this it is easy to obtain the mean and variance in terms of y_0, ρ and the tail of the normal distribution. These results, which may, of course, be easily obtained by direct integration, (see §7.15) are of considerable practical importance.

A similar argument can be used to obtain the conditional probability generating function when discrete random variables are involved. Thus suppose that X and Y are random variables which can only take the values 0, 1, 2,... so that $p_{r,s}$ is the probability that $X = r$, $Y = s$. Then the joint probability generating function is

$$P(z, w) = \sum_0^\infty \sum_0^\infty p_{rs} z^r z^s.$$

The probability that $X = r$, conditional on $Y = s$, is

$$p(r|s) = p_{r,s} \left(\sum_{r=0}^\infty p_{r,s} \right)^{-1}.$$

Then writing

$$P(z|s) = \sum_{r=0}^\infty z^r p(r|s)$$

we have

$$P(z|s) \sum_{r=0}^\infty p_{r,s} = \frac{1}{2\pi i} \int P(z, w) w^{-1-s} \, dw,$$

where the integral is taken round the origin in the complex plane. Using a unit circle as the path of integration we can therefore write ($z = e^{iu}$, $w = e^{iv}$)

$$P(z|s) = \frac{\int_0^{2\pi} e^{-isv} P(e^{iu}, e^{iv}) \, dv}{\int_0^{2\pi} e^{-isv} P(1, e^{iv}) \, dv}. \tag{6.34}$$

This can sometimes be evaluated by the method of steepest descents.

6.15. Conjugate distributions

Moment generating functions lead to another idea which has been found useful in a variety of applications. Suppose that $F(x)$ is a cumulative probability distribution and that

$$M(t) = \int_{-\infty}^{\infty} e^{tx} dF(x)$$

is finite in an interval I which contains the origin $t = 0$. Then for any value of t in this interval, the function

$$G(x, t) = M(t)^{-1} \int_{-\infty}^{x} e^{tu} dF(u) \tag{6.35}$$

is a probability distribution function which is said to be 'conjugate' to $F(x)$. Clearly $G(x, 0) = F(x)$.

Such a distribution can sometimes be given a direct probabilistic interpretation. As an example consider the growth of a population having specified age-specific birth and death rates (Lotka 1939, 1945). (I owe this example to Professor E. J. Williams.)

Suppose the population consists of $N(t)$ females at time t which are such that the probability of any female surviving to age x is $l(x)$ ($l(0) = 1$ and $l'(x) \leqslant 0$), whilst the probability of any female giving birth to one offspring between the age of x and $x+dx$ is $m(x)\,dx$. Then it can be proved that such a population increases, or decreases, asymptotically as $N_0\,e^{at}$, where N_0 is a constant, and a is a constant (which may be negative) known as the 'intrinsic rate of natural increase'. If by some artificial means the population had been forced to produce a constant rate of birth, the population size would be asymptotically constant, and the age distribution would be such that the probability of an individual chosen at random having an age lying in the range $(x, x+dx)$ would be

$$f(x)\,dx = \frac{l(x)\,dx}{\displaystyle\int_0^\infty l(x)\,dx}.$$

If, on the other hand, the population is allowed to increase at its natural rate, $N(t)$ will be asymptotically of the form $N_0\,e^{at}$ where a is given by

$$\int_0^\infty e^{-ax} l(x)m(x)\,dx = 1,$$

and the population will have asymptotically a different age distribution which we write as $g(x)$.

To find $g(x)$ write $N(t) = N_0\,e^{at}$, and suppose that the asymptotic form of the age distribution has been attained. The number of individuals in the age group $(x, x+dx)$ at time t will be equal to the number of individuals, born in the time interval $(t-x, t-x-dx)$, which have survived to time t. This is

$$N(t-x)l(x)\int_0^\infty g(u)m(u)\,du.$$

This will be proportional to $g(x)$ and hence, since $N(t) = N_0\,e^{at}$,

$$g(x) = \frac{e^{-ax}l(x)}{\displaystyle\int_0^\infty e^{-au}l(u)\,du},$$

which is a distribution conjugate to $f(x)$.

Returning to (6.35), the moment generating function of $G(x, t)$ is

$$M_1(\theta, t) = M(t)^{-1}\int_0^\infty e^{\theta x + tx}\,dF(x)$$

$$= M(t+\theta)M(t)^{-1}, \qquad (6.36)$$

and thus the characteristic function is

$$\phi_1(\theta, t) = M(t+i\theta)M(t)^{-1}$$

which exists for all θ, and all t such that (6.35) is a probability distribution.

In particular we see from (6.36) that the mean of the conjugate distribution is

$$M'(t)M(t)^{-1}.$$

As t varies, the mean varies and if τ is a value of t such that the mean is m, say, we have

$$e^{-m\tau}M'(\tau) - me^{-m\tau}M(\tau) = 0 \tag{6.37}$$

$$= \frac{d}{dt}\{e^{-mt}M(t)\} \quad (t = \tau).$$

In fact τ is the value of t for which

$$e^{-mt}M(t)$$

attains its unique minimum since

$$\frac{d^2}{dt^2}\{e^{-mt}M(t)\} = e^{-mt}\int_0^\infty e^{tx}(x-m)^2 \, dF$$

$$\geqslant 0.$$

From (6.36) we further see that if we have two distributions, $F_1(x)$ and $F_2(x)$, with corresponding conjugate distributions, $G_1(x, t)$ and $G_2(x, t)$, where t is such that these both exist, the conjugate distribution with parameter t of the convolution $F_1 * F_2$ has a characteristic function which is the product of the characteristic functions of $G_1(x, t)$ and $G_2(x, t)$, and is therefore the convolution $G_1(x, t) * G_2(x, t)$.

Taking the conjugate of a distribution introduces a new parameter t. Sometimes this adds essentially to the number of parameters in the distribution but in an interesting class of cases the original distribution is unchanged in form, only the parameters being changed. Thus for a probability distribution $f(x, \theta)$ $(x \geqslant 0)$, depending on a single parameter θ, which is of the form

$$f(x, \theta) = Q(\theta)R(x)e^{-\theta x}, \tag{6.38}$$

we see that the conjugate distributions are

$$\frac{e^{-(\theta-t)x}R(x)}{\displaystyle\int_0^\infty R(x)e^{-(\theta-t)x} \, dx},$$

which can be written as

$$Q(\theta - t)R(x)e^{-(\theta-t)x},$$

and are therefore of the same form because from (6.38) we have

$$\int_0^\infty R(x)e^{-\theta x} \, dx = Q(\theta)^{-1},$$

for all values of θ for which (6.38) is a distribution.

Another application of conjugate distributions occurs in the theory of mutation in genetics. Suppose that we have a class of individuals whose ages have a probability distribution with a density function $f(x)$, and assume that during their lifetimes mutations may occur in their germ cells. If these mutations are due to radiation it is often reasonable to suppose that the number of mutations which have occurred in an individual aged x has a Poisson distribution with mean λx. The distribution of the ages of all those members of the class which have not had mutations will then be

$$f_1(x) = \frac{e^{-\lambda x} f(x)}{\displaystyle\int_0^\infty e^{-\lambda u} f(u)\, du},$$

which is a distribution conjugate to $f(x)$.

6.16. The characteristic functional

An important generalization of the characteristic function is obtained when we consider random functions instead of random variables. Suppose that $X_0, \ldots, X_n$ are $n+1$ random variables with a joint distribution. Then their joint characteristic function is

$$\phi(\theta_0, \ldots, \theta_n) = E\left\{\exp i \sum_0^n \theta_i X_i\right\}. \tag{6.39}$$

It is a general principle in analysis that to theorems on sums there correspond theorems on integrals, and vice versa. If $X(t)$ is in some sense a 'random function' defined on an interval (a, b) we could consider the values

$$X(a + in^{-1}(b-a)), \quad i = 0, 1, \ldots, n$$

as $n+1$ random variables and $\theta_0, \ldots, \theta_n$, similarly as values of a function $\theta(t)$ at the same points. Then the integral analogue of (6.39) is seen to be

$$\phi(\theta(u)) = E\left\{\exp i \int_a^b \theta(t) X(t)\, dt\right\}, \tag{6.40}$$

which is known as the characteristic functional of the random function $X(t)$. The argument, $\theta(u)$, of the functional is not the value of $\theta(t)$ at some particular point but the whole form of the function θ, which is therefore written in (6.40) with a dummy variable u.

If $X(t)$ is defined in such a way that it is continuous with probability one, the joint characteristic function of $X(t_1), \ldots, X(t_k)$, for values $t_1, \ldots, t_k$ can be obtained heuristically from (6.40) by letting $\theta(u)$ converge to the 'function'

$$\sum_{j=1}^k \theta_j\, \delta(u - t_j),$$

where $\delta(u)$ is Dirac's delta 'function' which is equal to zero everywhere, except at $u = 0$ where it is infinity in such a way that the integral of $\delta(u)$ is unity. Thus $\phi\{\theta(u)\}$ contains all the information needed to find the joint distribution of any variates based on $X(t)$ for $a \leqslant t \leqslant b$.

To define (6.40) it is necessary to define the sense in which $X(t)$ is to be a 'random function'. In the general theory of random processes, with which this book is not concerned, this is done by setting up a probability measure in the space of such functions, but here it will be sufficient to consider functions which depend on a finite or enumerably infinite sequence of random variables. Thus we could consider functions

$$X(t, \varepsilon_1, \varepsilon_2, \ldots),$$

where the ε_i are random variables which we can take as independent. The expectation in (6.40) is then taken over the joint distribution of the ε_i. This definition suffices for many purposes.

We now consider two illustrations of this theory. Suppose (Bartlett 1955) that

$$X(t) = \varepsilon_1 \cos \lambda t + \varepsilon_2 \sin \lambda t, \qquad (6.41)$$

where λ is a constant, and ε_1, ε_2 are independently and normally distributed with zero means and unit standard deviations. Then

$$\phi\{\theta(u)\} = E\left\{\exp i \int_a^b \theta(t) X(t) \, dt\right\}$$

$$= E \exp i\left\{\varepsilon_1 \int_a^b \theta(t) \cos \lambda t \, dt + \varepsilon_1 \int_a^b \theta(t) \sin \lambda t \, dt\right\}$$

$$= \exp{-\frac{1}{2}}\left[\left\{\int_a^b \theta(t)\cos \lambda t \, dt\right\}^2 + \left\{\int_a^b \theta(t)\sin \lambda t \, dt\right\}^2\right]$$

$$= \exp{-\frac{1}{2}} \int_a^b \int_a^b \theta(s)\theta(t)\cos \lambda(s-t) \, ds \, dt. \qquad (6.42)$$

As a second example consider a Poisson process occurring on the line $t \geqslant 0$. This is defined by interpreting t as time and writing $N(t)$ for the number of events occurring in the interval $(0, t)$. We suppose that for each value of t, $N(t)$ is a Poisson variate with mean λt, and we have shown that this provides a consistent definition. $N(t)$ is a random function defined ln the half-line $t > 0$. However, for most purposes we do not need to use measure theory in function-space to define its distribution since for any finite interval $(0, t)$ we can define $N(t)$ as a Poisson variate with mean λt, and $N(u)$ $(0 < u < t)$ can be defined by the values $x_1, \ldots, x_n$ at which $N(u)$ has jumps of unit size. Conditional on $N(t)$, $x_1, \ldots, x_n$ can be regarded as independently distributed in rectangular distributions on the interval $(0, t)$.

We could, as before, define the characteristic functional as

$$E \exp \left\{ i \int_0^t \theta(t) N(t) \, dt \right\}$$

but in this case it is more convenient to take

$$\Phi\{\theta(u)\} = E \exp i \int_0^t \theta(x) \, dN(x)$$

$$= E \exp i \{\theta(x_1) + \ldots + \theta(x_n)\}$$

$$= \sum_{n=0}^{\infty} e^{-\lambda t} \frac{(\lambda t)^n}{n!} \, E \exp i \{\theta(x_1) + \ldots + \theta(x_n)\},$$

where the expectation in the last expression is taken over the possible values of the x_i conditional on n. Then

$$\Phi\{\theta(u)\} = \sum_{n=0}^{\infty} e^{-\lambda t} \frac{(\lambda t)^n}{n} \left\{ t^{-1} \int_0^t \exp i\theta(x) \, dx \right\}^n$$

$$= \exp \left\{ -\lambda t + \lambda \int_0^t \exp i\theta(x) \, dx \right\}. \tag{6.43}$$

From this various properties of the random function $N(t)$ can be obtained. In particular if $\theta(u)$ is put equal to θ, we get

$$\exp\{\lambda t (e^{i\theta} - 1)\}$$

which is the characteristic function of $N(t)$.

Expressions such as (6.40) are known as 'product integrals'. This is a mathematical concept which is related to a product of a number of factors in the same way that an integral is related to a sum. Thus just as an integral

$$\int_a^b f(x) \, dx$$

is approximated by a sum

$$n^{-1} \sum_{s=0}^{n} f\{a + s n^{-1}(b-a)\},$$

and behaves in many ways like a sum, so the product integral

$$\exp \int_a^b \log f(x) \, dx$$

is approximated by the expression

$$\exp n^{-1} \sum_{s=0}^{n} \log f\{a + s n^{-1}(b-a)\},$$

which can be written as a product

$$\left[\prod_{s=0}^{n} f\{a + s n^{-1}(b-a)\} \right]^{1/n}.$$

Thus product integrals will occur as soon as we consider the analogue of (6.39) obtained by replacing the sum by an integral. For a general survey of product integration in analysis see Birkhoff (1937). The application of characteristic functionals in the theory of random processes is discussed at some length by Bartlett and Kendall (1951), and by Arley (1943). The characteristic functional of the distribution of the ages of individuals in a simple birth and death process (§ 3.22) has been found by D. G. Kendall (1950b). The general theory of characteristic functionals and the generalization of Theorem 6.19 is considered in Chapter VI of Grenander (1963).

Bibliographical notes

A full account of the theory of characteristic functions is given in Lukacs (1960, 1961) and Lukacs and Laha (1964). Wintner (1947) gives a more detailed study from the point of view of a pure mathematician. The reader should also consult Cramér (1962). LaSalle (1962) gives a useful account of recent Russian work.

Exercises

Exercise 6.1. Prove that if $F(x)$ is the integral of its derivative, $\phi(t) \to 0$ as $|t| \to \infty$.

Exercise 6.2. Using Theorem 6.2 prove that the binomial distribution, suitably scaled, tends to the normal distribution as its index increases.

Exercise 6.3. If $n > 0$ prove that $\Gamma(n+i\theta)\Gamma(n)^{-1}$ is the characteristic function of a distribution and find the latter.

Exercise 6.4. Prove that $(\cosh \pi x)^{-1}$ is a probability density and obtain its characteristic function.

Exercise 6.5. Using Carleman's criterion (6.12), and Theorem 6.18, prove that a multivariate distribution is uniquely determined by its moments if

$$\sum_{j=0}^{\infty} (\lambda_{2j})^{1/2j}$$

is divergent, where λ_{2j} is the sum of the marginal moments of order $2j$.

Exercise 6.6. If

$$\phi(t) = \sum_{0}^{\infty} m_k (it)^k / k!$$

is a characteristic function of a distribution which is analytic in the neighbourhood of $t = 0$, prove that

$$1 + \sum_{k=1}^{\infty} m_{s+k} m_s^{-1} (it)^k / k!$$

and

$$1 + m_1 it + \sum_{k=2}^{\infty} m_{s+k} m_1^k m_s^{k-1} m_{s+1}^{-k} (it)^k / k!$$

are characteristic functions for $s = 1, 2, \ldots$.

7. Special Continuous Distributions

7.1. The rectangular distribution

A RANDOM variable X has a rectangular distribution if it is uniformly distributed over some finite interval (a, b), i.e. the probability density is $(b-a)^{-1}$ inside, and zero outside, this interval. Since the more general case can be obtained by translation and rescaling we shall only consider intervals of unit length, and the most useful case is that where the interval is $(0, 1)$. Then by simple integration we find

$$\mu_n = (n+1)^{-1}, \tag{7.1}$$

and

$$\phi(t) = \int_0^1 e^{itx}\, dx$$
$$= (e^{it} - 1)/it. \tag{7.2}$$

These are obviously the limits of the corresponding results for the discrete rectangular distribution considered in §2.7 when the latter is rescaled so that the variate takes the values k^{-1}, $2k^{-1}, \ldots, (k-1)k^{-1}$, 1 with probabilities k^{-1}, and k tends to infinity.

It is sometimes convenient to shift the origin of this distribution to the point $\tfrac{1}{2}$ so that X is uniformly distributed over the interval $(-\tfrac{1}{2}, \tfrac{1}{2})$ and the mean is zero. The characteristic function is then

$$\int_{-\frac{1}{2}}^{\frac{1}{2}} e^{itx}\, dx = 2(\sin \tfrac{1}{2}it)/it, \tag{7.3}$$

and the moments are given by

$$m_{2r+1} = \mu_{2r+1} = 0, \tag{7.4}$$

$$m_{2r} = \mu_{2r} = \frac{1}{(2r+1)2^{2n}}. \tag{7.5}$$

If X is any random variable having a cumulative distribution $F(x)$ which is absolutely continuous, $Y = F(X)$ will be rectangularly distributed. The most interesting problem connected with the rectangular distribution is that of determining the distribution of a sum, $Z = X_1 + \ldots + X_n$, of independent random variables each distributed in the same rectangular distribution which it is now more convenient to suppose uniform on the interval $(0, 1)$. In this case we have in fact that Z has the probability density

$$\frac{1}{(n-1)!}\left\{x^{n-1}-\binom{n}{1}(x-1)^{n-1}+\binom{n}{2}(x-2)^{n-1}-\ldots\right\}, \qquad (7.6)$$

the series ending with a term in $(x-[x])^{n-1}$ where $[x]$ is the integral part of x. The cumulative distribution is easily found by integration.

The first proof of this result was apparently that of Laplace (1776), but the corresponding result for the discrete distribution of §2.7 had been already found by de Moivre and (7.6) can be found from this by a simple limiting process. The result is remarkable in that it has probably been independently rederived by more writers than almost any other result in the history of probability theory (a list of such proofs is given by Seal (1950–1)), and a wide variety of different methods have been employed. Thus Packer (1950) uses mathematical induction on n, Hall (1927) uses an elegant geometric argument, and Irwin (1927) uses characteristic functions, the latter two proofs being given in Kendall and Stuart, Vol. I (1958).

We give Irwin's proof as an illustration of the use of characteristic functions in finding the distribution of sums or random variables. From (7.2) the characteristic function of Z is

$$\left(\frac{e^{it}-1}{it}\right)^n,$$

and the distribution of Z clearly has a probability density given by the inversion formula (6.7) which becomes

$$f(x) = \frac{1}{2\pi}\int_{-\infty}^{\infty} e^{-itx}\left(\frac{e^{it}-1}{it}\right)^n dt, \qquad (7.7)$$

the integral being taken along the real axis from $-\infty$ to ∞. The integrand is everywhere analytic but this is no longer so if the binomial expansion is used to represent it as the sum of $n+1$ integrals. We therefore choose $\delta > 0$ and small, and distort the path of integration to a path consisting of the real axis from $-\infty$ to $-\delta$, the upper half of the circle $|t| = \delta$, and the real axis from δ to ∞. Then

$$f(x) = \frac{1}{2\pi}\sum_{j=0}^{n}\binom{n}{j}(-1)^j\int\frac{e^{it(j-x)}}{(-it)^n} dt,$$

where each integral is taken over the above path. Completing such a path by a large semi-circle either above or below the real axis we verify that

$$\int\frac{e^{iat}}{t^n} dt$$

taken over the original path is equal to zero if a is real and positive, whilst if a is real and negative it is equal to

$$-\frac{2\pi(-1)^{n/2}a^{n-1}}{(n-1)!}.$$

From this the result (7.6) follows. The cumulative distribution is

$$F(x) = \frac{1}{n!}\left\{x^n - \binom{n}{1}(x-1)^n + \binom{n}{2}(x-2)^n + \ldots\right\} \qquad (7.8)$$

and is tabulated to six decimal places for $n = 1(1)\ 12$, $x = 0\cdot0(0\cdot5)\ 6\cdot0$ by Packer (1951). The approach to normality is fairly rapid.

The above distribution has been used to discuss the effect of rounding-off errors in numerical calculations (Fisher and Wishart 1927, Lowan and Ledermann 1939). In this connexion Wouk (1961) has considered the related problem of the distribution of the nth digit in the decimal representation of a random variable.

It is also interesting to determine the distribution of the product of a number of different random variables each with a rectangular distribution. Suppose $X_1, \ldots, X_n$ are independently distributed uniformly on $(0, 1)$. If we put $Y_i = -\log X_i$, it is obvious that the distribution of Y_i is given by

$$e^{-Y_i}dY_i \qquad (0 \leqslant Y_i < \infty). \qquad (7.9)$$

This is a particular example of a 'gamma-type' distribution already used in various previous sections of this book and which will be considered in detail in §7.17. The fact that Y_i has the distribution (7.9) can also be expressed by saying that $-\frac{1}{2}Y_i$ has the 'χ^2 distribution with two degrees of freedom'. It follows from §7.17 that

$$S_n = Y_1 + \ldots + Y_n = -\log(X_1 \ldots X_n)$$

has the distribution

$$\Gamma(n)^{-1}S_n^{n-1}e^{-S_n}dS_n, \qquad (7.10)$$

and thus $T_n = X_1 \ldots X_n$ has the distribution

$$\Gamma(n)^{-1}(-\log T_n)^{n-1}dT_n. \qquad (7.11)$$

This also gives the distribution of $(X_1 \ldots X_n)^k$ (k any positive number) by a simple transformation (see also the work of Kolmogorov described in §7.16).

7.2. The normal distribution

In §2.9 we have already defined the normal distribution with mean zero and standard deviation σ to be the distribution with density

$$(2\pi\sigma^2)^{-\frac{1}{2}}e^{-(x^2/2\sigma^2)} \qquad (-\infty < x < \infty). \qquad (7.12)$$

In §6.9 we have obtained its moments and its characteristic function. From the latter it follows that the cumulant generating function is

$$K(t) = -\tfrac{1}{2}\sigma^2 t^2. \tag{7.13}$$

The moments, given by (6.17), satisfy the criterion of Theorem 6.12 and the normal distribution is therefore uniquely determined by them. This has the important consequence (Theorem 6.16) that if a sequence of distributions is such that their moments converge to those of a normal distribution, the sequence must converge to a normal distribution. It often happens in particular problems that the characteristic function of some complicated quantity cannot be found but that good asymptotic estimates of its moments are available, and these can be used to prove that the distribution of the quantity tends to normality.

One of the most important problems in probability theory is the determination of the conditions under which a sum,

$$S_n = X_1 + \ldots + X_n \tag{7.14}$$

of independent random variables X_i, has a distribution which tends to normality as n increases. We study this problem in detail in Chapter 8 but here we consider the simplest case because of its many applications.

THEOREM 7.1. *If X_1, X_2,... are a sequence of independently distributed random variables with the same distribution which has a finite second moment, the distribution of*

$$S_n = X_1 + \ldots + X_n,$$

suitably scaled, tends to normality as n increases.

Proof. We can take the mean and second moment of the distribution of X_i to be 0 and σ^2 respectively. Let the characteristic function of the distribution of X_i be $\phi(t)$. Then by Theorem 6.11 we have

$$\phi(t) = 1 - \tfrac{1}{2}\sigma^2 t^2 + t^2 K(t),$$

where K is a function of t which is bounded when t lies in any finite interval $(-a, a)$, and tends to zero as $t \to 0$. The characteristic function of $n^{-\frac{1}{2}}S_n$ is then

$$\{1 - \tfrac{1}{2}\sigma^2 n^{-1} t^2 + t^2 n^{-1} K(t n^{-\frac{1}{2}})\}^n,$$

and $K(t n^{-\frac{1}{2}})$ is bounded for $t n^{-\frac{1}{2}}$ in $(-a, a)$ and tends to zero as $t n^{-\frac{1}{2}} \to 0$. Fixing $a > 0$, and t, and letting n tend to infinity, the characteristic function of $n^{-\frac{1}{2}}S_n$ tends to

$$\exp -\tfrac{1}{2}\sigma^2 t^2,$$

for every fixed t, and uniformly in any finite interval. From Theorem 6.2 it follows that the distribution of $n^{-\frac{1}{2}}S_n$ tends to that of a normal distribution with zero mean and standard deviation σ.

7.3. The evaluation of the normal distribution

The evaluation of the integral

$$1 - \Phi(x) = (2\pi)^{-\frac{1}{2}} \int_x^\infty e^{-\frac{1}{2}u^2} \, du \tag{7.15}$$

is somewhat awkward and a variety of interesting formulae have been developed for this purpose. We first consider the following identity. Putting $v = \frac{1}{2}u^2$,

$$1 - \Phi(x) = (2\pi)^{-\frac{1}{2}} \int_{\frac{1}{2}x^2}^\infty (2v)^{-\frac{1}{2}} e^{-v} \, dv$$

$$= (2\pi)^{-1} \int_{\frac{1}{2}x^2}^\infty \int_{-\infty}^\infty e^{-v(1+w^2)} \, dw \, dv$$

$$= (2\pi)^{-1} \int_{-\infty}^\infty (1+w^2)^{-1} e^{-\frac{1}{2}x^2(1+w^2)} \, dw,$$

so that

$$(2\pi)^{\frac{1}{2}} e^{\frac{1}{2}x^2} \{1 - \Phi(x)\} = (2\pi)^{-\frac{1}{2}} \int_{-\infty}^\infty \frac{e^{-\frac{1}{2}x^2 w^2}}{1+w^2} \, dw. \tag{7.16}$$

The expression on the left-hand side of this equation is known as Mill's ratio. This formula can be used in two different ways. Changing the variable of integration to $t = \frac{1}{2}x^2 w^2$, we get

$$(2\pi)^{\frac{1}{2}} e^{\frac{1}{2}x^2} \{1 - \Phi(x)\} = \frac{x}{2\sqrt{\pi}} \int_0^\infty \frac{e^{-t} t^{-\frac{1}{2}}}{\frac{1}{2}x^2 + t} \, dt$$

$$= \frac{1}{x\sqrt{\pi}} \int_0^\infty e^{-t} t^{-\frac{1}{2}} A(t) \, dt, \tag{7.17}$$

where

$$A(t) = 1 - \left(\frac{2t}{x^2}\right) + \left(\frac{2t}{x^2}\right)^2 - \left(\frac{2t}{x^2}\right)^3 + \ldots + (-1)^n \left(\frac{2t}{x^2}\right)^n + \frac{(-2tx^{-2})^{n+1}}{1+2tx^{-2}},$$

n being any positive integer.

Integrating we obtain

$$(2\pi)^{\frac{1}{2}} e^{\frac{1}{2}x^2} \{1 - \Phi(x)\} = \frac{1}{x} - \frac{1}{x^3} + \frac{1.3}{x^5} - \ldots + (-)^n \frac{1.3.5 \ldots (2n-1)}{x^{2n+1}}$$

$$+ (-)^{n+1} \frac{x^{-2n-3}}{\sqrt{\pi}} \int_0^\infty \frac{t^{n+\frac{1}{2}} e^{-t} dt}{1+2tx^{-2}}. \tag{7.18}$$

The last integral can be easily shown to be less in absolute value than the last term in the sum if $x > 1$. We can therefore write

$$(2\pi)^{\frac{1}{2}} e^{\frac{1}{2}x^2} \{1 - \Phi(x)\} = \frac{1}{x} - \frac{1}{x^3} + \frac{1.3}{x^5} - \ldots, \tag{7.19}$$

the infinite series not converging, but being asymptotic in the sense the error after taking n terms is less in absolute value than the last term of the series. This means that if the series is truncated at any fixed point, the error, both absolute and relative, tends to zero as x increases. It is also clear that the left-hand side is always bounded above by the sum to n terms when n is odd, and bounded below when n is even. The series is therefore said to be 'enveloping' in the sense used by Pólya (1949a), who also gives a similar series for the bivariate normal distribution.

These results are useful for calculating $\Phi(x)$ when x is large and they also show that for x large $1 - \Phi(x)$ is asymptotically equal to

$$(2\pi)^{-\frac{1}{2}} x^{-1} e^{-\frac{1}{2}x^2},$$

and in fact we have the useful inequalities

$$(2\pi)^{-\frac{1}{2}} e^{-\frac{1}{2}x^2} (x^{-1} - x^{-3}) < 1 - \Phi(x) < (2\pi)^{-\frac{1}{2}} e^{-\frac{1}{2}x^2} x^{-1}. \tag{7.20}$$

Equation (7.16) can, however, be used in a different way for calculating $\Phi(x)$ (Moran 1956, Das 1956). If we have an integral over the range $(-\infty, \infty)$ of the form

$$I = \int_{-\infty}^{\infty} f(x)\, dx, \tag{7.21}$$

it is natural to approximate it numerically by the sum

$$S = h \sum_{n=-\infty}^{\infty} f(\delta + nh), \tag{7.22}$$

where $h > 0$. S is a periodic function of δ and when $f(x)$ satisfies certain conditions it is a very close approximation to I (Moran 1958). Thus if

$$f(x) = (2\pi)^{-\frac{1}{2}} e^{-\frac{1}{2}x^2},$$

and $h = 1$, S does not differ from $I = 1$ by more than 0.52×10^{-8} (Yates 1948). The accuracy is not quite so great for (7.16) but is high enough to be useful. Moreover, since S is a periodic function of δ it must lie between its maximum and minimum, and in this way Das (1956) has shown that

$$1 - \Phi(x) > (2\pi)^{-1} x h e^{-\frac{1}{2}x^2} \sum_{n=-\infty}^{\infty} \frac{e^{-\frac{1}{2}h^2(n+\frac{1}{2})^2}}{x^2 + h^2(n+\frac{1}{2})^2}, \tag{7.23}$$

and

$$1 - \Phi(x) < (2\pi)^{-1} x h e^{-\frac{1}{2}x^2} \sum_{n=-\infty}^{\infty} \frac{e^{-\frac{1}{2}h^2 n^2}}{x^2 + h^2 n^2}. \tag{7.24}$$

These inequalities are very sharp and moreover the left-hand side is very nearly equal to the mean of the two expressions on the right. For other inequalities see Exercises 7.3 and Shenton (1954).

In the construction of tables the key values have usually not been calculated by formulae of the above types but by using continued fraction
20

expansions, the earliest of which was found by Laplace. A detailed discussion of this and another more rapidly convergent continued fraction is given by Shenton (1954). (See also Ruben 1962b.)

Exact bounds for the normal integral have been given by a number of writers such as Pólya (1949a), Williams (1946), and Chu (1955). Bounds for Mill's ratio (7.16) are given by Birnbaum (1942), Sampford (1953), and Tate (1953).

7.4. Tables and approximations for the normal distribution

A very large number of tables of the normal distribution in various forms have been published, so many in fact that all that can be done here is to refer to Greenwood and Hartley's (1962) *Guide to Tables in Mathematical Statistics*, and to the National Bureau of Standards' *Guide to Tables of the Normal Probability Integral* (1952). Tables also exist for complex x.

There are a considerable number of approximate formulae for use in various situations. In high-speed electronic calculators it is very wasteful to have to store tables of functions and in order to avoid this Hastings (1955) has devised a series of formulae which represent $\Phi(x)$ and $\Phi'(x)$ as rational functions (ratios of polynomials) with numerical coefficients which are chosen empirically to give highly accurate approximations. As the coefficients can be easily stored in the computer this provides an effective substitute for large tables. Thus, for example,

$$\Phi'(x) = (2\pi)^{-\frac{1}{2}}e^{-\frac{1}{2}x^2}$$

is given by

$$(b_0 + b_2 x^2 + b_4 x^4 + b_6 x^6 + b_8 x^8)^{-1}, \tag{7.25}$$

to an accuracy better than 0·0008 in the range $(0, \infty)$ when $b_0 = 2\cdot511261$, $b_2 = 1\cdot172801$, $b_4 = 0\cdot494618$, $b_6 = -0\cdot063417$, and $b_8 = 0\cdot029461$. He also gives a result of a somewhat similar type for $\Phi(x)$.

7.5. Cramér's theorem

We have already seen that if X and Y are independent random variables with normal distributions, $Z = X + Y$ is also normally distributed. It is a remarkable fact that a converse exists to this result in the following form.

THEOREM 7.2 (Cramér's theorem). *If X and Y are two independent random variables such that $Z = X + Y$ has a normal distribution, so also have X and Y, it being possible, of course, that one of the latter has a degenerate distribution in which all the probability is concentrated into one point.*

Proof. Suppose that Z has zero mean and unit standard deviation, and that the characteristic functions of X and Y are $\phi_1(z)$ and $\phi_2(z)$. Then

$$\phi_1(z)\phi_2(z) = e^{-\frac{1}{2}z^2}. \tag{7.26}$$

Let the distribution functions of X and Y be $F(x)$ and $G(x)$. We can suppose the X and Y shifted, if necessary so that the median of X is zero. Then

$$\text{pr}(Z \leqslant x) \geqslant \text{pr}(X \leqslant 0)\text{pr}(Y \leqslant x)$$

$$\geqslant \tfrac{1}{2}\text{pr}(Y \leqslant x),$$

so that

$$G(x) = \text{pr}(Y \leqslant x) \leqslant Ke^{-\frac{1}{2}x^2}, \tag{7.27}$$

where K is a positive constant. Similarly

$$1 - G(x) = \text{pr}(Y > x) \leqslant Ke^{-\frac{1}{2}x^2},$$

and a similar inequality holds for $F(x)$, so that we can take the same K in all cases. From (6.4) we have

$$\phi_1(z) = 1 - iz \int_{-\infty}^{0} e^{izx} F(x)\, dx + iz \int_{0}^{\infty} e^{izx}\{1 - F(x)\}\, dx, \tag{7.28}$$

and using the above inequalities for the distribution of X, we get

$$|\phi_1(z)| \leqslant 1 + 4K|z| \int_{-\infty}^{\infty} e^{|z|x - \frac{1}{2}x^2}\, dx,$$

$$\leqslant 1 + 4K(2\pi)^{\frac{1}{2}}|z|\, e^{\frac{1}{2}|z|^2}$$

$$\leqslant K_1 e^{A|z|^2}, \tag{7.29}$$

where $K_1 > 0$, $A > 0$. Similarly

$$|\phi_2(z)| \leqslant K_2 e^{A|z|^2}.$$

Therefore

$$|\phi_1(z)| = \left| \frac{e^{-\frac{1}{2}z^2}}{\phi_2(z)} \right|$$

$$\geqslant K_3 e^{-Bz^2} > 0, \tag{7.30}$$

with $K_3 > 0$, $B > 0$. $\phi_1(2)$ is therefore an analytic function with no zeros or poles, and satisfies (7.29) and (7.30). $\text{Log}\,\phi_1(z)$ is therefore also an analytic function with no poles in the whole plane. From (7.29) and (7.30), it must satisfy an inequality of the form

$$|\log \phi_1(z)| \leqslant A + B|z|^2. \tag{7.31}$$

From a well-known theorem on analytic functions (Titchmarsh 1932, p. 85) it follows that $\log \phi_1(z)$ is a polynomial of the second order at most, so that

$$\phi_1(z) = e^{a + bz + cz^2}. \tag{7.32}$$

Since $\phi_1(0) = 1$, we must have $a = 0$, and since $\phi(-z) = \phi(\bar{z})$ we have

$$-bz + cz^2 = \bar{b}\bar{z} + \bar{c}\bar{z}^2.$$

Thus b must be purely imaginary, and c real and negative, so that $\phi(z)$ is the characteristic function of a normal distribution.

Lévy (1937a) has proved an extension of this theorem to the limits of sequences of distributions in the following form.

THEOREM 7.3. *If $\{X_n\}$, $\{Y_n\}$ are sequences of random variables such that X_n and Y_n are independent, and if the distributions of $Z_n = X_n + Y_n$ converge to some normal distribution, then the distributions of X_n and Y_n, after reduction to zero means and unit standard deviations, converge to normal distributions provided the ratios of the dispersions of X_n and Y_n to that of Z_n are bounded below by a non-zero constant.*

Proof. The last condition is necessary in order to avoid the situation where the contribution of Y_n (say) to $X_n + Y_n$ becomes relatively negligible for in this case we could take X_n as normally distributed and Y_n to have any sort of distribution whose dispersion tends to zero.

From Theorem 5.22 the dispersions of X_n and Y_n cannot exceed that of Z_n, and are therefore bounded. We can, if necessary, shift the distributions of X_n and Y_n in opposite directions in such a way that zero is a median of the distribution of X_n. The set of all the distributions of X_n is therefore compact (see the proof of Theorem 6.4) and we can extract a subsequence, n_k ($k = 1, 2, \ldots$), such that the distributions of X_{n_k} converge to a proper distribution, $F(x)$, which can be chosen to be non-normal if the theorem is false. Similarly we can ensure that the subsequence, n_k, is such that the sequence of distributions of Y_{n_k} converges to a distribution $G(x)$. From the last condition of the theorem neither $F(x)$ nor $G(x)$ can be degenerate and $F(x) * G(x)$ is normal, so that by Cramér's theorem both $F(x)$ and $G(x)$ must be non-degenerate normal distributions, which proves the result.

Sapagov (1951) has given another proof of this result in which an upper bound for the rate of convergence to normality is given in terms of the rate of convergence to normality of the distribution of Z_n. It is remarkable that the only proofs known of Cramér's theorem involve the use of the theory of integral functions and no purely real variable proof has been found.

7.6. Distributions dependent on the normal distribution

If X is distributed in a normal distribution with zero mean and unit standard deviation, $Z = \frac{1}{2}X^2$ is clearly distributed with the probability density (using the fact that $\Gamma(\frac{1}{2}) = \pi^{\frac{1}{2}}$)

$$\Gamma(\tfrac{1}{2})^{-1} e^{-z} z^{-\frac{1}{2}} \, dz. \tag{7.33}$$

As we shall prove later (§ 7.17), the sum of n independent random variables all having the distribution (7.33) will have a gamma type distribution,

$$\Gamma(\tfrac{1}{2}n)^{-1} e^{-z} z^{\frac{1}{2}n-1} \, dz. \tag{7.34}$$

The quantity $X_1^2 + \ldots + X_n^2$ is then said to have the 'χ^2 distribution with n degrees of freedom' if $\frac{1}{2}(X_1^2 + \ldots + X_n^2)$ has the distribution (7.34).

Not only does $\frac{1}{2}(X_1^2 + \ldots + X_n^2)$ have this distribution when the true mean of the X_i distribution is zero but it is also possible to show that whatever the mean of the X_i, the quantity

$$T = \frac{1}{2} \sum_{i=1}^{n} \left(X_i - n^{-1} \sum_{j=1}^{n} X_j \right)^2 \tag{7.35}$$

has a distribution like (7.34) but with n replaced by $n-1$.

This may be proved in a variety of ways but the most intuitive manner is to use n-dimensional geometry. Consider the X_i as jointly distributed in a space of n dimensions. Their joint distribution is a multivariate normal distribution with density

$$(2\pi)^{-\frac{1}{2}n} \exp\{ -\tfrac{1}{2}(X_1^2 + \ldots + X_n^2) \} \, dX_1 \ldots dX_n. \tag{7.36}$$

This is spherically symmetric about the origin and is therefore invariant under any orthogonal transformation. Consider the 'sample point' $X = (X_1, \ldots, X_n)$, and the line $x_1 = x_2 = \ldots = x_n$, where the x_i are running coordinates. The perpendicular from $(X_1, \ldots, X_n)$ on to this line will meet it at the point

$$M = (n^{-1} \textstyle\sum X_i, n^{-1} \sum X_i, \ldots, n^{-1} \sum X_i),$$

which is distant

$$n^{-\frac{1}{2}} \textstyle\sum X_i \tag{7.37}$$

from the origin O. Take any orthogonal transformation to new rectangular axes such that the new coordinates are $Y_1, \ldots, Y_n$ where Y_1 is measured along the line OM. Then $Y_1 = n^{-\frac{1}{2}} \sum X_i$ will be distributed normally with zero mean and unit standard deviation (we have proved this before), and the distance XM is such that XM^2 will be distributed as $Y_2^2 + \ldots + Y_n^2$. Since the Y_i are independent normal variates, $\frac{1}{2}(Y_2^2 + \ldots + Y_n^2)$ has the distribution (7.34) with n replaced by $n-1$.

However, XM^2 is

$$\sum_{i=1}^{n} \left(X_i - n^{-1} \sum_{i=1}^{n} X_i \right)^2,$$

and this proves the result.

It also follows from the above that (7.37), which is $n^{\frac{1}{2}}$ times the mean of the X_i, and T are distributed independently. This property uniquely characterizes the normal distribution as was first proved by Geary (1936) for the class of all distributions with finite moments, and generalized to the class of all distributions by Lukacs (1942). Lancaster (1960) gives another proof together with a series of other characterization theorems of a similar kind.

We have shown that

$$\bar{X} = n^{-1} \sum_{i=1}^{n} X_i \quad \text{and} \quad s^2 = \sum_{i=1}^{n} (X_i - \bar{X})^2$$

are independent. In statistical theory the ratio

$$t = \bar{X} n^{\frac{1}{2}} s^{-1} \tag{7.38}$$

is said to have 'Student's t-distribution' with $n-1$ 'degrees of freedom'. $\bar{X}$ is distributed on the range $(-\infty, \infty)$ and s on the range $(0, \infty)$ so putting (7.38) in the product of the two distributions and integrating out s, we immediately obtain

$$\frac{\Gamma(n/2)}{\pi^{\frac{1}{2}}\Gamma\{(n-1)/2\}} \frac{dt}{(n-1)^{\frac{1}{2}}\{1+t^2(n-1)^{-1}\}^{n/2}}, \qquad (7.39)$$

for the probability density of t over the range $(-\infty, \infty)$. This distribution is considered later (§ 7.20). The distribution of the product of two normal variates has been considered by Irwin (1946), Craig (1936), and Aroian (1947). For the distribution of the ratio XY^{-1} see Geary (1930).

7.7. Distributions of quadratic forms in normal variates

Suppose that $X_1, \ldots, X_n$ are independent random variables distributed normally with zero means and unit standard deviations. Consider the distribution of the quadratic form

$$Q = \mathbf{x}'\mathbf{A}\mathbf{x}, \qquad (7.40)$$

where $\mathbf{x}'$ is the row vector $(X_1, \ldots, X_n)$, and $\mathbf{A}$ is a symmetric matrix. If $\mathbf{A}$ were not symmetric we could consider the distribution of $\frac{1}{2}\mathbf{x}'(\mathbf{A}+\mathbf{A}')\mathbf{x}$. Since $\mathbf{A}$ is symmetric there exists an orthogonal transformation

$$\mathbf{x} = \mathbf{T}\mathbf{y},$$

such that

$$\mathbf{T}'\mathbf{A}\mathbf{T}$$

is a diagonal matrix

$$\begin{pmatrix} \lambda_1 & & & & & \\ & \lambda_2 & & & & \\ & & \cdot & & & \\ & & & \cdot & & \\ & & & & \cdot & \\ & & & & & \cdot \\ & & & & & & \lambda_n \end{pmatrix}, \qquad (7.41)$$

where the λ_i are the characteristic roots of the matrix $\mathbf{A}$. Then

$$\mathbf{x}'\mathbf{A}\mathbf{x} = \mathbf{y}'\mathbf{T}'\mathbf{A}\mathbf{T}\mathbf{y} = \sum_1^n \lambda_i Y_i^2, \qquad (7.42)$$

where $\mathbf{y}' = (Y_1,\ldots,Y_n)$, and the Y_i are independent normal variates with zero means and unit standard deviations. We already know that the characteristic function of Y_i^2 is

$$(1-2it)^{-\frac{1}{2}}, \tag{7.43}$$

so that the characteristic function of $\lambda_i\,Y_i^2$ is

$$(1-2i\lambda_i t)^{-\frac{1}{2}}, \tag{7.44}$$

and that of $\mathbf{x}'\mathbf{A}\mathbf{x}$ must be

$$\prod_{i=1}^{n}(1-2i\lambda_i t)^{-\frac{1}{2}} \tag{7.45}$$

$$=|\mathbf{1}-2it\,\mathbf{A}|^{-\frac{1}{2}}. \tag{7.46}$$

$\mathbf{x}'\mathbf{A}\mathbf{x}$ will then have a gamma-type distribution if and only if all the non-zero λ_i are equal. By taking the logarithm of (7.45) and expanding it is easy to find all the cumulants of the distribution in terms of the λ_i. The calculation of the actual distribution of $\mathbf{x}'\mathbf{A}\mathbf{x}$ in the general case is discussed in detail by Plackett (1960) and Imhof (1961) who give useful references.

Next suppose we have two quadratic forms $Q_1 = \mathbf{x}'\mathbf{A}\mathbf{x}$, and $Q_2 = \mathbf{x}'\mathbf{B}\mathbf{x}$, in the same set of random normal variables. Then their joint characteristic function is

$$E\{\exp(it_1\mathbf{x}'\mathbf{A}\mathbf{x}+it_2\mathbf{x}'\mathbf{B}\mathbf{x})\}. \tag{7.47}$$

Suppose that t_1 and t_2 are real. Then the above is

$$E\{\exp i\mathbf{x}'(t_1\mathbf{A}+t_2\mathbf{B})\mathbf{x}\},$$

where $t_1\mathbf{A}+t_2\mathbf{B}$ is a real symmetric matrix. It follows that for real t_1, t_2, the joint characteristic function is obtained from (7.46) by replacing $t\mathbf{A}$ by $t_1\mathbf{A}+t_2\mathbf{B}$, and is therefore

$$|\mathbf{1}-2it_1\mathbf{A}-2it_2\mathbf{B}|^{-\frac{1}{2}}. \tag{7.48}$$

By analytical continuation this must hold for all values of t_1, t_2. From this result we can deduce some important criteria for the independence of the distributions of Q_1 and Q_2. Clearly for this to hold it is necessary and sufficient that

$$|\mathbf{1}-2it_1\mathbf{A}||\mathbf{1}-2it_2\mathbf{B}| = |\mathbf{1}-2it_1\mathbf{A}-2it_2\mathbf{B}|, \tag{7.49}$$

a result first obtained by Cochran. Since

$$(\mathbf{1}-2it_1\mathbf{A})(\mathbf{1}-2it_2\mathbf{B}) = \mathbf{1}-2it_1\mathbf{A}-2it_2\mathbf{B}-4t_1 t_2\mathbf{A}\mathbf{B},$$

it is sufficient that

$$\mathbf{A}\mathbf{B} = 0. \tag{7.50}$$

It is a remarkable fact that this is also a necessary condition (Craig 1943). If (7.49) holds we have

$$-\tfrac{1}{2}\log|\mathbf{1}-2it_1\mathbf{A}|-\tfrac{1}{2}\log|\mathbf{1}-2it_2\mathbf{B}| = -\tfrac{1}{2}\log|\mathbf{1}-2it_1\mathbf{A}-2it_2\mathbf{B}|.$$

To obtain the coefficient of t_1 on the left-hand side we use (7.45) from which it follows that the rth cumulant of the distribution of $\mathbf{x}'\mathbf{A}\mathbf{x}$ is

$$k_r = 2^{r-1}(r-1)! \sum_{i=1}^{n} \lambda_i^r. \tag{7.51}$$

However, it is known that if $\lambda_1,\ldots, \lambda_n$ are the roots of $\mathbf{A}$, $\lambda_1^r,\ldots, \lambda_n^r$ are the roots of $\mathbf{A}^r$. Furthermore, the sum of the roots of a symmetric matrix $\mathbf{A}$ is equal to the sum of the elements of the leading diagonal, the latter being known as the trace of $\mathbf{A}$ and written as $\operatorname{tr}\mathbf{A}$. Thus from the above

$$k_r = 2^{r-1}(r-1)! \operatorname{tr}\mathbf{A}^r. \tag{7.52}$$

It follows from this that in the joint cumulant generating function

$$\log E\{\exp i\mathbf{x}'(t_1\mathbf{A}+t_2\mathbf{B})\mathbf{x}\},$$

the coefficient of $i_1^k t_2^l$ $(k, l \geqslant 1)$ must be equal to zero, and in particular so must the coefficient of $t_1^2 t_2^2$. This coefficient is best found by replacing $\mathbf{A}$ in (7.52) by $t_1\mathbf{A}+t_2\mathbf{B}$ so that the coefficient of $t_1^2 t_2^2$ is obtained by picking out the coefficient of $t_1^2 t_2^2$ in

$$2^3(3!) \operatorname{tr}(t_1\mathbf{A}+t_2\mathbf{B})^4.$$

This is

$$2^3(3!) \operatorname{tr}(\mathbf{A}^2\mathbf{B}^2+\mathbf{A}\mathbf{B}\mathbf{A}\mathbf{B}+\mathbf{A}\mathbf{B}^2\mathbf{A}+\mathbf{B}\mathbf{A}^2\mathbf{B}+\mathbf{B}\mathbf{A}\mathbf{B}\mathbf{A}+\mathbf{B}^2\mathbf{A}^2). \tag{7.53}$$

Now for any two matrices $\operatorname{tr}(XY) = \operatorname{tr}(YX)$ and the trace of a sum is equal to the sum of the traces. Hence the trace in (7.53) is equal to

$$\operatorname{tr}(2\mathbf{A}^2\mathbf{B}^2+2\mathbf{A}\mathbf{B}\mathbf{A}\mathbf{B}+2\mathbf{A}\mathbf{B}^2\mathbf{A}) = \operatorname{tr}\{2\mathbf{B}\mathbf{A}^2\mathbf{B}+2(\mathbf{A}\mathbf{B})^2+2\mathbf{B}\mathbf{A}^2\mathbf{B}\}$$

$$= 2\operatorname{tr}(\mathbf{A}\mathbf{B})^2+4\operatorname{tr}(\mathbf{B}\mathbf{A}\mathbf{A}\mathbf{B})$$

$$= 2\operatorname{tr}\{(\mathbf{A}\mathbf{B})^2\}+4\operatorname{tr}\{(\mathbf{A}\mathbf{B})'\mathbf{A}\mathbf{B}\}. \tag{7.54}$$

Write $\mathbf{C} = \mathbf{A}\mathbf{B}$, and C_{ij} for its typical element. Then the above is equal to

$$2\operatorname{tr}\mathbf{C}^2+4\operatorname{tr}\mathbf{C}'\mathbf{C} = 2\sum_{ij}(C_{ij}C_{ji}+2C_{ij}^2). \tag{7.55}$$

Since $C_{ij}^2+C_{ij}C_{ji}+C_{ji}^2 \geqslant 0$, it must follow that each equals zero and therefore $C_{ij}^2 = 0$ for every pair (i,j). Thus $\mathbf{A}\mathbf{B} = 0$. This proof is due to Kawada (1950). (See also Lancaster 1954.) Another proof is given by Aitken (1950).

These results are of particular interest when $\mathbf{A}$ and $\mathbf{B}$ are such that all their roots are either zero or unity. It follows from (7.42) that in this case the quadratic forms have distributions of the form (7.34).

Kac (1945) has given a similar condition for the independence of the distribution of a quadratic and a linear form in the same variates.

7.8. Cochran's theorems

An important set of theorems has been given by Cochran (1934) which are concerned with sums of quadratic forms in normal random variables. (See also James 1952.)

THEOREM 7.4. *If $X_1,..., X_n$ are independent normal variates with zero means and unit standard deviations and*

$$Q = \sum_1^n X_i^2 = \sum_1^k Q_j = \sum_1^k \mathbf{x}'\mathbf{A}_j\mathbf{x}, \qquad (7.56)$$

where the Q_j are quadratic forms $\mathbf{x}'\mathbf{A}_j\mathbf{x}$ with $\mathbf{x}' = (X_1,..., X_n)$, and $\mathbf{A}_j$ is symmetric with rank r_j, then the following conditions are such that each implies the other two:

(1) $\sum_j r_j = n$;

(2) *for each j, Q_j is distributed in a χ^2 distribution,*

(3) *the Q_j are distributed independently.*

Proof. It is to be noticed first that an essential condition of this theorem is that it is known that the Q_j are quadratic forms in normal variates. If this restriction is removed the result is no longer true (James 1952).

Suppose (1) is true and choose an orthogonal transformation of the X_i to new variates Y_i such that Q_1 is diagonalized to become

$$Q_1 = \lambda_1 Y_1^2 + ... + \lambda_{r_1} Y_{r_1}^2 \quad (\lambda_i > 0).$$

Then

$$Q - Q_1 = (1 - \lambda_1) Y_1^2 + ... + (1 - \lambda_{r_1}) Y_{r_1}^2 + Y_{r_1+1}^2 + ... + Y_n^2, \qquad (7.57)$$

which must be a quadratic form of rank not greater than $n - r_1$. Thus $\lambda_1 = ... = \lambda_{r_1} = 1$, and Q_1 must be distributed as χ^2 with r_1 degrees of freedom independently of $Q - Q_1$. From this it does not follow directly that Q_1 is independent of each of the other Q_i's, or independent of them jointly. This is, however, true and follows when we consider the representation of the Q_i ($i > 1$) as quadratic forms in the Y_i. For consider Q_2. We know that it also is distributed as χ^2 and is therefore a positive definite or semi-definite form. If it contains any term involving Y_1 it must have a term of the form aY_1^2 ($a > 0$). This is true for all the Q_i with $i > 1$ but we know that their sum has no term in Y_1^2, and thus arguing in a similar manner Q_2 can contain no terms in $Y_2,..., Y_{r_1}$, and must therefore be independent of Q_1. It also follows that Q_1 is independent of $Q_2,..., Q_k$ taken jointly. Thus (2) and (3) follow from (1).

Now assume (2) and make an orthogonal transformation to variates Y_i, so that Q_1 becomes

$$\lambda_1 Y_1^2 + ... + \lambda_{r_1} Y_{r_1}^2$$

as before. This has characteristic function

$$\prod_{i=1}^{r_1} (1 - 2\lambda_i t)^{-k/2}$$

which must be of the form

$$(1-2it)^{-\frac{1}{2}s}$$

identically, and hence $\lambda_1 = \ldots = \lambda_{r_1}$, and $s = r_1$. Thus each Q_i is distributed as χ^2 with r_i degrees of freedom. Then

$$Q_1 = Y_1^2 + \ldots + Y_{r_1}^2$$

and

$$Q - Q_1 = Y_{r_1+1}^2 + \ldots + Y_n^2.$$

Arguing as before the Q_i must be distributed independently. Thus (2) implies (1) and (3).

If (3) is true we again make the orthogonal transformation leading to (7.57), so that $Q_1 = \lambda_1 Y_1^2 + \ldots + \lambda_{r_1} Y_{r_1}^2$.

The joint characteristic function of Q_1 and $Q - Q_1$ is

$$E \exp\{it_1 Q_1 + it_2(Q - Q_1)\}$$

$$= \left[\prod_{i=1}^{r_1} \{1 - 2it_1 \lambda_i - 2it_2(1 - \lambda_i)\} \right]^{-\frac{1}{2}} (1 - 2it_2)^{-\frac{1}{2}(n-r_1)}. \qquad (7.58)$$

If Q_1 and $Q - Q$ are distributed independently their joint characteristic function must also be of the form

$$\left[\prod_{i=1}^{r_1} (1 - 2it_1 \lambda_i) \prod_{j=1}^{r_1} \{1 - 2it_2(1 - \lambda_j)\} \right]^{-\frac{1}{2}} (1 - 2it_2)^{-\frac{1}{2}(n-r_1)}. \qquad (7.59)$$

Identifying this expression with (7.58) we have

$$\prod_{i=1}^{r_1} \{1 - 2it_1 \lambda_i - 2it_2(1 - \lambda_i)\} = \prod_{i=1}^{r_1} (1 - 2it_1 \lambda_i)\{1 - 2it_2(1 - \lambda_i)\}. \qquad (7.60)$$

This holds for all t_1 and t_2. Equating coefficients we see that this can only happen when $\lambda_1 = \ldots = \lambda_{r_1} = 1$. It then follows that Q_1 is distributed as χ^2 with r_1 degrees of freedom. As the similar result holds for each of the other Q_i the rest of the theorem follows. This theorem is of key importance in the statistical theory of the Analysis of Variance.

7.9. Ratios of quadratic forms

Another problem which is of great practical importance is to determine the distribution of a ratio,

$$T = \frac{\mathbf{x}'\mathbf{A}\mathbf{x}}{\mathbf{x}'\mathbf{B}\mathbf{x}}, \qquad (7.61)$$

of quadratic forms in a set of independent normal variates given by $\mathbf{x}' = (X_1, \ldots, X_n)$, where the X_i have zero means and the same standard deviation. In all the cases that have been investigated $\mathbf{x}'\mathbf{B}\mathbf{x}$ is positive definite. It is then possible, in principle, to write down infinite series for the distribution of T and it is also possible to obtain approximations by the method of steepest descent (see especially Daniels 1954, 1956) .One case of particular interest occurs when $\mathbf{B}$ has only positive unit roots. By a change of coordinates, if necessary, we can then write T in the form

$$T = \frac{\mathbf{x}'\mathbf{Ax}}{\sum X_i^2}. \tag{7.62}$$

If we regard $(X_1,\dots, X_n)$ as a point in n-dimensional space we see that it has a distribution which is spherically symmetric. It follows that $\sum X_i^2$, the square of the distance of this point from the origin, and T, a function depending only on the direction of the vector from the origin to $(X_1,\dots, X_n)$, must be distributed independently.

Then to obtain the moments of T we have

$$E(T^k) = \frac{E\{(\mathbf{x}'\mathbf{Ax})^k\}}{E\left\{\left(\sum_1^n X_i^2\right)^k\right\}}. \tag{7.63}$$

The numerator and denominator are often easy to find for small k. The distribution of expressions such as (7.62) has been found, exactly or approximately, in a number of special cases. Plackett (1960) gives a good survey of this problem (see also Robbins 1948c, and Grenander, Pollak, and Slepian 1959).

7.10. Other characterizations of the normal distribution

We have already characterized the normal distribution above by showing that it is the unique distribution for which certain properties of random variables obtain. There are a number of other such theorems which are particularly elegant and some will be here considered (for a survey with references see Lancaster 1960). We first prove the following remarkable result.

THEOREM 7.5. *Suppose that* $X_1,\dots, X_n$ *are n independent random variables* $(n \geqslant 2)$. *Let*

$$Y_1 = a_1 X_1 + \dots + a_n X_n, \tag{7.64}$$

$$Y_2 = b_1 X_1 + \dots + b_n X_n, \tag{7.65}$$

where the a_i *and* b_i *are given constants which are non-zero. If* Y_1, Y_2 *are independent, then each of the* X_i *are normally distributed.*

Notice that the existence of moments is not assumed.

Proof. We follow Lancaster's (1960) proof (earlier proofs were given by Darmois 1951, Skitovitch 1954, and Lukacs and King 1954). There is no restriction in supposing that all the a_i and b_j have absolute values less than unity. Let a be the smallest non-zero such absolute value. Choose ε so

$0 < \varepsilon < \gamma n^{-3}$ where $\gamma < 1$, and α so that

$$\mathrm{pr}(|X_j| > \alpha) < \varepsilon, \quad \text{all } j. \tag{7.66}$$

Then

$$\mathrm{pr}(|a_j X_j| > \alpha) < \varepsilon,$$

and

$$\mathrm{pr}(|b_j X_j| > \alpha) < \varepsilon.$$

$|Y_1| > n\alpha$ is only possible if at least one of the $|X_i| > \alpha$ and the events $(|X_i| > \alpha)$ are not mutually exclusive. Nevertheless, the probability of at least one of these happening is, by Boole's inequality (1.2), not greater than $n\varepsilon$. A similar argument holds for Y_2. However, $|Y_1|$ and $|Y_2|$ will both exceed $n\alpha$ if

$$|aX_j| > (2n-1)\alpha, \quad \text{for some } j,$$

and

$$|X_k| \leqslant \alpha, \quad \text{for all } k \neq j.$$

Thus

$$\mathrm{pr}(|Y_1| > n\alpha, |Y_2| > n\alpha) \geqslant \mathrm{pr}\{|a_j X_j| > (2n-1)\alpha\} \, \mathrm{pr}(|X| \leqslant \alpha),$$

$$\geqslant (1-\varepsilon)^{n-1} \mathrm{pr}\{|a_j X_j| > (2n-1)\alpha\}. \tag{7.67}$$

The left-hand side of (7.67) is equal to

$$\mathrm{pr}(|Y_1| > n\alpha)\mathrm{pr}(|Y_2| > n\alpha) \leqslant n^2\varepsilon^2.$$

Then

$$(1-\varepsilon)^{n-1} \mathrm{pr}\{|a_j X_j| > (2n-1)\alpha\} \leqslant n^2\varepsilon^2. \tag{7.68}$$

We have chosen $\varepsilon < n^{-3}$ and it can be shown by a simple calculation that this implies that $n > (1-\varepsilon)^{1-n}$. Hence

$$\mathrm{pr}\{|X_j| > (2n-1)\alpha a^{-1}\} \leqslant n^3\varepsilon^2. \tag{7.69}$$

Put $\varepsilon_1 = n^3\varepsilon^2 < \gamma\varepsilon$. This argument holds for all j, and putting

$$c = (2n-1)a^{-1} > 1$$

we conclude from (7.69) that

$$\mathrm{pr}(|X_j| > c\alpha) \leqslant \gamma\varepsilon, \quad \text{all } j. \tag{7.70}$$

Replacing (7.66) by (7.70) repeatedly we get

$$\mathrm{pr}(|X_j| > c^s\alpha) \leqslant \gamma^s\varepsilon, \quad s = 1, 2, \dots. \tag{7.71}$$

Putting this in the formula

$$\phi(t) = 1 - it \int_{-\infty}^{0} e^{itx} F(x) \, dx + it \int_{0}^{\infty} e^{itx}\{1 - F(x)\} \, dx,$$

as given in (6.4), writing $F_j(x)$ for the distribution of X_j, and remembering that γ can be chosen to be arbitrarily small provided it is less than unity, we see that each X_j has moments and cumulants of all orders and a characteristic function which is analytic in a non-zero circle around the origin.

Now suppose some of the ratios $a_i b_i^{-1}$ in (7.65) are equal. By redefinition of the X_i by taking new random variables which are sums of the corresponding $a_i X_i$ we can ensure that all the $a_i b_i^{-1}$ are unequal. (Some must be unequal in any case since otherwise Y_1 and Y_2 would not be independent.) The normality of the distributions of the individual terms will follow from that of the sums by Cramér's theorem. We can now rewrite (7.65) as

$$Y_1 = X_1 + X_2 + \ldots + X_n,$$
$$Y_2 = r_1 X_1 + r_2 X_2 + \ldots + r_n X_n, \tag{7.72}$$

where the r_i are unequal to each other and to zero. Let K_s^j be the sth cumulant of X_j, and $K_s(W)$ the sth cumulant of any random variable W, so that $K_s^j = K_s(X_j)$.

Consider the random variable $\lambda Y_1 + \mu Y_2$, where λ, μ are any two constants. Since the cumulants of sums of independent random variables are the sums of the cumulants we have

$$K_s(\lambda Y_1 + \mu Y_2) = \lambda^s K_s(Y_1) + \mu^s K_s(Y_2), \tag{7.73}$$

and the left-hand side is also equal to

$$K_s\{\sum_j (\lambda X_j + \mu r_j X_j)\} = \sum_j (\lambda + \mu r_j)^s K_s^j$$
$$= \sum_j (\lambda^s + \mu^s r_j^s) K_s^j, \tag{7.74}$$

from (7.73). Take $s > n$ and equate the coefficients of $\lambda^{s-1}\mu$, $\lambda^{s-2}\mu^2, \ldots, \lambda^{s-n}\mu^n$. We then obtain n equations in $K_1^s, \ldots, K_n^s$ which can be written

$$\begin{pmatrix} r_1 & r_2 & \cdots & r_n \\ r_1^2 & r_2^2 & \cdots & r_n^2 \\ & & \vdots & \\ r_1^n & r_2^n & \cdots & r_n^n \end{pmatrix} \begin{pmatrix} K_1^s \\ \vdots \\ K_n^s \end{pmatrix} = \begin{pmatrix} 0 \\ \vdots \\ 0 \end{pmatrix}$$

The determinant of the matrix on the left-hand side is necessarily non-zero because all the r_i are unequal and it therefore follows that

$$K_1^s = K_2^s = \ldots = K_n^s = 0. \tag{7.75}$$

Hence all the cumulants with suffices greater than n are zero and the characteristic function of each random variable must be of the form

$$e^{-P(t)},$$

where $P(t)$ is a polynomial of order n or less. By Marcinkiewicz's theorem (Theorem 6.23) $P(t)$ must be a quadratic and therefore the distributions of all the X_i must be normal. For another proof see Linnik (1956).

This theorem has some interesting corollaries. If X_1 and X_2 are independent random variables, and $X_1 + X_2$, $X_1 - X_2$ are distributed independently, then

X_1 and X_2 have moments of all orders and both are distributed normally. This striking result is known as Bernstein's theorem but was originally proved under the further restriction that the second moments were known to be finite.

A generalization of this result is that if the joint distribution of n independent random variables is spherically symmetric, the joint distribution must be normal. This follows at once from Theorem 7.5 on taking two components of an arbitrary orthogonal transformation of the variates.

In Lancaster's (1960) paper a number of other theorems are given which characterize the normal distribution in terms of the independence of quadratic and linear forms.

7.11. The truncation of a normal distribution

A distribution which is often important in practice, especially in genetics, is the truncated normal distribution. If we put

$$1 - F(x) = \frac{1}{\sqrt{(2\pi\sigma^2)}} \int_x^\infty \exp\left\{-\frac{(x-u)^2}{\sigma^2}\right\} dx,$$

this distribution is defined to be

$$\{F(x) - F(h)\}\{1 - F(h)\}^{-1} \quad (h \leqslant x \leqslant \infty), \tag{7.76}$$

where h is the point of truncation. The lower moments are then easily found. For example transforming, for simplicity, to the case where $\mu = 0$, $\sigma = 1$, we have

$$m_1 = \int_h^\infty x e^{-\frac{1}{2}x^2} dx \left(\int_h^\infty e^{-\frac{1}{2}x^2} dx\right)^{-1}$$

$$= (2\pi)^{-\frac{1}{2}} e^{-\frac{1}{2}h^2} \{1 - F(h)\}^{-1}. \tag{7.77}$$

Similarly

$$m_2 = (2\pi)^{-\frac{1}{2}} \int_h^\infty x^2 e^{-\frac{1}{2}x^2} dx \{1 - F(h)\}^{-1}$$

$$= 1 + (2\pi)^{-\frac{1}{2}} h \{1 - F(h)\}^{-1} e^{-\frac{1}{2}h^2}$$

$$= 1 + h m_1. \tag{7.78}$$

The distribution of a sum of independent random variables with a truncated normal distribution is discussed by Francis (1946).

7.12. The multivariate normal distribution

This has already been defined in § 5.3 and its characteristic function obtained in § 6.9. The moments are best obtained from the latter. We give the lower moments up to those of the fourth order and it is clearly sufficient to consider a four-variate distribution with zero means, unit variances, and a correlation matrix

$$\begin{pmatrix} 1 & \rho_{12} & \rho_{13} & \rho_{14} \\ \rho_{21} & 1 & \rho_{23} & \rho_{24} \\ \rho_{31} & \rho_{32} & 1 & \rho_{34} \\ \rho_{41} & \rho_{42} & \rho_{43} & 1 \end{pmatrix}, \tag{7.79}$$

where $\rho_{ij} = \rho_{ji}$. Then using an obvious notation we have

$$m_{1000} = 0.$$

$$m_{2000} = 1, \quad \mu_{1100} = \rho_{12}.$$

$$m_{3000} = 0, \quad \mu_{2100} = 0, \quad \rho_{1110} = 0$$

$$m_{4000} = 3, \quad \mu_{3100} = 3\rho_{12},$$

$$m_{2200} = 1 + 2\rho_{12}^2, \quad \mu_{2110} = \rho_{23} + 2\rho_{12}\rho_{13},$$

$$m_{1111} = \rho_{12}\rho_{34} + \rho_{13}\rho_{24} + \rho_{14}\rho_{23}.$$

Note that all the fourth-order moments can be derived from the last by identifying two or more of the variates. Moments of the sixth order are given by Wicksell (1919).

Any section of a multivariate normal distribution, i.e. any distribution conditional on fixing some of the variates, is clearly again a multivariate normal distribution as is obvious from the form of the probability density. Moreover, this conditional distribution is independent of the values of the fixed variates since the coefficients of the second-order terms in the remaining variables in the index of the exponential are unaltered. If $X_1, \dots, X_n$ have a joint multinormal distribution and $Y_1, \dots, Y_n$ are obtained from the X_i by a non-singular linear transformation it is clear that the Y_i also have a multinormal distribution. From this it follows that the conditional distribution of the X_i given m independent linear relations $(m < n)$,

$$\sum_{j=1}^{n} b_{ij} X_j = c_i \quad (i = 1, \dots, m),$$

is a degenerate multinormal distribution, i.e. the n random variables can be expressed linearly in terms of $n - m$ random variables with a multinormal distribution.

Let the distribution of the X_i have the density function

$$\frac{|A|^{\frac{1}{2}}}{(2\pi)^{\frac{1}{2}}} \exp -\tfrac{1}{2} x' A x, \tag{7.80}$$

where $x' = (x_1, \dots, x_n)$ and $A = (a_{ij})$. Then the expression

$$Q = \sum_{ij} a_{ij} X_i X_j \tag{7.81}$$

has the χ^2 distribution with n degrees of freedom. To prove this we use the fact that there exists an orthogonal transformation $\mathbf{x} = \mathbf{T}\mathbf{y}$ such that $\mathbf{T}'\mathbf{A}\mathbf{T}$ is a diagonal matrix whose elements are σ_i^{-2}, say. We then have

$$Q = \sum_{ij} a_{ij} X_i X_j = \sum_i \sigma_i^{-2} Y_i^2,$$

and thus Q is distributed in the χ^2 distribution with n degrees of freedom.

7.13. The evaluation of the multivariate normal integral

Suppose we have n random variables $X_1, \ldots, X_n$ with a joint multivariate normal distribution such that the means are all zero, the standard deviations unity, and the correlations $\rho_{ij} = E(X_i X_j)$. For many purposes it is necessary to evaluate the probability, $\mathrm{pr}(X_1 > x_1, \ldots, X_n > x_n)$. Consider first the simplest case in which $n = 2$ and $x_1 = x_2 = 0$.

We then have to evaluate the integral

$$P = \frac{1}{2\pi(1-\rho^2)^{\frac{1}{2}}} \int_0^\infty \int_0^\infty \exp -\left\{ \frac{x^2 - 2\rho xy + y^2}{2(1-\rho^2)} \right\} dx\, dy. \tag{7.82}$$

We change to new variables u, v defined by

$$u = x,$$
$$v = (y - \rho x)(1-\rho^2)^{-\frac{1}{2}}. \tag{7.83}$$

Since the exponent in the exponential can be written as

$$-\{x^2(1-\rho^2) + (y-\rho x)^2\}/2(1-\rho^2),$$

and the Jacobian of the transformation is

$$(1-\rho^2)^{-\frac{1}{2}},$$

the problem is transformed into evaluating the integral

$$\frac{1}{2\pi} \int\!\!\int \exp -\tfrac{1}{2}(u^2 + v^2)\, du\, dv \tag{7.84}$$

over the region of the plane,

$$\{u > 0,\ v > -\rho(1-\rho^2)^{-\frac{1}{2}}u\}. \tag{7.85}$$

The density function in (7.84) is circularly symmetrical and so the above integral must equal $\theta(2\pi)^{-1}$, where θ is the angle of the wedge defined by (7.85). Then $\theta = \frac{1}{2}\pi + \alpha$, where $\sin \alpha = \rho$ from (7.83). Hence

$$P = \tfrac{1}{4} + \tfrac{1}{2}\pi^{-1}\sin^{-1}\rho$$
$$= \tfrac{1}{2} - \tfrac{1}{2}\pi^{-1}\cos^{-1}\rho. \tag{7.86}$$

This is known as Sheppard's formula (Sheppard 1898).

Next consider the probability that $(X_1 > 0, \ X_2 > 0, \ X_3 > 0)$ where $E(X_i X_j) = \rho_{ij}$ and the distribution is trivariate normal with zero means and unit variances.

This is most simply obtained by using (1.60). Write A_1, A_2, and A_3 for the events $(X_1 > 0)$, $(X_2 > 0)$, and $(X_3 > 0)$. Then

$$\text{pr}(X_1 > 0, X_2 > 0, X_3 > 0) = \text{pr}(X_1 < 0, X_2 < 0, X_3 < 0)$$

$$= 1 - \text{pr}(\text{at least one of } X_1, X_2, X_3 > 0)$$

$$= 1 - \sum \text{pr}(A_i) + \sum \text{pr}(A_i A_j) - \text{pr}(A_1 A_2 A_3).$$

Hence

$$2\,\text{pr}(A_1 A_2 A_3) = 1 - \sum \text{pr}(A_i) + \sum \text{pr}(A_i A_j)$$

$$= 1 - \tfrac{3}{2} + (\tfrac{3}{4} + \tfrac{1}{2}\pi^{-1} \sum \sin^{-1} \rho_{ij}),$$

and thus

$$\text{pr}(A_1 A_2 A_3) = \tfrac{1}{8} + (4\pi)^{-1}(\sin^{-1} \rho_{12} + \sin^{-1} \rho_{13} + \sin^{-1} \rho_{23}). \qquad (7.87)$$

This can be given an instructive interpretation by an alternative proof. Write

$$X_1 = Z_1,$$

$$X_2 = \rho_{12} Z_1 + (1 - \rho_{12}^2)^{\frac{1}{2}} Z_2,$$

$$X_3 = \rho_{13} Z_1 + a Z_2 + b Z_3, \qquad (7.88)$$

where we have written for convenience

$$a = (\rho_{23} - \rho_{12}\rho_{13})(1 - \rho_{12}^2)^{-\frac{1}{2}},$$

$$b = \left\{1 - \rho_{13}^2 - \frac{(\rho_{23} - \rho_{12}\rho_{13})^2}{1 - \rho_{12}^2}\right\}^{\frac{1}{2}}.$$

It is easy to verify that this is a transformation to independent normal variates Z_1, Z_2, Z_3, and the condition that all $X_i > 0$ becomes

$$Z_1 > 0,$$

$$\rho_{12} Z_1 + (1 - \rho_{12}^2)^{\frac{1}{2}} Z_2 > 0,$$

$$\rho_{13} Z_1 + a Z_2 + b Z_3 > 0. \qquad (7.89)$$

Consider the region of space defined by these inequalities and its intersection by the unit sphere. Since the sums of the squares of the coefficients in (7.88) are equal to unity, the Z_i all have unit variances. (7.89) defines a triangle on the surface of the unit sphere whose area is 4π times the probability that $X_1 > 0, \ X_2 > 0, \ X_3 > 0$.

The area of a spherical triangle on a unit sphere whose angles are α, β, γ is known to be $\alpha + \beta + \gamma - \pi$, and in the present case these angles, from (7.89),

21

are given by

$$\cos(\pi - \alpha) = \rho_{12},$$

$$\cos(\pi - \beta) = \rho_{13},$$

$$\cos(\pi - \gamma) = \rho_{12}\rho_{13} + (1 - \rho_{12}^2)^{\frac{1}{2}}a,$$

$$= \rho_{23}.$$

Inserting these we find that the probability is

$$(4\pi)^{-1}(2\pi - \cos^{-1}\rho_{12} - \cos^{-1}\rho_{13} - \cos^{-1}\rho_{23}), \tag{7.90}$$

which is equal to (7.87).

The corresponding argument for more than three variates does not lead to any useful formulae since it requires the evaluation of the volume of an $n-1$ dimensional curvilinear simplex in the $n-1$ dimensional surface of a sphere in n-dimensions. This problem has an extensive history starting with Schläfli (1858). For further references see Ruben (1954, 1960, 1961a, b, c; 1962a). Kendall (1941) has given a method of obtaining the probability in terms of an infinite series in the ρ_{ij} and for the case of four variates this has been examined by Moran (1948b). Better methods are considered by Plackett (1954), David and Mallows (1961), and McFadden (1960).

When there are relations between the ρ_{ij} the problem sometimes becomes simpler. Thus if all the ρ_{ij} are equal to ρ, say, we can express the probability in terms of an integral which can be easily numerically evaluated. Suppose that $Y_1,..., Y_{n+1}$ are $n+1$ independent normal variates with zero means and unit variances. Write

$$X_i = Y_i - aY_{n+1} \quad (i = 1,...,n). \tag{7.91}$$

Then the X_i are distributed normally with zero means and all intercorrelations equal to $a^2(1+a^2)^{-1}$, which we take as equal to ρ. The probability that $(X_1 > 0,..., X_n > 0)$ is the probability that $(Y_1 > aY_{n+1},..., Y_n > aY_{n+1})$, which is thus

$$I = (2\pi)^{-(n+1)/2} \int_{-\infty}^{\infty} e^{-\frac{1}{2}x^2} \left(\int_{ax}^{\infty} e^{-\frac{1}{2}u^2} du \right)^n dt. \tag{7.92}$$

This integral also occurs elsewhere in probability theory. It can be evaluated most simply by replacing the outer integration by a sum and the error can then be made remarkably small (Moran 1956). Notice also that in the particular case where all the $\rho_{ij} = \frac{1}{2}$ the probability is $(n+1)^{-1}$.

In some of the equations (7.91) we can replace a by $-a$ and in this way we can extend the above method to the case where ρ_{ij} is of the form $\rho(-1)^{u_i+v_j}$, where u_i, v_j are integers depending on i and j.

Equally important in practice is the evaluation of $\text{pr}(X_1 > x_1,..., X_n > x_n)$ when the x_i are not all zero. The tabulation of such a quantity requires an entry for each x_i and for each value of the $\frac{1}{2}n(n-1)$ correlation coefficients, and therefore requires a classification for each of $\frac{1}{2}n(n+1)$ different para-meters, i.e. 3 for $n = 2$, 6 for $n = 3$, and so on. It is therefore not practicable to prepare tables for $n > 2$. For $n = 2$ tables have, however, been published. Pearson's *Tables for Statisticians and Biometricians*, Vol. II (1931), gives $\text{pr}(X_1 > x_1, X_2 > x_2)$ for $\rho = 0(0\cdot05)\,1$, and $x_1,\ x_2 = 0(0\cdot1)\,2\cdot6$, to $6D$; for $x_1,\ x_2$ in the same range and $-\rho = 0(0\cdot05)\,0\cdot6\,(7D)$, $0\cdot65(0\cdot05)\,0\cdot75\,(8D)$, $0\cdot80(0\cdot05)\,0\cdot95\,(7D)$. The National Bureau of Standard's *Tables of the Bivariate Normal Distribution* (1959) give a list of errors in Pearson's tables and $\text{pr}(X_1 > x_1, X_2 > x_2)$ for $\rho = 0(0\cdot05)\,0\cdot95(0\cdot01)\,1$ and $x_1,\ x_2 = 0(0\cdot1)\,4$ to $6D$, and for $-\rho = 0(0\cdot05)\,0\cdot95(0\cdot01)\,0\cdot99$ and $x_1,\ x_2 = 0(0\cdot1)\,4$ to $7D$. This volume also contains a useful introduction and bibliography. Owen (1962) gives tables of functions in terms of which the bivariate and trivariate normal integral can be calculated. Steck (1958a) gives tables of a function in terms of which the trivariate normal distribution can be found. For further related tables see also Greenwood and Hartley (1962).

The evaluation of the bivariate normal integral has been considered by Nicholson (1943) and Cadwell (1951).

Multivariate normal distributions which are truncated in one or more variates also have useful applications (see Tallis 1961; Curnow 1961; Rosenbaum 1961; Finney 1962, 1964; Nabeya 1961; Kamat 1958a, b).

7.14. The circular bivariate normal distribution

Consider a bivariate normal distribution with means zero, variances unity, and $\rho = 0$. This distribution has circular symmetry. For some applications it is desirable to be able to calculate the integral inside a circle of radius r and centre at the point (D, O). This is

$$p(r, D) = \frac{1}{2\pi} \iint e^{-\frac{1}{2}(x^2+y^2)}\,dx\,dy \qquad (7.93)$$

over the region $(x - D)^2 + y^2 \leqslant r$. $p(r, D)$ is known as the offset circle proba-bility function. A number of papers have been published on the evaluation of this integral and references to these are given in Owen (1962) who also gives a table to $3D$ for $r - D = -3\cdot9(0\cdot1)\,4\cdot0$, $D = 0(0\cdot1)\,6\cdot0(0\cdot5)\,10(1)\,20$.

7.15. Truncation and selection

If X and Y are jointly normally distributed with zero means and unit variances, and correlation ρ, the conditional distribution of Y given that $X = x$ is a normal distribution with mean ρx and variance $(1 - \rho^2)$. The conditional distribution of Y, when X is merely restricted to be not less than x, is, however, not normal and is of interest in situations, such as occur in genetics, where we wish to select higher values of Y but are only given values of X (Cochran 1951). Writing, as before,

$$1 - \Phi(x) = (2\pi)^{-\frac{1}{2}} \int_x^\infty e^{-\frac{1}{2}u^2}\, du,$$

the conditional distribution of Y, given $X > x$, will have the probability density

$$f(y \,|\, X > x) = [2\pi\{1 - \Phi(x)\}(1 - \rho^2)^{\frac{1}{2}}]^{-1} \int_x^\infty \exp\left\{ -\frac{1}{2}\left(\frac{u^2 - 2\rho u y + y^2}{1 - \rho^2} \right) \right\}\, du.$$

$$(7.94)$$

This is a skewed distribution whose skewness increases with x. The lower moments are easily found by repeated integration by parts. Thus

$$E(Y \,|\, X > x) = [2\pi\{1 - \Phi(x)\}(1 - \rho^2)^{\frac{1}{2}}]^{-1}$$

$$\times \int_{-\infty}^\infty y \int_x^\infty \exp\left\{ -\frac{1}{2}y^2 - \frac{1}{2}\left(\frac{(u - \rho y)^2}{1 - \rho^2} \right) \right\}\, du\, dy$$

$$= \frac{\rho e^{-\frac{1}{2}x^2}}{(2\pi)^{\frac{1}{2}}\{1 - \Phi(x)\}},$$

$$(7.95)$$

from which the increase in the expected mean of Y, resulting from truncation of X, can be found.

In a similar way it can be shown that

$$\text{var}(Y \,|\, X > x) = 1 - \rho^2 W(W - x),$$

$$(7.96)$$

where

$$W = \frac{e^{-\frac{1}{2}x^2}}{[(2\pi)^{\frac{1}{2}}\{1 - \Phi(x)\}]}.$$

$$(7.97)$$

Similarly in the joint distribution of X and Y conditional on $X > x$, the correlation of X and Y is

$$\rho\left(\frac{1 - W(W - x)}{1 - \rho^2 W(W - x)} \right)^{\frac{1}{2}}.$$

$$(7.98)$$

Cochran gives values ($2D$) of the mean, standard deviation, and correlation, for values of x such that $\Phi(x) = 0.5$, 0.25, 0.1, 0.05, and 0.01, and $\rho = 0.5$, 0.8, and 0.95, together with references to further development in this type of selection formulae (see also Weiler 1959, Finney 1964). This distribution can also be studied by using the conditional characteristic function defined in §6.14.

7.16. The logarithmic normal distribution

The normal distribution has often been used as an approximation to other distribution because of the simplicity of its properties and the ease of access to tables. In order to do this a transformation of the random variable being considered is sometimes used to improve the approximation and this leads directly to the idea of generating classes of distributions by a transformation $Y = f(X)$, where Y is to have a normal distribution, and $X = f^{-1}(Y)$ is the variate being studied. An endless variety of distributions can be generated in this way, and perhaps the most frequently used is the log-normal distribution. This is the distribution on the range $(0, \infty)$ obtained by supposing that $\log X$, where X is the variate, has a normal distribution with mean m and variance σ^2. Its probability density can therefore be written as

$$f(x)\, dx = \{\sigma x \sqrt{(2\pi)}\}^{-1} \exp\left\{-\frac{1}{2\sigma^2}(\log x - m)^2\right\} dx, \qquad (7.99)$$

where x lies in the range $(0, \infty)$.

Many biological measurements of quantities which are necessarily non-negative have distribution of this form and its usefulness can be further extended by supposing that $X - C$, where C is a constant, has a log-normal distribution, thus introducing a third parameter.

Using the form (7.99) the moments are easily found. If $n > 0$ we have

$$E(X^n) = \sigma^{-1}(2\pi)^{-\frac{1}{2}} \int_0^\infty x^{n-1} \exp\left\{-\frac{1}{2\sigma^2}(\log x - m)^2\right\} dx$$

$$= \sigma^{-1}(2\pi)^{-\frac{1}{2}} \int_{-\infty}^\infty \exp\left\{ny - \frac{1}{2\sigma^2}(y - m)^2\right\} dy$$

$$= \exp(mn + \tfrac{1}{2}n^2 \sigma^2). \qquad (7.100)$$

Hence the mean and variance are

$$m_1 = \exp(m + \tfrac{1}{2}\sigma^2), \qquad (7.101)$$

$$\mu_2 = \exp(2m + \sigma^2)(\exp\sigma^2 - 1). \qquad (7.102)$$

The third and fourth moments are given in Kendall and Stuart (1958).

Tables of the coefficient of variation $\mu_2^{\frac{1}{2}} m_1^{-1}$, the coefficient of skewness $\mu_3 \mu_2^{-\frac{3}{2}}$, and the coefficient of kurtosis $\mu_4 \mu_2^{-2} - 3$, for values of $\sigma = 0.0(0.05)$ $1.0(0.1)\,4$ are given in Aitchison and Brown (1957), which also contains a detailed history of this distribution and an account of the theory of its statistical fitting and applications. (See also Finney 1941.)

Since the distribution is obtained by a simple transformation from the normal distribution, the properties of the latter immediately translate into properties of logarithmic-normal distribution. For example, if $Z = XY$ is distributed in a logarithmic-normal distribution so also are X and Y if they are independent, and conversely.

A more interesting application is derived from the Central Limit Theorem which is considered in detail in the next chapter. We have, however, already proved in § 7.2 a special case of this theorem when we showed that if X_1, X_2,... is a sequence of independent and identically distributed random variables with zero means and a finite variance σ^2, the distribution of $n^{-\frac{1}{2}}(X_1 + ... + X_n)$ converges to normality. This result is sufficient for our purpose here.

It follows that if X_1, X_2,... is a sequence of independent random variables with the same distribution on the range $(0, \infty)$ such that $E(\log X_i)^2$ is finite, the product $T_n = X_1 X_2 ... X_n$ is such that

$$\frac{\log T_n - nE(\log X_i)}{\{n \operatorname{var}(\log X_i)\}^{\frac{1}{2}}} \tag{7.103}$$

is asymptotically distributed in a normal distribution.

In particular suppose that the X_i have a rectangular distribution on $(0, 1)$. Then $Y_i = -\log X_i$ has the exponential distribution

$$e^{-y} \, dy,$$

on the range $(0, \infty)$, so that $E(Y_i) = 1$, $\operatorname{var}(Y_i) = 1$. Then

$$U = (T_n e^{-n})^{n^{-1/2}} \tag{7.104}$$

will have asymptotically a distribution with density

$$f(u) \, du = \{u \sqrt{(2\pi)}\}^{-1} \exp\{-\tfrac{1}{2}(\log u)^2\} \, du. \tag{7.105}$$

(The exact distribution is obtained below in (7.108).)

The fact that (7.103) has the logarithmic-normal distribution asymptotically can be used as a plausible model for the occurrence of this distribution in biological measurements. If the size or weight of some organism can be regarded as the result of a large number of different causes during growth, each of which has an effect which can be represented as multiplying the final weight by a factor near unity, the joint effect will be that the final weight can be represented by a constant multiplied by a factor of the form X_1, $X_2 ... X_n$, and hence can be expected to have a logarithmic-normal distribution. Notice that if the shapes and density of biological objects are constant, a logarithmic-normal distribution in their dimensions will result in a similar distribution for their weights and volumes.

Kolmogorov (1941) has given a different application to explain the occurrence of the logarithmic-normal distribution in the distribution of the sizes (or weights) of small particles. (See also Halmos 1944.) Suppose that these particles have been produced by the repeated splitting of a rock whose initial size may be taken as unity. Let this be split into two parts of size X_1

and $1 - X_1$, where X_1 has a distribution on $(0, 1)$ which is symmetric about the point $\frac{1}{2}$ and such that $\log X_1$ has a finite second moment. If each of the two pieces is again split into two smaller pieces by the same rule we shall, after n splittings, have 2^n pieces. The weights of each of these can be written in the form $X_1 X_2 \ldots X_n$, and as n increases will tend to have logarithmic-normal distributions.

This seems to fit some classes of measurements quite well, but in the case of the sands from North African deserts, which have been extensively studied by R. A. Bagnold (1941), the logarithm of the particle sizes appears to be better fitted by a distribution whose probability density is of the form

$$f(x) = K\exp\{-\alpha x - (\beta x^2 + r)^{\frac{1}{2}}\}. \tag{7.106}$$

Such probability distributions have not yet been studied and arise by replacing the parabolic form of $\log f(x)$, when $f(x)$ is the density of a normal distribution, by a hyperbola.

The logarithmic normal distribution also occurs in the theory of the scattering of neutrons by protons (Condon and Breit 1936, Bethe 1937). Consider a neutron of initial energy E_0 moving in a field of protons of same energy and such that its velocity is large compared with that of the protons. After a collision it may be assumed to have an energy E uniformly distributed on the range $(0, E_0)$. Thus so long as its energy remains large compared with that of the protons its energy after n collisions will be that of T_n in (7.104) approximately.

Actually in this case it is possible to find the exact distribution of T_n (Halmos 1944). We have $T_n = X_1 \ldots X_n$, where each X_i is distributed uniformly on $(0, 1)$, and $\log T_n = \log X_1 + \ldots + \log X_n$. Each $\log X_i$ is distributed in the negative exponential distribution

$$\exp(\log X_i)\, d(-\log X_i) \quad (0 < -\log X_i < \infty),$$

and hence $Z = -\log T_n$ has the distribution

$$\Gamma(n+1)^{-1} e^{-Z} Z^n\, dZ, \tag{7.107}$$

by the reproductive property of the gamma-type distribution (§6.9 and §7.17). Thus T_n has the distribution

$$\Gamma(n+1)^{-1}(-\log T_n)^n\, dT_n. \tag{7.108}$$

The logarithmic-normal distribution also has the remarkable property of not being determined by its moments. (This was first observed by C. C. Heyde (1963).) Let k satisfy $0 < k < 1$. Then the distribution on $(0, \infty)$, with probability density

$$(2\pi)^{-\frac{1}{2}}\exp\{-\tfrac{1}{2}(\log y)^2\}\{1 + k\sin(2\pi \log y)\}, \tag{7.109}$$

has moments independent of k. To prove this, it is sufficient to show that

$$\int_0^\infty y^n \exp\{-\tfrac{1}{2}(\log y)^2 - \tfrac{1}{2}\}\sin(2\pi \log y)\, dy = 0,$$

for $n = 0, 1, 2, \ldots$. Writing $y = e^x$, we obtain

$$\int_{-\infty}^\infty \exp\{(n+1)x - \tfrac{1}{2}x^2\}\sin 2\pi x \, dx$$

$$= \exp\{\tfrac{1}{2}(n+1)^2\} \int_{-\infty}^\infty \exp\{-\tfrac{1}{2}(x-n-1)^2\}\sin 2\pi x \, dx$$

$$= \exp\{\tfrac{1}{2}(n+1)^2\} \int_{-\infty}^\infty \exp(-\tfrac{1}{2}x^2)\sin 2\pi x \, dx = 0,$$

because the integrand is odd.

By forming a mixture of a distribution of the form (7.109) with a similar one on the range $(-\infty, 0)$ with a different k, we obtain an example of a non-symmetric distribution whose odd moments are all zero.

Since the logarithmic-normal distribution is distributed on the range $(0, \infty)$, and has two parameters, it is particularly useful as a model of naturally occurring frequency distributions of non-negative quantities, a property shared by the gamma distribution (§ 7.17), and the first passage-time distribution (§ 7.23). The extent to which these distributions can be used to approximate each other has not been investigated.

Furthermore, just as we have used the gamma distribution in § 2.14 to compound a Poisson distribution by allowing the parameter of the latter to vary, so also can we use the logarithmic normal (Preston 1948, Anscombe 1950). Suppose we have a Poisson distribution defined by

$$p_n = (n!)^{-1}\lambda^n e^{-\lambda} \quad (n = 0, 1, \ldots),$$

and let $\log \lambda$ have a normal distribution with mean ξ and variance σ^2. Then

$$P_n = \{\sigma n! (2\pi)^{\frac{1}{2}}\}^{-1} \int_0^\infty \lambda^{n-1} \exp\{-\lambda - \tfrac{1}{2}(\log \lambda - \xi)^2 \sigma^{-2}\}\, d\lambda. \qquad (7.110)$$

This has not been evaluated but the lower moments are easily found by taking expectations conditional on λ fixed, and then using the moments given by (7.100). Some numerical results on (7.110) are given by Grundy (1951).

7.17. The gamma distribution

This is a distribution on the range $(0, \infty)$ with the continuous probability density

$$\frac{\lambda^m}{\Gamma(m)} e^{-\lambda x} x^{m-1}, \qquad (7.111)$$

where $\lambda > 0$, $m > 0$. When $m = 1$, this is the 'negative exponential' distribution. In Pearson's system of distributions it is also known as the type III distribution. We have already shown (§2.13, §3.14) that the integral of this distribution can be used to express the tails of the Poisson distribution in a closed form, and in §6.9 we have shown that the characteristic function is

$$\lambda^m(\lambda - it)^{-m}. \tag{7.112}$$

It follows from the form of this characteristic function that if X and Y are independent random variables which have gamma-type distributions with the same λ, and parameters m_1 and m_2, $X + Y$ has a gamma distribution with parameters λ, and $m_1 + m_2$. This result could also be shown by direct integration of the convolution integral.

The distribution of $X - Y$ is given by Kullback (1936).

In the theory of statistics it is often necessary to consider a random variable, χ^2, which is such that $\frac{1}{2}\chi^2$ has the distribution (7.111) with $\lambda = 1$, $m = \frac{1}{2}n$, where n is a non-negative integer. χ^2 is then said to have the 'χ^2-distribution with n degrees of freedom' and is sometimes written $\chi^2_{(n)}$. The sum of two independent random variables which have χ^2-distributions with m and n degrees of freedom thus has a χ^2-distribution with $m + n$ degrees of freedom.

If X has a normal distribution with mean zero and variance σ^2, we have seen in §7.6 that $\frac{1}{2}X^2\sigma^{-2}$ has the χ^2-distribution with one degree of freedom, and if $X_1, \ldots, X_n$ are n such random variables, $\Sigma(X_i - \bar{X})^2\sigma^{-2}$, where $\bar{X}$ is the mean of the X_i, has the χ^2-distribution with $n - 1$ degrees of freedom. In §7.6 we have also considered various quadratic forms in normal variates which have χ^2-distributions.

Returning to (7.111) we see from (7.112), or by direct integration, that the moments are given by

$$m_n = \lambda^{-n}\Gamma(m+n)\Gamma(m)^{-1}, \tag{7.113}$$

and the cumulants by

$$\kappa_n = m\lambda^{-n}(n-1)! \tag{7.114}$$

From the nature of the moments it is clear that they determine the gamma distribution uniquely, and by rescaling (7.114) it also follows that if X has the distribution (7.111) so that $E(X) = m\lambda^{-1}$, $E(X^2) = m(m+1)\lambda^{-2}$,

$$(\lambda X - m)m^{-\frac{1}{2}}$$

will have a distribution which converges to normality with zero mean and unit variance as m increases, a fact which is otherwise obvious from the central limit theorem in the case where m is integral. This convergence is fairly slow and for some practical purposes it is useful to use approximations derived from the fact that $X^{1/n}$ (for certain positive integral values of n) may have a distribution closer to normality than that of X. The accuracy of this approximation for $n = 2$, 3 is considered in detail in Kendall and Stuart

(1958). (See also Severo and Zelen 1960.) The cumulants of $\frac{1}{2}\log(n^{-1}\chi^2)$ are given by Wishart (1947).

There are a large number of tables of the tails of the distribution (7.111) in various forms. Pearson (1922) tabulates

$$\Gamma(p+1)^{-1} \int_0^x u^p e^{-u}\, du \qquad (7.115)$$

to $7D$ for $p = -0.95(0.05)\ 0(0.1)\ 5(0.2)\ 50$, for values of $x = u\sqrt{(p+1)}$ such that $u = 0(0.1)$ until the integral equals unity to within 0.00000005. (See also Harter 1964.) More useful tables are to be published by S. H. Khamis.

In the χ^2 form tables are more accessible. Thus Pearson and Hartley (1956) give the distribution of $\chi^2_{(n)}$ to $5D$ for $n = 1(1)\ 30(2)\ 70$, and $\chi^2 = 0.001(0.001)$ $0.01(0.01)\ 0.1(0.1)\ 2(0.2)\ 10(0.5)\ 20(1)\ 40(2)\ 70$. (See also Khamis 1964.) Inverse tables for selected values of the probability are given by Fisher and Yates (1953), Harter (1964), and Pearson and Hartley (1956). (A nomogram is given by Smirnov and Potapov 1957.)

Owing to the relationship between the tail of a Poisson distribution and that of a gamma distribution (as in equation (2.72)), tables of the Poisson distribution and the latter are to some extent interchangeable.

Let X and Y be independent gamma variates having distributions (7.111) with $\lambda = 1$, and the other parameter equal to m_1 and m_2. We have already discussed the distribution of $X + Y$ and $X - Y$. The distributions of $U = Y(X + Y)^{-1}$ and $V = X^{-1}Y$ are more interesting.

Consider first that of $V = X^{-1}Y$. Writing v for $x^{-1}y$, the probability density of the distribution function of the random variable V is, integrating out the variable x,

$$\Gamma(m_1)^{-1}\Gamma(m_2)^{-1} \int_0^\infty e^{-x(1+v)} x^{m_1-1} (xv)^{m_2-1} x\, dx = \frac{\Gamma(m_1+m_2)}{\Gamma(m_1)\Gamma(m_2)} \frac{(1+v)^{m_1+m_2}}{v^{m_2-1}}.$$

$$(7.116)$$

Since $U = V(V+1)^{-1}$, $V = U(1-U)^{-1}$, U is distributed on the range $(0, 1)$ with probability density

$$\frac{\Gamma(m_1+m_2)}{\Gamma(m_1)\Gamma(m_2)} u^{m_2-1}(1-u)^{m_1-1}, \qquad (7.117)$$

(7.116) and (7.117) will be considered in more detail in § 7.21 and § 7.22, both these distributions being of great importance in statistical theory.

Garti and Consoli (1954) have given explicit formulae for the distribution of the product of two, and approximations for the product of three, gamma-type variates. (See also Wells, Anderson, and Cell 1962.)

Cramér's theorem shows that if X and Y are independent random variables and $X + Y$ has a normal distribution, X and Y are also normally distributed.

A similar theorem does not hold for the gamma-type distribution and Mauldon (1956) has constructed a counter-example.

Since x ranges from 0 to ∞ we can define a new variate $Z = X^{-1}$ with the same range and (7.111) becomes (putting $\lambda = 1$ for simplicity),

$$\Gamma(m)^{-1} e^{-z^{-1}} z^{-1-m}\, dz. \tag{7.118}$$

This is known as the reciprocal gamma distribution. Its mean and variance are

$$m_1 = (m-1), \qquad\qquad m \neq 1, \tag{7.119}$$

$$\mu_2 = (m-1)^{-2}(m-2), \quad m \neq 1, 2. \tag{7.120}$$

Multivariate and truncated gamma distributions have also been considered. For the former see, for example, David and Fix (1961), and Lukacs and Laha (1964).

Gamma distributions have also been used to provide approximations to distributions on the range $(0, \infty)$, which are otherwise difficult to calculate, and the most convenient way to do this is to use a mixture of the form

$$a_1 f_1(x) + \ldots + a_k f_k(x), \tag{7.121}$$

where $a_i > 0$, $\sum a_i = 1$, and the $f_i(x)$ are probability densities of the form (7.111) with differing parameters λ_i and m_i. The advantage of this representation is that if the a_i, λ_i, and m_i are known the probability distribution can be quickly calculated from known tables. This approach has been used by Robbins and Pitman (1949) to obtain series for the distribution of quadratic forms in normal variates, whose characteristic roots are all non-negative.

7.18. The non-central χ^2-distribution

Suppose that $X_1, \ldots, X_n$ are n random variables which have normal distributions with zero means and unit variances. Then the quantity

$$S = X_1^2 + \ldots + X_n^2 \tag{7.122}$$

has the χ^2-distribution with n degrees of freedom and we have seen in §7.17 that $\frac{1}{2}\chi^2$ is distributed in the gamma distribution (7.111) with $\lambda = 1$ and $m = \frac{1}{2}n$. We now abandon the restriction that the means are zero and suppose that $E(X_i) = a_i$, $(i = 1, \ldots, n)$, the variances remaining the same. The quantity χ^2 in (7.121) is then said to have a non-central χ^2-distribution. This has been studied and occasionally rediscovered by a number of writers (Fisher 1928, Patnaik 1949, Quenouille 1949a, and Tang 1938).

Write

$$a^2 = a_1^2 + \ldots + a_n^2.$$

The joint distribution of the X_i is a spherically symmetric normal distribution about the point $P = (a_1, \ldots, a_n)$, and the distribution of S is that of the distance of $(X_1, ..., X_n)$ from the origin. There is clearly no alteration in this if we assume that P is the point $(a, 0, ..., 0)$. Hence we can put

$$S = Y_1^2 + (Y_2^2 + ... + Y_n^2)$$
$$= U + V, \tag{7.123}$$

where V has the χ^2-distribution with $n-1$ degrees of freedom, and U has a distribution with probability density $(0 \leqslant u < \infty)$

$$\tfrac{1}{2}(2\pi u)^{-\frac{1}{2}}\{\exp-\tfrac{1}{2}(u^{\frac{1}{2}}-a)^2 + \exp-\tfrac{1}{2}(u^{\frac{1}{2}}+a)^2\}. \tag{7.124}$$

The joint distribution of U and V, which are independent, therefore has density

$$\tfrac{1}{4}\Gamma\left(\frac{n-1}{2}\right)^{-1}(2\pi u)^{-\frac{1}{2}}\exp(-\tfrac{1}{2}v)(\tfrac{1}{2}v)^{(n-3)/2}\{\exp-\tfrac{1}{2}(u^{\frac{1}{2}}-a)^2 + \exp-\tfrac{1}{2}(u^{\frac{1}{2}}+a)^2\}.$$

Squaring the terms $(u^{\frac{1}{2}}-a)$ and $(u^{\frac{1}{2}}+a)$ we can write this as

$$2^{-\frac{1}{2}n}\Gamma\left(\frac{n-1}{2}\right)^{-1}(\pi u)^{-\frac{1}{2}}v^{(n-3)/2}\exp-\tfrac{1}{2}(u+v+a^2)\cosh au^{\frac{1}{2}}.$$

The range of u and v is $(0, \infty)$, so we can make the transformation,

$$x = u+v,$$
$$y = v(u+v)^{-1}, \tag{7.125}$$

into variates which have the ranges $(0, \infty)$ and $(0, 1)$ respectively. The Jacobian of the transformation is $(u+v)^{-1}$ and its inverse is

$$u = x(1-y),$$
$$v = xy. \tag{7.126}$$

Hence the distribution becomes

$$2^{-\frac{1}{2}n}\Gamma\left(\frac{n-1}{2}\right)^{-1}\Gamma(\tfrac{1}{2})^{-1}e^{-\frac{1}{2}(x+a^2)}x^{\frac{1}{2}(n-2)}y^{\frac{1}{2}(n-3)}(1-y)^{-\frac{1}{2}}\cosh\{ax^{\frac{1}{2}}(1-y)^{\frac{1}{2}}\}.$$

Next the cosh term is expanded in an infinite series and the distribution is integrated with respect to y over the range $(0, 1)$. The result is

$$2^{-\frac{1}{2}n}\Gamma\left(\frac{n-1}{2}\right)^{-1}\Gamma(\tfrac{1}{2})^{-1}e^{-\frac{1}{2}(x+a^2)}x^{\frac{1}{2}(n-2)}\sum_{r=0}^{\infty}(2r!)^{-1}\int_0^1(ax^{\frac{1}{2}})^{2r}y^{(n-3)/2}(1-y)^{r-\frac{1}{2}}dy$$

$$= 2^{-\frac{1}{2}n}\Gamma(\tfrac{1}{2})^{-1}e^{-\frac{1}{2}(x+a^2)}x^{\frac{1}{2}(n-2)}\sum_{r=0}^{\infty}(2r!)^{-1}(a^2x)^r\frac{\Gamma(r+\tfrac{1}{2})}{\Gamma\{(n+2r)/2\}}. \tag{7.127}$$

That this can also be expressed in terms of an infinite series of Bessel functions was shown by Fisher and Quenouille. For another more elegant proof see Kerridge (1965).

If a is kept fixed and n increases the distribution clearly tends to normality and from the representations (7.123) the lower moments are easily found. A much more detailed study of approximations to this distribution has been given by Patnaik (1949). Johnson (1959) has shown that the tail of a non-central χ^2-distribution can be related to the tail of the distribution of the difference between two Poisson variates and hence, by (2.76), as a sum of Bessel functions. This result is a generalization of (2.72).

Critical values in a certain sense are given in Owen (1962). General tables of the distribution are given by E. Fix (1949).

7.19. Cauchy-type distributions

Any non-negative integrable function can serve as a probability density after multiplying by a constant to make its integral unity, and one of the simplest such functions is $(1+x^2)^{-1}$, or, more generally, $(1+x^2)^{-m}$, where m must exceed $\frac{1}{2}$ to ensure that the integral over $(-\infty, \infty)$ is finite. To obtain the normalizing constant we have to evaluate

$$\int_{-\infty}^{\infty} \frac{dx}{(1+x^2)^m}.$$ (7.128)

Putting $y = (1+x^2)^{-1}$, this becomes

$$\int_0^1 y^{m-\frac{3}{2}}(1-y)^{-\frac{1}{2}} dy = \frac{\Gamma(m-\frac{1}{2})\Gamma(\frac{1}{2})}{\Gamma(m)},$$

and hence the normalized distribution is

$$\frac{\Gamma(m)}{\Gamma(m-\frac{1}{2})\Gamma(\frac{1}{2})} \cdot \frac{1}{(1+x^2)^m},$$ (7.129)

on the range $(-\infty, \infty)$. This clearly has no integral moment of order greater than the integral part of $2m-2$, but moments of lower order can be obtained in terms of gamma functions.

The particular case $m = 1$ is known as Cauchy's distribution although it had been considered before Cauchy. (7.129) then becomes

$$\frac{1}{\pi(1+x^2)},$$ (7.130)

and its characteristic function is

$$\Phi(t) = e^{-|t|},$$ (7.131)

as shown in §6.9. The characteristic function of (7.129), which is a good deal more complicated, is given by Kendall and Stuart (1958), Vol. I, p. 74. All these results are easily modified for any change in scale and origin of the distribution.

(7.131) has an important immediate application. Suppose that $X_1,..., X_n$ are a sequence of independent random variables with the distribution determined by (7.130). Then using the characteristic function we see that

$$\bar{X} = n^{-1}(X_1 + ... + X_n)$$

will have the characteristic function

$$\exp - n^{-1}(n|t|) = \exp - |t|. \tag{7.132}$$

Thus $\bar{X}$ will also have the distribution (7.130), and the mean of a sample of n values will be no more accurate than a single observation. This phenomenon is clearly connected with the non-existence of the second moment and we shall study it in detail in Chapter 9.

The distribution (7.130) can also be given a geometrical interpretation. Let θ be a random angle which is uniformly distributed over the range $(-\tfrac{1}{2}\pi, \tfrac{1}{2}\pi)$ and therefore has the probability density π^{-1}. The quantity $X = \tan \theta$ then obviously has the distribution (7.130). If P is the point $(0, 1)$,

$$\frac{dx}{\pi(1 + x^2)} \tag{7.133}$$

will be the angle subtended at P by the interval $(x, x + dx)$. (7.130) is therefore proportional to the light intensity falling on a plane from a uniform infinitely long strip-light at a unit distance from the plane. It can also be shown to be proportional to the charge induced in an infinite earthed conducting plane by an infinitely long wire parallel to the plane and uniformly charged. (7.129) can also be related to Riemann's elliptic geometry (Wintner 1957). Bivariate Cauchy-type distributions are considered by Ferguson (1962).

7.20. The t-distribution

Suppose that $X_1,..., X_n$ are independent random variables with zero means and standard deviation σ. In mathematical statistics the mean,

$$\bar{X} = n^{-1}(X_1 + ... + X_n), \tag{7.134}$$

and the sample standard deviation $(s \geqslant 0)$,

$$s^2 = (n-1)^{-1}(\textstyle\sum X_i^2 - n\bar{X}^2), \tag{7.135}$$

are used as estimates of the true mean (here zero but assumed unknown) and the true variance σ^2. Since $\bar{X}$ is distributed normally about the true mean with standard deviation of $\sigma/n^{\frac{1}{2}}$ it is useful to know the distribution of the quantity

$$t = \bar{X}n^{\frac{1}{2}}s^{-1}, \tag{7.136}$$

which is known as Student's t. Since t is scale-invariant its distribution is independent of σ, and is known as 'Student's t-distribution with $n-1$ degrees of freedom'. We have already shown that $\bar{X}$ and s are independent and thus derived the distribution of t already given in (7.39), where we obtained it by a geometrical argument.

Extensive tables of this distribution exist of which the most accessible is that of Pearson and Hartley (1956). This gives the distribution function to $5D$ for $t = 0(0\cdot1)\,8$ and $n-1 = 1(1)\,24$, 30, 40, 60, 120, and ∞. Inverse tables for particular values of the probability are given in Fisher and Yates (1953), and more extensive ones in Federighi (1959). These tables could have been calculated from the known tables of the incomplete beta function (Pearson 1948) or more conveniently from an asymptotic expansion given by Fisher. (See Kendall and Stuart (1958), Vol. I, p. 376.) A nomogram for t is given by James-Levy (1956).

As n increases the distribution of t tends to the normal distribution. This can be seen at once by observing that the denominator S converges in probability to σ whilst $n^{\frac{1}{2}}\bar{X}$ is normally distributed with variance σ^2. The convergence is, however, not very fast. Thus the values x for which

$$P(X > x) = 0\cdot975,\ 0\cdot995,$$

when X is normally distributed with zero mean and unit variance, are $1\cdot9600$ and $2\cdot5758$, whereas for the t-distribution the corresponding values are $2\cdot0211$, $2\cdot7045$ for $n = 40$ and $2\cdot2281$, $3\cdot1693$ for $n = 10$. For $n \geqslant 5$ (or 7, in the second case) a much more accurate approximation *at these particular probability levels* is obtained by regarding (respectively)

$$t_1 = t(1+0\cdot613|t|n^{-1})^{-1} \quad \text{and} \quad t_2 = t_1 - 0\cdot8n^{-1} \tag{7.137}$$

as normally distributed with zero mean and unit variance.

Bivariate and multivariate analogues of the t-distribution have also been considered. Another useful generalization results in Hotelling's 'T-distribution'. For distributions of this kind and the corresponding multivariate analogue of χ^2 and F-distributions the reader may consult Kendall and Stuart (1967), Vol. 3, Roy (1957), and Anderson (1958).

Just as the probability integral of the χ^2-distribution is related to that of the Poisson distribution, so also can a relationship be established between the integral of the t-distribution and the tail of a certain negative binomial (Fisher 1935). Let P be the probability that $t > t_0$, where $t_0 = n^{\frac{1}{2}}\tan\alpha$, and t has n degrees of freedom. Then

$$P = \int_{t_0}^{\infty} \frac{\Gamma\{(n+1)/2\}}{\pi^{\frac{1}{2}}\Gamma(n/2)} \frac{dt}{n^{\frac{1}{2}}(1+t^2n^{-1})^{\frac{1}{2}(n+1)}}$$

$$= \int_{\alpha}^{\frac{1}{2}\pi} \frac{\Gamma\{(n+1)/2\}}{\pi^{\frac{1}{2}}\Gamma(n/2)} \cos^{n-1}\theta\,d\theta, \tag{7.138}$$

on putting $t^2 n^{-1} = \tan^2 \theta$. Integrating the latter repeatedly by parts we find that

$$P = \tfrac{1}{2} - \tfrac{1}{2} \sin \alpha \left\{ 1 + \tfrac{1}{2} \cos^2 \alpha + \frac{1.3}{2.4} \cos^2 \alpha + \dots + \frac{1.3 \dots (n-3)}{2.4 \dots (n-2)} \cos^{n-2} \alpha \right\},$$

(7.139)

when n is even, and

$$P = \tfrac{1}{2} - \pi^{-1} \alpha - \pi^{-1} \sin \alpha \left\{ \cos \alpha + \tfrac{2}{3} \cos^3 \alpha + \dots + \frac{2.4 \dots (n-3)}{1.3 \dots (n-2)} \cos^{n-2} \alpha \right\},$$

(7.140)

when n is odd. In the case where n is even we can put $q = \cos^2 \alpha$ and obtain

$$1 - 2P = (1-q)^{\frac{1}{2}} \left\{ 1 + \tfrac{1}{2} q + \frac{1.3}{2.4} q^2 + \dots + \frac{1.3 \dots n-3}{2.4 \dots n-2} q^{\frac{1}{2}(n-2)} \right\}, \quad (7.141)$$

which expresses the tail of the t-distribution in terms of the partial sums of a negative binomial distribution. When n is odd we get a similar expansion except that we now need $\alpha = \tan^{-1} t_0 n^{-\frac{1}{2}}$.

In § 7.22 we shall see that the distribution can be related to a special case of the F-distribution, and hence expressed in terms of the beta distribution (§ 7.21).

7.21. The beta distribution of the first kind

This is a distribution on the range $(0, 1)$, with two parameters, and probability density

$$B(l, m)^{-1} x^{l-1} (1-x)^{m-1}, \tag{7.142}$$

where $l > 0$, $m > 0$, and $B(l, m)$ is the beta function which is equal to

$$\Gamma(l) \Gamma(m) \Gamma(l+m)^{-1}. \tag{7.143}$$

We have already used this distribution in § 7.17. As is well known,

$$B(l, m) = \int_0^1 x^{l-1} (1-x)^{m-1} \, dx, \tag{7.144}$$

and hence the moments are easily found to be

$$m_s = B(l+s, m) B(l, m)^{-1}. \tag{7.145}$$

The characteristic function is given in (6.23).
The mean is

$$m_1 = l(l+m)^{-1}, \tag{7.146}$$

and the second moment about the mean is

$$\mu_2 = lm(l+m)^{-2}(l+m+1)^{-1}. \tag{7.147}$$

Thus if $l(l+m)^{-1}$ is kept fixed whilst l and m tend to infinity, a random variable X having this distribution will converge in probability to m_1. In fact $(X-m_1)\mu_2^{-\frac{1}{2}}$ has a distribution which converges to normality with zero mean and unit standard deviation. This can be shown directly by rather awkward asymptotic analysis but follows more easily from the relationship with the gamma distribution described in the next paragraph.

Suppose that X and Y are independent random variables with gamma distributions

$$\Gamma(l)^{-1}e^{-x}x^{l-1}\,dx, \quad \Gamma(m)^{-1}e^{-y}y^{m-1}\,dy, \tag{7.148}$$

$(l, m > 0)$ respectively. Then $U = X(X+Y)^{-1}$ will have the distribution (7.142) above. To prove this we change to variables

$$u = x(x+y)^{-1} \quad (0 \leqslant u \leqslant 1),$$
$$v = x+y \quad (0 \leqslant v < \infty).$$

The Jacobian of this transformation is easily seen to be

$$\frac{\partial(u, v)}{\partial(x, y)} = v^{-1}, \tag{7.149}$$

and the distribution of U has a probability density

$$\Gamma(l)^{-1}\Gamma(m)^{-1}\int_0^\infty e^{-v}v^{l+m-1}(1-u)^{m-1}u^{l-1}\,dv = \frac{\Gamma(l+m)}{\Gamma(l)\Gamma(m)}u^{l-1}(1-u)^{m-1}, \tag{7.150}$$

as required.

We now want to show U is asymptotically normally distributed with mean (7.146) and variance (7.147). This is equivalent to saying that

$$T = \left(\frac{X}{X+Y} - \frac{l}{l+m}\right)\left\{\frac{(l+m)^2(l+m+1)}{lm}\right\}^{\frac{1}{2}}$$

is asymptotically normally distributed with zero mean and unit variance, when l and m both tend to infinity. However,

$$T = \frac{mX-lY}{\{lm(l+m)\}^{\frac{1}{2}}}\left\{\frac{(l+m)(l+m+1)}{(X+Y)^2}\right\}^{\frac{1}{2}}. \tag{7.151}$$

The first factor in T is clearly asymptotically normally distributed with zero mean and unit variance. The second factor 'converges in probability' to unity in the sense that given any small number ε, the probability that it is different from unity by more than ε tends to zero when l and m both increase. Although the two factors are not independent, it follows that T is also asymptotically normally distributed with zero mean and unit variance. In fact the mean and variance of T are also exactly equal to zero and unity respectively.

We now prove a somewhat unexpected result on the product of beta and gamma variates. Suppose that U has a beta distribution with parameters

22

(l, m) and let V have an independent gamma distribution with parameter $l+m$. Then $Z = UV$ has a gamma distribution with parameter l. This is not at all immediately obvious for if U is represented as a quotient $X(X+Y)^{-1}$ where X, Y have gamma distributions with parameters l and m, $X+Y$ and V are different and independent variates. The proof is, however, easily obtained by direct integration and Stuart (1962) also gives an elegant proof using characteristic functions. The simplest proof is obtained by returning to (7.150) and transforming to u and v. The joint distribution of U and $V_1 = X+Y$ is then, as before,

$$\Gamma(l)^{-1}\Gamma(m)^{-1}e^{-v_1}v_1^{l+m-1}u^{l-1}(1-u)^{m-1}\,du\,dv_1, \qquad (7.152)$$

and hence U and V_1 are distributed independently. The random variable with a gamma distribution with parameter $l+m$ can therefore be taken as V_1 and the result follows. In the particular case when $2l$ and $2m$ are integers, U and V can be represented as proportional to sums of squares of independent random variables having normal distributions with means zero and unit variances. Using the same kind of geometric representation as in §7.9 $U(U+V)^{-1}$ depends on the direction of a sample point in a space of $2(l+m)$ dimensions, and $U+V$ is the square of the distance of this point from the origin. They are therefore independent and the result again follows.

The previous argument shows, however, that the result is also true for any $l, m > 0$. This theorem was apparently discovered first by Sawkins (1940). (See also Weatherburn 1946 and Aitchison 1963.)

We have already seen (§2.8) that the beta distribution can be used to express the tail sums of the binomial and this is one way of obtaining tables of this distribution, that of the binomial being already extensively calculated. It can also be used to represent the tail sum of a negative binomial (see Exercise 7.14).

The beta distribution is also related to the t-distribution. In (7.39) put

$$x = \{1+t^2(n-1)^{-1}\}^{-1}, \qquad (7.153)$$

and the distribution density becomes (using the fact that $\Gamma(\tfrac{1}{2}) = \sqrt{\pi}$)

$$\frac{\Gamma(n/2)}{\Gamma(\tfrac{1}{2})\Gamma\{(n-1)/2\}}x^{(n-3)/2}(1-x)^{-\tfrac{1}{2}}, \qquad (7.154)$$

over the range $(0, 1)$. Thus tables of the incomplete beta function could be used to provide the distribution of t.

K. Pearson (1934) gives 7-decimal tables of

$$I_x(l, m) = \frac{\Gamma(l+m)}{\Gamma(l)\Gamma(m)}\int_0^x u^{l-1}(1-u)^{m-1}\,du \qquad (7.155)$$

for $l \geqslant m$, l, $m = 0.5(0.5)\,11(1)\,50$, $x = 0(0.01)\,1$. For values of l, m greater than 50, Wishart (1927) gives auxiliary tables to facilitate the calculation.

Harter (1964) and Vogler (1964) give percentage points. Inverse tables giving x for specified values of l, m, and $I_x(l, m)$ can be found in C. M. Thompson (1941), and the theory of the calculation of such inverse tables is discussed, with extensive references, by Wise (1946, 1950), and Aroian (1950). Wise (1960) also gives an approximate normalization and another approximation is given by Kimball and Leach (1959).

The beta distribution can be used to compound a parameter of another distribution which is to be allowed to vary over a finite interval. In §2.10 we compounded the binomial distribution by letting the probability p vary in a beta distribution and in this way obtained the negative hypergeometric. Another application which occurs in practice is that to the Poisson distribution (Moran 1954) and may be illustrated by a biological example. Suppose that an animal is inoculated by a dose of suspension of infective particles such as those of a virus. Let the number of such particles in the dose follow a Poisson distribution with mean λ, and let the probability of each particle forming a lesion by infecting the animal be p. Then the number of lesions will be a Poisson distribution with mean λp. If the animal is chosen at random out of a population of animals in which the distribution of p is given by a density function $f(p)$ on the range $(0, 1)$, the probability of k lesions will be

$$(k!)^{-1} \int_0^1 e^{-\lambda p}(\lambda p)^k f(p)\, dp, \tag{7.156}$$

which is not a Poisson distribution unless $f(p)$ collapses to a concentration at a single point. If $f(p)$ is the density of a beta distribution, (7.156) can be evaluated in terms of Whittaker's confluent hypergeometric distribution by (6.23) but the resulting distribution is usually too complicated to be useful.

The distribution of $S = X_1 + \ldots + X_n$ where the X_i are independent random variables with the same beta distribution has been obtained by Irwin (1927).

The beta distribution can be generalized in various ways, for example in the theory of multivariate analysis (Foster and Rees 1957). The k roots $\theta_1, \ldots, \theta_k$ of certain determinantal equations with random elements are known to lie in the interval $(0, 1)$, and if arranged in their order of magnitude, so that

$$0 < \theta_1 < \ldots < \theta_k < 1, \tag{7.157}$$

are known to have the joint distribution

$$K \prod_{i=1}^{k} \theta_i^{p-1}(1 - \theta_i)^{q-1} \prod_{i>j} (\theta_i - \theta_j), \tag{7.158}$$

where K, p, and q are certain constants. Then the distribution function of θ_k is

$$I_x(k; p, q) = K \int_0^x d\theta_k \int_0^{\theta_k} d\theta_{k-1} \ldots \int_0^{\theta_2} d\theta_1 \prod_{i=1}^{k} \theta_i^{p-1}(1 - \theta_i)^{q-1} \prod_{i>j} (\theta_i - \theta_j). \tag{7.159}$$

This can be regarded as a generalization of the beta distribution.

Another generalization which occurs in genetics (Moran 1962) arises when we assume that the probability density is proportional to an expression of the form

$$e^{ax+bx^2}x^{l-1}(1-x)^{m-1}. \tag{7.160}$$

When $b = 0$ the integral of this expression can be found (6.23) but for $b \neq 0$ is apparently not known. This distribution was originally found by S. Wright, and for $b = 0$ has already occurred above (7.156).

A 'non-central' beta distribution has also been defined (Nicholson 1954, Hodges 1955). For further work on generalizations of the beta distribution see Mauldon (1959).

7.22. The F-distribution

The 'beta distribution of the second kind' is defined to be a distribution on $(0, \infty)$ with the probability density (compare (7.116))

$$\frac{\Gamma(l+m)}{\Gamma(l)\Gamma(m)} \cdot \frac{x^{l-1}}{(1+x)^{l+m}}. \tag{7.161}$$

Put $y = (1+x)^{-1}$ and this becomes, after introducing the factor for the change of the differential element,

$$\frac{\Gamma(l+m)}{\Gamma(l)\Gamma(m)} y^{m-1}(1-y)^{l-1} dy, \tag{7.162}$$

which is the probability density of a beta distribution of the first kind. Thus the integral of (7.161) is unity and the distribution exists whenever $l > 0$, $m > 0$.

The F-distribution is a transformation of (7.161) which is obtained when we replace x by $lm^{-1}F$, and then l, m by $\frac{1}{2}n_1$, $\frac{1}{2}n_2$, so that (7.161) becomes

$$\frac{\Gamma(\frac{1}{2}n_1+\frac{1}{2}n_2)}{\Gamma(\frac{1}{2}n_1)\Gamma(\frac{1}{2}n_2)} \frac{n_1^{\frac{1}{2}n_1} n_2^{\frac{1}{2}n_2} F^{\frac{1}{2}(n_1-2)}}{(n_2+n_1 F)^{\frac{1}{2}(n_1+n_2)}} \tag{7.163}$$

for the probability distribution of F $(0 \leqslant F < \infty)$. This is said to be the 'F-distribution with n_1 and n_2 degrees of freedom'. In fact (7.163) is the distribution of

$$F = n_2 X(n_1 Y)^{-1}, \tag{7.164}$$

where X, Y are independent random variables with χ^2 distributions with n_1, n_2 degrees of freedom each. The form of the distribution is not dependent on the assumption that n_1 and n_2 are integers. In fact (7.161) is the distribution of a multiple of the ratio of two gamma variates for any $l, m > 0$. This has already been shown in (7.116). In particular from (7.39) we see that t^2 with $n-1$ degrees of freedom has the distribution (7.163) with $n_1 = 1$, $n_2 = n-1$.

In earlier work on this subject Fisher considered the distribution of the quantity $z = \frac{1}{2}\log F$ (obtained by a simple transformation of (7.163)) because it is more easily approximated by a normal distribution for n_1, n_2 large. A better approximation is given by Wishart (1957).

From the relationship between (7.161) and (7.162) we see that any variate having an F-distribution can be transformed into one having a beta distribution, and vice versa. This is sometimes useful when the beta-distribution is used to approximate to some distribution on a finite interval. Notice also that if F has the F-distribution with n_1 and n_2 degrees of freedom, F^{-1} has a similar distribution with n_2, n_1 degrees of freedom.

It is clear from the form of (7.161) and (7.163) that the moments do not always exist. By using the fact that the integral of (7.163) is unity it follows that the mean and variance of (7.163) are

$$m_1 = n_2(n_2-2)^{-1},$$
$$\mu_2 = 2n_2^2(n_1+n_2-2)\{n_1(n_2-2)^2(n_2-4)\}^{-1}, \tag{7.165}$$

provided $n_2 > 2$ in the first case and $n_2 > 4$ in the second. Moments for (7.161) are similar.

The characteristic function of F is apparently not known. On the other hand, the characteristic function of $Z = \frac{1}{2}\log F$ is

$$\frac{2n_1^{\frac{1}{2}n_1} n_2^{\frac{1}{2}n_2}}{B(\frac{1}{2}n_1, \frac{1}{2}n_2)} \int_{-\infty}^{\infty} \frac{e^{(it+n_1)x}}{(n_1 e^{2x} + n_2)^{\frac{1}{2}(n_1+n_2)}} dx$$

$$= \left(\frac{n_2}{n_1}\right)^{\frac{1}{2}it} \frac{\Gamma\{\frac{1}{2}(n_2-it)\}\Gamma\{\frac{1}{2}(n_1+it)\}}{\Gamma(\frac{1}{2}n_1)\Gamma(\frac{1}{2}n_2)}. \tag{7.166}$$

From this it follows that the first two moments of Z are approximately

$$m_1 = \frac{1}{2}(n_2^{-1} - n_1^{-1}), \tag{7.167}$$
$$\mu_2 = \frac{1}{2}(n_1^{-1} + n_2^{-1}), \tag{7.168}$$

when n_1 and n_2 are large.

Such moments are more easily obtained from the cumulant generating function (Wishart 1947)

$$\frac{1}{2}it \log(n_2 n_1^{-1}) + \log \Gamma\{\frac{1}{2}(n_2-it)\} + \log \Gamma\{\frac{1}{2}(n_1+it)\} - \log \Gamma(\frac{1}{2}n_1) - \log \Gamma(\frac{1}{2}n_2). \tag{7.169}$$

From the fact that a variate having the F-distribution can be represented as the ratio of two variates having gamma distributions, it follows that as n_1 and n_2 (or l and m) increase, the F-distribution converges to normality, and (7.165) can be used to rescale it. Other approximations for large n_1, n_2 are discussed in Kendall and Stuart (1958), Vol. I, pp. 380–. The distribution of Z, however, tends to normality faster than that of F.

Tables of the distribution of F in the direct form have not been calculated but could be easily found from the known tables of the beta distribution. There are, however, several tables in the inverse form, i.e. giving values of F corresponding to certain values of P and this is true also for the Z-distribution.

Thus Fisher and Yates (1938) give F to two decimals and Z to four decimals for $P = 0.2$, 0.1, 0.05, 0.01 and 0.001, for $n_1 = 1(1)$ 6, 8, 12, 24, and ∞, $n_2 = 1(1)$ 30, 40, 60, 120, and ∞. Merrington and Thompson (1943) give F to four decimals for $P = 0.5$, 0.25, 0.1, 0.05, 0.025, 0.01, and 0.005 for $n_1 = 1(1)$ 10, 12, 15, 20, 24, 30, 40, 60, 120, and ∞, and $n_2 = 1(1)$ 30, 40, 60, 120, and ∞. A similar table to $2D$ but omitting $P = 0.5$ is given in the *Biometrika Tables*, Vol. I (1956).

Just as for the gamma distribution there is no uniqueness theorem similar to Cramér's. It is possible (Mauldon 1956) to find independent variates X and Y which do not have gamma-type distribution but are such that XY^{-1} has an F-distribution.

In §7.18 we have generalized the χ^2-distribution to the non-central χ^2-distribution and a similar generalization is possible with F. This is obtained as the distribution of a ratio XY^{-1} where X and Y have non-central χ^2-distribution and is given in Kendall and Stuart (1961), Vol. II, p. 252. For statistical applications the interesting case is that where X has a non-central χ^2-distribution and Y the ordinary χ^2-distribution. This has been studied by Fisher (1928), Tang (1938), Patnaik (1949), and others. Tables calculated for special values are listed in Kendall and Stuart.

7.23. The Gaussian first passage-time distribution

This is sometimes known as the inverse Gaussian distribution and the reason for this name will appear in Chapter 9 where we show how it arises in the theory of the Brownian movement. It was in connexion with the latter that it was considered by Schrödinger (1915) and a fairly complete account of its properties, and statistical theory, is given by Tweedie (1957a, b).

The probability density of this distribution can be written

$$b\{2\pi x^3\}^{-\frac{1}{2}} \exp{-\frac{1}{2}\left\{\frac{(b-mx)^2}{x}\right\}}, \tag{7.170}$$

where $0 \leqslant x < \infty$, $b > 0$, and $m > 0$. There are in fact two independent parameters including the scale. To verify that the integral of (7.170) is unity we use some formulae on Bessel functions. For $v > 0$ and x real and non-zero, the Bessel function $K_n(x)$ satisfies

$$K_v(x) = \tfrac{1}{2}(\tfrac{1}{2}x)^v \int_0^\infty \exp(-t - \tfrac{1}{4}x^2 t^{-1}) t^{-1-v} \, dt. \tag{7.171}$$

This also holds for complex x when the real part of x^2 is strictly positive.

In the particular case $v = \frac{1}{2}$, we have for all $x \neq 0$, real or complex,

$$K_{\frac{1}{2}}(x) = \left(\frac{\pi}{2x}\right)^{\frac{1}{2}} e^{-x}. \tag{7.172}$$

Substituting in (7.170) after changing the parameters we verify that the integral of (7.170) is unity.

The shape of the distribution depends only on bm and curves for varying values of this are given by Tweedie. If we put $m = 0$, $b = 1$ in (7.170) we get as a density function

$$(2\pi x^3)^{-\frac{1}{2}} e^{-1/2x}, \tag{7.173}$$

which is an important special case. That the integral of this is unity follows by letting $m \to 0$ in the above integral and observing that the convergence is bounded, or more easily by putting $x^{-1} = u^2$.

The characteristic function of (7.170) is

$$\phi(t) = b \int_0^\infty (2\pi x)^{-\frac{3}{2}} \exp\{itx - \tfrac{1}{2}x^{-1}(b-mx)^2\} \, dx. \tag{7.174}$$

For $m > 0$, $\phi(t)$ is necessarily analytic in the neighbourhood of $t = 0$. Using (7.171) and (7.172) again we get

$$\phi(t) = \exp bm\left\{1 - \left(1 - \frac{2it}{m^2}\right)^{\frac{1}{2}}\right\}. \tag{7.175}$$

The important implication of this formula is that it shows that the sum of n independent random variables with the distribution (7.170) is a distribution of the same form with b replaced by nb.

Similarly the characteristic function of (7.173) becomes (putting $b = 1$ and letting $m \to 0$, or directly as in §6.9)

$$\exp -|t|^{\frac{1}{2}}(1 - i \operatorname{sgn} t), \tag{7.176}$$

so that the sum of n independent random variables, $X_1, \ldots, X_n$, with the distribution (7.173), has a probability density

$$n(2\pi x^3)^{-\frac{1}{2}} \exp -\frac{n^2}{2x}, \tag{7.177}$$

and thus $n^{-2}(X_1 + \ldots + X_n)$ has the same distribution as X_1. This is in sharp contrast to the normal distribution, and is clearly related to the fact that (7.173) has no second moment.

From the characteristic function the moments and cumulants can be found. Thus

$$m_1 = \frac{b}{m}$$

$$m_2 = \frac{b}{m^3}(1 + bm),$$

$$m_3 = \frac{b}{m^5}(3 + 3bm + b^2m^2),$$

$$m_4 = \frac{b}{m^7}(15 + 15bm + 6b^2m^2 + b^3m^3), \tag{7.178}$$

and in general

$$m_n = bm^{-1} \frac{K_{n-\frac{1}{2}}(bm)}{K_{\frac{1}{2}}(bm)}.$$ (7.179)

The cumulants, which are simpler, are given by Tweedie. The moments of (7.173) are all infinite.

If X is a random variable with the distribution (7.170) or (7.173), $Y = X^{-1}$ also has some interesting properties and has moments of all orders. Thus both of these distributions have moments of all negative orders.

7.24. Distributions on a circle

We consider the points on the circumference of a circle whose radius can be taken as unity. We can define each point by its angular position relative to some fixed direction so that the set of all points is equivalent to the set of points θ in the interval $0 \leqslant \theta < 2\pi$. On this space, probability distributions can be defined in the usual way, $F(x)$ being taken as the probability that a random direction satisfies $0 \leqslant \theta < x$, where $0 \leqslant x \leqslant 2\pi$.

We consider four particular distributions which occur in various contexts.

(1) *The uniform distribution.* In this case we assume that $F(x)$ has a probability density which is constant and therefore equal to $(2\pi)^{-1}$.

(2) *The circular normal distribution.* Here the probability density is taken as proportional to $\exp(k \cos \theta)$ where $k \geqslant 0$. $k = 0$ gives the uniform distribution, and $k < 0$ can be treated by replacing θ by $\pi - \theta$. Since

$$\int_0^{2\pi} \exp(k \cos \theta) \, d\theta = 2\pi I_0(k),$$ (7.180)

where $I_0(k)$ is the imaginary Bessel function of order zero, the probability density can be written

$$\frac{1}{2\pi I_0(k)} \exp(k \cos \theta).$$ (7.181)

(3) *The wrapped-up normal distribution.* The density is here

$$(2\pi\sigma^2)^{-\frac{1}{2}} \sum_{n=-\infty}^{\infty} \exp\left\{-\frac{1}{2\sigma^2}(\theta + 2\pi n)^2\right\}.$$ (7.182)

This may be regarded as resulting from wrapping a line with the probability density of the normal distribution an infinite number of times about the circumference of a circle. It can be shown (Stephens 1963) that this and (7.181) can be made to approximate to each other very closely.

(4) *The distribution of an offset normal distribution.* Circular distributions have been used to represent the probability distribution of wind directions. If fixed rectangular coordinates are taken at some point and the wind velocity resolved into components, X and Y, along these axes it seems very probable

that the best fit would be obtained by supposing these are jointly distributed in a bivariate normal distribution with means not necessarily zero, and arbitrary variances and correlation. The circular distribution is then that of the direction of the vector from zero to the point X, Y but is complicated and awkward.

Now consider the general distribution $F(x) = \text{pr}(0 \leqslant \theta < x)$. The natural analogue of the characteristic function is the set of Fourier–Stieltjes coefficients

$$c_n = a_n + ib_n = \int_0^{2\pi} e^{in\theta} \, dF(\theta), \tag{7.183}$$

where $n = 0, 1, \ldots$. Clearly $F(\theta)$ determines c_n uniquely. If $\sum |c_n|^2$ is convergent not only is the converse true but $F(\theta)$ is the integral of a probability density when $c_0 = 1$ if it is non-decreasing. The general theory is now that of Fourier–Stieltjes series (see Hardy and Rogosinski 1944 and Zygmund 1959). c_0 is necessarily unity, and $c_n = 0$ ($N \geqslant 1$) if and only if $F(\theta) = (2\pi)^{-1}\theta$, that is, if the distribution is uniform over the circle. We also need the theorem that if $c_n(m)$ ($m = 1, 2, \ldots$) (all n) is a sequence of Fourier–Stieltjes coefficients of distributions $F_m(\theta)$, the convergence of $F_m(\theta)$ to a distribution $F(\theta)$ implies $c_n(m) \to c_n$ (this is obvious), and the converse result that if the sequence $c_n(m)$ is convergent for each n to a number c_n, there exists an $F(\theta)$ such that $F_n(\theta) \to F(\theta)$ at all the continuity points of the latter which is therefore uniquely defined at these points. This result, which is the analogue of the theorems in §6.5, is proved in Zygmund (1959), p. 140. In particular if $c_n(m) \to 0$ ($m \to \infty$) then $F_m(\theta) \to (2\pi)^{-1}\theta$.

Given two random variables θ_1 and θ_2 we can define their sum in two different ways. We might take

$$\theta = \theta_1 + \theta_2 \pmod{2\pi}, \tag{7.184}$$

and this is the most natural procedure for most problems. Alternatively we might consider the angle between the direction $\theta = 0$ and the vector of components

$$(\cos\theta_1 + \cos\theta_2, \ \sin\theta_1 + \sin\theta_2). \tag{7.185}$$

Returning to (7.184) consider a sequence $\theta_1, \theta_2, \ldots$ of independent random directions all having the same distribution $F(\theta)$, and define

$$\Phi_N = \theta_1 + \ldots + \theta_N \pmod{2\pi}. \tag{7.186}$$

The asymptotic distribution of Φ_n will, in most cases, be the uniform distribution. This will certainly be so if $F(\theta)$ has any component which is the integral of some integrable function, for in this case each of the c_n ($n \geqslant 1$) will satisfy $|c_n| < 1$. The result then follows because

$$E(\exp Ni\Phi_n) = E\left(\prod_1^N \exp i\theta_i\right) = c_n^N, \tag{7.187}$$

which converges to zero. This is a consequence of the result that if θ_1 and θ_2 are angles with distributions $F_1(\theta)$, $F_2(\theta)$, and c_n, c_n' are correspondingly defined by (7.183), the Fourier–Stieltjes coefficients of the distribution of $\theta_1 + \theta_2 \pmod{2\pi}$ are $c_n c_n'$.

This convergence to uniformity is the analogue on a circle of the Central Limit Theorem. Clearly a necessary and suffiicien condition for this to hold is that all the c_n are such that $|c_n| < 1$. This leads to deep and unsolved problems in the theory of Fourier series. For a general account of the addition of random variables on a circle see Lévy (1939b). Dvoretsky and Wolfowitz (1951) have given an interesting discrete analogue (1951) of the above Central Limit Theorem.

The circular normal distribution given by (7.181) does not have all the properties that would be expected by analogy with the normal distribution on $(-\infty, \infty)$. Not only is it not the limiting distribution of a sum of angles but it does not have the reproductive property, i.e. if θ_1 and θ_2 have this distribution, the distribution of $\theta_1 + \theta_2 \pmod{2\pi}$ is not of the same form. To see this it is sufficient to calculate

$$c_n = \{2\pi I_0(k)\}^{-1} \int_0^{2\pi} \exp(in\theta + k \cos\theta)\, d\theta$$

$$= \frac{I_n(k)}{I_0(k)}, \tag{7.188}$$

so that c_n^2 is not of the same form as c_n. The limiting distribution of a sum of angles also depends on the manner in which it is defined. Thus in general if θ_1, θ_2,... are independent random angles with the same distribution (with all $|c_n| < 1$),

$$\theta_1 + ... + \theta_n \pmod{2\pi},$$

tends to the uniform distribution, whilst

$$n^{-\frac{1}{2}}(\theta_1 + ... + \theta_n) \pmod{2\pi} \tag{7.189}$$

tends to the wrapped-up normal distribution.

There is, however, one property of the circular normal distribution which is analogous to that of the normal distribution on $(-\infty, \infty)$, and that is the classical property which can be stated in the form that the 'mean of a sample is the value of the population mean which maximizes the joint probability of the sample values'. This was the property which Gauss used to characterize the normal distribution on $(-\infty, \infty)$. v. Mises (1918) used this property to look for a circular analogue. Suppose that $F(\theta)$ is a circular distribution which has everywhere a probability density $f(\theta)$ which is itself differentiable.

Write

$$c_1 = \int e^{i\theta} f(\theta)\, d\theta, \tag{7.190}$$

and suppose this is non-zero.

If $c_1 = |c_1| (\cos \alpha + i \sin \alpha)$, the mean is defined to be the direction α, and using now the vector definition of addition, the mean of a sample of values $\theta_1, \ldots, \theta_n$ is defined to be the direction, θ, of the point

$$\left(\sum_1^n \cos \theta_i, \sum_1^n \sin \theta_i \right). \tag{7.191}$$

This must satisfy the equation

$$\sin \theta \left(\sum_1^n \cos \theta_i \right) - \cos \theta \left(\sum_1^n \sin \theta_i \right) = \sum_1^n \sin(\theta - \theta_i) = 0. \tag{7.192}$$

Rotate the axes of coordinates so that the mean, α, is zero. Then the joint probability of the values $\theta_1, \ldots, \theta_n$, given that they come from a distribution of the form $f(x)$ with mean ϕ, is

$$\prod_1^n f(\theta_i - \phi),$$

which is to be maximized for variations in ϕ to obtain an estimator of the mean. At the maximum value, $\hat{\phi}$, of ϕ we must have

$$\sum_1^n \frac{f'(\theta_i - \hat{\phi})}{f(\theta_i - \hat{\phi})} = 0, \tag{7.193}$$

and this equation must define the way in which $\hat{\phi}$ depends on $\theta_1, \ldots, \theta_n$. Assuming that ϕ is also the value of θ satisfying (7.192), i.e.

$$\sum_1^n \sin(\phi - \theta_i) = 0, \tag{7.194}$$

the fact that equations (7.193) and (7.194) are equivalent implies that

$$f'(\theta)/f(\theta) = \text{constant}(\sin \theta),$$

and hence

$$f(\theta) = K \exp(k \cos \theta)$$

where K, k are constants. From this (7.181) follows. (See also Pólya 1930.)

The statistical theory of this distribution has been considered by Watson (1957a, b), Watson and Williams (1956), and Stephens (1962). Tables have been constructed by Gumbel, Greenwood, and Durand (1953), who give probabilities of the form

$$\text{pr}(-\alpha \leqslant \theta \leqslant \alpha),$$

for $k = 0(0.2) 4.0$, $\alpha = 0°(5°) 180°$, to five decimals with second and fourth central differences in α.

A number of other papers deal with circular distributions. References are given in Gumbel, Greenwood, and Durand (1953). The distribution of the sums of vectors of unit lengths which have uniformly distributed directions really belongs to the subject of random walks and will be treated in Chapter 10.

7.25. Distributions on spheres

Distribution functions on spheres have also been considered. Since any point on a sphere is determined by two coordinates we can take these as the co-latitude θ, and the azimuth angle ϕ, with respect to some given axis, 'the polar direction', and a direction for $\theta \neq 0$, π, such that $\phi = 0$. We can then as before set up a distribution function, $F(\theta, \phi)$, which we identify with the probability that a random vector of coordinates (Θ, Φ) satisfies $0 \leqslant \Theta \leqslant \theta$, $0 < \Phi \leqslant \phi$. This in turn defines a measure on the surface of the sphere. For the most part, however, attention has been confined to distributions which are independent of the azimuth.

Consider than a distribution independent of ϕ, and let (θ_1, ϕ_1) be a sample point. Addition could be defined by supposing that if (θ_2, ϕ_2) is another sample point, the sum is such that θ_2 and ϕ_2 are measured from (θ_1, ϕ_1) as the polar direction, ϕ_1 being measured from some null position whose definition is irrelevant since the probability distribution is uniform over ϕ. Moving from $(0, 0)$ to (θ_1, ϕ_1) and then to a new point distant θ_2 from the latter, and so on, defines a random walk on the sphere and it is not difficult to show that the asymptotic distribution is always the distribution with uniform density all over the surface of the sphere, the sole exception being the case where the probability density is entirely concentrated at the two points $\theta = 0$, $\theta = \pi$. Notice the simplicity of this case compared with the circle. If the distribution is not independent of ϕ, the theory is much more complicated.

Alternatively, addition can be defined in terms of vectors. If $(\theta_1, \phi_1), \ldots, (\theta_n, \phi_n), \ldots$ are a sequence of independent sample directions, all with respect to the same polar direction, we could consider their sum as defined by the direction of the sum of the random unit vectors to which they correspond. This definition is important in some statistical applications. If each vector has a direction with a distribution defined relative to a polar direction which is that of the preceding vector, independence is lost and the problem becomes more complicated. We shall consider it later in connexion with the theory of 'stiff chains' developed by H. Daniels (Chapter 10).

The spherical analogue of the circular normal distribution is a distribution on the sphere whose density depends only on θ, and is given by

$$\frac{k}{2 \sinh k} \exp(k \cos \theta), \tag{7.195}$$

it being easy to verify that the integral of this over the surface of the sphere is unity. k is the measure of dispersion, and when it is large the density is approximately proportional to

$$\exp(-\tfrac{1}{2}k\theta^2),$$

in the neighbourhood of the polar direction. Thus it is locally like a bivariate circular normal distribution, an analogous property holding for the circular normal.

This distribution also arises in statistical mechanics. If we consider a statistical assemblage of weakly interacting electric dipoles (such as polar molecules in a dielectric gas) which are subject to an external electric field, the density of the probability distribution of their potential energies will be proportional to

$$\exp\left(-\frac{U}{kT}\right),$$

where U is the potential energy, k is Boltzmann's constant, and T is the absolute temperature. Since the potential energy is proportional to the cosine of the angle θ between the dipole and the field, the distribution of θ will be such that (7.195) is the distribution of the vector direction of the dipole as represented on the surface of the sphere.

Fisher (1953) introduced this distribution in order to deal with the distributions of remanent magnetism in rock specimens in the study of paleomagnetism, and in many cases (Watson and Irving 1957) it provides a good fit (it had been used in other contexts before this). The distribution of the sum of n unit vectors having the directional distribution (7.195) has been considered in detail by Watson, Williams, and Stephens in the papers cited in § 7.24. For further work on distributions on a sphere the reader should consult some of the papers referred to in § 7.24 and Breitenberger (1963). Another symmetric distribution on the surface of the sphere is obtained from diffusion equation (see Roberts and Ursell 1960).

Bibliographic notes

An index of discrete and continuous distributions is given by Haight (1961). An extensive guide to tables of distributions is given by Greenwood and Hartley (1962). (See also Owen 1962.)

Two general classes of distributions are those of Pearson and Gram-Charlier. A good account of these, with references, is given in Chapter VI of Kendall and Stuart (1958). For the latter see also Boas (1949). Another system of distributions has been constructed by N. L. Johnson (1949a, b).

For the generation of random numbers from given distributions see Marsaglia (1961a, b), Box and Muller (1958), and the surveys in Tocher (1963) and Hammersley and Handscomb (1964). Quenouille (1959) has given tables of one thousand random variates derived from five typical analytical distributions including the log-normal and the two-sided exponential distributions.

Characterization theorems for distributions are discussed in Lukacs and Laha (1964) and in Prokhorov (1966). Multivariate distributions are indexed in Haight (1961), and these include multivariate generalizations of many of the distributions considered in this chapter.

Exercises

Exercise 7.1. Prove that the cumulants of the rectangular distribution on $(-\frac{1}{2}, \frac{1}{2})$ are given by

$$\kappa_{2r+1} = 0,$$

$$\kappa_{2r} = (2r)^{-1}B_{2r},$$

where B_n is the Bernoulli number of order n.

Exercise 7.2. The distribution (7.8) can be obtained in another manner. Consider the region of space for which $X_i \geqslant 0$, $i = 1,..., n$, and $\sum_1^n X_i \leqslant x$. $F(x)$ is the ratio of the volume of that part of this region cut off by the condition $X_i \leqslant 1$, $i = 1,..., n$, to the volume of the whole region. By considering n random variables with a joint distribution which is uniform over $X_i \geqslant 0$, $\sum_1^n X_i \leqslant x$, find the probability that all the X_i are less than unity by using the results of § 1.15 and hence prove (7.8).

Exercise 7.3. Represent the square of

$$A(x) = (2\pi)^{-\frac{1}{2}} \int_0^x e^{-\frac{1}{2}u^2} \, du$$

as an integral over a square in the plane. By comparing with integrals over inscribed and circumscribed quadrants prove that

$$(1-e^{-\frac{1}{2}x^2})^{\frac{1}{2}} < 2A(x) < (1-e^{-x^2})^{\frac{1}{2}}.$$

Prove also the stronger result

$$2A(x) < (1-e^{-2\pi^{-1}x^2})^{\frac{1}{2}}.$$

(Pólya 1949a).

Exercise 7.4. Using Theorem 6.18 (the Cramér–Wold theorem) prove that if the sum $(X_1 + Y_1,..., X_k + Y_k)$ of two sets $(X_1,..., X_k)$, $(Y_1,..., Y_k)$ of random variables is distributed in a multivariate normal distribution, and the two sets are independent, each is also distributed in multivariate normal distributions.

Exercise 7.5. Construct an example of a non-normal distribution of two random variables X_1 and X_2 which has an analytic density function and for which the marginal distributions of X and Y are normal.

Exercise 7.6. Construct an example of an analytic probability density function such that the conditional distribution of X for every given Y is normal, the conditional distribution of Y for every given X is normal, and the joint distribution is not bivariate normal.

Exercise 7.7. Let X and Y be independent normal variates with zero means and standard deviation σ. Show that XY^{-1} has the distribution (7.130). (This property does not characterize the normal distribution—see Steck 1958a and Laha 1959.)

Exercise 7.8. If X and Y are independent random variables uniformly distributed on $(0, 1)$ prove that

$$U = (-2 \log X)^{\frac{1}{2}} \cos(2\pi Y),$$

$$V = (-2 \log X)^{\frac{1}{2}} \sin(2\pi Y),$$

are independent random variables distributed normally with zero means and unit variances (Box and Muller 1958).

Exercise 7.9. $X_1, \ldots, X_n, Y_1, \ldots, Y_n$ are independent random variables normally distributed with zero means and unit standard deviations. Obtain the characteristic function of

$$\sum_1^n X_i Y_i$$

and its distribution for $n = 1$.

Exercise 7.10. Let $(X_1, \ldots, X_n)$ be a set of random variables with a joint multinormal distribution with zero means and unit variances. Let the correlation matrix (ρ_{ij}) be arbitrary and suppose $(Y_1, \ldots, Y_n)$ is a random permutation of $(X_1, \ldots, X_n)$. Show that in general the joint distribution of the Y_i is not multinormal.

Exercise 7.11. $X_1, \ldots, X_n$, are independent random variables each having the same negative exponential distribution. Prove that

$$X_1 + \frac{1}{2} X_2 + \ldots + \frac{1}{n} X_n$$

and

$$\max(X_1, \ldots, X_n)$$

have the same distribution (D. G. Kendall).

Exercise 7.12. Using the simple birth process (§ 3.20) interpret equation (2.72) relating the tails of the Poisson and χ^2-distributions.

Exercise 7.13. Let $X_1, X_2, \ldots$ be a sequence of independent random variables having the same negative-exponential distribution. Let $Y = X_1 + \ldots + X_n$, where n has a geometric distribution QP^{n-1} $(n = 1, 2, \ldots)$. Prove that Y has a negative exponential distribution.

Exercise 7.14. Convert equation (2.97) into a relationship between the tails of a negative binomial distribution and a beta distribution of the first kind.

Exercise 7.15. Prove that a non-central χ^2 with n degrees of freedom and a non-centrality parameter a^2 is the same as that of a random variable X whose distribution is the ordinary χ^2 distribution with $n + 2\kappa$ degrees of freedom, where κ has the Poisson distribution with mean $\frac{1}{2}a^2$ (James and Stein 1961, Kerridge 1965).

Exercise 7.16. Using a conditional characteristic function as in § 6.14, obtain formulae (7.96) and (7.98).

8. Sequences and Sums

8.1. IN THIS chapter we consider in a more general way the behaviour of sequences of random variables. We assume $X_1, X_2, \ldots$ is a sequence of random variables having proper distributions on the interval $(-\infty, \infty)$ and we begin by considering the various senses in which this sequence can be said to converge.

8.2. Convergence in distribution

We say the sequence $\{X_n\}$ converges in distribution if the distribution, $F_n(x)$, of X_n converges completely in the sense of § 5.5, or equivalently, with the Lévy metric. We have proved there that this type of convergence has the Cauchy property, so that if the sequence $\{F_n\}$ is mutually convergent, there exists a proper distribution, $F(x)$, to which $F_n(x)$ converges.

This type of convergence involves only the distributions of the individual X_n and does not require anything to be known about their mutual dependence or even that they have a joint distribution. It is therefore not a definition of the convergence of random variables themselves. In all the other definitions of convergence of random variables used in this chapter we shall have to assume that all the variables have a joint distribution.

8.3. Convergence in probability and convergence in distribution

The types of convergence of random variables which we now consider have all been already defined in principle in Chapter 4, but in order to keep the rest of the book as independent as possible of that chapter, and in order to provide a direct probability insight into what happens we shall redevelop the theory again here.

A sequence of random variables $\{X_n\}$, which we now suppose to have a joint distribution, is said to converge in probability to a random variable X, which has a joint distribution with the X_n, if

$$\mathrm{pr}(|X_n - X| > \varepsilon)$$

converges to zero as n increases for every $\varepsilon > 0$.

It is then sometimes convenient to write

$$X_n \xrightarrow{p} X.$$

As explained in Chapter 4, the idea of convergence in probability is a translation into probability language of the idea of 'convergence in measure'.

THEOREM 8.1. *Convergence in probability has the Cauchy property of convergence, i.e. if $\{X_n\}$ is a sequence of jointly distributed random variables such that given any $\varepsilon > 0$, $\eta > 0$, there exists n_0 such that*

$$\text{pr}(|X_n - X_m| > \varepsilon) < \eta, \tag{8.1}$$

for all $m, n \geqslant n_0$, then there exists a random variable X, jointly distributed with the X_n, such that

$$\text{pr}(|X_n - X| > \varepsilon)$$

converges to zero for every $\varepsilon > 0$.

It is simpler to delay the proof of this theorem until almost certain convergence has been considered.

THEOREM 8.2. *Convergence in probability implies convergence in distribution.*

This is almost obvious. Suppose the random variables, X_n, with distribution functions $F_n(x)$, converge in probability to a random variable, X, with a proper distribution function, $F(x)$. Then for every $\varepsilon > 0$, $\eta > 0$, there exists n_0 so that for all $n \geqslant n_0$,

$$\text{pr}(|X_n - X| > \varepsilon) < \eta,$$

and hence

$$F_n(x - \varepsilon) - \eta \leqslant F(x) \leqslant F_n(x + \varepsilon) + \eta.$$

Letting ε, $\eta \to 0$ the result follows. The converse is clearly false, since the X_n could, for example, be independent and identically distributed.

Convergence in distribution is an important property of a sequence of random variables, especially if a bound for the error is known, because it often provides a method of approximating to an unknown distribution in terms of a known one. The most frequently used method of proving such convergence is not by direct consideration of the distributions themselves but from their characteristic function. Another method, especially useful in combinatorial problems, is to prove convergence of moments to the moments of some known distribution, provided the latter is then uniquely determined.

We here prove one particular result which is occasionally useful in such problems.

THEOREM 8.3 (Bernstein's lemma) (Bernstein 1926–7). *Suppose that X_n, Y_n, and Z_n are random variables such that $X_n = Y_n + Z_n$, where Y_n and Z_n have some joint distribution. If Y_n has a distribution, $F_n(y)$, which converges to some proper distribution $F(x)$ with non-zero dispersion, and if the variance of Z_n, $\sigma^2(Z_n)$, tends to zero, the distribution of X_n also tends to $F(x)$ (the means being taken as zero).*

Proof. Using Tchebychev's inequality we have, given any $\varepsilon > 0$,

$$\mathrm{pr}(|Z_n| > \varepsilon) \to 0. \tag{8.2}$$

Then

$$\mathrm{pr}(X_n \leqslant x) \leqslant \mathrm{pr}(Y_n < x + \varepsilon) + \mathrm{pr}(|Z_n| > \varepsilon),$$

and the left-hand side converges to $F(x+\varepsilon)$ as n increases. Thus $\mathrm{pr}(X_n \leqslant x) \leqslant F(x)$ since ε is arbitrary. Similarly we can show that

$$\mathrm{pr}(X_n \geqslant x)$$

converges to a quantity not less than $F(x-0)$, so that the result follows.

8.4. Almost certain convergence

Let X_n be a sequence of random variables with a joint distribution. We say that X_n converges almost certainly to a random variable X, which has a joint distribution with the X_n, if the probability that the sequence $\{X_n\}$ converges to X (in the usual sense) is unity. This definition requires that we have a theory of probability measures on the space of infinite sequences such as is developed in Chapter 4. Assuming this theory, the definition is equivalent to asserting that, for every $\varepsilon > 0$,

$$\lim_m \mathrm{pr}(|X_n - X| > \varepsilon, \quad \text{for some } n \geqslant m)$$

$$= \lim_m \lim_N \{1 - \mathrm{pr}(|X_n - X| \leqslant \varepsilon, \quad \text{for all } n \text{ satisfying } m \leqslant n \leqslant N)\}$$

$$= 0. \tag{8.3}$$

When this happens we write

$$X_n \xrightarrow{\text{a.c.}} X. \tag{8.4}$$

Notice that this also corresponds to the concept in Chapter 4 of the convergence 'almost everywhere' of a sequence of functions on some measure space. Similarly, given the sequence of random variables $\{X_n\}$, with a joint distribution, we say that it is almost certainly convergent in the Cauchy sense if for every $\varepsilon > 0$,

$$\lim_m \mathrm{pr}(|X_{n_1} - X_{n_2}| > \varepsilon, \quad \text{for some } n_1, n_2 \geqslant m)$$

$$= \lim_m \lim_N \{1 - \mathrm{pr}(|X_{n_1} - X_{n_2}| \leqslant \varepsilon \quad \text{for } m \leqslant n_1 < n_2 \leqslant N)\}$$

$$= 0. \tag{8.5}$$

THEOREM 8.4. *Almost certain convergence has the Cauchy property, i.e. if* $\{X_n\}$ *is a sequence of jointly distributed random variables which converges with probability one, there is a random variable X jointly distributed with the X_n such that $X_n \xrightarrow{\text{a.c.}} X$, and conversely.*

Proof. Suppose that $\{X_n\}$ is an almost certainly convergent sequence. The X_i are defined by a joint probability distribution, and outside a set of probability measure zero they converge. Define X to be a quantity equal to their limit when they converge and to zero (say) when they do not. Then by the way X is defined it is a random variable jointly distributed with the X_n because this joint distribution can be defined by a limiting process. Moreover, it satisfies the condition that the X_n converge to it with probability one. The converse of the theorem is obvious from (8.3) and (8.5).

THEOREM 8.5. *If a sequence, $\{X_n\}$, converges almost certainly to a random variable X, it also converges in probability to X.*

This follows at once from (8.3).

THEOREM 8.6. *The converse of Theorem 8.5 is false, i.e. it is possible to have a sequence $\{X_n\}$ which converges in probability to a random variable X without converging almost certainly to X.*

Proof. The proof follows the example in § 4.12. Let Y be a random variable uniformly distributed on the half-open interval $[0, 1)$ (the point $Y = 1$ being excluded for simplicity only). For any positive integer n we define the half-open interval I_n in the following manner.

Let

$$n = n_0 + n_1 2 + n_2 2^2 + \ldots + n_{k-1} 2^{k-1} + 2^k, \tag{8.6}$$

where $n_i = 0, 1$, $k = 0, 1, \ldots$, and take I_n to be the half-open interval

$$\frac{n - 2^k}{2^k} \leqslant x < \frac{n + 1 - 2^k}{2^k}. \tag{8.7}$$

Thus $I_1, I_2, I_3, \ldots$ are the intervals

$$[0, 1), [0, \tfrac{1}{2}), [\tfrac{1}{2}, 1), \ldots.$$

Define the random variable X_n to be equal to unity when Y lies in I_n, and to zero otherwise. Then the sequence $\{X_n\}$ converges in probability to the 'random variable' X which is always equal to zero. This follows because if $2^k \leqslant n < 2^{k+1}$, the probability that $X_n \neq 0$ is 2^{-k} which tends to zero as n increases. On the other hand, given Y in $[0, 1)$, there are always an infinite number of values of n for which X_n is equal to either zero or one so that the sequence $\{X_n\}$ never converges in the ordinary sense.

Although the converse of Theorem 8.5 is not true in general there is a partial converse which is important.

THEOREM 8.7. *If a sequence $\{X_n\}$ converges in probability to a random variable X, it is always possible to pick out a sub-sequence $\{X_{n_k}\}$ $(k = 1, 2, \ldots)$ which converges almost certainly to X, and moreover it is possible to do this in such a way that*

$$\sum_{k=1}^{\infty} \mathrm{pr}(|X_{n_k} - X| > 2^{-k}) < \infty. \tag{8.8}$$

Proof. (This is equivalent to Theorem 4.11.) Since X_n converges in probability to X there exists a positive integer m_k corresponding to every positive integer k such that

$$\text{pr}(|X_m - X| > 2^{-k}) < 2^{-k}, \tag{8.9}$$

for every $m > m_k$. Define

$$n_1 = m_1, \quad n_2 = \max(n_1 + 1, m_2), \ldots, n_{k+1} = \max(n_k + 1, m_{k+1}),$$

and so on.

Given any $\varepsilon > 0$, define N to be an integer such that $2^{-N+1} < \varepsilon$. Then since for all $k > N$,

$$\text{pr}(|X_{n_k} - X| > 2^{-k}) < 2^{-k},$$

$$\sum_{k=N}^{\infty} \text{pr}(|X_{n_k} - X| > 2^{-k}) < 2^{-N+1} < \varepsilon, \tag{8.10}$$

so that

$$\text{pr}(|X_{n_k} - X| \leqslant 2^{-k} \text{ for all } k > N) > 1 - \varepsilon. \tag{8.11}$$

Thus the probability that X_{n_k} converges in the ordinary sense to X is greater than $1 - \varepsilon$ for every $\varepsilon > 0$, and so X_{n_k} converges almost certainly to X, whilst (8.8) is satisfied because of (8.10).

Proof of Theorem 8.1. Suppose that $\{X_n\}$ is a sequence of random variables which converge in probability in the Cauchy sense, i.e. given any $\varepsilon > 0, \eta > 0$ there exists n_0 so that

$$\text{pr}(|X_m - X_n| > \varepsilon) < \eta, \tag{8.12}$$

for all $m, n > n_0$. We pick out a subsequence of these which converges almost certainly. For each k there is an integer m_k such that

$$\text{pr}(|X_{m_k} - X_n| > 2^{-k}) < 2^{-k}, \tag{8.13}$$

for all $n > m_k$. As before put $n_1 = m_1$, $n_{k+1} = \max(n_k + 1, m_{k+1})$. Then the sequence $\{X_{n_k}\}$ satisfies the inequality

$$\text{pr}(|X_{n_k} - X_{n_l}| > 2^{-k}) < 2^{-k}, \tag{8.14}$$

for $l > k$, and hence

$$\text{pr}(|X_{n_k} - X_{n_{k+1}}| \leqslant 2^{-k} \text{ for all } k > k_0)$$

$$\geqslant 1 - \sum_{k_0}^{\infty} 2^{-k} = 1 - 2^{1-k_0}. \tag{8.15}$$

The sequence $\{X_{n_k}\}$ therefore converges almost certainly in the Cauchy sense, and hence there exists a random variable, X, to which it converges almost certainly. Hence $\{X_{n_k}\}$ converges in probability to X, and using (8.1) again, $\{X_n\}$ converges in probability to X.

One of the most important methods of establishing almost certain convergence and convergence in probability is to obtain bounds on the above probabilities by using moments, and this immediately leads to the idea of convergence in mean.

8.5. Convergence in rth mean $(r > 0)$

X_n, a sequence of random variables, is said to converge in rth mean $(r > 0)$ to a random variable X, if $E|X_n|^r$ and $E|X|^r$ are finite and

$$E(|X_n - X|^r) \to 0. \tag{8.16}$$

We write this $X_n \xrightarrow{r} X$.

THEOREM 8.8. *If $\{X_n\}$ converges to X in rth mean for some $r > 0$, it converges to X in probability.*

Proof. This follows at once from Tchebychev's inequality (Theorem 5.20),

$$\mathrm{pr}(|X_n - X| > \varepsilon) \leqslant \varepsilon^{-r} E(|X_n - X|^r). \tag{8.17}$$

THEOREM 8.9. *Convergence in rth mean satisfies the Cauchy criterion.*

Proof. If $X_n \xrightarrow{r} X$, then by Theorem 5.12

$$E(|X_n - X_m|^r) \leqslant c_r E(|X_n - X|^r + |X_m - X|^r), \tag{8.18}$$

where $c_r = 1$ if $0 < r \leqslant 1$, and $c_r = 2^{r-1}$ if $r > 1$. Hence

$$E(|X_n - X_m|^r) \tag{8.19}$$

can be made arbitrarily small by making n and m large. Conversely suppose (8.19) tends to zero as n and m increase. Then by using an inequality of the form (8.17) with X replaced by X_m, it is clear that $\{X_n\}$ is convergent in probability so that there is a subsequence, $\{X_{n_k}\}$, which converges almost certainly to a random variable X which has a joint distribution with the X_n. Then for any fixed n, $X_n - X_{n_k}$ converges almost certainly to $X_n - X$.

The taking of an expectation is equivalent to integrating over the joint distribution of the X_n and X. Hence using the Fatou–Lebesgue theorem (Theorem 4.14),

$$E|X_n - X|^r = E \lim_k |X_n - X_{n_k}|^r,$$

$$\leqslant \lim_k E|X_n - X_{n_k}|^r, \tag{8.20}$$

and the right-hand side tends to zero as n increases. Hence $X_n \xrightarrow{r} X$, and $E|X|^r$ is finite.

THEOREM 8.10. *If $X_n \xrightarrow{r} X$, then*

$$E|X_n|^r \to E|X|^r. \tag{8.21}$$

Proof. This follows from inequalities such as

$$E|X_n|^r \leqslant E(|X_n - X|^r + E|X|^r), \tag{8.22}$$

when $r \leqslant 1$, for then

$$|E|X_n|^r - E|X|^r| \leqslant E|X_n - X|^r.$$

For $r > 1$ the result follows from a similar use of Minkowski's inequality (5.66).

THEOREM 8.11. *If* $X_n \xrightarrow{r} X$, *then*

$$X_n \xrightarrow{s} X,$$

for any s satisfying $0 < s < r$.

Proof. This follows at once from the fact (Theorem 5.16) that

$$s^{-1} \log E|X|^s \leqslant r^{-1} \log E|X|^r, \quad \text{if } s \leqslant r.$$

Although convergence in rth mean implies convergence in probability, it does not imply almost certain convergence, nor is it implied by the latter and we state this as a theorem.

THEOREM 8.12. *It is possible that*

$$X_n \xrightarrow{r} X \quad \text{but not} \quad X_n \xrightarrow{a.c.} X,$$

and also that

$$X_n \xrightarrow{a.c.} X \quad \text{but not} \quad X_n \xrightarrow{r} X \text{ for any } r.$$

Proof. Define X_n to be a sequence of independent random variables which take the values 0, $n^{1/2r}$, with probabilities $1 - n^{-1}$, n^{-1}. Take X to be the degenerate random variable always equal to zero. Then $E|X_n|^r = n^{-\frac{1}{2}} \to 0$ but

$$\text{pr(all } X_n = 0, \text{ for } m \leqslant n \leqslant N) = \prod_{n=m}^{N} (1 - n^{-1}) \qquad (8.23)$$

which diverges to zero as N increases for all values of m. Using Theorem 8.8 this provides another proof of Theorem 8.6.

Conversely suppose X_n takes the values 0, e^n with probabilities $1 - n^{-2}$, n^{-2}. Then for any $r > 0$,

$$E|X_n|^r = n^{-2} e^{nr} \to \infty, \qquad (8.24)$$

so that X_n does not converge to zero in rth mean for any r. On the other hand,

$$\text{pr(all } X_n = 0, \text{ for } m \leqslant n \leqslant N) = \prod_{n=m}^{N} (1 - n^{-2}). \qquad (8.25)$$

As N increases this infinite product converges to a non-zero quantity which itself tends to unity as m increases. Thus X_n converges to zero almost certainly.

For simplicity the two above examples have been constructed so that convergence occurs to a degenerate random variable. A more general case can be obtained by adding the same random variable to all the X_n and to X.

Since almost certain convergence implies convergence in probability, the above example also shows that convergence in probability does not imply convergence in rth mean of any order.

Just as the class of all distribution functions can be made into a metric space by introducing a distance function so, too, a class of random variables which are jointly defined over some probability space (or, in other words, have a joint distribution) can be made into a metric space by introducing a distance defined by

$$E|X-Y|^r, \quad \text{if } 0 < r \leqslant 1, \tag{8.26}$$

or by

$$(E|X-Y|^r)^{1/r}, \quad \text{if } r \geqslant 1, \tag{8.27}$$

provided that $E|X|^r$ is finite for each of the random variables. The fact that (8.26) and (8.27) are distance functions follows from (5.65) and (5.66). The fact that the sum of two random variables with finite rth moment also has a finite rth moment shows that these metric spaces are also linear.

8.6. Theorems of Tauberian type

In analysis it is often necessary to consider theorems involving the convergence of some element, Y_n say, to an element Y, where 'convergence' is defined in various different senses. Suppose that Y_n converges to Y in two senses which we call A and B, and write the convergence as $Y_n \xrightarrow{A} Y$, $Y_n \xrightarrow{B} Y$. If for the class of objects, $\{Y\}$, under consideration, $Y_n \xrightarrow{A} Y$ always implies $Y_n \xrightarrow{B} Y$ we say that convergence (A) is a 'stronger' condition than convergence (B) (or 'at least as strong'). A common type of theorem in analysis asserts that some type of convergence (A) is stronger than some other type (B) and such theorems are known as 'Abelian' theorems since they were first given this name in the study of the relative strengths of various methods of summability of series such as Abel's. If convergence (A) is strictly stronger than convergence (B), $Y_n \xrightarrow{B} Y$ does not imply that $Y_n \xrightarrow{A} Y$, and so the converse of the Abelian theorem is false in general.

However, in particular cases it is often useful to draw such conclusions and we can sometimes do so by imposing further conditions on the elements Y_n. Such a converse is known as a Tauberian theorem (from Tauber's theorem in the theory of summability) and is therefore 'the corrected false converse of an Abelian theorem'.

In this way we can obtain various theorems on the implications of modes of convergence of random variables.

If a random variable X is such that $\mathrm{pr}(|X| > k) = 0$ for some finite k it is said to be bounded almost certainly. We say that a sequence, $\{X_n\}$, is almost certainly bounded if this is true for each X_n with the same k.

THEOREM 8.13. *If $\{X_n\}$ is a sequence of random variables which is almost certainly bounded, and $X_n \xrightarrow{p} X$, then $X_n \xrightarrow{r} X$ for every $r > 0$.*

Proof. Since $X_n \xrightarrow{p} X$, the distribution of X, which is independent of n, cannot include any positive probability outside the range $(|X| \leqslant k)$. Then for any $r > 0$, $\varepsilon > 0$

$$E|X_n - X|^r \leqslant \varepsilon^r \mathrm{pr}(|X_n - X| \leqslant \varepsilon) + (2k)^r \mathrm{pr}(|X_n - X| > \varepsilon).$$

By choosing ε small and n large this can be made arbitrarily small and hence

$$E|X_n - X|^r \to 0.$$

This theorem is a special case of the following more general result.

THEOREM 8.14. *If* $\{X_n\}$ *is a sequence of random variables such that* $|X_n| < Y$ *for each* n, *where* Y *is another random variable such that* $E|Y|^r = k < \infty$ *($r > 0$), then* $X_n \xrightarrow{p} X$ *implies* $X_n \xrightarrow{r} X$, *and* $E|X|^r < \infty$.

Proof. Since $X_n \xrightarrow{p} X$, there exists a subsequence n_k ($k = 1, 2, \ldots$) such that X_{n_k} converges almost certainly to X, and therefore $|X| \leqslant Y$ with unit probability, so that $E|X|^r \leqslant k < \infty$.

Let $F(x_n, x)$ be the joint distribution of X_n and X. Given $\varepsilon > 0$, we can choose x_0 so that

$$\int_{\Omega - E} \int (|x|^r + |x_n|^r) \, dF(x_n, x) < \varepsilon,$$

where E is the set where $|x| < x_0$, $|x_n| < x_0$, and Ω is the whole plane. This is possible because the corresponding integral, taken over the whole plane, is convergent. Then

$$E\{|X_n - X|^r\} = \int \int |x_n - x|^r \, dF(x_n, x)$$

$$= \int_E \int |x_n - x|^r \, dF(x_n, x) + \int_{\Omega - E} \int |x_n - x|^r \, dF(x_n, x)$$

$$\leqslant \int_E \int |x_n - x|^r \, dF(x_n, x) + c_r \varepsilon,$$

where c_r is the constant of Theorem 5.12. The last integral has a bounded integrand, and since $X_n \to X$ in probability, can be made as small as desired by increasing n. Thus $E(|X_n - X|^r) \to 0$.

Another theorem of similar type which assumes a little less and also has a weaker conclusion is the following.

THEOREM 8.15. *If*

$$\sup E(|X_n|^r) = k < \infty \quad (r > 0), \tag{8.28}$$

and $X_n \xrightarrow{p} X$, *then* $X_n \xrightarrow{r'} X$ *for every* r' *in* $0 < r' < r$.

Proof. Let $F(x_n, x)$ be the joint distribution of X_n and X, and let E be any measurable set in the plane. Then for $a > 0$,

$$\int_E \int |x_n|^r dF(x_n, x) = \int_E \int_{|x_n|>a} |x_n|^r dF(x_n, x) + \int_E \int_{|x_n|\leqslant a} |x_n|^r dF(x_n, x)$$

$$\leqslant \int_E \int_{|x_n|>a} |x_n|^{r'-r}|x_n|^r dF(x_n, x) + \int_E \int_{|x_n|\leqslant a} a^{r'} dF(x_n, x)$$

$$\leqslant a^{r'-r} \int_E \int_{|x_n|>a} |x_n|^r dF(x_n, x) + \int_E \int_{|x_n|\leqslant a} a^{r'} dF(x_n, x).$$
$$(8.29)$$

Given $\varepsilon > 0$, choose a so $a^{r'-r}k < \frac{1}{2}\varepsilon$. Then for any E such that

$$\int_E \int dF(x_n, x) < \eta = \frac{1}{2}\varepsilon a^{-r'}$$

we have

$$\int_E \int |x_n|^r dF(x_n, x) < \varepsilon. \tag{8.30}$$

Since $X_n \overset{p}{\longrightarrow} X$, there exists a subsequence, X_{n_k}, which converges to X almost certainly. Then

$$E\{|X|^r\} \leqslant \varliminf_k E\{|X_{n_k}|^r\} \leqslant k,$$

by the Fatou–Lebesgue theorem (Theorem 4.14). Arguing as above we see that given any $\varepsilon > 0$, there exists $\eta > 0$ so that if E is any set such that

$$\int_E \int dF(x_n, x) < \eta$$

then

$$\int_E \int |x|^r dF(x_n, x) < \varepsilon.$$

Using the inequality of Theorem 5.12 we see that if

$$\int_E \int dF(x_n, x) < \eta,$$

then

$$\int_E \int |x_n - x|^r dF(x_n, x) < \kappa\varepsilon, \tag{8.31}$$

where κ depends on r' only. Let E be the set where either $|x_n| \geqslant c > 0$ or $|x| \geqslant c > 0$, where c is some prescribed constant. Then

$$E(|X_n - X|^r) = \int_E \int |x_n - x|^r dF(x_n, x) + \int_{\Omega-E} \int |x_n - x|^r dF(x_n, x). \tag{8.32}$$

By choosing c sufficiently large we can make the first integral arbitrarily small. In the second integral x_n and x are bounded and since $X_n \xrightarrow{p} X$ this integral can be made arbitrarily small by making n large. Thus $E\{|X_n - X|^r\}$ tends to zero as n increases.

8.7. The Slutsky–Fréchet theorem

Suppose that $f(x)$ is a continuous function for all values of x. It is sometimes useful to consider the convergence of the sequence $f(X_n)$, where $\{X_n\}$ is a sequence of random variables. This is dealt with by the following theorem which was proved by Slutsky (1925) in a slightly more special case, and generalized by Fréchet (1937, p. 178).

THEOREM 8.16. *If $f(x)$ is continuous everywhere and $\{X_n\}$ converges to X in probability, or almost certainly, so does $f(X_n)$ to the random variable $f(X)$.*

Proof. Given any $\varepsilon > 0$, choose $a > 0$ so that

$$\mathrm{pr}(|X| > a) < \varepsilon.$$

In any finite range $f(x)$ is uniformly continuous, so that given $\eta > 0$ there exists $\xi > 0$ such that

$$|f(y) - f(x)| < \eta$$

for every pair of values x, y which are such that $|y - x| < \xi$, and x lies in $(-a, a)$. Then

$$\mathrm{pr}(|f(X_n) - f(X)| < \eta, |X| \leqslant a) \geqslant \mathrm{pr}(|X_n - X| < \xi, |X| \leqslant a), \quad (8.33)$$

and by choosing n sufficiently large we can make this not less than $1 - 2\varepsilon$ if X_n converges to X in probability. The proof for almost certain convergence is similar. Notice that if the probability distribution of X is concentrated at a single point $X = x$, it is only necessary to assume that $f(x)$ is continuous at x.

8.8. Sequences of events

The most important application of the above ideas of convergence are to the convergence of the partial sums of sequences of random variables. Thus if $X_1, X_2,\ldots$ is a sequence of random variables with a joint distribution we write

$$S_n = X_1 + \ldots + X_n, \quad (8.34)$$

and we consider the convergence of the sequence $\{S_n\}$.

Before considering sequences of random variables in general it is useful to consider sequences of events $E_1, E_2,\ldots$ which may be independent or not. For example, we may suppose that these events are the results of successive independent trials which have two possible outcomes, 'success' with probability p_n, and 'failure' with probability $1 - p_n$. If, as in Chapter 1, we define indicator variables, X_n, to equal 1 or 0 according as E_n occurs or not, the number of successes in the first n trials is $X_1 + \ldots + X_n$, so that this can be regarded as a sequence of random variables.

Consider then a sequence of events, E_n, with probabilities p_n, and indicator variates X_n, so that

$$\mathrm{pr}(X_n = 1) = p_n = 1 - \mathrm{pr}(X_n = 0). \tag{8.35}$$

THEOREM 8.17 (Borel 1909, Cantelli 1917). *If the events in the sequence are independent the probability that an infinite number of them occur is 0, or 1, according as $\sum p_n$ converges or not.*

Proof. First suppose that $\sum p_n$ converges and that α is the probability that an infinite number of events occur. Put $\alpha_{n,N}$ for the probability that at least one of the events $E_n, E_{n+1}, \ldots, E_N$ occurs. Then

$$\alpha = \lim_n \lim_N \alpha_{nN}. \tag{8.36}$$

Using Boole's inequality (1.2) we have

$$\alpha_{nN} \leqslant p_n + \ldots + p_N, \tag{8.37}$$

and thus

$$\alpha \leqslant \lim_n \sum_n^\infty p_i = 0,$$

if $\sum p_n$ converges.

Conversely suppose $\sum p_n$ diverges. Then the product

$$\prod_{i=n}^{N} (1 - p_i) \tag{8.38}$$

tends to zero as N increases by the usual theory of infinite products. But since the events are independent, this is the probability that none of the events $E_n, \ldots, E_N$ occur, and hence is equal to $1 - \alpha_{n,N}$. Then

$$\lim_N \alpha_{nN} = 1,$$

so that $\alpha = 1$. It follows that the probability that an infinite number of events occur can only take the values 0 or 1.

The first part of the theorem can also be proved by using Tchebychev's inequality. Write $X_n = 1, 0$ according as E_n occurs or not. Then

$$E(S_n) = \sum_1^n p_i < K, \quad \text{say.} \tag{8.39}$$

Hence

$$\mathrm{pr}(\text{at least } k \text{ events occur in trials } 1, \ldots, n) \leqslant k^{-1}K.$$

Since n is arbitrary, and does not occur on the right-hand side,

$$\mathrm{pr}(\text{more than } k \text{ events occur at all}) \leqslant k^{-1}K. \tag{8.40}$$

The probability of an infinite number of events is therefore zero.

(8.36) and (8.39) both remain true if the events are not independent and hence both proofs of the first part of the theorem remain valid. This result is so important that we state it as a theorem.

THEOREM 8.18. *Whether or not the events are independent, the convergence of* Σp_n *implies that with probability one, only a finite number occur.*

Theorems 8.17 and 8.18 are known as the Borel–Cantelli lemmas.

The most important application of these theorems is to the probabilities of each of a sequence of random variables exceeding certain values in absolute value, so that we have the following result.

THEOREM 8.19. *If* $\{X_n\}$ *is a sequence of random variables, and* $\{a_n\}$ *a sequence of constants, the probability of an infinite number of realizations of the events*

$$|X_n| > a_n$$

is zero if the series

$$\Sigma \operatorname{pr}(|X_n| > a_n) \tag{8.41}$$

is convergent. If the random variables are also independent, the divergence of this series implies that the probability is unity.

Notice that in Theorems 8.17–8.19 we are tacitly assuming a theory of probability measure on the space of all sequences of events.

8.9. Convergence of sums

Consider a sequence of independent random variables X_1, X_2,.... In this section we shall begin to study the convergence of the sequence of random variables, $\{S_n\}$, where

$$S_n = X_1 + \ldots + X_n, \tag{8.42}$$

the convergence being taken in the various senses defined in § 8.3–§ 8.5.

This is not the only property of sums S_n which we have to study. Let $\{a_n\}$, $\{b_n\}$ be sequences of non-negative qualities and consider the random variables

$$T_n = a_n^{-1} S_n - b_n. \tag{8.43}$$

If T_n converges to zero in one of the above senses we say that the sum S_n is 'stable' relative to the sequences $\{a_n\}$, $\{b_n\}$, in the corresponding sense. This is equivalent to asserting that T_n converges to a random variable which is degenerate at zero, i.e. such that the probability of it being zero is unity. To generalize this idea suppose that the distribution of T_n converges to some known distribution when $\{a_n\}$ and $\{b_n\}$ are suitably chosen. Then S_n converges in distribution, and the conditions under which this occurs, especially when the limiting distribution is normal, are of great importance, and will be studied later in this chapter.

The convergence and stability of sums, S_n, of independent random variables, $\{X_n\}$, are of special interest because the assumption of independence in the X_n introduces a special structure into the sequence of random variables S_n, and hence theorems on the convergence of the S_n can be proved which are stronger than those established previously for arbitrary sequences.

The principal analytical techniques for dealing with such problems are based on the ideas of centring, truncating, and symmetrizing the random variables X_n.

To centre a random variable X_n we replace it by $X_n - c_n$ where c_n is a suitably chosen constant which is usually the mean (if it exists), or in more general cases the median, or one of the medians if there is more than one.

X_n is said to be truncated at a value $c > 0$, if it is replaced by a random variable Y_n defined by

$$Y_n = X_n, \quad \text{if } |X_n| < c,$$

$$= 0, \quad \text{if } |X_n| \geqslant c. \tag{8.44}$$

Symmetrization will be dealt with in § 8.12.

The value of truncation arises from the following result which we state as a theorem.

THEOREM 8.20. *Given two sequences $\{X_n\}$, $\{Y_n\}$, defined on the same probability space, the convergence of the infinite series*

$$\sum_n \mathrm{pr}(X_n \neq Y_n)$$

implies that the probability that the two series $\sum X_n$, $\sum Y_n$ do not converge or diverge together, is zero.

This is obvious from Theorem 8.18.

The general problem of the convergence of $\sum X_n$ can be most easily dealt with by relating it to the simpler special case where the X_n have finite second moments. Defining the variance as the second moment about the mean we have

$$\mathrm{var}\left(\sum_{n=m}^{N} X_n\right) = \sum_{n=m}^{N} \mathrm{var}(X_n). \tag{8.45}$$

Applying Tchebychev's inequality we now have

$$\mathrm{pr}\left[\left|\sum_{m}^{N}\{X_n - E(X_n)\}\right| > \varepsilon\right] \leqslant \varepsilon^{-2} \sum_{m}^{N} \mathrm{var}(X_n), \tag{8.46}$$

and we immediately deduce the following theorem.

THEOREM 8.21. *If $\sum \mathrm{var}(X_n)$ converges,*

$$\sum \{X_n - E(X_n)\}$$

converges in probability.

By Theorem 8.24 we shall later be able to conclude that this series also converges almost certainly

The following example is of particular importance historically (see Kac 1959 for an interesting account). Suppose that θ_n is a sequence of independent random variables distributed on the interval $(0, 2\pi)$ in such a way that

$$E(e^{i\theta_n}) = 0. \tag{8.47}$$

Then if $\{c_n\}$ is a sequence of real numbers,

$$\Sigma c_n e^{i\theta_n}$$

converges in probability to a random variable if

$$\Sigma c_n^2$$

converges. This follows on putting $X_n = c_n e^{i\theta_n}$ and observing that $\text{var}(X_n) = E(X_n^2) \leqslant c_n^2$. The result then follows. In particular the series

$$\sum_{1}^{\infty} \pm a_n, \tag{8.48}$$

where the signs are taken as $+$ and $-$ with probabilities $\tfrac{1}{2}$, converges in probability if Σa_n^2 converges. Later (Theorems 8.24 and 8.26) we shall also see that this series then converges almost certainly and if Σa_n^2 diverges, it diverges with probability one.

8.10. The Kolmogorov inequalities

In order to strengthen the above results we prove two remarkable inequalities due to Kolmogorov (1928, 1930).

THEOREM 8.22. *Suppose that $\{X_n\}$ is a sequence of independent random variables with finite variances, and*

$$S_n = \sum_{1}^{n} X_i. \tag{8.49}$$

Then for every $\varepsilon > 0$,

$$\text{pr}\left\{ \max_{1 \leqslant i \leqslant n} |S_i - E(S_i)| \geqslant \varepsilon \right\} \leqslant \varepsilon^{-2} \sum_{1}^{n} \text{var}(X_i). \tag{8.50}$$

Proof. For simplicity suppose $E(X_i) = 0$, which imposes no real restriction. Write

$$U_k = 1, \quad \text{if } \max_{1 \leqslant i \leqslant k} |S_i| \leqslant \varepsilon,$$

$$= 0, \quad \text{otherwise}; \tag{8.51}$$

$$V_k = 1, \quad \text{if } |S_1| \leqslant \varepsilon,..., |S_{k-1}| \leqslant \varepsilon, |S_k| > \varepsilon,$$

$$= 0, \quad \text{otherwise}, \tag{8.52}$$

$$= U_{k-1} - U_k, \quad \text{for } k \geqslant 2.$$

Writing $U_0 = 1$, we also have $V_1 = U_0 - U_1$.

Then for $n \geq k$,

$$E(S_n^2 V_k) = E(S_k^2 V_k) + E\{(S_n - S_k)^2 V_k\}$$

$$\geq E(S_k^2 V_k)$$

$$\geq \varepsilon^2 E(V_k).$$

Summing this inequality from $k = 1$ to $k = n$, we have

$$\sum_1^n \mathrm{var}(X_i) = E(S_n^2)$$

$$\geq \sum_{k=1}^n E(S_n^2 V_k)$$

$$\geq \varepsilon^2 \sum_{k=1}^n E(V_k) = \varepsilon^2 E(U_0 - U_n).$$

From the above definitions we have

$$E(U_0 - U_k) = 1 - \mathrm{pr}\left(\max_{1 \leq i \leq k} |S_i| \leq \varepsilon \right)$$

$$= \mathrm{pr}\left(\max_{1 \leq i \leq k} |S_i| > \varepsilon \right),$$

and thus

$$\mathrm{pr}\left(\max_{1 \leq i \leq n} |S_i| > \varepsilon \right) \leq \varepsilon^{-2} \sum_1^n \mathrm{var}(X_i).$$

Since the right-hand side of this inequality is a continuous function of ε for $\varepsilon > 0$ it follows that (8.50) is also true.

By introducing a further condition it is possible to get an inequality in the opposite direction.

THEOREM 8.23 (Kolmogorov's second inequality). *If* $\{X_n\}$ *is a sequence of independent random variables such that* $|X_i| \leq c > 0$ *for all i, then*

$$\mathrm{pr}\left\{ \max_{1 \leq i \leq n} |S_i - E(S_i)| \geq \varepsilon \right\} \geq 1 - (\varepsilon + 2c)^2 \left\{ \sum_1^n \mathrm{var}(X_i) \right\}^{-1}. \qquad (8.53)$$

Proof. It is sufficient to suppose $E(X_i) = 0$ and replace c by $2c$ since $|X_i - E(X_i)| \leq 2c$. We make the convention that $S_0 = 0$. Then with the notation of the previous theorem we have

$$S_{k-1} U_{k-1} + X_k U_{k-1} = S_k U_{k-1} = S_k U_k + S_k V_k.$$

Since $U_k V_k = 0$, and

$$E(S_{k-1} X_k U_{k-1}) = 0,$$

we have

$$E(S_{k-1}^2 U_{k-1}) + E(X_k^2) E(U_{k-1}) = E(S_k^2 U_k) + E(S_k^2 V_k). \qquad (8.54)$$

Furthermore, we have

$$|S_k V_k| \leqslant |S_{k-1} V_k| + |X_k V_k| \leqslant (\varepsilon + 2c) V_k,$$

so that

$$E(S_k^2 V_k) = E(S_k^2 V_k^2) \leqslant (\varepsilon + 2c)^2 E(V_k).$$

Using this in (8.54), and the fact that $U_n \leqslant U_{k-1}$, we get

$$E(S_{k-1}^2 U_{k-1}) + \text{var}(X_k) E(U_n) \leqslant E(S_k^2 U_k) + (\varepsilon + 2c)^2 E(V_k).$$

Sum this over $k = 1, 2,..., n$, and we obtain

$$E(U_n) \sum_1^n \text{var}(X_k) \leqslant E(S_n^2 U_n) + (\varepsilon + 2c)^2 \sum_1^n E(V_k)$$

$$\leqslant \varepsilon^2 \text{pr}\left(\max_{1 \leqslant i \leqslant n} |S_i| \leqslant \varepsilon \right) + (\varepsilon + 2c)^2 \sum_1^n E(V_k)$$

$$\leqslant (\varepsilon + 2c)^2.$$

Then

$$\text{pr}\left(\max_{1 \leqslant i \leqslant n} |S_i| > \varepsilon \right) = 1 - E(U_n)$$

$$\geqslant 1 - (\varepsilon + 2c)^2 \left\{ \sum_1^n \text{var}(X_i) \right\}^{-1},$$

and again observing that the right-hand side is continuous in ε, (8.53) follows.

8.11. Almost certain convergence of sums

As before we consider sequences, $\{X_n\}$, of independent random variables. We first prove a strengthened form of Theorem 8.21.

THEOREM 8.24. *If $\sum \text{var}(X_n)$ converges then,*

$$\sum \{X_n - E(X_n)\} \tag{8.55}$$

converges almost certainly.

Proof. Using Theorem 8.22 we have, for every $\varepsilon > 0$,

$$\text{pr}\left\{ \max_{m,n > N} |S_m - S_n - E(S_m) + E(S_n)| \geqslant \varepsilon \right\} \geqslant \varepsilon^{-2} \sum_N^\infty \text{var}(X_i). \tag{8.56}$$

Given any fixed ε we can choose N large enough to make the left-hand side as small as desired and hence (8.55) converges.

Using Theorem 8.23 we conclude in a similar manner that if $|X_n| < c$ for all n, where c is some positive constant, the divergence of $\sum \text{var}(X_i)$ implies the almost certain divergence of $\sum \{X_i - E(X_i)\}$.

THEOREM 8.25. *If there exists $c > 0$ such that $|X_n| < c$ for all n, the almost certain convergence of $\sum X_n$ implies that the series $\sum \text{var}(X_n)$ and $\sum E(X_n)$ both converge in the ordinary sense.*

Proof. Introduce another sequence of independent random variables, $\{Y_n\}$, which are independent of the $\{X_n\}$ but such that Y_n has the same distribution as X_n for every n. Then $|X_n - Y_n| < 2c$, $E(X_n - Y_n)$ exists and is zero, and $\text{var}(X_n - Y_n) = 2\,\text{var}(X_n)$. The series $\Sigma(X_n - Y_n)$ must converge almost certainly and, the terms being bounded, it follows from the above remark that $\Sigma\,\text{var}(X_n - Y_n) = 2\Sigma\,\text{var}(X_n)$ must converge. Then since ΣX_n and $\Sigma\{X_n - E(X_n)\}$ both converge almost certainly, their difference, $\Sigma E(X_n)$, must converge in the ordinary sense.

It may seem that the study of random variables which are bounded is not sufficiently general to provide a method of obtaining useful general theorems on the convergence of random series, but this is not true. A great many theorems in this subject are most easily established by a method of truncation. Supposing that X_n is a random variable of general distribution. We choose $c > 0$ and define two new random variables Y_n and Z_n dependent on X_n by

$$Y_n = X_n, \quad \text{if}\,|X_n| < c,$$
$$= 0, \quad \text{if}\,|X_n| \geqslant c.$$
$$Z_n = X_n, \quad \text{if}\,|X_n| \geqslant c,$$
$$= 0, \quad \text{if}\,|X_n| < c,$$
$$= X_n - Y_n.$$

If c is chosen to depend on n (and is therefore written as c_n) it is often possible to choose the sequence $\{c_n\}$ in such a way that the sequence $\{Z_n\}$ has, with probability one, only a finite number of non-zero members, and therefore converges almost certainly. $\{Y_n\}$, on the other hand, is a sequence of bounded (and sometimes uniformly bounded) random variables and hence each X_n has a finite variance which enables us to use the much simpler theorems on sequences of random variables with finite variances. This train of ideas leads to the following important result, due to Kolmogorov (1928).

THEOREM 8.26 (The Three Series Theorem). *If $\{X_n\}$ is a sequence of independent random variables, and Y_n is X_n truncated to $c > 0$, the series*

$$\Sigma X_n$$

converges almost certainly, if and only if the three series

$$\Sigma\,\text{pr}(|X_n| \geqslant c), \tag{8.57}$$
$$\Sigma\,\text{var}(Y_n), \tag{8.58}$$
$$\Sigma E(Y_n), \tag{8.59}$$

all converge in the ordinary sense.
   This result holds for any $c > 0$.

24

Proof. If (8.57) converges, then by Theorem 8.20 the series $\sum X_n$ and $\sum Y_n$ must either converge together or diverge together unless an event of probability zero occurs. Then the convergence of (8.58) and (8.59) implies the almost certain convergence of $\sum Y_n$ and hence that of $\sum X_n$.

If $\sum X_i$ converges almost certainly, then the sequence $\{X_n\}$ must converge almost certainly to zero, so that given any number $c > 0$, the series $\sum \text{pr}(|X_n| \geqslant c)$ must converge by Theorem 8.17, so that (8.57) converges.

Thus $\sum Y_n$ converges almost certainly and using Theorem 8.25, (8.58) and (8.59) must converge also.

In discussing the almost certain convergence of infinite series of random variables we have, of course, tacitly assumed the theory of measure on an infinite dimensional product space that was developed in Chapter 4. We now show that if a series, $\sum X_n$, of independent random variables does not converge almost certainly, it must diverge almost certainly. This 'zero–one' law requires that we consider the infinite dimensional space explicitly.

This space consists of all sequences $(X_1, X_2, ...)$ and on it is defined a measurable field of sets in the manner considered in §4.9. Let E be the event that the series converges. This corresponds to a measurable set in this field. Let F be the class of all measurable sets A such that

$$P(AE) = P(A)P(E).$$

Since the convergence of the series does not depend on $(X_1, ..., X_n)$ for any finite n, every cylinder set defined on the space of values of $(X_1, ..., X_n)$ belongs to F. The class of all such sets for given n is a Borel field F_n contained in F. Hence F also contains the limit of every monotone sequence of such sets and therefore the set E. Hence $P(E) = P(E)^2$ and $P(E) = 0$ or 1. This proves the result.

In particular we see that the series (8.48) diverges almost certainly if $\sum a_n^2$ diverges.

8.12. Centring and symmetrization

So far we have centred our random variables at their means and hence have assumed the latter to exist. To deal with more general cases we need to centre at a median and having done this, symmetrization is a further tool.

In §5.14 we have defined the median of the distribution of a random variable X, with a distribution $F(x)$, to be any number, written as $x = \text{med}(X)$, such that

$$F(x-0) \leqslant \tfrac{1}{2} \leqslant F(x).$$

There may be more than one such median, but this will not cause any trouble if we make the convention that the median of a random variable such as cX (c a positive constant) will always be taken equal to $c\,\text{med}(X)$.

If X is any random variable we define a corresponding but independent symmetrized random variable W to have the distribution of $X - X'$, where X' is a random variable independent of X and having the same distribution. The distribution of W is then symmetric in the sense that

$$\text{pr}(W \leqslant a) = \text{pr}(W \geqslant -a) \qquad (8.60)$$

THEOREM 8.27. *Given $x > 0$ and a real,*

$$\tfrac{1}{2}\text{pr}\{X - \text{med}(X) \geqslant x\} \leqslant \text{pr}(W \geqslant x), \qquad (8.61)$$

$$\tfrac{1}{2}\text{pr}(|X - \text{med } X| \geqslant x) \leqslant \text{pr}(|W| \geqslant x) \leqslant 2\,\text{pr}(|X - a| \geqslant \tfrac{1}{2}x). \quad (8.62)$$

Proof.

$$\begin{aligned}
\text{pr}(W \geqslant x) &= \text{pr}\{X - \text{med}(X) - X' + \text{med}(X) \geqslant x\} \\
&\geqslant \text{pr}\{X - \text{med}(X) \geqslant x,\ X' - \text{med}(X) \leqslant 0\} \\
&\geqslant \text{pr}\{X - \text{med}(X) \geqslant x\}\text{pr}\{X' - \text{med}(X) \leqslant 0\} \\
&\geqslant \tfrac{1}{2}\text{pr}\{X - \text{med}(X) \geqslant 0\}.
\end{aligned}$$

The median is valuable here because $\text{med}(X)$ is a median of X', and the use of the median as a centred constant enables a probability $\tfrac{1}{2}$ to be introduced.

Adding to (8.60) the similar inequality for $-X$ we get the left-hand side of (8.62). The right-hand side follows from

$$\begin{aligned}
\text{pr}(|W| \geqslant x) &= \text{pr}\{|X - \text{med}(X) - X_1 + \text{med}(X)| \geqslant x\} \\
&\leqslant 2\,\text{pr}\{|X - \text{med}(X)| > \tfrac{1}{2}x\}.
\end{aligned}$$

We next establish inequalities similar to (8.61) and (8.62) involving not W and X, but the largest out of sequences of random variables related to W and X.

THEOREM 8.28. *If we have a sequence $\{X_n\}$ of random variables together with their related symmetrical variables $\{W_n\}$, if $\varepsilon > 0$, and if $\{a_i\}$ is a sequence of real numbers, then*

$$\tfrac{1}{2}\text{pr}\left[\max_{1 \leqslant j \leqslant n}\{X_j - \text{med}(X_j)\} \geqslant \varepsilon\right] \leqslant \text{pr}\left(\max_{1 \leqslant j \leqslant n} W_j \geqslant \varepsilon\right), \qquad (8.63)$$

and

$$\tfrac{1}{2}\text{pr}\left\{\max_{1 \leqslant j \leqslant n}|X_j - \text{med}(X_j)| \geqslant \varepsilon\right\} \leqslant \text{pr}\left(\max_{1 \leqslant j \leqslant n}|W_j| \geqslant \varepsilon\right), \qquad (8.64)$$

and the right-hand side is not greater than

$$2\,\text{pr}\left(\max_{1 \leqslant j \leqslant n}|X_j - a_j| \geqslant \tfrac{1}{2}\varepsilon\right). \qquad (8.65)$$

This result also holds for n infinite.

Proof. Write X'_i for a variate having the distribution of X_i, and $W_i = X_i - X'_i$. Then put

$$U_i = 0, \quad \text{if } X_i - \text{med}(X_i) < \varepsilon,$$
$$= 1, \quad \text{if } X_i - \text{med}(X_i) \geqslant \varepsilon.$$
$$V_i = 0, \quad \text{if } X'_i - \text{med}(X_i) > 0,$$
$$= 1, \quad \text{if } X'_i - \text{med}(X_i) \leqslant 0.$$

Then when $U_i V_i = 1$ we certainly have $W_i \geqslant \varepsilon$. From the definition of V_i we have

$$E(V_i) = \text{pr}\{X'_i - \text{med}(X_i) \leqslant 0\} \geqslant \tfrac{1}{2}.$$

Then

$$E\left\{1 - \prod_i (1 - U_i V_i)\right\} \geqslant \tfrac{1}{2} E\left\{1 - \prod_i (1 - U_i)\right\}. \tag{8.66}$$

To prove this consider the series

$$E\{U_1 V_1 + U_2 V_2 (1 - U_1 V_1) + U_3 V_3 (1 - U_2 V_2)(1 - U_1 V_1) + \ldots\} \tag{8.67}$$

$$= E[1 + (1 - U_1 V_1)\{-1 + U_2 V_2 + U_3 V_3 (1 - U_2 V_2) + \ldots\}]$$

$$= E[1 + (1 - U_1 V_1)(1 - U_2 V_2)\{-1 + U_3 V_3 + U_4 V_4 (1 - U_3 V_3) + \ldots\}]$$

$$= E\left\{1 - \prod_i (1 - U_i V_i)\right\},$$

which is the left-hand side of (8.66). On the other hand (8.67) is not less than

$$E\{U_1 V_1 + U_2 V_2 (1 - U_1) + U_3 V_3 (1 - U_2)(1 - U_1) + \ldots\}, \tag{8.68}$$

because $1 - U_i V_i \geqslant 1 - U_i$.

Moreover, since the V_i are distributed independently of the U_i, (8.68) is not less than

$$\tfrac{1}{2} E\{U_1 + U_2 (1 - U_1) + U_3 (1 - U_1)(1 - U_2) + \ldots\} = \tfrac{1}{2} E\left\{1 - \prod_i (1 - U_i)\right\},$$

so that (8.66) is proved.

Taking n infinite, if necessary, in (8.63), we have

$$\tfrac{1}{2} \text{pr}\left[\max_j \{X_j - \text{med}(X_j)\} \geqslant \varepsilon\right] = \tfrac{1}{2} \text{pr}[\text{at least one of } \{X_j - \text{med}(X_j)\} \geqslant \varepsilon]$$

$$= \tfrac{1}{2} - \tfrac{1}{2} \text{pr}[\text{all } \{X_j - \text{med}(X_j)\} < \varepsilon]$$

$$= \tfrac{1}{2} E\left\{1 - \prod_i (1 - U_i)\right\}.$$

Similarly

$$pr\left(\max_j W_j \geq \varepsilon\right) = 1 - pr(\text{all } W_j < \varepsilon)$$

$$\geq 1 - \prod_j \{1 - pr(W_j \geq \varepsilon)\}$$

$$\geq 1 - \prod_j [1 - pr\{X_j - \text{med}(X_j) \geq \varepsilon, X_j' - \text{med}(X_j) \leq 0\}]$$

$$\geq 1 - \prod_j E(1 - U_j V_j)$$

and thus (8.63) follows.

(8.64) follows by adding (8.63) to the similar inequality with the signs changed. (8.65) is obtained by observing that

$$pr\left(\max_j |W_j| \geq \varepsilon\right) = pr\left(\max_j |X_j - a_j - X_j' + a| \geq \varepsilon\right)$$

$$= 1 - pr(\text{all } |X_j - a - X_j' + a| < \varepsilon)$$

$$\leq 1 - pr(\text{all } |X_j - a_j| < \tfrac{1}{2}\varepsilon)pr(\text{all } |X_j' - a_j| < \tfrac{1}{2}\varepsilon)$$

$$\leq 1 - \{pr(\text{all } |X_j - a_j| < \tfrac{1}{2}\varepsilon)\}^2$$

$$\leq 2\{1 - pr(\text{all } |X_j - a_j| < \tfrac{1}{2}\varepsilon)\}, \tag{8.69}$$

because $1 - x^2 \leq 2(1-x)$ when $0 \leq x \leq 1$. Finally, the last expression in (8.69) is equal to (8.65).

From these inequalities we deduce the following theorem due to P. Lévy.

THEOREM 8.29. *If $\{X_n\}$ is a sequence of independent random variables and*

$$S_n = \sum_1^n X_i,$$

we have, for all $\varepsilon > 0$,

$$pr\left[\max_{1 \leq k \leq n} \{S_k - \text{med}(S_k - S_n)\} \geq \varepsilon\right] \leq 2\,pr(S_n \geq \varepsilon), \tag{8.70}$$

and the similar inequality with moduli.

Proof. The result with moduli follows by adding (8.70) to the similar inequality with signs reversed.

To prove (8.70) we have

$$2\,pr(S_n \geq \varepsilon) \geq 2 \sum_{k=1}^n pr\{S_j - \text{med}(S_j - S_n) < \varepsilon \text{ for } j = 1, ..., k-1,$$

$$\text{and } S_k - \text{med}(S_k - S_n) \geq \varepsilon\}pr\{S_n - S_k - \text{med}(S_n - S_k) \geq 0\}$$

$$\geq pr\left[\max_{1 \leq k \leq n} \{S_k - \text{med}(S_k - S_n)\} \geq \varepsilon\right].$$

As we have remarked before, the independence of the X_i gives the sequence of random variables S_n a special structure and by using the above inequalities we can now prove the following remarkable theorem which provides a converse to Theorem 8.5 for sums of independent variates.

THEOREM 8.30. *If the sequence $\{X_n\}$ is independent and*

$$S_n = \sum_1^n X_k,$$

the convergence in probability of the sequence $\{S_n\}$ implies that it converges almost certainly, and the two modes of convergence are therefore equivalent for sums of independent variates.

Proof. Let S be the random variable to which S_n converges in probability. Then by Theorem 8.7 there exists a subsequence S_{n_k} which converges almost certainly to S, and does so in such a way that

$$\sum_k \mathrm{pr}(|S_{n_k} - S_{n_{k+1}}| \geqslant 2^{-k}) \tag{8.71}$$

is convergent. $|S_{n_k} - S|$ thus converges almost certainly to zero.

Let n be an integer satisfying $n_k < n \leqslant n_{k+1}$. Applying Theorem 8.29 with absolute values to the sum of X_i from S_{n_k} to $S_{n_{k+1}}$ we have

$$\mathrm{pr}\left\{ \max_n |S_n - S_{n_k} - \mathrm{med}(S_n - S_{n_{k+1}})| \geqslant 2^{-k} \right\} \leqslant 2\,\mathrm{pr}(|S_{n_k} - S_{n_{k+1}}| \geqslant 2^{-k}),$$

and hence

$$\sum_k \mathrm{pr}\left\{ \max_n |S_n - S_{n_k} - \mathrm{med}(S_n - S_{n_{k+1}})| \geqslant 2^{-k} \right\}$$

is convergent, where in each case we have $n_k < n \leqslant n_{k+1}$. Using this we have

$$\max_{n_k < n \leqslant n_{k+1}} |S_n - S - \mathrm{med}(S_n - S_{n_{k+1}})|$$

$$\leqslant \max_{n_k < n \leqslant n_{k+1}} |S_n - S_{n_k} - \mathrm{med}(S_n - S_{n_{k+1}})| + |S - S_{n_k}|.$$

These two expressions on the right both converge almost certainly to zero so that $S_n - \mathrm{med}(S_n - S_{n_{k+1}})$ converges almost certainly to S. But S_n converges in probability to S, and $\mathrm{med}(S_n - S_{n_{k+1}})$, which is not a random variable, must converge to zero in the ordinary sense. Thus finally S_n converges to S almost certainly.

8.13. The convergence of distributions of sums

We have already seen in 8.3 that the convergence of the distributions of a sequence of random variables, $\{X_n\}$, in no way implies their convergence in probability or almost certainly. It is a remarkable fact that in the particular case of sums, S_n, of independent random variables this conclusion does follow.

THEOREM 8.31. *If $\{X_n\}$ is a sequence of independent random variables,*

$$S_n = \sum_{i=1}^{n} X_i,$$

and the distribution of S_n converges to a proper distribution, the sequence $\{S_n\}$ converges in probability and almost certainly.

Proof. We first prove this when the distributions of the X_i are symmetric. Let $\phi_i(t)$ be the characteristic function of X_i and $\phi(t)$ that of the limiting distribution of S_n. Then

$$\prod_{1}^{n} \phi_i(t) \to \phi(t), \tag{8.72}$$

for all real t, and $\phi_i(t)$, $\phi(t)$ are real. Let $(-T, T)$ be an interval in which $\phi(t) > K > 0$, for some K. Write $F_i(x)$ for the distribution of X_i. Then

$$\prod_{1}^{\infty} \phi_i(t) = \prod_{1}^{\infty} \{1 - (1 - \phi_i(t))\} \tag{8.73}$$

is a convergent infinite product converging to a non-zero continuous function in $(-T, T)$.

Hence it converges uniformly, and so also does

$$\sum_{1}^{\infty} |1 - \phi_i(t)|. \tag{8.74}$$

From Theorem 6.26, we have

$$\mathrm{pr}(|X_i| \geq T^{-1}) \leq 7T^{-1} \int_{0}^{T} \{1 - \phi_i(t)\} \, dt, \tag{8.75}$$

and

$$\int_{-T^{-1}}^{T^{-1}} x^2 \, dF_i(x) \leq 3T^{-2}\{(1 - \phi_i(T)\}. \tag{8.76}$$

It follows that there exists a $T > 0$ such that

$$\sum \mathrm{pr}(|X_i| \geq T^{-1})$$

is convergent, and if X_i' is X_i truncated at T^{-1},

$$\sum \mathrm{var}(X_i')$$

is convergent. From Theorem 8.26 it follows that $\sum X_n$ converges almost certainly because $E(X_i) = 0$.

Now consider the more general case in which the distributions of the X_i are not symmetric. To each X_i we associate an independent random variable X_i' with the same distribution, and a symmetrized random variable $W_i = X_i - X_i'$. Then the distribution of

$$T_n = \sum_{1}^{n} W_i$$

also converges to a proper distribution, and $\sum W_i$ must converge almost certainly. By Theorem 8.27 we have for every $x > 0$

$$\text{pr}\{|X_i - \text{med}(X_i)| \geqslant x\} \leqslant 2\,\text{pr}(|W_i| \geqslant x).$$

Truncate $X_i - \text{med}(X_i)$ at $\pm x$, and call the truncated random variable Z_i, so that

$$Z_i = X_i - \text{med}(X_i), \quad \text{if } |X_i - \text{med}(X_i)| < x, \qquad (8.77)$$
$$= 0, \qquad\qquad \text{otherwise.}$$

Similarly let U_i be W_i truncated at x so that

$$U_i = W_i, \quad \text{if } |W_i| < x,$$
$$= 0, \quad \text{otherwise.} \qquad (8.78)$$

Then using (8.62) and integrating by parts, we find

$$E(Z_i^2) \leqslant 2E(U_i^2) + 2x^2\,\text{pr}(|W_i| \geqslant x).$$

Now applying the previous argument to the series W_i we see that there exists an $x > 0$ such that

$$\sum \text{pr}(|W_i| \geqslant x),$$

and

$$\sum \text{var}(U_i),$$

are both convergent, and hence

$$\sum \text{pr}\{|X_i - \text{med}(X_i)| \geqslant x\},$$
$$\sum E(Z_i^2),$$

and

$$\sum E(Z_i)$$

are convergent, so that $\sum X_i$ converges almost certainly by Theorem 8.26.

8.14. Application to stationary processes

THEOREM 8.32. *Consider a sequence of independent random variables, $\{X_n\}$, where n now ranges over the values, ..., $-2, -1, 0, 1, 2,...$. We suppose that all the X_n have the same distribution with a zero mean and finite variance. Then the series*

$$Y_n = \sum_0^\infty a_k X_{n-k} \qquad (8.79)$$

converges almost certainly if

$$\sum_0^\infty a_k^2$$

is convergent.

Proof. The sequence of random variables $\{a_k X_{n-k}\}$ are such that

$$\sum_{k=0}^{\infty} E(a_k^2 X_{n-k}^2) < \infty,$$

and the infinite series (8.79) converges in mean square and therefore almost certainly from Theorem 8.24. Y_n is thus a well-defined random variable with a zero mean, and a variance equal to $\sum_0^{\infty} a_n^2 \operatorname{var}(X_n)$.

The sequence of random variables $\{Y_n\}$ is said to form a stationary random process. In the study of such processes a key part is played by serial correlations, the correlations between Y_n and Y_{n+s} ($s = \pm 1, \pm 2,...$). It is more convenient, however, to deal with serial covariances, $c_s = E(Y_n Y_{n+s})$, since if these are known for all integral s the correlations can be found by dividing by $E(Y_n^2)$, which is the serial covariance of order zero.

Suppose the conditions above are satisfied, and write σ^2 for $E(X_n^2)$. Then

$$c_s = E(Y_n Y_{n-s}) = E(Y_n Y_{n+s}) = \sum_{k=0}^{\infty} a_k a_{k+s} E(X_n^2)$$

$$= \sigma^2 \sum_{k=0}^{\infty} a_k a_{k+s}. \tag{8.80}$$

If $\sum_0^{\infty} a_n$ is absolutely convergent so is $\sum_0^{\infty} a_n^2$ and the condition used in Theorem 8.32 above is weaker than the assumption that $\sum_0^{\infty} |a_n|$ is finite. It is now convenient to use this further restriction.

THEOREM 8.33. *If $\sum_0^{\infty} |a_n|$ is convergent, $\sum_0^{\infty} |c_s|$ is convergent.*

Proof. We have

$$|c_s| = \left| \sigma^2 \sum_{k=0}^{\infty} a_k a_{k+s} \right| \leqslant \lim_{N \to \infty} \sigma^2 \sum_{k=0}^{N} |a_k| |a_{k+s}|.$$

Then

$$\sum_{-n}^{n} |c_s| \leqslant \lim_{N \to \infty} \sigma^2 \sum_{k=0}^{N} \sum_{s=-n}^{n} |a_k| |a_{k+s}|$$

$$\leqslant \lim_{N \to \infty} \sigma^2 \sum_{k=0}^{N} |a_k| \sum_{-\infty}^{\infty} |a_s|$$

and the right-hand side is bounded.

THEOREM 8.34. *If a sequence of identically distributed but not necessarily independent random variables $\{X_n\}$ with zero means and finite variance σ^2 is such that*

$$c_s = E(X_n X_{n+s})$$

is independent of n, and $\sum_0^{\infty} c_s$ is absolutely convergent, then

$$\sum_0^{\infty} a_n X_n$$

converges in mean square if $\sum_0^{\infty} |a_n|$ converges.

Proof. We have, for $n \leqslant m$,

$$E\left(\sum_{k=n}^{m} a_k X_k\right)^2 = \sum_{k=n}^{m} \sum_{l=n}^{m} a_k a_l c_{k-l}$$

$$\leqslant \frac{1}{2} \sum_{k=n}^{m} \sum_{l=n}^{m} c_{k-l}(a_k^2 + a_l^2)$$

$$\leqslant \sum_{-\infty}^{\infty} |c_s| \sum_{k=n}^{m} a_k^2,$$

and this can be made arbitrarily small by making n large.

Thus $\{Y_n\} = (\sum_0^\infty a_k X_{n+k})$ is a well-defined random variable and its serial covariances, which we write as c_k', are given by

$$c_k' = E\left\{\left(\sum_0^\infty a_s X_{n+s}\right)\left(\sum_0^\infty a_r X_{n+k+r}\right)\right\}$$

$$= \sum_{s=0}^{\infty} \sum_{r=0}^{\infty} a_s a_r c_{s-k-r}. \tag{8.81}$$

In these circumstances $\sum_0^\infty c_k'$ is itself absolutely convergent, for if

$$\sum_0^\infty |c_k| < K < \infty,$$

$$\sum |c_k'| \leqslant \sum_{s=0}^{\infty} \sum_{r=0}^{\infty} |a_s| |a_r| \sum_0^\infty |c_k|$$

$$\leqslant K\left(\sum_{s=0}^{\infty} |a_s|\right)^2 < \infty. \tag{8.82}$$

It follows that any random process $\{X_n\}$ with zero means and equal finite standard deviations such that

$$c_k = E(X_n X_{n+k})$$

is independent of n, and $\sum |c_k|$ convergent, can be operated on by a 'moving average'

$$\sum_{s=0}^{\infty} a_s X_{n-s} \tag{8.83}$$

to give a new random process satisfying the same conditions.

The effect of such a moving average on the serial covariances is best represented by using generating functions. If $\{c_k\}$ are the serial covariances of such a process we write

$$C(z) = \sum_{k=-\infty}^{\infty} c_k z^k, \tag{8.84}$$

and call it the serial covariance generating function. Multiplying equation (8.81) by z^k and summing we see that the effect of taking a moving average such as (8.83) is to produce a new random process with serial covariance generating function

$$C'(z) = \left(\sum_0^\infty a_n z^n \right) \left(\sum_0^\infty a_n z^{-n} \right) C(z). \tag{8.85}$$

Finally, we should remark that Theorem 8.32 can be strengthened in the following way.

THEOREM 8.35. *If* $\{X_n\}$ *($n = 0, \pm 1, \pm 2,...$) is a sequence of independent random variables with zero means, and the same distribution, the series*

$$Y_n = \sum_0^\infty a_k X_{n-k}$$

converges almost certainly for each n if

$$\sum_0^\infty a_k$$

converges absolutely.

This result is, in one respect, more general than Theorem 8.32 since it is not assumed that $E(X_i^2)$ is finite.

Proof. Since the mean $E(X_n)$ exists, $E(|X_n|)$ is finite. Then for $m > n$, and any $\varepsilon > 0$,

$$\mathrm{pr}\left(\left| \sum_n^m a_k X_{n-k} \right| \geqslant \varepsilon \right) \leqslant \varepsilon^{-1} E \left| \sum_n^m a_k X_{n-k} \right|$$

$$\leqslant \varepsilon^{-1} \sum_n^m |a_k| E(|X_n|). \tag{8.86}$$

For given ε this can be made as small as we like, independently of $m > n$, by making n sufficiently large. The series for Y_n therefore converges in probability and by Theorem 8.30 it also converges almost certainly, since the random variables $\{a_k X_{n-k}\}$ are independent.

8.15. The stability of sums and the strong law of large numbers

As before we consider partial sums, S_n, of a sequence of random variables. If S_n converges to a proper random variable and $\{b_n\}$ is any sequence of positive numbers increasing to infinity, the distribution of $S_n b_n^{-1}$ collapses into the singular distribution with all its probability concentrated at zero. When S_n does not converge in any of the senses used above it is, however, often still possible to prove that when suitably rescaled and relocated it converges to zero.

We therefore say that a sequence of partial sums

$$S_n = \sum_1^n X_k$$

of random variables $\{X_n\}$ is 'stable in probability', or 'almost certainly stable' if, given two sequences of real numbers, $\{a_n\}$ and $\{b_n\}$, such that b_n tend to infinity,

$$S_n b_n^{-1} - a_n$$

converges to zero in probability, or almost certainly, respectively. Such sequences always exist, and the interest is in their nature.

The typical example of this situation is that of repeated binomial trials with probability p. We have already shown in § 1.9 that if $X_i = 1, 0$ according to success or failure at the ith trial, $n^{-1}\sum X_i$ converges in probability to p. In fact it also converges almost certainly to p. These are the simplest cases of the weak and strong laws of large numbers and form a pattern for the consideration of stability in general.

If the independent random variables X_n have finite variances the weak law is easily generalized and in fact, if b_n is any sequence of positive numbers such that

$$b_n \left\{ \sum_1^n \mathrm{var}(X_i) \right\}^{-\frac{1}{2}}$$

tends to infinity, then

$$b_n^{-1} \sum_1^n \{ X_i - E(X_i) \} \tag{8.87}$$

converges in probability to zero.

This follows at once by applying Tchebychev's inequality (§ 1.9). Note in particular that if the variances are equal, b_n can be taken equal to n^α, $\alpha > \frac{1}{2}$.

Our principal aim is to strengthen this theorem by dropping the assumption that the variances are finite and by considering almost certain convergence. In fact we prove the following result which is known as Kolmogorov's Strong Law of Large Numbers.

THEOREM 8.36. *Let $\{X_n\}$ be a sequence of independent random variables with the same distribution. Then*

$$T_n = n^{-1} \sum_1^n X_i$$

converges almost certainly to a finite constant m, if and only if $E(X_i)$ exists (i.e. $E|X_i| < \infty$), and then $m = E(X_i)$.

Proof. Suppose first that T_n converges almost certainly to m where m is finite. We introduce indicator functions U_n such that

$$U_n = 1, \quad \text{if } |X| \geq n,$$
$$\quad = 0, \quad \text{if } |X| < n, \tag{8.88}$$

where X has the distribution of the X_i. Then

$$\mathrm{pr}(n \leqslant |X| < n+1) = E(U_n - U_{n+1}),$$

so that

$$E(|X|) < \sum_{n=0}^{\infty} (n+1)\mathrm{pr}(n \leqslant |X| < n+1)$$

$$< \sum_{n=0}^{\infty} (n+1)E(U_n - U_{n+1})$$

$$< 1 + \sum_{1}^{\infty} E(U_n). \tag{8.89}$$

Similarly

$$E(|X|) \geqslant \sum_{1}^{\infty} E(U_n). \tag{8.90}$$

Now

$$n^{-1}X_n = n^{-1}\sum_{1}^{n} X_i - n^{-1}\sum_{1}^{n-1} X_i$$

$$= T_n - \left(\frac{n-1}{n}\right)T_{n-1}.$$

Since T_n converges almost certainly to m, $n^{-1}X_n$ must converge almost certainly to zero, and the series

$$\sum_{1}^{\infty} E(U_n) = \sum_{n=1}^{\infty} \mathrm{pr}(|X_n| \geqslant n) \tag{8.91}$$

is convergent. From (8.89) this implies that $E(|X|)$ is finite.

The fact that $m = E(|X|)$ will follow if we now show that if $E(|X|)$ is finite, T will converge almost certainly to $E(X)$.

Define Y_n to be the truncated random variable,

$$Y_n = X_n, \quad \text{if } |X_n| < n,$$
$$= 0, \quad \text{if } |X_n| \geqslant n.$$

Since

$$\sum \mathrm{pr}(|X_n| \geqslant n) = \sum_{n} E(U_n) \leqslant E|X|, \tag{8.92}$$

by (8.90),

$$\sum \mathrm{pr}(X_n \neq Y_n)$$

is finite, and it is sufficient to show that

$$T'_n = n^{-1}\sum_{1}^{n} Y_i \tag{8.93}$$

converges almost certainly to $E(X)$. Let $F(x)$ be the distribution of X. The integral

$$E(|X|) = \int_{-\infty}^{\infty} |x|\, dF(x)$$

exists and therefore by dominated convergence,

$$E(Y_n) = \int_{-n}^{n} x \, dF(x)$$

converges to $E(X)$. Then

$$E(T_n') = n^{-1} \sum_{1}^{n} E(Y_i),$$

which is the Cesàro mean of the sequence $\{E(Y_i)\}$, must also converge to $E(X)$ by the standard theorem by which ordinary convergence implies Cesàro convergence. Thus it is sufficient to show that

$$T_n' - E(T_n') \tag{8.94}$$

converges to zero almost certainly. We now have

$$\sum_{n=1}^{\infty} n^{-2} \operatorname{var}(Y_n) \leqslant \sum_{n=1}^{\infty} n^{-2} E(Y_n^2)$$

$$\leqslant \sum_{n=1}^{\infty} n^{-2} E\{X_n^2(1 - U_n)\}$$

$$\leqslant \sum_{n=1}^{\infty} n^{-2} E[X_n^2\{(U_0 - U_1) + (U_1 - U_2) + \ldots + (U_{n-1} - U_n)\}]$$

$$\leqslant E \sum_{n=1}^{\infty} n^{-2} \sum_{k=1}^{n} k^2 (U_{k-1} - U_k)$$

$$\leqslant E \sum_{k=1}^{\infty} k^2 (U_{k-1} - U_k) \sum_{n=k}^{\infty} \frac{1}{n^2}$$

$$\leqslant E \sum_{k=1}^{\infty} (U_{k-1} - U_k)\left(1 + k^2 \int_{k}^{\infty} x^{-2} \, dx\right)$$

$$\leqslant E \sum_{k=1}^{\infty} (U_{k-1} - U_k)(2 + k - 1)$$

$$\leqslant 2 + E \sum_{k=1}^{\infty} (k - 1)(U_{k-1} - U_k)$$

$$\leqslant 2 + E \sum_{k=1}^{\infty} U_k$$

$$\leqslant 2 + E(|X|). \tag{8.95}$$

Write

$$S_n' = \sum_{j=1}^{n} \frac{Y_j - E(Y_j)}{j}. \tag{8.96}$$

Then since

$$\sum_{n=1}^{\infty} n^{-2} \operatorname{var}(Y_n)$$

is bounded, S'_n converges almost certainly. Putting $S'_0 = 0$,

$$T'_n - E(T'_n) = n^{-1} \sum_{j=1}^{n} j(S'_j - S'_{j-1})$$

$$= n^{-1}(-S'_0 - S'_1 - \ldots - S'_{n-1} + nS'_n)$$

$$= S'_n - n^{-1} \sum_{j=1}^{n-1} S'_j. \tag{8.97}$$

Since S'_n converges almost certainly to a random variable with a proper distribution, we apply again the Cesàro limit theorem to conclude that

$$n^{-1} \sum_{j=1}^{n-1} S'_j = \frac{(n-1)}{n}(n-1)^{-1} \sum_{j=1}^{n-1} S'_j$$

converges almost certainly to the same random variable, and hence $T'_n - E(T'_n)$ converges almost certainly to zero. Thus T_n converges almost certainly to $E(X_i)$ which proves the second part of the theorem, and shows that in the first part m must equal $E(|X_i|)$.

This theorem is remarkable for the fact that the only assumption made about the distribution of the X_i is that its mean exists. Under some circumstances it is possible to weaken the restriction on independence. It is also possible to give a much simpler proof of the sufficiency part of the theorem by using characteristic functions.

Further more general theorems on stability will be found in Chapter V of Loève (1963).

8.16. The law of the iterated logarithm

Consider a sequence $\{X_n\}$ of independent random variables with zero means and the same distribution. Writing, as before,

$$S_n = \sum_{i=1}^{n} X_i,$$

we have shown that $n^{-1}S_n$ converges almost surely to zero. The graph of $n^{-1}S_n$ must thus tend to zero with probability one. A more detailed description of the way in which this happens can be given if we can find a decreasing sequence of positive constants b_n such that $|S_n| > b_n$ for more than a finite number of values of n with zero probability. Thus if we plot $y = n^{-1}b_n$ and $y = -n^{-1}b_n$ against n we obtain a 'funnel' which converges to the line $y = 0$ as n increases, and with probability one the graph of $n^{-1}S_n$ will lie inside this funnel except for a finite number of values of n. It is a remarkable fact that under fairly wide conditions a sequence b_n can be constructed in such a way that if $\delta > 0$ the sequence $(1+\delta)b_n$ always satisfies these conditions whilst if $\delta < 0$ an infinite number of the events

$$|S_n| > (1-\delta)b_n$$

will occur with probability one.

Clearly some restriction on the distributions of the X_i will be necessary. We suppose that their second moments are finite and that the X_i are bounded in the manner determined by (8.100) below but we do not assume that they all have the same distribution. We first prove Theorem 8.37 (due in this form to Kolmogorov 1929) in which the X_i are bounded in this way, and then extend the result to more general cases.

THEOREM 8.37 (Law of the Iterated Logarithm). *Let $\{X_n\}$ be a sequence of independent random variables with zero means. Write*

$$S_n = \sum_1^n X_n, \tag{8.98}$$

$$s_n^2 = \mathrm{var}(S_n)$$

$$= \sum_1^n \mathrm{var}(X_n). \tag{8.99}$$

Suppose that $s_n^2 \to \infty$ and $X_n = o\{s_n(\log\log s_n^2)^{-\frac{1}{2}}\}$, so that we can choose K arbitrarily small so that

$$|X_n| \leqslant K s_n (\log\log s_n^2)^{-\frac{1}{2}}, \tag{8.100}$$

for all sufficiently large n, where K is a positive constant and the logarithms are natural logarithms. Then the probability of an infinite number of the events

$$|S_n| > (1+\delta)s_n(2\log\log s_n^2)^{\frac{1}{2}} \tag{8.101}$$

is zero if $\delta > 0$, and unity if $\delta < 0$.

At first sight the restriction (8.100) appears to spoil the useful applicability of the theorem because many of the distributions to which we would like to apply it are distributions of unbounded random variables. However, as we shall see in Theorem 8.39, the result can be extended to many unbounded random variables by truncation. The proof of this theorem is long and complicated. We give it here as an example of the very careful analytical discussion which is necessary in general theorems of this type in the theory of probability. This proof is that given by Kolmogorov (1929) and similar expositions are given in Loève (1963) and Khintchine (1933).

For the first part of the theorem it is clearly sufficient to consider the probability that the inequality

$$S_n > (1+\delta)s_n t_n \tag{8.102}$$

is satisfied infinitely often, where $t_n = (2\log\log s_n^2)^{\frac{1}{2}}$ and $\delta > 0$.

We first prove that the probability that (8.102) occurs for an infinite number of values of n is zero. s_n^2 tends to infinity and by (8.100), $\mathrm{var}(X_n)s_n^{-2}$ tends to zero. Therefore given any constant $c > 1$, we can choose an increasing subsequence n_k such that s_{n_k} is asymptotically equal to c^k, and $s_{n_k}s_{n_{k-1}}^{-1}$ tends to c. In particular we take $c = 1 + \frac{1}{2}\delta$, so that s_{n_k} is asymptotically equal to

$$(1+\tfrac{1}{2}\delta)^k. \tag{8.103}$$

Write

$$T_k = \max_{1 \leqslant n \leqslant n_k} S_n, \tag{8.104}$$

and consider the probability that

$$T_k > (1+\delta_1)s_{n_k}t_{n_k}, \tag{8.105}$$

where

$$1 < 1+\delta_1 < \frac{1+\delta}{1+\frac{1}{2}\delta}. \tag{8.106}$$

Then

$$\mathrm{pr}\{S_n > (1+\delta)s_n t_n \text{ infinitely often}\}$$
$$\leqslant \mathrm{pr}\{T_k > (1+\delta)s_{n_{k-1}}t_{n_{k-1}} \text{ infinitely often}\}, \tag{8.107}$$

because $S_n > (1+\delta)s_n t_n$ for some n satisfying $n_{k-1} < n < n_k$ implies

$$T_k > (1+\delta)s_{n_{k-1}}t_{n_{k-1}}.$$

But by (8.103), and the fact that $t_{n_k}t_{n_{k-1}}^{-1} \to 1$,

$$(1+\delta)s_{n_{k-1}}t_{n_{k-1}}$$

is asymptotically equal to

$$\left(\frac{1+\delta}{1+\frac{1}{2}\delta}\right)s_{n_k}t_{n_k}$$

and this is greater than

$$(1+\delta_1)s_{n_k}t_{n_k}. \tag{8.108}$$

Thus the right-hand side of the inequality (8.107) is not greater than

$$\mathrm{pr}\{T_k > (1+\delta_1)s_{n_k}t_{n_k} \text{ infinitely often}\}.$$

Using the Borel–Cantelli lemma (Theorem 8.17) we see that it is sufficient to prove that

$$\sum \mathrm{pr}\{T_k > (1+\delta_1)s_{n_k}t_{n_k}\} \tag{8.109}$$

is a convergent series.

From Theorem 8.29 we have

$$\mathrm{pr}\left[\max_{1 \leqslant k \leqslant n} \{S_k - \mathrm{med}(S_k - S_n)\} \geqslant \varepsilon\right] \leqslant 2\,\mathrm{pr}(S_n \geqslant \varepsilon), \tag{8.110}$$

for every $\varepsilon > 0$. By Tchebychev's inequality, and remembering that the mean

25

of S_n is zero,

$$\text{med}(S_k - S_n) \leqslant \sqrt{\{2 \, \text{var}(S_k - S_n)\}}$$
$$\leqslant \sqrt{(2 \, \text{var} \, S_n)}. \tag{8.111}$$

Replace ε in (8.110) by $\varepsilon - \sqrt{(2 \, \text{var} \, S_n)}$ and we have, for $\varepsilon > \sqrt{(2 \, \text{var} \, S_n)}$,

$$\text{pr}\left(\max_{1 \leqslant k \leqslant n} S_k \geqslant \varepsilon\right) \leqslant \text{pr}\left[\max_{1 \leqslant k \leqslant n} \{S_k - \text{med}(S_k - S_n)\} \geqslant \varepsilon - \sqrt{(2 \, \text{var} \, S_n)}\right]$$
$$\leqslant 2 \, \text{pr}\{S_n \geqslant \varepsilon - \sqrt{(2 \, \text{var} \, S_n)}\}. \tag{8.112}$$

Put

$$\varepsilon = (1 + \delta_1)s_{n_k} t_{n_k},$$

and then

$$\text{pr}\{T_k > (1 + \delta_1)s_{n_k} t_{n_k}\} \leqslant 2 \, \text{pr}\left\{S_{n_k} > \left(1 + \delta_1 - \frac{\sqrt{2}}{t_{n_k}}\right) s_{n_k} t_{n_k}\right\}.$$

We can choose k sufficiently large so that

$$1 + \delta_1 - \frac{\sqrt{2}}{t_{n_k}} > 1 + \delta_2,$$

where $\delta_2 > 0$. Thus it is sufficient to show that

$$\sum \text{pr}\{S_{n_k} > (1 + \delta_2)s_{n_k} t_{n_k}\} \tag{8.113}$$

is convergent where $\delta_2 > 0$. To do this we use the following lemma.

LEMMA 8.1 (Kolmogorov 1929). *Let the independent random variables* $\{X_i\}$ *have zero means and variances* $\sigma_i^2 = E(X_i^2)$, *and be such that* $|X_i| \leqslant d < \infty$. *Let* t *be a real number such that* $0 < td \leqslant s_n$, *where* $s_n^2 = \text{var}(S_n) = \sum \sigma_i^2$. *Then for* $S_n = \sum_1^n X_i$, *and any* $\varepsilon > 0$,

$$\text{pr}(S_n > \varepsilon s_n) < \exp\{-t\varepsilon + \tfrac{1}{2}t^2(1 + \tfrac{1}{2}t \, d s_n^{-1})\}. \tag{8.114}$$

Proof. Let X be a random variable with mean zero and variance σ_i^2. Then

$$E(X^n) = E(X^2 X^{n-2}) \leqslant d^{n-2}\sigma_i^2 \quad (n > 2).$$

The series

$$E(e^{tX}) = 1 + \tfrac{1}{2}t^2 E(X^2) + \frac{1}{3!} t^3 E(X^3) + \ldots \tag{8.115}$$

is absolutely convergent. Since $0 < td \leqslant 1$ it does not exceed

$$1 + \tfrac{1}{2}t^2\sigma_i^2\left\{1 + \tfrac{1}{3}td + \frac{1}{3.4}t^2 d^2 + \ldots\right\} = 1 + \tfrac{1}{2}t^2\sigma_i^2\left[1 + \tfrac{1}{3}td\left\{1 + \tfrac{1}{4} + \frac{1}{4.5} + \ldots\right\}\right]$$
$$< 1 + \tfrac{1}{2}t^2\sigma_i^2(1 + \tfrac{1}{2}td)$$
$$< \exp\{\tfrac{1}{2}t^2\sigma_i^2(1 + \tfrac{1}{2}td)\}.$$

Now if $s_n^2 = \mathrm{var}(S_n)$, $t > 0$,

$$\mathrm{pr}\{S_n > \varepsilon s_n\} \leqslant e^{-t\varepsilon} E\{\exp(t S_n s_n^{-1})\}$$

$$\leqslant e^{-t\varepsilon} \prod_{i=1}^{n} E\{\exp(t X_n s_n^{-1})\}.$$

We now use the above inequality for $E(e^{tX_i})$ with t replaced by ts_n^{-1}. Then so long as

$$t s_n^{-1} d \leqslant 1$$

we have

$$\mathrm{pr}(S_n > s_n) < \exp\left\{-t\varepsilon + \sum_{i=1}^{n} \tfrac{1}{2} t^2 s_n^{-2} \sigma_i^2 (1 + \tfrac{1}{3} t d s_n^{-1})\right\}$$

$$< \exp\{-t\varepsilon + \tfrac{1}{2} t^2 (1 + \tfrac{1}{3} t d s_n^{-1})\},$$

as required.

We now apply this inequality using

$$\varepsilon = (1 + \delta_2) t_n,$$

and, from (8.100),

$$d = K s_n (\log\log s_n^2)^{-\frac{1}{2}},$$

where K is small. Putting $t = \varepsilon$ we have

$$\mathrm{pr}(S_n > s_n) \leqslant \exp\{-\tfrac{1}{2}\varepsilon^2(1 - \varepsilon d s_n^{-1})\}.$$

Using this for $n = n_k$, and noticing that $\varepsilon d s_{n_k}^{-1}$ can be made as small as we please by choosing K small and n_k large, we have, for δ_3 arbitrarily small and positive,

$$\mathrm{pr}(S_{n_k} > \varepsilon s_{n_k}) \leqslant \exp-\{\tfrac{1}{2}(1-\delta_3)\varepsilon^2\}$$

$$\leqslant \exp-\tfrac{1}{2}(1-\delta_3)(1+\delta_2)^2(2\log\log s_{n_k}^2).$$

The right-hand side is asymptotically equal to

$$\{2k\log(1+\tfrac{1}{2}\delta)\}^{-(1-\delta_3)(1-\delta_2)^2}. \tag{8.116}$$

Since δ_3 can be made arbitrarily small, we can make

$$(1-\delta_3)(1+\delta_2)^2 > 1 + \delta_4,$$

where $\delta_4 > 0$. Then (8.116) is the term of a convergent series, so that the event (8.102) occurs infinitely often only with probability zero. This proves the first part of the theorem.

We now consider the sequence of events

$$S_n > (1-\delta)s_n t_n, \tag{8.117}$$

and show that they occur infinitely often. It is clearly sufficient to show that this happens in the subsequence n_k.

Write

$$u_k^2 = s_{n_k}^2 - s_{n_{k-1}}^2.$$

Then since $s_{n_k}^2$ is asymptotically equal to c^{2k}, u_k^2 is asymptotically equal to

$$c^{2k}(1-c^{-2}).$$

Furthermore,

$$v_k = (2\log\log u_k^2)^{\frac{1}{2}}$$

is then asymptotically equal to $(2\log\log s_{n_k}^2)^{\frac{1}{2}} = t_{n_k}$.

Consider the event

$$S_{n_k} - S_{n_{k-1}} > (1-\delta)u_k v_k, \tag{8.118}$$

where $\delta > 0$. We first show this event occurs infinitely often with probability one. To do this we have to use another lemma also due to Kolmogorov (1929) which provides a lower bound instead of the upper bound in Lemma 8.1. For convenience we write B for s_n^2.

LEMMA 8.2. *Let the X_i be independent random variables with zero means and variances σ_i^2, and such that*

$$|X_i| \leqslant d < \infty.$$

If

$$B = \sum\sigma_i^2, \quad w = \frac{xd}{B} < \frac{1}{256}, \quad \lambda = \frac{x^2}{B} > 512,$$

then

$$\mathrm{pr}(S_n > x) > \exp\left\{-\frac{1}{2}\frac{x^2}{B}(1+3\varepsilon)\right\}, \tag{8.119}$$

where

$$\varepsilon = \max\left\{32\left(\frac{\log\lambda}{\lambda}\right)^{\frac{1}{2}}, \ 64w^{\frac{1}{2}}\right\}.$$

Proof. Put $\delta = \dfrac{\varepsilon}{8}$, so that

$$\delta^2 = \max(64w, 16\log\lambda/\lambda).$$

Then since

$$w < \tfrac{1}{256},$$

and

$$16\log\lambda/\lambda < 9\log 2/2^5 = 0{\cdot}1949 < \tfrac{1}{4},$$

we have

$$\delta^2 < \tfrac{1}{4}, \quad \delta < \tfrac{1}{2}, \quad \delta > 2\delta^2.$$

Write

$$a = \frac{x}{B(1-\delta)}.$$

Then, since $\delta < \tfrac{1}{2}$,

$$\frac{x}{B} < a < \frac{2x}{B}$$

and

$$ad < 2w < \tfrac{1}{128}.$$

Moreover,

$$a^2 B = \frac{x^2}{B(1-\delta)^2} > \lambda > 512.$$

By differentiation it is easy to show that if $u > 0$,

$$1 + u > e^{u(1-u)}.$$

From this it follows, using the same kind of argument as before, that

$$E(e^{aX_i}) \geq 1 + \frac{a^2 \sigma_i^2}{2} \left(1 - \frac{ad}{3} - \frac{a^2 d^2}{12} - \cdots\right)$$

$$\geq 1 + \frac{a^2 \sigma_i^2}{2}(1 - \tfrac{1}{2}ad)$$

$$\geq \exp \tfrac{1}{2}a^2 \sigma_i^2 (1 - \tfrac{1}{2}ad - \tfrac{1}{2}a^2 \sigma^2)$$

$$\geq \exp \tfrac{1}{2}a^2 \sigma_i^2 (1 - ad),$$

where we have used the fact that $ad < 1$. From this it follows that

$$E(e^{aS_n}) \geq \exp\left\{\frac{\Sigma a^2 \sigma_i^2}{2}(1 - ad)\right\}$$

$$\geq \exp\left\{\frac{a^2 B}{2}(1 - \tfrac{1}{4}\delta^2)\right\},$$

since $ad < 2w < \tfrac{1}{4}\delta^2$. Now write

$$W(x) = \mathrm{pr}(S_n > x).$$

Then

$$E(e^{aS_n}) = -\int_{-\infty}^{\infty} e^{ax}\, dW(x)$$

$$= a\int_{-\infty}^{\infty} e^{ax} W(x)\, dx,$$

which is a convergent integral for real a because S_n is bounded. We split this into the sum of five integrals in the following manner.

$$E(e^{aS_n}) = a\left(\int_{-\infty}^{0} + \int_{0}^{aB(1-\delta)} + \int_{aB(1-\delta)}^{aB(1+\delta)} + \int_{aB(1+\delta)}^{8aB} + \int_{8aB}^{\infty}\right)$$

$$= a(J_1 + J_2 + J_3 + J_4 + J_5), \quad \text{say.}$$

We must now estimate each of these separately.
  Since $W(x) \leq 1$, we have

$$aJ_1 \leq a\int_{-\infty}^{0} e^{ax}\, dx = 1.$$

We next show that for $8aB \leqslant x$,

$$W(x) < e^{-2ax}.$$

To show this consider again (8.114). This inequality holds so long as $tB^{-\frac{1}{2}}d \leqslant 1$. If $xd = \varepsilon B^{\frac{1}{2}}d \leqslant B$, this inequality will hold if we put $t = xB^{-\frac{1}{2}}$, and we then get

$$P(S_n > x) < \exp\left\{-\frac{x^2}{B} + \frac{x^2}{2B}\left(1 + \frac{xd}{2B}\right)\right\}$$

$$< \exp\left\{-\frac{x^2}{2B}\left(1 - \frac{xd}{2B}\right)\right\}$$

$$< \exp-\frac{x^2}{4B}.$$

Conversely if $xd = \varepsilon B^{\frac{1}{2}}d \geqslant B$, we put $t = B^{\frac{1}{2}}d^{-1}$, and obtain

$$P(S_n > x) < \exp\left\{-\frac{x}{d} + \frac{B}{2d^2}(1 + \tfrac{1}{2})\right\}$$

$$< \exp-\frac{x}{4d}.$$

Using these results, we have for $8aB \leqslant x \leqslant Bd^{-1}$,

$$W(x) < \exp-\frac{x^2}{4B}$$

$$< \exp-2ax,$$

since $x > 8aB$. On the other hand, if $x \geqslant Bd^{-1}$ we have

$$W(x) < \exp-\frac{x}{4d}$$

$$< \exp-2ax,$$

because $ad < \frac{1}{128} < \frac{1}{8}$. We therefore have

$$aJ_5 < a\int_{8aB}^{\infty} e^{-2ax}\, dx < 1.$$

We have already shown that

$$E(e^{aS_n}) \geqslant \exp\{\tfrac{1}{2}a^2B(1 - \tfrac{1}{4}\delta^2)\},$$

and using the facts that

$$\delta^2 < \tfrac{1}{4}, \quad a^2B > 512,$$

we have

$$E(e^{aS_n}) > 8,$$

so that

$$aJ_1 + aJ_5 < \tfrac{1}{4}E(e^{aS_n}).$$

Now consider J_2 and J_4. So long as $0 \leqslant x \leqslant 8aB$ we have

$$xd < 8adB < B,$$

because $ad < \frac{1}{128}$. However, we have already shown above that

$$\text{pr}(S_n > x) < \exp\left\{-\frac{x^2}{2B}\left(1-\frac{xd}{2B}\right)\right\},$$

so long as $xd < B$. Furthermore, we then also have

$$\frac{xd}{2B} \leqslant 4ad < \tfrac{1}{8}\delta^2.$$

Thus

$$\text{pr}(S_n > x) < \exp\left\{-\frac{x^2}{2B}(1-\tfrac{1}{8}\delta^2)\right\}$$

and

$$a(J_2+J_4) \leqslant \left(\int_0^{aB(1-\delta)} + \int_{aB(1+\delta)}^{8aB}\right)\exp\left\{ay-\frac{y^2}{2B}(1-\tfrac{1}{8}\delta^2)\right\}dy.$$

The expression

$$f(y) = ay-\frac{y^2}{2B}(1-\tfrac{1}{8}\delta^2),$$

is a quadratic which reaches a maximum at the point

$$y_0 = aB(1-\tfrac{1}{8}\delta^2)^{-1}.$$

We have

$$aB(1-\delta) < aB < y_0 < aB(1+\delta),$$

and therefore in the two ranges $\{0, aB(1-\delta)\}$, $\{aB(1+\delta), 8aB\}$, $f(y)$ is never greater than its value at $y = aB(1+\delta)$. Its value there is

$$\tfrac{1}{2}a^2B\{1-\delta^2+\tfrac{1}{8}\delta^2(1+\delta)^2\} \leqslant \tfrac{1}{2}a^2B(1-\tfrac{1}{2}\delta^2),$$

because $\delta < \tfrac{1}{2}$. We therefore have

$$a(J_2+J_4) < a\int_0^{8aB}\exp\{\tfrac{1}{2}a^2B(1-\tfrac{1}{2}\delta^2)\}\,dy$$

$$< 8a^2B\exp\{\tfrac{1}{2}a^2B(1-\tfrac{1}{2}\delta^2)\}.$$

Since $\lambda > 512$, we have, using the above inequalities,

$$\log(32\lambda) < \log(\lambda^2) = 2\log\lambda \leqslant \tfrac{1}{8}\lambda\delta^2.$$

Thus

$$\log 32u < \tfrac{1}{8}u\delta^2$$

when $u = \lambda$, and differentiating we readily verify that this remains true for $u > \lambda$. Since $a^2B > \lambda$, we have

$$\log 32a^2B < \tfrac{1}{8}a^2B\delta^2,$$

and hence

$$a(J_2+J_4) < \tfrac{1}{4}\exp\{\tfrac{1}{2}a^2B(1-\tfrac{1}{4}\delta^2)\}$$

$$< \tfrac{1}{4}E(e^{aS_n}).$$

Finally, we have

$$aJ_3 = E(e^{aS_n}) - a(J_1+J_2+J_4+J_5)$$

$$> \tfrac{1}{2}E(e^{aS_n})$$

$$> \tfrac{1}{2}\exp\{\tfrac{1}{2}a^2B(1-\tfrac{1}{4}\delta^2)\}$$

$$> \tfrac{1}{2}\exp\{\tfrac{1}{2}a^2B(1-\delta)\},$$

because $\delta > \tfrac{1}{4}\delta^2$. On the other hand, $W(x)$ is non-increasing so that

$$aJ_3 = a\int_{aB(1-\delta)}^{aB(1+\delta)} e^{ay} W(y)\,dy$$

$$< 2a^2B\,\delta W\{aB(1-\delta)\}\exp\{a^2B(1+\delta)\}$$

$$< 2a^2BW\{aB(1-\delta)\}\exp\{a^2B(1+\delta)\},$$

because $\delta < 1$. Thus, for $x = aB(1-\delta)$,

$$W(x) = W\{aB(1-\delta)\} > \frac{1}{4a^2B}\exp\{-\tfrac{1}{2}a^2B(1+3\delta)\}.$$

Arguing as before, it is easy to show that

$$\log 4a^2B < \tfrac{1}{2}a^2B\delta,$$

so that

$$W(x) > \exp\{-\tfrac{1}{2}a^2B(1+4\delta)\}$$

$$> \exp\left\{-\frac{x^2(1+4\delta)}{2B(1-\delta)^2}\right\}$$

$$> \exp\left\{-\frac{x^2}{2B}(1+3\varepsilon)\right\},$$

since, for $0 < \delta < \tfrac{1}{2}$,

$$1+3\varepsilon = 1+24\delta > \frac{1+4\delta}{(1-\delta)^2}.$$

This completes the proof.

We can now rephrase the conclusions of the lemma. If we have a sequence of identically distributed independent random variables, X_i, such that $|X_i| \leqslant d < \infty$, and $B = nE(X_i^2)$, $S_n = X_1+\ldots+X_n$, then for any arbitrarily small number ε

$$\mathrm{pr}(S_n > x) > \exp\left\{-\frac{x^2}{2B}(1+3\varepsilon)\right\}$$

so long as x lies in the range determined by the inequalities

$$\frac{xd}{B} < \frac{\varepsilon^2}{4096},$$

$$\frac{xd}{B} < \frac{1}{256},$$

$$x > 16(2B)^{\frac{1}{4}},$$

$$\frac{\log x^2 B^{-1}}{x^2 B^{-1}} < \frac{\varepsilon^2}{1024}.$$

We now use this to prove that the event (8.118) happens infinitely often and to do this we must show that the series

$$\sum_k \text{pr}\{S_{n_k} - S_{n_{k-1}} > (1-\delta)u_k v_k\} \tag{8.120}$$

is divergent.

$$B = u_k^2 = s_{n_k}^2 - s_{n_{k-1}}^2$$

is asymptotically equal to $c^{2k}(1-c^{-2})$, V_k is $(2\log\log B)^{\frac{1}{2}}$, x is

$$(1-\delta)B^{\frac{1}{2}}(2\log\log B)^{\frac{1}{2}},$$

and d is

$$KB^{\frac{1}{2}}(\log\log B)^{-\frac{1}{2}}.$$

x^2/B and $x/B^{\frac{1}{2}}$ are decreasing functions of n and xd/B can be made arbitrarily small by choosing n so large that K in (8.100) is sufficiently small. Thus the conditions of the lemma are satisfied and we conclude that given ε arbitrarily small we can choose k_0 so large that

$$\text{pr}\{S_{n_k} - S_{n_{k-1}} > (1-\delta)u_k v_k\} < \exp\{-\tfrac{1}{2}(1-\delta)^2 v_k^2(1+3\varepsilon)\},$$

for all $k > k_0$. Choosing 3ε to be equal to δ and using the fact that $(1-\delta^2) < 1$, the above expression is not less than

$$\exp\{-(1-\delta)\log\log u_k^2\} = \{(\log u_k^2)^{1-\delta}\}^{-1}, \tag{8.121}$$

which is asymptotically equal to

$$\{2k\log c + \log(1-c^{-2})\}^{\delta-1}.$$

This is the term of a divergent series. Thus (8.120) is a divergent series. Now consider the events

$$|S_{n_{k-1}}| < 2s_{n_{k-1}}t_{n_{k-1}}. \tag{8.122}$$

Using the first part of the theorem, with probability one there exists n_0 so that (8.122) is true for all $n_{k-1} > n_0$. Thus the combined event

$$\{S_{n_k} - S_{n_{k-1}} > (1-\delta)u_k v_k | S_{n_{k-1}}| < 2s_{n_{k-1}}t_{n_{k-1}}\}$$

must occur infinitely often. If both these events occur,

$$S_{n_k} > (1-\delta)u_k v_k - 2s_{n_{k-1}} t_{n_{k-1}},$$

and as k increases the right-hand side is asymptotically equal to

$$\{(1-\delta)(1-c^{-2})^{\frac{1}{2}} - 2c^{-1}\}s_{n_k} t_{n_k}.$$

Given $\delta_1 > 0$ we can choose δ and c so that this is not less than

$$(1-\delta_1)s_{n_k} t_{n_k}$$

so that the theorem is proved.

As the theorem stands it applies immediately to sequences of X_i which are obtained by centring a series of binomial trials with constant probability. Thus if this probability is p we write

$$X_i = -p, \quad \text{with probability } q,$$
$$\quad = q, \quad \text{with probability } p.$$

The variance is then pq and $s_n^2 = npq$, and thus

$$2\log\log s_n^2$$

is asymptotically equal to

$$2\log\log n.$$

Then clearly $|X_n| = o(s_n t_n^{-\frac{1}{2}})$. In this way we get a very much strengthened form of the Strong Law of Large Numbers for repeated binomial trials and we therefore state it as a separate theorem.

THEOREM 8.38. *Let S_n be the number of successes in the first n of an infinite sequence of independent binomial trials with probability p. Then for $\delta > 0$, the events*

$$|S_n - np| > (1+\delta)(2npq\log\log n)^{\frac{1}{2}} \tag{8.123}$$

will occur an infinite number of times with probability zero, and with probability one, the events

$$|S_n - np| > (1-\delta)(2npq\log\log n)^{\frac{1}{2}} \tag{8.124}$$

will occur infinitely often.

It follows from this that if we choose any number $K > 0$ and carry out such a sequence of trials, the event

$$|S_n - np| > K(npq)^{\frac{1}{2}}, \quad \text{for some } n, \tag{8.125}$$

will occur with probability one (and, indeed, infinitely often). This provides a useful and convenient method for faking experiments. A statistician informed that in n trials of an experiment, the number of successes exceeded $\frac{1}{2}n$ (where n is the number of trials) by more than $3n^{\frac{1}{2}}$, would decisively reject the hypothesis that $p = \frac{1}{2}$. The above theorem shows that even if p really is $\frac{1}{2}$ such results can certainly be obtained by continuing the experiment long enough.

How long it is necessary to continue the experiment to obtain a given significance level is a problem recently studied by N. E. Day.

Theorem 8.38 can be proved directly in a somewhat simpler manner than Theorem 8.37, since direct estimates of the probabilities are available from the known results on the normal approximation to the binomial distribution (see Feller 1957b, p. 191).

Theorem 8.37 has only been proved for bounded distributions of the X_i. To extend the range of applicability we use the method of truncation, but we must now assume that all the X_i have the same distribution. We can then obtain the following theorem.

THEOREM 8.39. *The results of Theorem 8.37 remain correct if we replace condition (8.100) by the condition that for some $\varepsilon > 0$,*

$$E|X_i|^{2+\epsilon}$$

is bounded by some finite constant K, provided all the X_i have the same distribution (with mean zero).

Proof. Define truncated random variables, Y_n, by

$$Y_n = X_n, \quad \text{if } |X_n| < n^{\frac{1}{2}}(\log\log n)^{-\frac{3}{2}}$$
$$= 0, \qquad \text{if } |X_n| \geqslant n^{\frac{1}{2}}(\log\log n)^{-\frac{3}{2}}.$$

It is convenient to put

$$a_n = n^{\frac{1}{2}}(\log\log n)^{-\frac{3}{2}}.$$

Then if $F(x)$ is the distribution function of X_n,

$$|E(Y_n)| = |E(Y_n) - E(X_n)| \leqslant \int_{a_n}^{\infty} u\, dF(u) + \int_{-\infty}^{-a_n} |u|\, dF(u)$$

$$\leqslant a_n^{-1-\epsilon}\left\{\int_{a_n}^{\infty} u^{2+\epsilon}\, dF(u) + \int_{-\infty}^{-a_n} |u|^{2+\epsilon}\, dF(u)\right\}$$

$$\leqslant a_n^{-1-\epsilon}K. \tag{8.126}$$

$$\leqslant K(\log\log n)^{\frac{3}{2}(1+\epsilon)}n^{-\frac{1}{2}(1+\epsilon)}. \tag{8.127}$$

From this it follows that

$$E\left(\sum_1^n Y_i\right) = E\left(\sum_1^n Y_i - \sum_1^n X_i\right) = o(s_n). \tag{8.128}$$

Furthermore ($F(u)$ being continuous to the right),

$$E(Y_n^2) - E(X_n^2) = \int_{a_n}^{\infty} u^2\, dF(u) + \int_{-\infty}^{-a_n} u^2\, dF(u)$$

$$\leqslant a_n^{-\epsilon}K$$

$$\leqslant K(\log\log n)^{\frac{3}{2}\epsilon}n^{-\frac{1}{2}\epsilon}. \tag{8.129}$$

Combining this with (8.127) we see that

$$\text{var}(Y_n) - \text{var}(X_n) = o(n^{-\frac{1}{2}\epsilon}). \tag{8.130}$$

We can now write $(s_n')^2$ for the variance of the sum

$$\sum_1^n Y_i$$

and from (8.130), $(s_n')^2$ is asymptotically equal to s_n^2, since the latter increases as n. Thus the bounded random variables, $|Y_n|$, satisfy

$$|Y_n| = o\,[s_n'\{\log\log(s_n')^2\}^{-\frac{1}{2}}], \tag{8.131}$$

so that the conditions of Theorem 8.37 apply to the Y_n and it is sufficient to consider the probability of an infinite number of the events

$$\left|\sum_1^n Y_i\right| > (1+\delta)s_n'\{2\log\log(s_n')^2\}^{-\frac{1}{2}}$$

$(\delta > 0, \delta < 0)$, since the series

$$\sum \text{pr}(X_i \neq Y_i)$$

is bounded by

$$\sum\left\{\int_{a_i}^\infty dF(u) + \int_{-\infty}^{-a_i} dF(u)\right\} \leqslant \sum Ka_i^{-2-\epsilon}, \tag{8.132}$$

which is convergent.

This proves the theorem. By a more elaborate method of truncation Hartman and Wintner (1941) have shown that the conclusions of Theorem 8.39 remain valid when we only assume that $E|X_n|^2$ is finite.

A number of papers have been published which extend and sharpen the above results in various directions. (See Feller 1943a, 1946.)

8.17. The central limit problem

Under a wide variety of conditions sums of random variables,

$$S_n = \sum_1^n X_i, \tag{8.133}$$

have distributions which, when suitably centred and scaled, tend to the normal distribution. We have already encountered examples of this in the study of sums of binomial and Poisson variates, and an example where it does not occur—that of the Cauchy distribution. We have also proved in §7.2 that if the X_i are independently distributed in the same distribution with mean zero and variance $\sigma^2 < \infty$, $S_n(n\sigma^2)^{-\frac{1}{2}}$ has a distribution which converges to the normal distribution with zero mean and unit variance. It is this type of theorem that we wish to generalize. When the X_i have the same distribution, the finiteness of the variance is not a necessary condition, as we shall see later.

If we drop the condition that all the X_i have the same distribution we shall have to impose some condition that ensures that no particular individual term in the sum (8.133) has a real influence on the asymptotic distribution. Thus suppose that the dispersion of S_n remains bounded and the distribution of S_n converges to some distribution without rescaling. Then by Cramér's theorem this distribution cannot be normal unless the distribution of each X_i is also normal.

We begin with cases where the individual X_i have finite variances.

8.18. The case of bounded variance

The first main progress in extending Theorem 7.1 was achieved by Liapounov who proved the following result,

THEOREM 8.40 (Liapounov's theorem). *Suppose that $\{X_i\}$ is a sequence of independent random variables with zero means and variances σ_i^2. If*

$$s_n^2 = \sigma_1^2 + \ldots + \sigma_n^2 \tag{8.134}$$

and, for some $\delta > 0$,

$$s_n^{-2-\delta} \sum_1^n E|X_i|^{2+\delta} \to 0, \tag{8.135}$$

then the distribution of

$$s_n^{-1} \sum_1^n X_i \tag{8.136}$$

converges to normality with mean zero and unit standard deviation.

Proof. We first show that if $\delta > 1$ we can reduce the result to the case $\delta = 1$. After this we can assume $0 < \delta \leqslant 1$.

Consider the mixture $\dfrac{1}{n} \sum_1^n F_i(x)$, where $F_i(x)$ is the distribution of X_i. For any positive number r we have

$$E\left(\frac{1}{n} \sum_1^n |X_i|^r\right) = \frac{1}{n} \int_{-\infty}^{\infty} |x|^r \sum_1^n dF_i(x)$$
$$= E(|Y|^r), \tag{8.137}$$

where Y is distributed in the mixed distribution.

From Theorem 5.15, $\log E(|Y|^r)$ $(r > 0)$ is a function which is convex from below and hence, for $\delta > 1$,

$$\delta \log E(|Y|^3) \leqslant (\delta-1)\log E(|Y|^2) + \log E(|Y|^{2+\delta}) \tag{8.138}$$

so that if (8.135) is satisfied for $\delta > 1$,

$$s_n^{-3} \sum_1^n |X_i|^3 \to 0. \tag{8.139}$$

We can therefore now assume $0 < \delta \leqslant 1$. We know from Theorem 5.16 that for $r > 0$

$$(E|X|^r)^{1/r}$$

is a non-decreasing function of r. Hence

$$\max_{k \leqslant n} \frac{\sigma_k^{2+\delta}}{s_n^{2+\delta}} \leqslant \max_{k \leqslant n} \frac{E(|X_k|^{2+\delta})}{s_n^{2+\delta}}$$

and the right-hand side tends to zero by (8.135).

We now introduce a lemma which is an improvement of Theorem 6.10.

LEMMA 8.3. *If $\phi(t)$ is the characteristic function of a random variable Y such that $E(|Y|^{2+\delta}) < \infty$,*

$$\phi(t) = 1 + itE(Y) - \tfrac{1}{2}t^2 E(Y^2) + \theta|t|^{2+\delta} 2^{1-\delta} E(|Y|^{2+\delta}) \qquad (8.140)$$

for t real, and some θ such that $0 \leqslant |\theta| \leqslant 1$.

Proof. By integration by parts we see easily that

$$e^{iy} = 1 + iy - \tfrac{1}{2}y^2 - y^2 \int_0^1 (e^{iyt} - 1)(1 - t)\,dt. \qquad (8.141)$$

For any real a we have

$$|e^{ia} - 1| \leqslant 2|a| \leqslant 4\left|\frac{a}{2}\right|^\delta, \quad \text{if } |\tfrac{1}{2}a| \leqslant 1,$$

and

$$|e^{ia} - 1| \leqslant 2 < 4\left|\frac{a}{2}\right|^\delta, \quad \text{if } |\tfrac{1}{2}a| \geqslant 1.$$

Putting yt for a, the last term in (8.141) does not exceed in absolute value

$$2^{1-\delta}y^{2+\delta}.$$

Replacing y by the random variable tY in (8.141) and taking expectations we get (8.140).

Let $\phi_i(t)$ be the characteristic function of X_i, and $\Phi(t)$ that of $s_n^{-1}S_n$. Then

$$\log \Phi(t) = \sum_1^n \log \phi_i(ts_n^{-1})$$

$$= \sum_1^n \log \left\{ 1 - \tfrac{1}{2}t^2 \sigma_i^2 s_n^{-2} + |t|^{2+\delta}\theta_i \frac{E(|X_i|^{2+\delta})}{s_n^{2+\delta}} \right\}$$

$$= -\tfrac{1}{2}t^2\{1 + o(1)\} + \theta'|t|^{2+\delta} \frac{\sum E(|X_i|^{2+\delta})}{s_n^{2+\delta}}\{1 + o(1)\},$$

where $|\theta_i| \leqslant 2$ and $|\theta'| \leqslant 2$. As n tends to infinity this tends to $-\tfrac{1}{2}t^2$, and by Theorem 6.2 the result is proved. This theorem was first given by Liapounov in (1900, 1901). Clearly this settles a large proportion of the practical cases in which the Central Limit Theorem is likely to be required (for example, the distribution of inversions considered in §1.22). Yet it assumes the existence of moments of a higher order than occur in the statement of the conclusions of the theorem, and can in fact be strengthened to the following theorem.

THEOREM 8.41 (Lindeberg 1922). *If* $\{X_i\}$ *is a sequence of independent random variables with zero means and finite variances* σ_i^2, *and*

$$s_n^2 = \sigma_1^2 + \ldots + \sigma_n^2,$$

then the distribution of

$$s_n^{-1} \sum_1^n X_i$$

tends to a normal distribution with zero mean and unit variance if, for every $\varepsilon > 0$,

$$\lim_{n \to \infty} s_n^{-2} \sum_{i=1}^n \int_{|x| \geqslant \varepsilon s_n} x^2 \, dF_i(x) = 0, \tag{8.142}$$

where $F_i(x)$ *is the distribution of* X_i.

Later, in Theorem 8.42, we shall see that the condition (8.142) is also necessary for the conclusion.

Proof of Theorem 8.41. The condition (8.142) is clearly designed to make truncation useful. Furthermore, it implies that s_n^2 tends to infinity, and that

$$\sigma_i s_n^{-1}$$

tends to zero uniformly in i for $i \leqslant n$. To prove the latter remark write $g_n(\varepsilon)$ for the expression in (8.142) which tends to zero. We have $g_n(\varepsilon) \to 0$ for every $\varepsilon > 0$. We first show that we can choose a sequence, $\{\varepsilon_n\}$, converging to zero such that $\varepsilon_n^{-2} g_n(\varepsilon_n)$ converges to zero. To do this consider the sequence $k = 1, 2, \ldots$ and define an increasing sequence, $\{n_k\}$, such that $g_n(k^{-1}) < k^{-3}$ for $n \geqslant n_k$. Then for any n satisfying $n_k \leqslant n < n_{k+1}$ (all k) put $\varepsilon_n = k^{-1}$. For such an n

$$\varepsilon_n^{-2} g_n(\varepsilon_n) = k^2 g_n(k^{-1}) < k^{-1}, \tag{8.143}$$

and thus

$$\varepsilon_n^{-2} g_n(\varepsilon_n) \to 0.$$

We then have

$$s_n^{-2} \max_{1 \leqslant k \leqslant n} \sigma_k^2 = s_n^{-2} \max_{1 \leqslant k \leqslant n} \int_{-\infty}^{\infty} x^2 \, dF_k(x)$$

$$\leqslant s_n^{-2} \left\{ \varepsilon_n^2 s_n^2 + \max_{1 \leqslant k \leqslant n} \int_{|x| \geqslant \varepsilon_n s_n} x^2 \, dF_k(x) \right\}$$

$$\leqslant \varepsilon_n^2 + g_n(\varepsilon_n). \tag{8.144}$$

The latter expression tends to zero since $g_n(\varepsilon_n)$ does also, and hence

$$\max_{1 \leqslant k \leqslant n} \frac{\sigma_k^2}{s_n^2}$$

tends to zero so that $\sigma_i s_n^{-1}$ tends to zero uniformly in i for $i \leqslant n$. This provides the necessary condition mentioned above that ensures that none of the X_i make an individual contribution to S_n which is not in the limit negligible by itself.

We now introduce truncated variates in a more sophisticated manner than we have done so far, in that we consider sum of truncated variates whose point of truncation depends on the number of terms in the sum. To do this we write, for $1 \leqslant i \leqslant n$,

$$Y_{ni} = X_i, \quad \text{if } |X_i| < \varepsilon_n s_n,$$

$$= 0, \quad \text{if } |X_i| \geqslant \varepsilon_n s_n. \tag{8.145}$$

Put

$$S_{nn} = \sum_{i=1}^{n} Y_{ni}.$$

Then

$$\mathrm{pr}(S_{nn} \neq S_n) \leqslant \sum_{i=1}^{n} \mathrm{pr}(Y_{ni} \neq X_i)$$

$$\leqslant \sum_{i=1}^{n} \int_{|x| \geqslant \varepsilon_n S_n} dF_i(x)$$

$$\leqslant \sum_{i=1}^{n} \varepsilon_n^{-2} \int_{|x| \geqslant \varepsilon_n s_n} s_n^{-2} x^2 \, dF_i(x)$$

$$\leqslant \varepsilon_n^{-2} g(\varepsilon_n). \tag{8.146}$$

In order to prove that the sequence of variates $s_n^{-1} S_n$ converges in distribution to a normal variate it is therefore sufficient to prove the same result for $s_n^{-1} S_{nn}$.

By truncating the X_i we have, in general, shifted their means from zero and altered their variances. We must therefore next show that this has a negligible effect on the distribution of $s_n^{-1} S_{nn}$.

We have, for $k \leqslant n$,

$$|E(Y_{nk})| = \left| \int_{|x| < \varepsilon_n s_n} x \, dF_k(x) \right|$$

$$= \left| \int_{|x| \geqslant \varepsilon_n s_n} x \, dF_k(x) \right|$$

$$\leqslant \varepsilon_n^{-1} s_n^{-1} \int_{|x| \geqslant \varepsilon_n s_n} x^2 \, dF_k(x). \tag{8.147}$$

Then

$$s_n^{-1} |E(S_{nn})| \leqslant s_n^{-1} \sum_{k=1}^{n} |E(Y_{nk})|$$

$$\leqslant \varepsilon_n^{-1} g_n \varepsilon_n, \tag{8.148}$$

which tends to zero. Hence the asymptotic distribution of $s_n^{-1} S_{nn}$ is that of $s_n^{-1}\{S_{nn} - E(S_{nn})\}$. Write

$$s_{nn} = \text{var}(S_{nn})$$

$$= \sum_{i=1}^{n} \text{var}(Y_{ni}). \qquad (8.149)$$

Then, using (8.148),

$$1 - s_n^{-2} s_{nn}^2 = s_n^{-2}\left\{\sum_{i=1}^{n} \text{var}(X_i) - \sum_{i=1}^{n} \text{var}(Y_i)\right\}$$

$$= s_n^{-2}\left[\sum_{i=1}^{n} \int_{|x| \geqslant \varepsilon_n s_n} x^2 \, dF_i(x) + \{E(S_{nn})\}^2\right]$$

$$\leqslant g_n(\varepsilon_n) + \varepsilon_n^{-2} g_n(\varepsilon_n)^2, \qquad (8.150)$$

and this also converges to zero. It is therefore sufficient to consider the distribution of

$$s_{nn}^{-1}\{S_{nn} - E(S_{nn})\}. \qquad (8.151)$$

Consider the third absolute moments of the Y_{ni} about their means. We have

$$s_{nn}^{-3} \sum_{k=1}^{n} E|Y_{nk} - E(Y_{nk})|^3 \leqslant 2s_{nn}^{-3} \varepsilon_n s_n \sum_{k=1}^{n} E|Y_{nk} - E(Y_{nk})|^2$$

$$\leqslant 2s_{nn}^{-1} \varepsilon_n s_n. \qquad (8.152)$$

Since $s_n s_{nn}^{-1}$ converges to unity and ε_n to zero, this converges to zero.

If we now look at the proof of Theorem 8.40 it will be seen that the conditions remain sufficient if the distributions of the X_i in the sum S_n are allowed to depend on n. This can be seen by following through the details of the proof. (8.152) therefore implies that the distribution of (8.151) converges to a normal distribution with mean zero and unit variance. The proof of Theorem 8.41 is therefore complete.

It is a remarkable fact that the condition (8.142) is also necessary if the other assumptions of the theorem hold. We state this result as Theorem 8.42 which is due to Feller (1936, 1937a).

THEOREM 8.42. *If $\{X_n\}$ is a sequence of independent random variables with zero means and finite variances σ_i^2, and*

$$s_n^{-1} \sum_{1}^{n} X_i$$

has a distribution which converges to normality, and if $\max_{1 \leqslant k \leqslant n} \sigma_i s_n^{-1} \to 0$ *uniformly in k, then for every $\varepsilon \to 0$,*

$$s_n^{-2} \sum_{1}^{n} \int_{|x| \geqslant \varepsilon s_n} x^2 \, dF_i(x)$$

converges to zero.

26

Proof. Let $\phi_k(t)$ be the characteristic function of X_i. The ratios $\sigma_k s_n^{-1}$ $(k = 1,...,n)$ are bounded above by a quantity which tends to zero with n. We have, from Theorem 6.11,

$$\phi_k(ts_n^{-1}) = 1 - \tfrac{1}{2}t^2 \frac{\sigma_k^2}{s_n^2}\, \theta_k,$$

where $|\theta_k| \leqslant 1$, so that

$$\max_{1 \leqslant k \leqslant n} |\phi_k(ts_n^{-1}) - 1| \leqslant \tfrac{1}{2}t^2 s_n^{-2} \max_{1 \leqslant k \leqslant n} \sigma_k^2, \qquad (8.153)$$

and the left-hand side must tend to zero as n increases. Furthermore,

$$\sum_{k=1}^{n} |\phi_k(ts_n^{-1}) - 1|^2 \leqslant \tfrac{1}{4}t^4 \sum_{k=1}^{n} \left(\frac{\sigma_k}{s_n}\right)^4 \leqslant \tfrac{1}{4}t^4 \sum_{k=1}^{n} \frac{\sigma_k^2}{s_n^2} \max_{1 \leqslant i \leqslant n} \frac{\sigma_i^2}{s_n^2}, \qquad (8.154)$$

and these three expressions also tend to zero as n increases.

From (8.153) it follows that for n large enough

$$\log \phi_k(ts_n^{-1})$$

exists, and that

$$E(\exp its_n^{-1} S_n) = \prod_{1}^{n} \phi_k(ts_n^{-1})$$

$$= \exp\left\{ \sum_{1}^{n} \log \phi_k(ts_n^{-1}) \right\}$$

converges to

$$\exp -\tfrac{1}{2}t^2$$

as n increases for each fixed t. We also know that we can write

$$\log z = z - 1 + \theta(z-1)^2, \qquad (8.155)$$

where θ tends to zero as z tends to unity. Using (8.153) and (8.154) it follows that

$$\left| \sum_{1}^{n} \{1 - \phi_k(ts_n^{-1})\} - \tfrac{1}{2}t^2 \right|$$

tends to zero as n increase. We can write this as

$$\left| \tfrac{1}{2}t^2 - \sum_{k=1}^{n} \int \{1 - \exp(its_n^{-1} x)\} \, dF_k(x) \right| = o(1).$$

Taking the real part by itself we have

$$\left| \tfrac{1}{2}t^2 - \sum_{k=1}^{n} \int \{1 - \cos(ts_n^{-1}x)\} \, dF_k(x) \right| = o(1),$$

and splitting the integral into two parts we get for any $\varepsilon > 0$,

$$\sum_{k=1}^{n} \int_{|x| \geqslant \varepsilon s_n} \{1 - \cos(t s_n^{-1} x)\} \, dF_k(x)$$

$$= \tfrac{1}{2} t^2 - \sum_{k=1}^{n} \int_{|x| < \varepsilon s_n} \{1 - \cos(t s_n^{-1} x)\} \, dF_k(x) + o(1). \qquad (8.156)$$

Now $0 \leqslant 1 - \cos u \leqslant \tfrac{1}{2} u^2$, for $0 \leqslant u < \infty$, and the sum of the integrals on the right-hand side of (8.156) is non-negative and not greater than

$$\tfrac{1}{2} t^2 s_n^{-2} \sum_{k=1}^{n} \int_{|x| < \varepsilon s_n} x^2 \, dF_k(x) = \tfrac{1}{2} t^2 s_n^{-2} \left\{ s_n^2 - \sum_{k=1}^{n} \int_{|x| \geqslant \varepsilon s_n} x^2 \, dF_k(x) \right\}$$

$$= \tfrac{1}{2} t^2 \{ 1 - g_n(\varepsilon) \}. \qquad (8.157)$$

We also have

$$\sum_{k=1}^{n} \int_{|x| \geqslant \varepsilon s_n} \{1 - \cos(t s_n^{-1} x)\} \, dF_k(x) \leqslant 2 \sum_{k=1}^{n} \int_{|x| \geqslant \varepsilon s_n} dF_k(x)$$

$$\leqslant 2 \varepsilon^{-2}, \qquad (8.158)$$

by Tchebychev's inequality. (8.156) therefore implies that

$$\tfrac{1}{2} t^2 g_n(\varepsilon) \leqslant 2 \varepsilon^{-2} + o(1), \qquad (8.159)$$

where the $o(1)$ term tends to zero with n. Thus

$$g_n(\varepsilon) \leqslant 2 t^{-2} \{ 2 \varepsilon^{-2} + o(1) \}$$

so that

$$\overline{\lim} \, g_n(\varepsilon) \leqslant 2 t^{-2} \varepsilon^{-2}. \qquad (8.160)$$

Since t has been chosen arbitrarily it follows that $g_n(\varepsilon) \to 0$ as n increases for every fixed positive ε, which proves the required result.

Much research has been devoted to the problem of setting bounds to the rate of convergence to the normal distribution in the central limit problem. For this see in particular Berry (1941), Esseen (1945), and Gnedenko and Kolmogorov (1954).

The above theorems can also be easily extended in a natural way to sums of independent vector variates, thus arriving at a multivariate distribution for the components of normed sums which is asymptotically multivariate normal. The easiest way to obtain such theorems is to apply the above results to a linear form in the components of the vector variates. For each arbitrarily chosen set of coefficients, the linear form has an asymptotically normal distribution and, using a uniformity argument, the multivariate result follows from the Cramér–Wold theorem (Theorem 6.18).

8.19. Independent variates with unbounded variances

We now drop the assumption that the variances of the X_i are bounded, and we want conditions on the distributions of the X_i which will ensure the existence of two sequences, $\{a_n\}$ and $\{b_n\}$, such that

$$\frac{S_n - a_n}{b_n}$$

has a distribution which converges to the normal distribution, where $S_n = X_1 + \ldots + X_n$.

In order to discuss this we first consider a more general problem. Suppose that S_n is the sum of n independent random variables, X_{ni}, whose distributions depend on n. We write this

$$S'_n = X_{n1} + \ldots + X_{nn}. \tag{8.161}$$

We let n increase and consider conditions under which S'_n has a distribution which tends to a normal distribution with finite mean and variance. This includes the degenerate case where the variance is zero.

We say that a particular term, X_{nk}, regarded as dependent on n for a fixed k, is 'asymptotically negligible' if for every $\varepsilon > 0$,

$$\operatorname{pr}(|X_{nk}| > \varepsilon)$$

tends to zero as n increases.

If the distribution of S'_n tends to a normal distribution with finite mean and finite non-zero variance, the distribution of X_{nk} for any particular k will either have to tend to normality or be asymptotically negligible. This follows from Cramér's theorem (Theorem 7.2).

However, this is not sufficient. The fact that any prescribed X_{nk}, in the sum (8.161), is individually negligible does not mean that the largest X_{nk} in S'_n is also negligible. Write

$$\alpha_{ni} = \operatorname{pr}(|X_{ni}| > \varepsilon), \tag{8.162}$$

and

$$\eta_n = \sum_{i=1}^{n} \alpha_{ni}. \tag{8.163}$$

Then the number of terms, X_{ni}, which are such that

$$|X_{ni}| > \varepsilon,$$

will be a random variable with mean η_n which need not be small even if all the α_{ni} are small. However, by using Boole's inequality (§ 1.3) we see that

$$\operatorname{pr}\left(\max_{1 \leqslant i \leqslant n} |X_{ni}| > \varepsilon\right) \leqslant \sum_{i=1}^{n} \alpha_{ni} = \eta_n. \tag{8.164}$$

The X_{ni} are said to be 'uniformly asymptotically negligible' if

$$\max_{1 \leqslant k \leqslant n} \mathrm{pr}(|X_{nk}| > \varepsilon) = \max_{1 \leqslant k \leqslant n} \alpha_{nk}$$

tends to zero as n increases, for every $\varepsilon > 0$. This is sometimes known as the UAN condition. We shall, however, use the rather stronger condition that η_n tends to zero as n increases, and we then say that the largest term in S'_n is asymptotically individually negligible.

THEOREM 8.43. *If the distribution of S'_n is such that its dispersion (as defined in § 5.14) lies between two non-zero finite limits, a sufficient condition for this distribution to tend to the normal distribution is that the largest term in the sum is asymptotically individually negligible, i.e.*

$$\eta_n \to 0, \quad for\ every\ \varepsilon > 0. \tag{8.165}$$

In defining the dispersion we take some probability p where $0 < p < 1$ and consider the dispersion L corresponding to this probability. We can take L as the greatest lower bound of numbers l such that p is not less than the least upper bound of $F_S(x+l) - F_S(l)$ for all x, where $F_S(x)$ is the distribution of S'_n. This is merely a device for ensuring that the sum S'_n is suitably scaled.

In order to prove Theorem 8.43 we use a strengthened form of Theorem 8.40.

THEOREM 8.44. *Suppose that $\{X_i\}$ is a sequence of independent random variables such that $|X_i| \leqslant a$, $E(X_i) = 0$, $\sigma_i^2 = E(X_i^2)$, $s_n^2 = \sigma_1^2 + \ldots + \sigma_n^2$. Then the distance of the distribution of $s_n^{-1} S_n$, where*

$$S_n = X_1 + \ldots + X_n,$$

from the normal distribution with zero mean and unit standard deviation can be bounded by a quantity depending only on as_n^{-1} and tending to zero with the latter.

Proof. Following the proof of Theorem 8.40, let $\phi_i(t)$ be the characteristic function of X_i, and $\Phi_n(t)$ that of $s_n^{-1} S_n$. Then

$$\phi_i(t) = 1 - \tfrac{1}{2}\sigma_i^2 t^2 + \tfrac{1}{6}\theta_i a^3 t^3,$$

where $|\theta_i| \leqslant 1$ and θ_i may depend on t and i. Then

$$\log \Phi_n(t) = \sum_1^n \{1 - \tfrac{1}{2}\sigma_i^2 t^2 s_n^{-2} + \tfrac{1}{6}\theta_i(as_n^{-1})^3 t^3\}.$$

For each fixed t the difference of this expression from $-\tfrac{1}{2}t^2$ can be bounded by a quantity depending only on as_n^{-1} and tending to zero with it uniformly in any finite interval in t. From the convergence theorem (Theorem 6.3) we see that the distance of this distribution from a normal distribution with zero mean and unit standard deviation can be similarly bounded.

Proof of Theorem 8.43. Choose some $\varepsilon > 0$, dependent on n, and write

$$X'_{ni} = X_{ni}, \quad \text{if} \, |X_{ni}| \leqslant \varepsilon,$$
$$= 0, \qquad \text{if} \, |X_{ni}| > \varepsilon,$$
$$\alpha_{ni} = \text{pr}(|X_{ni}| > \varepsilon).$$

Then if

$$S''_n = \sum_1^n X'_{ni},$$

$$\text{pr}(S'_n \neq S_n) \leqslant \sum_{i=1}^n \alpha_{ni}, \qquad (8.166)$$

which can be made arbitrarily small by making n large. Hence it is sufficient to prove that the distribution of

$$S''_n - E(S''_n)$$

tends to normality with a zero mean when suitably scaled. Write

$$(s''_n)^2 = \sum_1^n \text{var}(X'_{ni}). \qquad (8.167)$$

If s''_n were to tend to zero with n, the dispersion of S''_n would tend to zero and this is not possible because of (8.166) and the fact that the dispersion of S'_n is bounded below. Hence as $\varepsilon \to 0$, s''_n is bounded below. Applying Theorem 8.44 to the sum S''_n we see that its distribution tends to normality as $\varepsilon \to 0$ and n increases. This proves the result. Notice that in this proof we have made the point of truncation depend on n.

Lévy (1937a) has given a converse to this theorem in the form that if each term is asymptotically individually negligible and the distribution of S'_n tends to normality, the largest term, after centring at medians, must be asymptotically negligible.

8.20. The case of identically distributed variates

Suppose now that

$$S_n = X_1 + \ldots + X_n,$$

where the X_i are independently distributed in the same distribution, $F(x)$, which is otherwise arbitrary. We say that $F(x)$ belongs to the 'domain of attraction' of the normal distribution if there exist functions of n, A_n, and B_n such that the distribution of

$$\frac{S_n - A_n}{B_n} \qquad (8.168)$$

tends to the normal distribution with zero mean and unit variance.

We have already seen in Theorem 7.1 that if the distribution $F(x)$ has a finite variance, σ^2, it belongs to the domain of attraction of the normal law and we can take

$$A_n = nE(X_i), \tag{8.169}$$

$$B_n = n^{\frac{1}{2}}\{\text{var}(X_i)\}^{\frac{1}{2}}. \tag{8.170}$$

We look for a general necessary and sufficient condition for $F(x)$ to be attracted to the normal distribution. Write

$$A(x) = \text{pr}(|X| > x)$$

$$= \int_{-\infty}^{-x-0} dF(x) + \int_x^\infty dF(x), \tag{8.171}$$

and

$$B(x) = \int_{-x-0}^x x^2 \, dF(x)$$

$$= E(X'^2), \tag{8.172}$$

where X' is X if $|X| \leqslant x$, and zero otherwise. If the limit of $B(x)$ as x tends to infinity is finite, the variance of X is finite and so $F(x)$ belongs to the domain of attraction of the normal distribution. We therefore suppose that

$$\lim_{x \to \infty} B(x) \tag{8.173}$$

is infinite.

THEOREM 8.45. *In order that $F(x)$ belongs to the domain of attraction of the normal distribution it is necessary and sufficient that*

$$\lim_{x \to \infty} \frac{x^2 A(x)}{B(x)} = 0. \tag{8.174}$$

Proof. Note first that if (8.173) is finite the denominator in (8.174) tends to a finite non-zero limit, and since

$$\int_{-\infty}^\infty x^2 \, dF(x)$$

is a convergent integral, the numerator in (8.174) tends to zero, so that (8.174) is satisfied, and as stated above we can eliminate this case. Thus we assume $F(x)$ not to have a finite variance.

Let

$$X' = |X|, \quad \text{if } |X| \leqslant x,$$

$$= 0, \quad \text{if } |X| > x,$$

and let X'' be X' centred at its median. We first show that

$$\text{var}(X'') = \text{var}(X')$$

is asymptotically equal to $B(x)$, i.e. that

$$\text{var}(X')B(x)^{-1} \to 1 \tag{8.175}$$

as $x \to \infty$. Since

$$\text{var}(X') = B(x) - E(X')^2,$$

it is sufficient to show that

$$E(X')^2 B(x)^{-1} \to 0. \tag{8.176}$$

Given any $\varepsilon > 0$ choose $\alpha > 0$ so that

$$\int_{\alpha}^{\infty} dF(x) + \int_{-\infty}^{-\alpha-0} dF(x) < \varepsilon, \tag{8.177}$$

and then $\beta > \alpha$ so that

$$\int_{-\alpha-0}^{\alpha} t \, dF(t) < \varepsilon B(x)^{\frac{1}{2}}$$

for $x > \beta$. Then for $x > \beta$

$$E(X') = \int_{-\alpha-0}^{\alpha} t \, dF(t) + \int_{-x-0}^{-\alpha-0} t \, dF(t) + \int_{\alpha}^{x} t \, dF(t),$$

so that

$$|E(X')| \leqslant \varepsilon B(x)^{\frac{1}{2}} + \int_{-x-0}^{-\alpha-0} |t| \, dF(t) + \int_{\alpha}^{x} |t| \, dF(t)$$

$$\leqslant \varepsilon B(x)^{\frac{1}{2}} + \left\{ \int_{-x-0}^{-\alpha-0} t^2 \, dF(t) \right\}^{\frac{1}{2}} \left\{ \int_{-x-0}^{-\alpha-0} dF(t) \right\}^{\frac{1}{2}}$$

$$+ \left\{ \int_{\alpha}^{x} t^2 \, dF(t) \right\}^{\frac{1}{2}} \left\{ \int_{\alpha}^{x} dF(t) \right\}^{\frac{1}{2}}$$

$$\leqslant B(x)^{\frac{1}{2}} \{ \varepsilon + \varepsilon^{\frac{1}{2}} + \varepsilon^{\frac{1}{2}} \}. \tag{8.178}$$

Since there exists such a β for every $\varepsilon > 0$, (8.176) is proved.

If the distribution of S_n tends to normality after possible rescaling, the contribution of each X_i must be individually negligible after such rescaling. Suppose that this is not so and let S_n be rescaled so that the dispersion of $k_n S_n$ for some p satisfying $0 < p < 1$ is bounded between two non-zero constants. Then if X_1 is not individually negligible,

$$\text{pr}(|k_n X_1| > \varepsilon)$$

does not tend to zero. It follows that

$$k_n S_n = k_n X_1 + k_n (X_2 + \ldots + X_n)$$

is the sum of two independent random variables and its distribution tends to normality with finite second moment as n increases. Since $k_n X_1$ has a distribution whose dispersion does not tend to zero it must be normally distributed by Cramér's theorem. But this is ruled out by hypothesis and hence each X_i is individually negligible. By Theorem 8.43 (and Lévy's converse) the necessary and sufficient condition that the distribution of S_n tends to normality is that the largest term is asymptotically negligible.

Now

$$nA(x) = n\operatorname{pr}(|X_i| > x)$$

is the expected number of the X_i whose moduli are greater than x. We therefore have

$$1 - \{1 - A(X)\}^n \leqslant \operatorname{pr}\left(\max_{1 \leqslant k \leqslant n} |X_i| > x\right) \leqslant nA(x). \tag{8.179}$$

Hence if we can choose x dependent on n in such a way that

$$nA(x)$$

tends to zero, the largest term will be asymptotically negligible. We also had to choose x so that the dispersion of S_n', as measured by its variance

$$n\operatorname{var}(X'), \tag{8.180}$$

is asymptotically large compared with x^2 in order for Theorem 8.44 to be applicable. Thus both

$$nA(x) \quad \text{and} \quad n^{-1}x^2 B(x)^{-1} \tag{8.181}$$

must be small. Write $D(x)$ for the product

$$x^2 A(x) B(x)^{-1}. \tag{8.182}$$

Then if the expressions (8.181) are arbitrarily small so also is $D(x)$. The converse is also true for if $D(x)$ tends to zero as x increases, associate each value of x with the number n defined by

$$n^2 \leqslant \frac{x^2}{A(x)B(x)} < (n+1)^2, \tag{8.183}$$

and then both expressions in (8.181) tend to zero as n increases.

The scaling factor in the above theorem is $n^{\frac{1}{2}}\operatorname{var}(X')$ which is asymptotically equal to $n^{\frac{1}{2}}B(x)$. Unless the variance of X_i is finite this increases faster than $n^{\frac{1}{2}}$. Lévy has shown, however, that $B(x)$ increases more slowly than any power of n.

8.21. Intermittent convergence

Consider the behaviour of

$$D(x) = \frac{x^2 A(x)}{B(x)}$$

as x tends to infinity. We have already seen that if the upper limit of $D(x)$ is zero, the distribution of S_n tends to normality when suitably scaled. On the other hand, if the lower limit of $D(x)$ is greater than zero the distribution of

S_n does not tend to normality however it is scaled. We may have circumstances in which the lower limit is zero and the upper limit non-zero. Thus suppose the upper limit is $K > 0$, and the lower limit is zero.

Then we can find a sequence of values $x_1, x_2, \ldots$ such that $D(x_i) \to 0$. With each of these we associate, knowing $A(x_i)$ and $B(x_i)$, the integer n_i defined by

$$n_i^2 \leqslant \frac{x_i^2}{A(x_i)B(x_i)} < (n_i + 1)^2. \qquad (8.184)$$

The distribution of S_{n_i} then tends to normality. On the other hand, there must also exist a sequence $x_1', x_2', \ldots$ tending to infinity, and a corresponding set of increasing integers n_i' such that $D(x_i') > \tfrac{1}{2}K$ for all i, and thus the distribution of $S_{n_i'}$ does not tend to normality. As n increases the distribution of S_n will tend towards normality and away again alternately. This is known as 'intermittent convergence' (P. Lévy 1937a).

It is interesting to construct an example where such intermittent convergence occurs. Consider a sequence of positive values, α_k ($k = 1, 2, \ldots$) where $\alpha_1 = 1$, and the sequence decreases so quickly that $\alpha_k - \alpha_{k+1}$ is practically the same size as α_k, and such that α_{k+1}/α_k is to converge to zero. For example, we might take

$$\alpha_1 = 1, \quad \alpha_2 = 10^{-6}, \quad \alpha_3 = \alpha_2^{\alpha_2^{-1}}, \ldots, \quad \alpha_{k+1} = \alpha_k^{\alpha_k^{-1}}, \quad \text{etc.}$$

Suppose that the random variable X takes the values $x_1, x_2, \ldots$ with probabilities $\alpha_1 - \alpha_2, \alpha_2 - \alpha_3, \ldots$ and choose the x_k to increase so fast that

$$x_{k+1}^2 \alpha_{k+1} > Kx_k^2 \alpha_k, \qquad (8.185)$$

where $K > 1$. The probability that X takes the value x_k is therefore very nearly α_k, and $B(x)$ will be a step function which increases at the points x_k by amounts very nearly equal to $x_k^2 \alpha_k$. $B(x)$ is constant in the interval $(x_k + 0, x_{k+1} - 0)$, and since $K > 1$, it is in this interval of about the same order as $x_k^2 \alpha_k$. Now consider the variation of

$$D(x) = \frac{x^2 A(x)}{B(x)}$$

as x increases from $x_k + 0$ to $x_{k+1} - 0$. When x is just above x_k it is of the same order as

$$\frac{x_k^2 \alpha_{k+1}}{x_k^2 \alpha_k} = \frac{\alpha_{k+1}}{\alpha_k},$$

which is very small. On the other hand, when x is just below x_{k+1}, $D(x)$ is of the order of

$$\frac{x_{k+1}^2 \alpha_{k+1}}{x_k^2 \alpha_k} > K > 1.$$

Hence there is here intermittent convergence to normality.

The intuitive reasons for this behaviour may be explained as follows. Consider the sum S_n as n increases. The chance that any of the X_i in $X_1 + \ldots + X_n$ will exceed the value x_k is about $n\alpha_{k+1}$. Hence as n increases this will remain very small in comparison with the dispersion of S_n for a long time after the time when the probability of at least one of the X_i taking the value x_k (this is about $n\alpha_k$) becomes large. Thus the largest of the X_i will be small compared with the dispersion of S_n and the distribution of S_n will be nearly normal. However, as n increases still further the probability that one or more of the X_i will take the value x_{k+1} will become non-negligible and the distribution will become very non-normal again.

Thus the property of intermittent convergence is related to the behaviour of the tails of the distribution. Lévy has further clarified this dependence by proving the following theorem.

THEOREM 8.46. *For identically distributed independent random variables,* $\{X_i\}$, *the condition that for all α in the range $0 < \alpha < 2$,*

$$E|X_i|^\alpha < \infty,$$

is necessary but not sufficient for convergence of the distribution of S_n to normality, and is sufficient but not necessary for the distribution to show at least intermittent convergence.

Using this theorem and the previous results it follows that, for identically distributed X_i, each of the conditions below implies, but is not implied by, the condition beneath it.

(1) $E|X_i|^2 < \infty$.
(2) The distribution of S_n tends to normality.
(3) $E|X_i|^\alpha < \infty$ for all α satisfying $0 < \alpha < 2$.
(4) The distribution of S_n converges at least intermittently to normality.

The idea of a 'domain of attraction' can be extended, with some modification, to other distributions and a detailed account of such possibilities is given in Loève (1963).

8.22. Some generalizations

Central Limit Theorems for sums

$$S_n = X_1 + \ldots + X_n,$$

where the X_i are random variables which are not independent, are occasionally useful. In order to make the distribution of the sums tend to normality fairly stringent conditions are necessary and these are usually such that ensure that the dispersion of the sum, S_n, increases at least as fast as some power of n, and that terms X_i, X_j which are sufficiently far apart are independent or nearly independent. The first theorems of this type were given by Bernstein (1926–7). A number of other such theorems have also been given by Loève (1950); Diananda (1953, 1954, 1955); Hoeffding and Robbins (1948); Marsaglia (1954); Lomnicki and Zaremba (1957); and Rosenblatt (1961). These theorems are often useful in the theory of random processes.

In all the above Central Limit Theorems the final conclusion has been of the form that a distribution function, $F_n(x)$, converges as n increases to a normal distribution function. If $F_n(x)$ is the integral of its derivative it is in many cases interesting to try to prove the stronger result that the derivative converges to the density function of the normal distribution. Even when the distribution of the individual components of the sum are discrete so that $F_n(x)$ is a discrete distribution it is often possible to show that the ordinates, when suitably scaled, converge to the normal density function. We have already proved a theorem of this type for the binomial distribution, and the general theory has occasioned an extensive literature for which Gnedenko and Kolmogorov (1954) may be consulted. See also Haden (1947) and Hammersley (1952).

Limit theorems when the number of random variables in the sum is itself a random variable are given by Robbins (1948b). The limiting behaviour of repeated convolutions of more general measures than probability measures has been studied by Bergström (1963).

Bibliographic note

The most comprehensive treatment of the subjects of this chapter will be found in Loève (1963), but the reader should also consult Khintchine (1933), Lévy (1937a), and Gnedenko and Kolmogorov (1954). Fréchet (1937) gives an exhaustive discussion of modes of convergence, and Linnik (1961) should be consulted for work on the asymptotic theory of large deviations in sums of random variables.

9. The Arithmetic of Distributions and the Brownian Movement

9.1. WE HAVE already seen that given two distributions, $F_1(x)$ and $F_2(x)$, there always exists another distribution, $F_1 * F_2$, the convolution of F_1 and F_2, which is the distribution of the sum of two independent random variables having the distribution functions $F_1(x)$ and $F_2(x)$. The convolution operator, '*', is thus analogous to multiplication and the analogy is strengthened by the fact that the characteristic function of $F_1 * F_2$ is the product of the characteristic functions of $F_1(x)$ and $F_2(x)$. However, division is obviously not possible in general and in fact even when an analogue of division is possible, the result is not necessarily unique.

We say that a characteristic function, $\phi(t)$, is 'decomposable' if it can be written as

$$\phi(t) = \phi_1(t)\phi_2(t), \tag{9.1}$$

where $\phi_1(t)$, $\phi_2(t)$ are the characteristic functions of two non-degenerate distributions.

We begin by considering discrete distributions but before doing this it is useful to prove a result which relates the distribution $F_1(x) * F_2(x)$ to the two distributions of which it is composed (for some other theorems on convolutions see § 5.8).

We say that a point, x_0, is a point of increase of $F(x)$ if for every $\varepsilon > 0$,

$$F(x_0+\varepsilon)-F(X_0-\varepsilon) > 0. \tag{9.2}$$

Clearly, if we have further that the left-hand side does not tend to zero with ε, x_0 is also a point of discontinuity. Given a distribution, $F(x)$, the set of all points of increase is said to be the 'support' of the distribution. Notice that it is possible for the support of a discrete distribution to consist of the whole line so that there is not a one-to-one correspondence between the discontinuity points of $F(x)$ and the points of the support of purely discrete distributions. In fact the support is the 'closure' of the set of discontinuities, if the distribution is purely discrete. This can be illustrated by constructing a purely discrete distribution whose discontinuity points $x_1, x_2,...$ form an enumerable set which is dense in the whole line $(-\infty, \infty)$.

THEOREM 9.1. *If $F(x) = F_1(x) * F_2(x)$, the support of $F(x)$ contains all points which can be written as $x_1 + x_2$, where x_1 belongs to the support of $F_1(x)$, and x_2 to that of $F_2(x)$.*

Proof. This follows at once from the fact that

$$F(x+h) - F(x-h) = \int \{F_1(x-y+h) - F_1(x-y-h)\} \, dF_2(y)$$

$$\geq \{F_1(x-y_0+\tfrac{1}{2}h) - F_1(x-y_0-\tfrac{1}{2}h)\}\{F_2(y_0+\tfrac{1}{2}h) - F_2(y_0-\tfrac{1}{2}h)\},$$

$$(9.3)$$

where y_0 is any given point in the range of y.

Let l and r be the least and greatest points of increase of $F(x)$, or $-\infty, \infty$ if there is no least or no greatest. Similarly define l_1, r_2, for $F_1(x)$, and l_2, r_2 for $F_2(x)$. Then

$$l = l_1 + l_2,$$

$$r = r_1 + r_2,$$

and

$$l - r = (l_1 - r_1) + (l_2 - r_2). \tag{9.4}$$

From this it follows that a distribution containing all its probability concentrated at two points is indecomposable.

The most interesting discrete distributions are those in which the probability is concentrated at points which can all be represented in the form $a + nh$ (a finite, $h > 0$, $n = 0, \pm 1, ...$). Such a distribution is said to be a 'lattice distribution' and the most frequently occurring lattice distributions have $a = 0$, $h = 1$, and $n \geq 0$. It is then convenient to treat such distributions by using their probability generating functions (§ 2.1), such as

$$P(z) = \sum_{n=0}^{\infty} p_n z^n. \tag{9.5}$$

rather than their characteristic functions $P(e^{i\theta})$. The corresponding distribution is then decomposable if

$$P(z) = P_1(z)P_2(z),$$

where $P_1(z)$, $P_2(z)$ are finite or infinite series with non-negative coefficients. Since $P(1) = 1$, we can arrange that the sum of the coefficients in $P_1(z)$ and $P_2(z)$ each add to unity.

Even when $P(z)$ is a finite polynomial the problem of its decomposability is difficult. If $P(z)$ is indecomposable we can say it is 'prime'. A theorem of unique decomposition into prime factors, such as holds for positive integers, is not then true. This may be illustrated by the following simple example.

Consider the distribution with generating function

$$P(z) = \tfrac{1}{6}(1+z+z^2+z^3+z^4+z^5)$$
$$= \{\tfrac{1}{3}(1+z+z^2)\}\{\tfrac{1}{2}(1+z^3)\}$$
$$= \{\tfrac{1}{3}(1+z^2+z^4)\}\{\tfrac{1}{2}(1+z)\}. \qquad (9.6)$$

This gives us two factorizations and we can show that the factors are indecomposable. First we see from the remark following (9.4) that $\tfrac{1}{2}(1+z)$ and $\tfrac{1}{2}(1+z^3)$ are indecomposable. Consider $\tfrac{1}{3}(1+z+z^2)$. If this is decomposable it must decompose into two factors of the form $(q+pz)(Q+Pz)$ where p, q, P, Q are real and positive. But the roots of $1+z+z^2 = 0$ are complex and hence this is impossible. From Theorem 9.1, $\tfrac{1}{3}(1+z^2+z^4)$ cannot have a factor of the form $(Q+Pz)$. If it is decomposable it must factorize as $(q+pz^2)(Q+Pz^2)$ which is again impossible. Thus (9.6) gives us two different compositions into prime factors. This problem can be generalized as follows.

THEOREM 9.2. *The class of distributions,* $\{P_n(z)\}$, *where*

$$P_n(z) = \frac{1}{n}(1+z+\ldots+z^{n-1}), \qquad (9.7)$$

is indecomposable if n is prime.

A proof of this is given by Krasner and Ranulac (1937). On the other hand, if n is not prime suppose it equals lm where l, m are integers greater than unity. Then

$$P_n(z) = P_l(z)P_m(z^l)$$
$$= P_m(z)P_l(z^m).$$

Thus if

$$n = p_1^{l_1} p_2^{l_2} \ldots p_k^{l_k}$$

where $p_1, p_2, \ldots$ are primes, (9.7) can be decomposed into prime factors in

$$\frac{\Sigma l_i}{l_1! \ldots l_k!} \qquad (9.8)$$

different ways, it being easy to verify that since $P_n(z)$ is indecomposable, so also is $P_n(z^m)$, $m > 1$. Krasner and Ranulac (1937) have shown that these are the only decompositions of $P_n(z)$.

It is even possible to have an indecomposable $P(z)$ such that $P(z)^2$ is decomposable into factors $P_1(z)$, $P_2(z)$, which are not equal to $P(z)$ (Dugué 1957). For this we take

$$P(z) = \tfrac{1}{35}(1+2z+5z^2+12z^3+15z^4) = \tfrac{1}{35}(1-z+5z^2)(1+3z+3z^2),$$

and then

$$P(z)^2 = \tfrac{1}{7}(1+3z+3z^2)\tfrac{1}{175}(1+z+8z^2+17z^3+28z^4+45z^5+75z^6).$$

Examination of the roots of $P(z) = 0$ show that $P(z)$ is indecomposable. This subject will clearly bear much further study.

In Cramér's theorem (Theorem 7.2) we showed that if the normal distribution is composed of two components, each of the latter must be normal also. In the theory of discrete distribution there is a theorem, due to D. A. Raikov (1938), which is analogous to this.

THEOREM 9.3 (Raikov's theorem). *If Z is a random variable whose distribution is a Poisson distribution with generating function*

$$P(z) = e^{\lambda(z-1)},\tag{9.9}$$

and Z = X + Y, where X and Y are independent random variables with non-degenerate distributions, then X and Y are also Poisson variates.

Proof. The support of Z is the set of integers 0, 1, 2,.... Consider the support of X. The distance between any two points of this set must be an integer. Thus the distribution of X must be confined to the points α, $\alpha \pm 1$, $\alpha \pm 2$,..., where α is some constant. Furthermore, the distribution of X must have a left extremity. Thus by repositioning there is no restriction in supposing that the distribution of X is confined to the points 0, 1, 2,..., zero being its true left extremity. The distribution of Y must also be confined to these points. We can then write

$$e^{\lambda(z-1)} = P_1(z)P_2(z),\tag{9.10}$$

where the $P_i(z)$ are power series which we write as

$$P_1(z) = a_0 + a_1 z + a_2 z^2 + \dots,\tag{9.11}$$

$$P_2(z) = b_0 + b_1 z + b_2 z^2 + \dots,\tag{9.12}$$

where $a_i \geqslant 0$, $b_i \geqslant 0$, and

$$\Sigma a_i = \Sigma b_i = 1.$$

Next we have

$$e^{-\lambda}\lambda^n(n!)^{-1} = a_0 b_n + \dots + a_n b_0 \quad (a_0 > 0, b_0 > 0),\tag{9.13}$$

so that

$$a_n \leqslant b_0^{-1}e^{-\lambda}\lambda^n(n!)^{-1}.$$

This implies that $P_1(z)$ is an integral function on the whole z-plane, and

$$|P_1(z)| \leqslant b_0^{-1}e^{\lambda(|z|-1)}, \quad \text{for all } z.\tag{9.14}$$

We also have

$$P_1(z) = e^{\lambda(z-1)}P_2(z)^{-1},$$

and using an inequality similar to (9.14) for $P_2(z)$ we have

$$|P_1(z)| \geqslant |e^{\lambda(z-1)}a_0 e^{\lambda(1-|z|)}|$$

$$\geqslant a_0|e^{\lambda(z-|z|)}|.$$

Thus

$$|\log P_1(z)| \leqslant \alpha + \beta|z|,\tag{9.15}$$

where α and β are real finite constants. $P_1(z)$ is never zero or infinite for finite

z, and hence $\log P_1(z)$ is an integral function, of order not exceeding unity. It follows from Liouville's theorem (Titchmarsh 1932) that $\log P_1(z)$ is of the form $a+bz$, and hence $P(z)$ must be the generating function of a Poisson distribution. A generalization of this result to convolutions of Poisson and normal distributions is given by Linnik (1957).

9.2. Absolutely continuous distributions

We now consider an example of the indecomposability of absolutely continuous distributions, i.e. distributions which are the integrals of their derivatives. Consider the distribution

$$f(x) = (2\pi)^{-\frac{1}{2}}x^2 e^{-\frac{1}{2}x^2}. \tag{9.16}$$

That this is a distribution function is easily verified by integration. Its characteristic function is

$$\phi(t) = (2\pi)^{-\frac{1}{2}}\int_{-\infty}^{\infty} x^2 e^{itx-\frac{1}{2}x^2}\,dx$$

$$= -\frac{d^2}{dt^2}(2\pi)^{-\frac{1}{2}}\int_{-\infty}^{\infty} e^{itx-\frac{1}{2}x^2}\,dx$$

$$= -\frac{d^2}{dt^2}e^{-\frac{1}{2}t^2}$$

$$= (1-t^2)e^{-\frac{1}{2}t^2}. \tag{9.17}$$

Suppose that $Z = X+Y$ has the distribution (9.16), where X and Y are independent random variables. By following the arguments used in the proofs of Cramér's theorem (Theorem 7.2) we see that the even moments of X and Y are bounded by those of Z, and hence that their characteristic functions must be integral functions of orders not exceeding that of (9.17).

If the distributions of X or Y had a normal component it is easy to see that the probability density of $f(x)$ would have to be non-zero everywhere, which is false. Hence the distributions of X and Y have no normal component. Furthermore, if either were of the form

$$(1-t^2)e^{-\frac{1}{2}\sigma^2 t^2} \quad (0 \leqslant \sigma^2 < 1), \tag{9.18}$$

the other would be of the form

$$e^{-\frac{1}{2}(1-\sigma^2)t^2}$$

which we have shown to be impossible. Hence the only factors of the characteristic function which can be integral functions of the correct order are factors of the forms

$$(1-t)e^{-\frac{1}{2}\sigma^2 t^2}, \tag{9.19}$$

$$(1+t)e^{-\frac{1}{2}\sigma^2 t^2}, \tag{9.20}$$

where $0 < \sigma^2 \leqslant 1$. These do not satisfy the conditions for characteristic functions (Theorem 6.19) and in particular $\phi(-t) \neq \overline{\phi(t)}$. Thus (9.16) is indecomposable.

9.3. Infinitely divisible distributions

To carry further the theory of the divisibility of distribution we now consider a class of distributions which are said to be 'infinitely divisible'.

A distribution function, $F(x)$, is said to be infinitely divisible if for every positive integer n, there exists a distribution, $F_1(x)$, such that $F(x) = F_1(x)^{n*}$, or equivalently, if $\phi(t)$ is the characteristic function of $F(x)$, $\phi(t)^{1/n}$ must be a characteristic function for every such n.

Clearly the normal distribution, the Poisson distribution, the gamma distribution, and the negative binomial distributions are infinitely divisible. Furthermore, if the sequence of distributions, $\{F_n(x)\}$, are each infinitely divisible, so also are $F_1(x) * F_2(x)$, and $\lim_n F_n(x)$, if the latter is a proper distribution. We also have the following simple result.

THEOREM 9.4. *If $\phi(t)$ is the characteristic function of an infinitely divisible distribution,*

$$\phi(t) \neq 0, \qquad (9.21)$$

for all real t.

Proof. $\phi(t)$ is continuous at the origin. Hence $\phi(t) \neq 0$ in the neighbourhood of the origin. Consider

$$\lim_{n \to \infty} \phi(t)^{1/n}. \qquad (9.22)$$

Each $\phi(t)^{1/n}$ is a characteristic function and as n increases their values for a given t converge to 0 or 1. In the neighbourhood of the origin (9.22) is a continuous function equal to unity. Since the conditions of Theorem 6.2 are satisfied, (9.22) is a characteristic function itself (possibly degenerate), and is therefore continuous everywhere so that (9.22) cannot equal zero for any real t. Thus (9.21) follows. The characteristic function $\phi(t) = \frac{1}{3}(2+e^{it})$ shows that the converse of Theorem 9.4 is not true, i.e. (9.21) is not a sufficient condition.

Another definition of infinitely divisible distributions was given by Lévy (1937a). In this definition $F(x)$ is said to be infinitely divisible if given any $\varepsilon > 0$, $\varepsilon' > 0$, it can be represented as a convolution

$$F_1(x) * F_2(x) \dots * F_k(x)$$

such that for each i,

$$1 + F_i(-\varepsilon) - F_i(\varepsilon - 0) < \varepsilon'. \qquad (9.23)$$

In this definition the $F_i(x)$ are not assumed to be the same.

These two definitions are equivalent. Assume that $F(x)$ satisfies the first. Then as n increases, $\phi(t)^{1/n}$ tends to unity uniformly in any finite interval on the t-axis. Now using the second part of Theorem 6.26 (equation (6.32)) we see that given $\varepsilon > 0$, $\varepsilon' > 0$ we can choose n so large that (9.23) is satisfied since we already know from the first definition that $\phi(t)^{1/n}$ is a characteristic function. For a proof of the converse result that the second definition implies the first see Doob (1953). We only use the first definition here.

It follows that if $\phi(t)$ is the characteristic function of an infinitely divisible distribution so also are $\phi(t)^\alpha$ and $|\phi(t)|^\alpha$ for every $\alpha > 0$, the latter result being obvious from the equation

$$|\phi(t)|^\alpha = \phi(t)^{\frac{1}{2}\alpha}\phi(-t)^{\frac{1}{2}\alpha}. \tag{9.24}$$

9.4. Homogeneous additive processes

Infinitely divisible processes can be looked at from another point of view. Let t be a real number ($-\infty < t < \infty$) which we interpret as time. Then we define a 'random process' to be a set of random variables, $X(t)$, where t ranges over some set in the interval $-\infty < t < \infty$. Such a process is 'additive' if given any set of values of t such as $t_1, \ldots, t_n$, where

$$t_1 < t_2 < \ldots < t_n, \tag{9.25}$$

the random variables

$$X(t_2) - X(t_1),\ X(t_3) - X(t_2), \ldots,\ X(t_n) - X(t_{n-1}) \tag{9.26}$$

are all independent. If, in addition, the distribution of

$$X(t_2) - X(t_1)$$

depends only on $t_2 - t_1$ the process is said to be 'homogeneous'.

Clearly the distribution of $X(t_2) - X(t_1)$ must be infinitely divisible if the process is additive and homogeneous, and conversely to every infinitely divisible distribution corresponds an additive homogeneous process. A good deal of the interest in the study of the latter is concerned with the joint distribution of the $X(t)$ for different points of time, and we shall study such questions later for some particular processes. Here we are concerned simply with the distribution of $X(t_2) - X(t_1)$.

Two examples which are fundamental in the whole theory are the Brownian process in which $X(t_2) - X(t_1)$ has a normal distribution with mean zero and variance proportional to $|t_2 - t_1|$, and the Poisson process in which $X(t_2) - X(t_1)$ has the Poisson distribution with parameter proportional to $t_2 - t_1$ (where $t_2 > t_1$).

In fact the general homogeneous additive process can be regarded as arising from the addition of a Brownian process and a process obtained by generalizing the Poisson process.

Consider an elementary Poisson process in which $N(t)$, the number of events occurring in the time interval $(0, t)$, has the Poisson distribution with mean λt, and such that the numbers of events occurring in non-overlapping intervals are distributed independently. The probability generating function of $N(t)$ is

$$P(z) = e^{\lambda t(z-1)}. \tag{9.27}$$

We now write u for the variable in the characteristic function to avoid confusion with t as used for time.

Suppose that with each event we associate a random variable, Z, having the distribution $F(z)$, and that the Z's with different events are independent. If $X(t)$ is the sum of the Z's for the events in the interval $(0, t)$, the characteristic function of $X(t)$ is

$$\exp \lambda t\{\psi(u) - 1\}, \tag{9.28}$$

where $\psi(u)$ is the characteristic function of the distribution $F(z)$. We may express this by saying that the second characteristic function is

$$\log \phi(u) = \lambda t \left\{ \int_{-\infty}^{\infty} e^{ixu} \, dF(x) - 1 \right\}. \tag{9.29}$$

Then the corresponding homogeneous additive process will be such that in any time interval $(0, t)$, $X(t)$ will be constant except for a finite number of jumps each of which has the distribution $F(x)$. The number of jumps of size lying in the interval $a < x \leqslant b$ will have a Poisson distribution with mean $\lambda t\{F(b) - F(a)\}$. We therefore expect that we can generalize (9.29) by letting the number of small jumps increase indefinitely provided we make them small enough for their sum to have a finite dispersion. The theorem of the next section shows that this can be done in such a way that by further adding a normal component, the general form of infinitely divisible distributions and homogeneous additive processes is obtained.

9.5. The canonical representation of an infinitely divisible characteristic function

THEOREM 9.5. $\phi(u)$ is the characteristic function of an infinitely divisible distribution if and only if $\log \phi(u)$ can be written in the form

$$iau - \tfrac{1}{2}\sigma^2 u^2 + \int_{-\infty}^{0-0} \left(e^{iux} - 1 - \frac{iux}{1+x^2} \right) dM(x)$$

$$+ \int_{0+0}^{\infty} \left(e^{iux} - 1 - \frac{iux}{1+x^2} \right) dN(x), \tag{9.30}$$

where a and σ are real numbers, $\sigma \geqslant 0$, $M(x)$ and $N(x)$ are non-decreasing in the intervals $(-\infty, 0)$ and $(0, \infty)$ respectively, and such that

$$\lim_{x \to \infty} M(-x) = \lim_{x \to \infty} N(x) = 0, \tag{9.31}$$

and

$$\int_{-\varepsilon}^{0} x^2 \, dM(x), \quad \int_{0}^{\varepsilon} x^2 \, dN(x), \tag{9.32}$$

are finite for every $\varepsilon > 0$.

Proof. We first prove that this is a sufficient condition, i.e. that if $\log \phi(u)$ is of the form (9.30), $\phi(u)$ is the characteristic function of an infinitely divisible distribution.

The effect of the term iau is simply to shift the mean of the distribution so that we can suppose $a = 0$. Similarly the effect of the term $-\frac{1}{2}\sigma^2 u^2$ is to incorporate a normally distributed component so that we can likewise put $\sigma = 0$.

We have already seen that (9.29) is the logarithm of the characteristic function of an infinitely divisible distribution and so also must be any function which is the limit of any sequence of functions of the form (9.29) provided this limit is uniform in every finite interval. From the condition that the integrals (9.32) exist we see that

$$\lim_{\varepsilon \to 0} \int_{-\varepsilon}^{0} x^2 \, dM(x) = \lim_{\varepsilon \to 0} \int_{0}^{\varepsilon} x^2 \, dN(x) = 0, \tag{9.33}$$

and hence that

$$\log \phi(u) = \lim_{\varepsilon \to 0} \int_{-\infty}^{-\varepsilon} \left(e^{iux} - 1 - \frac{iux}{1+x^2} \right) dM(x) + \lim_{\varepsilon \to 0} \int_{\varepsilon}^{\infty} \left(e^{iux} - 1 - \frac{iux}{1+x^2} \right) dN(x). \tag{9.34}$$

Consider the integral

$$I_1 = \int_{\varepsilon}^{\infty} \left(e^{iux} - 1 - \frac{iux}{1+x^2} \right) dN(x). \tag{9.35}$$

From the properties of $N(x)$ this is clearly an absolutely convergent Riemann–Stieltjes integral and is the limit of sums of terms of the form

$$\left(e^{iux_i} - 1 - \frac{iux_i}{1+x_i^2} \right) \{ N(x_{i+1}) - N(x_i) \}.$$

Each of these is the logarithm of the characteristic function of a Poisson distribution whose position has been shifted through a finite distance. I_1 is therefore the logarithm of a characteristic function, and so also must be

$$\lim_{\varepsilon \to 0} I_1,$$

which exists because the integrand in (9.35) is of order x^2 for x small, and the integral (9.32) is finite. A similar argument applies to the first integral in (9.34) and the proof is complete. Note that (9.30) is still the logarithm of a characteristic function if each of the four terms comprising it are multiplied by arbitrary positive constants. Thus

$$iaut - \tfrac{1}{2}\sigma^2 u^2 t + t \int_{-\infty}^{0-0} \left(e^{iux} - 1 - \frac{iux}{1+x^2} \right) dM(x) + t \int_{0+0}^{\infty} \left(e^{iux} - 1 - \frac{iux}{1+x^2} \right) dN(x)$$

is the logarithm of the characteristic function of the distribution of $X(t) - X(0)$ in a homogeneous additive process for all t in the interval $0 \leqslant t < \infty$.

We next have to prove that if $\phi(u)$ is the characteristic function of an infinitely divisible distribution, $\log \phi(u)$ can be written in the form (9.30).

We first show that any function of the form (9.30) can be written in the form

$$iau + \int_{-\infty}^{\infty} \left(e^{iux} - 1 - \frac{iux}{1+x^2} \right) \frac{1+x^2}{x^2} dG(x), \tag{9.36}$$

and conversely, where $G(x)$ is a bounded non-decreasing function such that $G(-\infty) = 0$. At $x = 0$ the integrand in (9.36) is undefined and we take it as defined by continuity so that it has the value $-\tfrac{1}{2}u^2$. We then suppose that $G(x)$ has a jump of magnitude σ^2 at $x = 0$, so that this makes a contribution of $-\tfrac{1}{2}u^2\sigma^2$, which corresponds to the normal component.

To prove the equivalence of (9.30) and (9.36) we therefore define $G(x)$ by the conditions

$$G(0+0) - G(0-0) = \sigma^2, \tag{9.37}$$

$$M(x) = \int_{-\infty}^{x} \frac{1+x^2}{x^2} dG(x), \quad x < 0, \tag{9.38}$$

$$N(x) = -\int_{x}^{\infty} \frac{1+x^2}{x^2} dG(x), \quad x > 0. \tag{9.39}$$

Because of the conditions (9.31) and (9.32) this is always possible, and conversely if $G(x)$ satisfies the above conditions, $M(x)$ and $N(x)$ as defined by (9.38) and (9.39) will satisfy the conditions of Theorem 9.5. The expression (9.36) is known as the Lévy–Khintchine canonical representation whereas (9.30) is known as the Lévy canonical representation. It is now clearly sufficient to prove that every infinitely divisible distribution has a second characteristic function of the form (9.36).

Let $\phi(u)$ be the characteristic function of an infinitely divisible distribution. We prove that there exists a sequence $\{\phi_n(u)\}$ of function which converge to $\phi(u)$ everywhere, and uniformly in every bounded interval, and whose logarithms are of the Lévy–Khintchine canonical form, i.e.

$$\log \phi_n(u) = ia_n u + \int_{-\infty}^{\infty} \left(e^{iux} - 1 - \frac{iux}{1+x^2} \right) \frac{1+x^2}{x^2} dG_n(x). \tag{9.40}$$

Write

$$\Phi(u) = \log \phi(u).$$

We know that $\phi(u)^{1/n}$ is a characteristic function so that there is a distribution function, $F_n(x)$, such that

$$\phi(u)^{1/n} = \int_{-\infty}^{\infty} e^{ixu}\, dF_n(x). \tag{9.41}$$

Furthermore, we have

$$n\{\phi(u)^{1/n} - 1\} = n\left\{\exp\frac{1}{n}\Phi(u) - 1\right\}$$

$$= n\left\{\frac{1}{n}\Phi(u) + O(n^{-2})\right\}$$

$$= \Phi(u) + O(n^{-1}). \tag{9.42}$$

Thus for every u,

$$\Phi(u) = \lim_n n\{\phi(u)^{1/n} - 1\}.$$

Then from (9.41),

$$n\{\phi(u)^{1/n} - 1\} = n\int_{-\infty}^{\infty} (e^{ixu} - 1)\, dF_n(x)$$

$$= niu\int_{-\infty}^{\infty} \frac{x}{1+x^2}\, dF_n(x) + n\int_{-\infty}^{\infty}\left(e^{iux} - 1 - \frac{iux}{1+x^2}\right) dF_n(x). \tag{9.43}$$

These integrals are convergent because $F_n(x)$ is a distribution function. We can then put

$$\log\phi_n(u) = ia_n u + \int_{-\infty}^{\infty}\left(e^{iux} - 1 - \frac{iux}{1+x^2}\right)\frac{1+x^2}{x^2}\, dG_n(x),$$

where $\phi_n(u)$ is a function defined by

$$\log\phi_n(u) = n\{\phi(u)^{1/n} - 1\}$$

$$= \Phi_n(u), \quad \text{say}, \tag{9.44}$$

and

$$a_n = n\int_{-\infty}^{\infty} \frac{x}{1+x^2}\, dF_n(x), \tag{9.45}$$

$$G_n(x) = n\int_{-\infty}^{x} \frac{y^2}{1+y^2}\, dF_n(y). \tag{9.46}$$

This is of the form (9.36) and $\phi_n(u)$ is therefore a characteristic function by the first part of the theorem, and converges to $\phi(u)$ as required.

We now show that there exists a finite constant, a, and a bounded non-decreasing function, $G(x)$, such that

$$a = \lim_n a_n,$$

$$G(x) = \lim_n G_n(x),$$

and

$$\int_{-\infty}^{\infty} dG(x) = \lim_n \int_{-\infty}^{\infty} dG_n(x).$$

Consider the integral

$$\lambda_n(u) = \int_0^1 \left\{ \log \phi_n(u) - \frac{\log \phi_n(u+h) + \log \phi_n(u-h)}{2} \right\} dh$$

$$= \int_0^1 dh \int_{-\infty}^{\infty} e^{iux}(1-\cos xh) \frac{1+x^2}{x^2} dG_n(x). \tag{9.47}$$

By bounded convergence we can reverse the order of integration so that we obtain

$$\int_{-\infty}^{\infty} e^{iux}\left(1 - \frac{\sin x}{x}\right) \frac{1+x^2}{x^2} dG_n(x) = \int_{-\infty}^{\infty} e^{iux} dH_n(x), \tag{9.48}$$

where

$$H_n(x) = \int_{-\infty}^{x} \left(1 - \frac{\sin y}{y}\right) \frac{1+y^2}{y^2} dG_n(y). \tag{9.49}$$

We know that $\phi(u)$ is a continuous function and therefore $\lambda_n(u)$ converges to a continuous function as n increases. Using the continuity theorem (Theorem 6.3) applied to (9.48) we conclude that $H_n(x)$ converges weakly to a bounded non-decreasing function, $H(x)$, at all the continuity points of the latter, and that

$$\lim_n \int_{-\infty}^{\infty} dH_n(x) = \int_{-\infty}^{\infty} H(x).$$

From Helly's second theorem (Theorem 6.5) it follows that

$$\lim_n G_n(x) = \int_{-\infty}^{x} \left(1 - \frac{\sin y}{y}\right)^{-1} \frac{y^2}{1+y^2} dH(y)$$

$$= G(x), \quad \text{say.}$$

Using the same theorem again,

$$\lim_n \int_{-\infty}^{\infty} \left(e^{iux} - 1 - \frac{iux}{1+x^2}\right) \frac{1+x^2}{x^2} dG_n(x)$$

$$= \int_{-\infty}^{\infty} \left(e^{iux} - 1 - \frac{iux}{1+x^2}\right) \frac{1+x^2}{x^2} dG(x). \tag{9.50}$$

This establishes the existence and properties of $G(x)$, and it then follows that $\lim_n a_n$ must exist also. This concludes the proof of Theorem 9.5.

THEOREM 9.6. *The representations* (9.30) *and* (9.36) *are unique.*

Proof. We use an argument similar to the above and write

$$\lambda(u) = \int_0^1 \left\{ \log \phi(u) - \frac{\log \phi(u+h) + \log \phi(u-h)}{2} \right\} dh$$

$$= \int_{-\infty}^{\infty} e^{iux} \left(1 - \frac{\sin x}{x} \right) \frac{1+x^2}{x^2} dG(x)$$

$$= \int_{-\infty}^{\infty} e^{iux} dH(x), \tag{9.51}$$

where

$$H(x) = \int_{-\infty}^{x} \left(1 - \frac{\sin y}{y} \right) \frac{1+y^2}{y^2} dG(y).$$

The integrand in this integral is bounded between two positive constants and hence the unique determination of $H(x)$ (in the weak sense) by equation (9.51) also uniquely defines $G(x)$. Once $G(x)$ is determined, a must be fixed also.

9.6. The case of bounded variance

Kolmogorov (1932) had earlier obtained a canonical representation for infinitely divisible distributions which have finite variances.

THEOREM 9.7 (the Kolmogorov canonical representation). *If $\phi(u)$ is the characteristic function of an infinitely divisible distribution which has finite variance, $\log \phi(u)$ can be written in the form*

$$iau + \int_{-\infty}^{\infty} (e^{iux} - 1 - iux) x^{-2} dK(x), \tag{9.52}$$

where a is a real constant, and $K(x)$ is a non-decreasing bounded function such that $K(-\infty) = 0$. Furthermore, this representation is unique.

Proof. The integrand in (9.52) does not exist at $x = 0$ and is defined there by continuity so that it equals $-\frac{1}{2}u^2$.

Since the second moment is finite, $\phi(u)$ and $\log \phi(u)$ can be differentiated twice. Then

$$\left| \lim_{h \to 0} h^{-2} \{ \log \phi(h) - 2 \log \phi(0) + \log \phi(-h) \} \right| < \infty. \tag{9.53}$$

Since the distribution is infinitely divisible we have

$$\log \phi(u) = ia_1 u + \int_{-\infty}^{\infty} \left(e^{iux} - 1 - \frac{iux}{1+x^2} \right) \frac{1+x^2}{x^2} dG(x). \tag{9.54}$$

Then from (9.53), after inserting (9.54), it follows that

$$\int_{-\infty}^{\infty} (1+x^2) \, dG(x)$$

is finite, and also

$$\int_{-\infty}^{\infty} x \, dG(x).$$

Put

$$K(x) = \int_{-\infty}^{x} (1+x^2) \, dG(x),$$

and

$$a = a_1 + \int_{-\infty}^{\infty} x \, dG(x).$$

Then (9.54) becomes

$$ia_1 u + \int_{-\infty}^{\infty} \left(e^{iux} - 1 - \frac{iux}{1+x^2} \right) x^{-2} \, dK(x)$$

$$= ia_1 u + \int_{-\infty}^{\infty} (e^{iux} - 1 - iux) x^{-2} \, dK(x) + \int_{-\infty}^{\infty} iux \, dG(x)$$

$$= iau + \int_{-\infty}^{\infty} (e^{iux} - 1 - iux) x^{-2} \, dK(x), \tag{9.55}$$

as required.

9.7. Examples of infinitely divisible distributions

The obvious case of the normal distribution will be treated in detail later in this chapter. Another particularly interesting example is the gamma distribution (§ 7.17) which we write as

$$f(x) = \Gamma(t)^{-1} e^{-x} x^{t-1} \quad (0 \leqslant x < \infty), \tag{9.56}$$

where $t > 0$. The characteristic function is

$$\phi(u) = (1 - iu)^{-t}. \tag{9.57}$$

The functions involved in the representations can then be found as above by considering the limiting behaviour of $\phi(u)^{1/n}$. In this way we get the Lévy representation (9.30) with

$$a = t \int_{0}^{\infty} \frac{e^{-y}}{1+y^2} \, dy, \tag{9.58}$$

$$\sigma = 0, \quad M(x) = 0 \quad (x < 0), \tag{9.59}$$

and

$$N(x) = -t \int_{x}^{\infty} \frac{e^{-y}}{y} \, dy \quad (x > 0). \tag{9.60}$$

The other representations are then easily derived. For this distribution there is no normal component and it can be shown that when regarded as a homogeneous additive process, the curve, $Y = X(t)$, has an infinite number of positive jumps or discontinuities in every interval.

The Cauchy, Poisson, and negative binomial are also obviously infinitely divisible. We consider only the last here. The characteristic function can be written

$$\phi(u) = \left(\frac{p}{1 - qe^{iu}}\right)^t, \tag{9.61}$$

where $0 < p = 1 - q < 1$, and $t > 0$. Then we can easily verify that the Lévy representation (9.30) is given by

$$a = t \sum_{s=1}^{\infty} \frac{q^s}{1 + s^2}, \tag{9.62}$$

$$\sigma = 0,$$

$$M(x) = 0 \quad (x < 0), \tag{9.63}$$

and

$$N(x) = t \sum_{s=1}^{\infty} \frac{q^s}{s} I(x - s) + t \log p, \tag{9.64}$$

where $I(x) = 1$ if $x > 0$, and is zero otherwise. It is instructive to observe that the gamma distribution can be obtained as the limit of the negative binomial after rescaling. This is most easily seen from the characteristic function. In (9.61) we put $u = vp$ and let $p \to 0$, $q \to 1$ so that the limit of (9.61) becomes

$$(1 - iv)^{-t}.$$

The same result can be obtained by considering the limits of (9.62) and (9.63).

The idea of infinite divisibility has also been extended to multivariate distributions (see Lévy 1937a, 1948b).

9.8. The Poisson process and Campbell's theorem

Before studying applications of the above theory of infinitely divisible distributions to the arithmetic of distributions we consider in more detail the two basic additive processes, the Poisson process, and normal or Brownian process.

The Poisson process may be defined by a random variable, $X(t)$, with the property that $X(t_2) - X(t_1)$ $(t_1 < t_2)$, has a Poisson distribution with mean $\lambda(t_2 - t_1)$ $(\lambda > 0)$, and that quantities such as $X(t_2) - X(t_1)$ for sets of non-overlapping intervals are independent. It is convenient to fix $X(0)$ as equal to zero and consider what happens when $t > 0$.

We have already (§ 9.4) generalized this idea by supposing that the jumps in the process $\{X(t)\}$ have a general distribution instead of unit values, and a further generalization of this gives the general infinitely divisible distribution, except for the normal component. We now consider a generalization in a different direction which has interesting applications in electrical theory.

We start from a Poisson process with $X(0) = 0$, and refer to the jumps in the graph of $X(t)$ as 'events'. The number of events occurring in an interval $(0, t)$ has a Poisson distribution with mean λt. If there are n of them denote their times of occurrence by $t_1,..., t_n$. If we suppose that they are numbered from 1 to n in a random manner independent of their time of occurrence we can suppose the quantities $\{t_i\}$ to be uniformly and independently distributed over the interval $(0, t)$.

Now suppose that when an event occurs at a time u it has an effect which lasts after it in such a way that the effect at time $v > u$ is $f(v-u)$, and furthermore that the effects of different events are additive. Then the total effect, $Z(t)$ say, at time t of the events at times $t_1,..., t_n$ will be

$$Z(t) = \sum_1^n f(t-t_i). \tag{9.65}$$

This can be written in the form

$$Z(t) = \int_0^t f(t-u)\,dX(u), \tag{9.66}$$

where $X(u)$ is the random function of jumps defined above. (9.66) is an example of a 'stochastic integral'.

We are interested in the distribution of $Z(t)$ and to ensure that it exists we assume that $f(t)$ is a bounded integrable function on the interval $(0, T)$. The characteristic function of $Z(T)$, conditional on there being n events, will be the nth power of the characteristic function of the effect due to a single event, at x say. It is therefore

$$\left\{T^{-1} \int_0^T e^{iuf(x)}\,dx\right\}^n. \tag{9.67}$$

Averaging over the distribution of n the characteristic function becomes

$$\phi(u) = \exp \lambda \int_0^T \{e^{iuf(x)} - 1\}\,dx. \tag{9.68}$$

The main interest in this theory arises when we consider the asymptotic distribution when T tends to infinity. To make this meaningful in a simple way we assume in addition that $f(x)$ tends to zero as x tends to infinity and that $|f(x)|$ is bounded and finitely integrable over the range $(0, \infty)$. The characteristic function is then

$$\phi(u) = \exp \lambda \int_0^\infty \{e^{iuf(x)} - 1\}\,dx. \tag{9.69}$$

The first and second moments can be obtained by differentiation of $\phi(u)$, and are

$$m_1 = \lim_{u \to 0} \lambda \int_0^\infty e^{iuf(x)} f(x)\, dx\ \phi(u)$$

$$= \lambda \int_0^\infty f(x)\, dx, \tag{9.70}$$

and

$$m_2 = \lim_{u \to 0} \lambda \int_0^\infty e^{iuf(x)} f(x)^2\, dx\ \phi(u) + \lim_{u \to 0} \left\{ \lambda \int_0^\infty e^{iuf(x)} f(x)\, dx \right\}^2 \phi(u)$$

$$= m_1^2 + \lambda \int_0^\infty f(x)^2\, dx. \tag{9.71}$$

(9.70) and (9.71) constitute 'Campbell's theorem' (Campbell 1909), and can obviously be derived directly. For other discussions of this result see Whittaker (1937, 1938), Rowland (1936, 1937), and Khintchine (1938).

It is worth pointing out that $\phi(u)$ in (9.69) must be the characteristic function of an infinitely divisible distribution. This follows from direct considerations, and (9.69) is not in one of the canonical forms for infinitely divisible distributions.

The process can be generalized somewhat by allowing the after-effect to have a random element. Thus we may suppose each jump to be associated with a random variable, V, which has some distribution and that the after-effect at t of a jump V at u is $Vf(t-u)$. If the jumps at $t_1, t_2, \ldots$ correspond to random variables $V_1, V_2, \ldots$ which are independent and have the same distribution, $F(v)$, the characteristic function is easily seen to be

$$\phi(u) = \exp \lambda \int_0^\infty \int_0^\infty \{e^{iuvf(x)} - 1\}\, dx\, dF(v), \tag{9.72}$$

from which the moments can be derived. In particular,

$$m_1 = \lambda E(V) \int_0^\infty f(x)\, dx, \tag{9.73}$$

$$\mu_2 = \lambda E(V^2) \int_0^\infty f(x)^2\, dx. \tag{9.74}$$

The formulae for the moments in this and the previous case can be summarized by the result that the nth cumulant is

$$\lambda E(V^n) \int_0^\infty f(x)^n\, dx. \tag{9.75}$$

Rice (1944) shows how the inverse transforms of (9.72) can be used to find an expansion of the distribution of $Z(t)$ in an Edgeworth series, and that this distribution tends to normality when λ increases and the other conditions remain fixed.

It is also interesting to consider the 'serial correlations' of the above process. For simplicity, we consider the case where each jump is of the same size. When t becomes very large we see that the process defined by (9.65) becomes stationary so that we can write

$$Z(t) = \sum_1^\infty f(t-t_i), \tag{9.76}$$

where $t > t_1 > t_2 > t_3 > ...$, and all events prior to time t are considered. It is not hard to show that under the assumptions made on $f(t)$, (9.76) is almost certainly convergent (first prove convergence in probability and then use Theorem 8.30). $Z(t)$ is then a stationary process in the sense that if $t, t_1, t_2,...$ are prescribed times, the joint distribution of

$$Z(t), Z(t_1),..., Z(t_n)$$

depends only on $t-t_1,..., t-t_n$. We have already found $E\{Z(t)\}$ and $E\{Z(t)^2\}$. Consider $E\{Z(t)Z(t-u)\}$ where $u > 0$. This could be obtained by considering the characteristic function of the joint distribution of $Z(t)$ and $Z(t-u)$. However, we obtain it here directly. Suppose first that we consider only the effects of events in the interval $(t-T, t)$ where $T > u$, and suppose that there are n events in this interval at instants $x_1,..., x_n$ $(t-T \leqslant x_i \leqslant t)$. Then making the convention that $f(x) = 0$ for $x < 0$, we have

$$Z(t) = \sum_1^n f(t-x_i),$$

$$Z(t-u) = \sum_1^n f(t-u-x_i),$$

and taking expectations conditional on n,

$$E(Z(t)Z(t-u)) = nE\{f(t-x_1)f(t-u-x_1)\} + n(n-1)E\{f(t-x_1)f(t-u-x_2)\},$$

where x_1, x_2 are uniformly distributed over the interval $(t-T, t)$. This is equal to

$$nT^{-1} \int_{t-T}^t f(t-x)f(t-u-x)\,dx$$

$$+ n(n-1)T^{-2} \int_{t-T}^t f(t-x)\,dx \int_{t-T}^t f(t-u-y)\,dy.$$

Taking the expectations of n and n^2, and letting T tend to infinity, we get

$$E\{Z(t)Z(t-u)\} = \lambda \int_0^\infty f(x)f(x-u)\,dx + \lambda^2 \left\{ \int_0^\infty f(x)\,dx \right\}^2. \tag{9.77}$$

It follows that the 'serial covariance' of the process is

$$C(u) = \lambda \int_{-\infty}^\infty f(x)f(x-u)\,dx, \tag{9.78}$$

this integral existing because of the restrictions placed on $f(x)$, and, as above, the convention being made that $f(x)$ is defined to be zero when $x < 0$.

(9.78) leads to a particular case of a very important result in the theory of stationary processes. Consider the Fourier transform of $f(x)$,

$$F(y) = (2\pi)^{-\frac{1}{2}} \int_{-\infty}^{\infty} e^{ixy} f(x) \, dx. \tag{9.79}$$

Then

$$f(x) = (2\pi)^{-\frac{1}{2}} \int_{-\infty}^{\infty} e^{-ixy} F(y) \, dy, \tag{9.80}$$

and by Parseval's theorem for Fourier transforms (Titchmarsh 1932)

$$\int_{-\infty}^{\infty} f(x)f(x-u) \, dx = \int_{-\infty}^{\infty} e^{iuy} F(y)^2 \, dy. \tag{9.81}$$

Except for a scale factor, the right-hand side of (9.81) is a characteristic function since it is the Fourier transform of a non-negative integrable function. This formula therefore shows that all serial correlation functions derived in the above way have the same form as characteristic functions, a fact which is true of stationary processes in general.

Campbell's theorem originally arose in the study of the 'shot effect' in thermionic vacuum tubes. If such a tube consists of a heated emitter and an anode, the electrons leaving the former cause an impulse on striking the anode. When the flow of electrons is very large there is a substantial number in the intervening space whose electric field has a cushioning effect. However, at smaller flows the emission and impact on the anode of an electron is approximately a random event independent of other such events so that the number of impacts on the anode in an interval has a Poisson distribution. The effect of these on the circuit depends on the latter's electrical characteristics which determine $f(x)$.

This subject has a very large literature. On the physical aspects there are extensive discussions and references in Middleton (1960), Davenport and Root (1958), and Bell (1960), and for the mathematical theory the reader should consult Blanc-Lapierre and Fortet (1953) and Fortet (1951).

The ideas of Campbell's theorem can be extended to a more general situation where $X(t)$ in (9.66) is a general homogeneous additive process. If $f(t) = e^{-\lambda t}$ ($\lambda > 0$), and $X(t)$ is a homogeneous additive process with only positive jumps, the distribution of $Z(t)$ in (9.66) can be interpreted as the distribution of the content of an infinite dam which is being fed by an input such that $X(t)$ is the total input up to time t, and the release is continuous at a rate proportional to the amount of water in the dam. The dam can then never run dry and the characteristic function of the distribution of $Z(t)$ is easily found. Note that the distribution of $Z(t)$ is then infinitely divisible but $\{Z(t)\}$ as a function of t is not a homogeneous additive process.

9.9. The Brownian process

The homogeneous additive process in which the distribution of $X(t_1) - X(t_2)$ is always a normal distribution is known by various names as the Gaussian, normal, Wiener, or Brownian process. It has been extensively studied not only for its own sake but because of the light it shows on many applied problems. We begin by considering its definition in more detail and its elementary properties. We then consider some of its applications and finally study its properties in detail.

We denote by $X(t)$ the random variable defining the state of the process at time t and we suppose that for any set of values $t_1, ..., t_n$, the corresponding random variables $X(t_1), ..., X(t_n)$ have a joint multinormal distribution which is such that $X(t_1) - X(t_2)$, $X(t_2) - X(t_3), ...,$ are independent if $(t_1, t_2), (t_2, t_3), ...$ are non-overlapping intervals (i.e. without common interior points). It is convenient to put $X(0) = 0$, and suppose that $X(t)$ $(t > 0)$ is normally distributed with mean zero and variance t.

The above definition involves random variables, $X(t)$, defined for arbitrary, but only finite sets of values of t. In practice we are usually interested in the properties of the whole set of values $X(t)$ for $t \geqslant 0$, and this is a 'random function'. A completely adequate treatment would therefore demand that we set up a probability measure in the space of all such functions on a suitably chosen σ-field of this space. We shall not follow this procedure here (see in particular Doob 1953), and consider instead the approach due to Lévy which defines the function $X(t)$ as the limit, in the probability sense, of a sequence of simpler functions.

Suppose that $0 < t_1 < t_2 < t_3$, and consider the joint distribution of $X(t_1)$, $X(t_2)$, and $X(t_3)$. From the definition we see, following the same kind of argument used in the theory of Markov chains in §3.9, that given $X(t_1)$ and $X(t_3)$, the distribution of $X(t_2)$ is independent of any values of $X(t)$ for $t < t_1$, or $t > t_3$. The joint distribution of $X(t_1)$, $X(t_2)$ and $X(t_3)$ is a trivariate normal distribution whose second moments are given by the matrix

$$\begin{pmatrix} t_1 & t_1 & t_1 \\ t_1 & t_2 & t_2 \\ t_1 & t_2 & t_3 \end{pmatrix}. \tag{9.82}$$

The distribution of $X(t_2)$, given $X(t_1)$ and $X(t_3)$, is normal with a mean, $m(t_2)$, and variance, σ_2^2, which can be easily shown to be

$$m(t_2) = \frac{(t_3 - t_2)X(t_1) + (t_2 - t_1)X(t_3)}{t_3 - t_1}, \tag{9.83}$$

and

$$\sigma_2^2 = \frac{(t_2 - t_1)(t_3 - t_2)}{(t_3 - t_1)}. \tag{9.84}$$

Thus we could write

$$X(t_2) = m(t_2) + \sigma_2 Z,$$

where Z is a random variable distributed normally and independently of $X(t_1)$ and $X(t_2)$, with mean zero and unit variance. We now define $X(t)$ in an interval which we may take as $[0, 1]$. We do this by taking an enumerable set of values of t which are dense in the interval and defining $X(t)$ by a limiting process. As such a sequence we take

$$0, 1, \tfrac{1}{2}, \tfrac{1}{4}, \tfrac{3}{4}, \tfrac{1}{8}, \tfrac{3}{8}, \tfrac{5}{8}, \tfrac{7}{8}, \ldots,$$

the rule of formation being obvious. For any positive integer n, the first $2^n + 1$ values in this sequence are the numbers $m2^{-n}$ ($m = 0, 1, \ldots, 2^n$) in a certain order. We suppose that $X(t)$ is defined for $x = m2^{-n}$. Define $X_n(t)$ to be equal to these random variables when $x = m2^{-n}$ and to be the linear interpolation between the two nearest values when $x \neq m2^{-n}$.

$X_n(t)$ is a random function composed of 2^n straight segments. We can get from $X_n(t)$ to $X_{n+1}(t)$ by bisecting each of these and adding, at the point of bisection, a random variable which is normally distributed with mean zero and variance 2^{-n-2}. The straight segment is then replaced by two new ones.

Consider the sequence of random functions, $X_n(t)$, on the interval $[0, 1]$. We prove that this converges uniformly with probability one. For each t, write

$$Z_n(t) = X_n(t) - X_{n-1}(t).$$

Then the sequence $\{Z_n(t)\}$ is independent and moreover

$$\max |Z_n(t)| = \max\left\{ Z_n\left(\frac{1}{2^n}\right), Z_n\left(\frac{3}{2^n}\right), \ldots, Z_n\left(\frac{2^n - 1}{2^n}\right) \right\}.$$

Each of the random variables on the right-hand side has zero mean and variance 2^{-n-2}. We therefore have

$$\mathrm{pr}\left(\max_t |Z_n(t)| > x \right) \leqslant \mathrm{pr}\left(\max_m \left| Z_n\left(\frac{m}{2^n}\right) \right| > x \right)$$

$$\leqslant 2^{n-1} \mathrm{pr}\left(\left| Z_n\left(\frac{1}{2^n}\right) \right| > x \right). \tag{9.85}$$

Now we know that if Y is a random variable normally distributed with zero mean and unit variance, and $\alpha \geqslant 0$,

$$\mathrm{pr}(|Y| > \alpha) = 2(2\pi)^{-\frac{1}{2}} \int_\alpha^\infty e^{-\frac{1}{2}u^2}\, du$$

$$\leqslant 2(2\pi\alpha^2)^{-\frac{1}{2}} e^{-\frac{1}{2}\alpha^2} \tag{9.86}$$

28

(as shown in §7.3). Thus the last expression in (9.85), with

$$x = c\left(\frac{n\log 2}{2^n}\right)^{\frac{1}{2}},$$

is not greater than

$$\frac{2^{n-1}}{c(\pi n\log 2)^{\frac{1}{2}}}\,2^{-c^2n}. \tag{9.87}$$

Taking $c > 1$, the series

$$\sum_n \frac{2^{n-1}}{c(\pi n\log 2)^{\frac{1}{2}}}\,2^{-c^2n}$$

is convergent and using Theorem 8.17 (the first Borel–Cantelli lemma) we have

$$\max_t |Z_n(t)| < c\left(\frac{n\log 2}{2^n}\right)^{\frac{1}{2}} \tag{9.88}$$

for all sufficiently large n, with probability one. Thus the series

$$\sum_n \max_t |Z_n(t)|$$

converges almost certainly, and uniformly in t. Since the limit of a uniformly convergent sequence of continuous functions is continuous we have the following striking theorem.

THEOREM 9.8. *With probability one, the realization of a Brownian motion is a continuous function.*

We shall see later that it is also almost certain that $X(t)$ is not differentiable at any given point, and in fact $X(t)$ is not a function of bounded variation in any finite interval.

9.10. The diffusion equation

Consider the distribution of $X(t_2)$ given that of $X(t_1)$. For simplicity replace these random variables by Y and X, respectively. Then the probability that Y lies in the range $(y, y+dy)$ given that $X = x$ is

$$P_t(y, x)\,dx = (2\pi t)^{-\frac{1}{2}}\exp\left\{\frac{-(y-x)^2}{2t}\right\}dy. \tag{9.89}$$

Differentiating $P(y, x)$ with respect to x, y, and t we see that it satisfies the two equations

$$\frac{\partial}{\partial t}P_t(y, x) = \frac{1}{2}\frac{\partial^2}{\partial y^2}P_t(y, x) \tag{9.90}$$

$$= \frac{1}{2}\frac{\partial^2}{\partial x^2}P_t(y, x). \tag{9.91}$$

These are Kolmogorov's forward and backward equations, and the forward equation is a special case of the Fokker–Planck equation which is of importance in physics, and in genetics (Moran 1962a).

We consider these equations of Kolmogorov in a more general manner (Kolmogorov 1931). Suppose that we have a Markov process in continuous time so that $X(t)$ is the value of the resulting random function at time t. The future development of the process from t onwards, conditional on $X(t)$, will be independent of the values of X before t. Suppose that the distribution of $Y = X(t_2)$ given $X(t_1) = x$ $(t_1 < t_2)$ is

$$F(y \mid x, t_2 - t_1)$$

and depends on $t_2 - t_1$ but not on t_1.

THEOREM 9.9. *If the above circumstances hold and if $F(y \mid x, t)$ has uniformly bounded fourth order derivatives with respect to x, y, and t when $t > 0$, and if*

$$\lim_{t \to 0} \frac{\int_{-\infty}^{\infty} |y - x|^3 \, dF(y \mid x, t)}{\int_{-\infty}^{\infty} |y - x|^2 \, dF(y \mid x, t)} = 0, \tag{9.92}$$

where both numerator and denominator are finite, then $f(y \mid x, t)$, the derivative of $F(y \mid x, t)$ with respect to y, satisfies the two equations

$$\frac{\partial f(y \mid x, t)}{\partial t} = A(x) \frac{\partial f(y \mid x, t)}{\partial x} + B(x) \frac{\partial^2 f(y \mid x, t)}{\partial x^2}, \tag{9.93}$$

and

$$\frac{\partial f(y \mid x, t)}{\partial t} = -\frac{\partial}{\partial y} \{A(y) f(y \mid x, t)\} + \frac{\partial^2}{\partial y^2} \{B(y) f(y \mid x, t)\}. \tag{9.94}$$

Here $A(x)$ and $B(x)$ are defined by the equations

$$A(x) = \lim_{t \to 0} t^{-1} \int_{-\infty}^{\infty} (y - x) \, dF(y \mid x, t), \tag{9.95}$$

$$B(x) = \lim_{t \to 0} \tfrac{1}{2} t^{-1} \int_{-\infty}^{\infty} (y - x)^2 \, dF(y \mid x, t), \tag{9.96}$$

and it is assumed that these limits exist.

Proof. From the hypotheses, $f(y \mid x, t)$ exists everywhere and has bounded third-order derivatives. Define

$$a(x, t) = \int_{-\infty}^{\infty} (y - x) f(y \mid x, t) \, dy, \tag{9.97}$$

$$b(x, t) = \int_{-\infty}^{\infty} (y - x)^2 f(y \mid x, t) \, dy, \tag{9.98}$$

and

$$c(x, t) = \int_{-\infty}^{\infty} |y - x|^3 f(y | x, t) \, dy. \tag{9.99}$$

$c(x, t)$ is finite by assumption and tends to zero as $t \to 0$. We now have, for $h > 0$,

$$f(y | x, h + t) = \int_{-\infty}^{\infty} f(z | x, h) f(y | z, t) \, dz$$

$$= \int_{-\infty}^{\infty} f(z | x, h) \left\{ f(y | x, t) + (z - x) \frac{\partial}{\partial x} f(y | x, t) \right.$$

$$\left. + \tfrac{1}{2}(z - x)^2 \frac{\partial^2}{\partial x^2} f(y | x, t) + \tfrac{1}{6}(z - x)^3 \frac{\partial^3}{\partial \xi^3} f(y | \xi, t) \right\} dz,$$

where ξ is some quantity lying in the interval (x, z). Integrating we obtain

$$f(y | x, h + t) = f(y | x, t) + a(x, h) \frac{\partial}{\partial x} f(y | x, t)$$

$$+ b(x, h) \frac{\partial^2}{\partial x^2} f(y | x, t) + \tfrac{1}{6} \theta c(x, t), \tag{9.100}$$

where θ is bounded. Since the limits (9.95) and (9.96) exist, using (9.92), (9.100) gives (9.93) on dividing by h and proceeding to the limit. Kolmogorov's proof is a little more general than the above. Notice that for the proof of (9.93) only the existence and boundedness of the fourth-order derivatives of $F(y | x, t)$ are needed.

The forward equation is best proved from the backward equation (9.93), using the stronger assumption of the existence and boundedness of the fifth-order derivatives of $F(y | x, t)$ (and therefore of the fourth-order derivatives of $f(y | x, t)$). It can then be shown that $A(x, t)$ and $B(x, t)$ have continuous derivatives of the second order, and that the limits in (9.92), (9.95), and (9.96) are uniform.

Consider a finite interval $a \leqslant y \leqslant b$, and on it define a function $R(y)$ which has a bounded third-order derivative and is such that

$$R(a) = R(b) = R'(a) = R'(b) = 0.$$

Outside this interval put $R(y) = 0$.

Then

$$\int_{-\infty}^{\infty} \frac{\partial}{\partial t} f(y\,|\,x,t) R(y)\,dy = \frac{\partial}{\partial t} \int_{-\infty}^{\infty} f(y\,|\,x,t) R(y)\,dy$$

$$= \lim_{h\to 0} h^{-1} \int_{-\infty}^{\infty} \{f(y\,|\,x,t+h) - f(y\,|\,x,t)\} R(y)\,dy$$

$$= \lim_{h\to 0} h^{-1} \left\{ \int_{-\infty}^{\infty} R(y) \int_{-\infty}^{\infty} f(z\,|\,x,t) f(y\,|\,z,h)\,dz\,dy \right.$$

$$\left. - \int_{-\infty}^{\infty} f(y\,|\,x,t) R(y)\,dy \right\}$$

$$= \lim_{h\to 0} h^{-1} \left[\int_{-\infty}^{\infty} f(z\,|\,x,t) \int_{-\infty}^{\infty} f(y\,|\,z,h)\Big\{(y-z)R'(z) \right.$$

$$\left. + \tfrac{1}{2}(y-z)^2 R''(z) + \tfrac{1}{6}(y-z)^3 R'''(\xi)\Big\} \right]\,dy\,dz$$

$$= \lim_{h\to 0} h^{-1} \left[\int_{-\infty}^{\infty} f(z\,|\,x,t)\Big\{R'(z)a(z,h) + \tfrac{1}{2}R''(z)b(z,h) \right.$$

$$\left. + \tfrac{1}{6}R'''(\xi)c(z,h)\Big\} \right]\,dz$$

$$= \int_{-\infty}^{\infty} f(z\,|\,x,t)\{R'(z)A(z) + R''(z)B(z)\}\,dz, \qquad (9.101)$$

the taking of the differentiation and limiting operations under the integral sign being easily justified. Integrating by parts, we have

$$\int_{-\infty}^{\infty} f(z\,|\,x,t)A(z)R'(z)\,dz = -\int_{-\infty}^{\infty} R(z)\frac{\partial}{\partial z}\{A(z)f(z\,|\,x,t)\}\,dz,$$

and similarly integrating by parts twice,

$$\int_{-\infty}^{\infty} f(z\,|\,x,t)B(z)R''(z)\,dz = \int_{-\infty}^{\infty} R(z)\frac{\partial^2}{\partial z^2}\{B(z)f(z\,|\,x,t)\}\,dz.$$

Inserting these in (9.101)

$$\int_{-\infty}^{\infty} R(z)\left[\frac{\partial}{\partial t} f(z\,|\,x,t) + \frac{\partial}{\partial z}\{A(z)f(z\,|\,x,t)\} - \frac{\partial^2}{\partial z^2}\{B(z)f(z\,|\,x,t)\}\right] dz = 0.$$

$$(9.102)$$

Since this holds for any $R(z)$ satisfying the prescribed conditions, (9.94) follows.

Feller (1937b) has shown, conversely, that given suitable initial conditions the two diffusion equations can be solved to give the transition probabilities of a Markov process. The main use of these equations in practical problems is to provide approximations for processes which are discrete or intrinsically more complicated. We discuss some such applications in the next chapter.

It is also often useful to consider equations for the characteristic function of the transition probability. If we write (using θ for the variable of the characteristic function since t is now the time variable)

$$\phi(\theta) = \int_{-\infty}^{\infty} e^{i(y-x)\theta} f(y \mid x, t) \, dy$$

we get from the backward equation

$$e^{ix\theta} \frac{\partial \phi(\theta)}{\partial t} = \int_{-\infty}^{\infty} e^{iy\theta} A(x) \frac{\partial f(y \mid x, t)}{\partial x} \, dy + \int_{-\infty}^{\infty} e^{iy\theta} B(x) \frac{\partial^2 f(y \mid x, t)}{\partial x^2} \, dy$$

$$= A(x) \frac{\partial}{\partial x} \{e^{ix\theta} \phi(\theta)\} + B(x) \frac{\partial^2}{\partial x^2} \{e^{ix\theta} \phi(\theta)\}. \tag{9.103}$$

Similarly the forward equation gives, using the behaviour of $f(y \mid x, t)$ at $y = \pm \infty$,

$$\frac{\partial \phi(\theta)}{\partial t} = \int_{-\infty}^{\infty} e^{i(y-x)\theta} \{i\theta A(y) - \theta^2 B(y)\} f(y \mid x, t) \, dy, \tag{9.104}$$

which is rather less convenient.

If we compare the conditions of Theorem 9.9 with the properties of an additive homogeneous process we see that the latter cannot be the solution of a diffusion process of the above type unless it consists solely of its normal component, i.e. if we use the representation (9.30), $M(x)$ and $N(x)$ must be identically zero.

9.11. Simple physical interpretation of the diffusion equation

A. Einstein was the first to give a physical interpretation of the diffusion equations of the Brownian movement which is capable of being compared with experiment. The 'Brownian movement' was first observed physically by R. Brown (1828), a botanist, who noticed that when pollen is dispersed in water the individual particles were in uninterrupted irregular motion. The first proper physical theory was given by Einstein in a series of papers (1905; 1906a, b; 1907; 1908) which have been subsequently published as a book (1956).

Consider spherical particles, of mass m and radius r, in a liquid. If these are not too large compared with the molecules which surround them, collisions with the latter will result in each particle performing a random walk in three dimensions. For our purposes it is only necessary to consider one of the three coordinates defining the position of the particle. Thus if these are (x, y, z) we consider only x. If the particle has position x_0 at time t_0, and position x_1 at a time $t_1 > t_0$ we can reasonably expect that $x_1 - x_2$ will be distributed normally with a variance proportional to $t_1 - t_0$. This will be so if the impulses received by collision with the molecules are sufficiently numerous during the interval (t_0, t_1), and the process will be Markovian if we further assume that the initial velocity can be neglected in comparison with the total changes in velocity due to all the impulses during (t_0, t_1).

If the particle starts from a position $x = 0$, and its x coordinate has the value x at a time t later, and if further there is no external force, or gradient of density in the liquid, the probability distribution of x, $f(x, t)$, will satisfy an equation of the form

$$\frac{\partial f}{\partial t} = D\frac{\partial^2 f}{\partial x^2}, \tag{9.105}$$

since we are in fact assuming that the central limit theorem implies that the cumulative effect of all the impulses will make $f(x, t)$ of the form

$$f(x, t) = \tfrac{1}{2}(\pi D t)^{-\frac{1}{2}} e^{-x^2/4Dt}.$$

We aim to calculate D from physical considerations.

Suppose that there is an external force K on each particle, and consider such a particle by itself in a homogeneous fluid. If the particle moved through the fluid with constant velocity, v, it would experience an average resisting force due to the molecules of the fluid. Assume that the particle, besides being spherical, is large enough for Stokes law to give a good estimate of the resisting force. This asserts that the resisting force is $6\pi\eta r v$, where η is the coefficient of viscosity and v is the velocity. Since this must equal K we have

$$v = \frac{K}{6\pi\eta r}. \tag{9.106}$$

Now suppose the liquid is in a finite container and that there are initially ν particles per unit volume each of which is acted on by a force K along the x-axis. When equilibrium is reached there will be a gradient of particle density along x so that we write $\nu(x)$ for the density of particles per unit volume at a point with x-coordinate equal to x.

In the absence of any force the net number of particles passing across unit area of the plane $x = \text{constant}$, in unit time, would be

$$-D\frac{\partial \nu(x)}{\partial x} \tag{9.107}$$

for this assumption leads, by a simple argument, to the equation (9.105) and is in fact the definition of the diffusion coefficient D. In fact since a state of equilibrium has been reached the net number of particles crossing unit area of the plane must be zero, and since if there were no gradient, the number crossing to the right would be

$$vv = \frac{Kv}{6\pi\eta r},$$
(9.108)

we must have

$$-D\frac{\partial v(x)}{\partial x} = \frac{Kv}{6\pi\eta r}.$$
(9.109)

We may look at this in another way. The particles exercised an osmotic pressure dependent on their density, and it is the gradient in this pressure which acts in opposition to the force K.

It is known that the osmotic pressure caused by a suspension of particles per unit volume is

$$\frac{RTv}{N},$$
(9.110)

where $R = 8{\cdot}31 \times 10^7$ is the constant of the gas equation, T is the absolute temperature in degrees, and N is the number of particles contained in one 'gramme-molecule' of the fluid, i.e. the number of particles contained in a number of grammes of fluid equal to the weight of one particle in units of a single hydrogen atom.

The force resisting K is then

$$\frac{RT}{N}\frac{\partial v(x)}{\partial x},$$
(9.111)

as is easily seen by considering the work done against osmotic pressure by moving a single particle through a distance x. Equating (9.111) to K we get

$$\frac{RT}{N}\frac{\partial v(x)}{\partial x} = \frac{6\pi\eta r D}{v}\frac{\partial v(x)}{\partial x},$$

so that

$$D = \frac{RTv}{6N\pi\eta r},$$
(9.112)

which is the required result. This result has important physical applications.

The assumptions of the above theory are somewhat rough and we shall consider a more exact theory in §9.15.

9.12. Further properties of the Brownian movement

We have already seen in Theorem 9.8 that with probability one the realization of the Brownian movement is a continuous function. On the other hand, we can also show that at any particular point, there is unit probability that $X(t)$ is not differentiable, and we state this as a theorem.

THEOREM 9.10. *Given any particular point t_0, there is probability one that $X(t)$ is not differentiable at t_0.*

Proof. Take $t > t_0$. We shall show that with probability one,

$$\limsup_{t \to t_0} \frac{X(t) - X(t_0)}{t - t_0} = \infty. \qquad (9.113)$$

If this is so, then by symmetry the lower limit is $-\infty$ with probability one, and no derivative can exist. Take $x > 0$, $\delta > 0$, and consider

$$\text{pr}\left\{\text{least upper bound}_{0 \leqslant t \leqslant \delta} \frac{X(t_0 + t) - X(t_0)}{t} \geqslant x\right\}. \qquad (9.114)$$

This is not less than

$$\text{pr}\left[\max_{0 \leqslant t \leqslant \delta} \{X(t_0 + t) - X(t_0)\} \geqslant x\delta\right]. \qquad (9.115)$$

With probability one $X(t)$ is continuous, and if

$$\max_{0 \leqslant t \leqslant \delta} \{X(t_0 + t) - X(t_0)\} \geqslant x\delta,$$

we must have

$$X(t_0 + t) - X(t_0) = x\delta,$$

for the first time at some point t_1. Having attained this value, the distribution of

$$X(t_0 + \delta) - X(t_0 - t_1)$$

is symmetric. Thus

$$\text{pr}\left[\max_{0 \leqslant t \leqslant \delta} \{X(t_0 + t) - X(t_0)\} \geqslant x\delta\right]$$

$$= 2\,\text{pr}\left[\max_{0 \leqslant t \leqslant \delta}\{X(t_0 + t) - X(t_0)\} \geqslant x\delta,\; X(t_0 + \delta) - X(t_0) \geqslant x\delta\right]$$

$$= 2\,\text{pr}\{X(t_0 + \delta) - X(t_0) \geqslant x\delta\}$$

$$= \frac{2}{(2\pi)^{\frac{1}{2}}} \int_{x\delta^{\frac{1}{2}}}^{\infty} e^{-\frac{1}{2}u^2}\,du,$$

which tends to unity as $\delta \to 0$. A similar result holds for the lower limit, and hence the derivative cannot exist.

This suggests that $X(t)$ is unlikely to be a function of bounded variation, as is shown by Theorem 9.11.

THEOREM 9.11. *With probability one, $X(t)$ for $0 \leqslant t \leqslant 1$ is not a function of bounded variation.*

Proof. Write $x_i = iN^{-1}$, $i = 0, 1, ..., N$, and consider the sum

$$S_N = \sum_{i=0}^{N-1} \{X(x_{i+1}) - X(x_i)\}^2. \qquad (9.116)$$

The variance of $X(t_{i+1}) - X(t_i) = t_{i+1} - t_i = N^{-1}$, and hence NS_N is distributed as χ^2 with N degrees of freedom. If time is rescaled so that $X(t) - X(0)$ has variance $\sigma^2 t$ and it was desired to estimate σ^2 the above result shows that from a single realization of the Brownian movement over some interval $(0, t)$ it is possible to estimate σ^2 with as great an accuracy as desired.

We exploit this fact in the following way. Instead of taking the above values of x_i we put $x_i = i2^{-N}$, $i = 0, 1,..., 2^N$. Then with $\text{var}\{(X(t) - X(0)\} = t$, for convenience, we have

$$2^N S_{2^N} = 2^N \sum_{i=0}^{2^N-1} \{X(x_{i+1}) - X(x_0)\}^2 \tag{9.117}$$

distributed as χ^2 with 2^N degrees of freedom. We show that

$$\lim_N S_{2^N}$$

exists with probability one and equals unity.

From (9.117) it follows at once that S_{2^N} converges in probability to unity as N increases. We can, however, show that it also converges almost certainly to unity as N increases. From the properties of the χ^2 distribution we have

$$E(S_{2^N}) = 1,$$

$$\text{var}(S_{2^N}) = 2^{-N}.$$

Hence for $\varepsilon > 0$,

$$\text{pr}(|S_{2^N} - 1| \geqslant \varepsilon) \leqslant \varepsilon^{-2} 2^{-N}, \tag{9.118}$$

by Tchebychev's theorem. Since this is the term of a convergent series, S_{2^N} must converge to unity almost certainly.

The quadratic variation of $X(t)$ over the time interval $[0, t]$ is therefore equal to t when defined with respect to a binary division of this interval. Lévy (1940) has proved that the same result holds for any dense division of the interval. (See also Doob 1953, p. 395.)

Let $t_0 = 0 < t_1 < ... < t_n = 1$ be any division of the interval $[0, 1]$. We have

$$\sum_{i=0}^{n-1} |X(t_{i+1}) - X(t_i)|^2 \leqslant \max_i |X(t_{i+1}) - X(t_i)| \sum_{i=0}^{n-1} |X(t_{i+1}) - X(t_i)|. \tag{9.119}$$

Since $X(t)$ is continuous with probability one, it must also be uniformly continuous with probability one, and hence

$$\sum_{i=0}^{n-1} |X(t_{i+1}) - X(t_i)| \tag{9.120}$$

cannot converge with increasing n to a finite quantity so that $X(t)$ is, with probability one, not a function of bounded variation. In the proof of this result it is clearly sufficient to take the t_i as the binary division points used above so Lévy's result is not required. Theorem 9.11 could also be proved by considering the convergence of the sum (9.120) directly.

9.13. The projective invariance of the Brownian motion

Lévy (1944a, b) discovered a remarkable result about the Brownian movement which is best described as a theorem of 'projective invariance' since it shows that, in a certain sense, the properties of $X(t)$ remain invariant when t is subjected to a transformation of the type used in projective geometry.

Consider the values of $X(t)$ at $t = t_1, t_2, t_3, t_4$, where $t_1 < t_2 < t_3 < t_4$. These four random variables are jointly distributed in a multivariate normal distribution in such a way that

$$X(t_4) - X(t_3), \; X(t_3) - X(t_2), \; X(t_2) - X(t_1)$$

are independent normal variates with zero means and variances $t_4 - t_3$, $t_3 - t_2$, $t_2 - t_1$ respectively. The variance–covariance matrix (§6.9) for $X(t_1)$, $X(t_2)$, $X(t_3)$, $X(t_4)$ is thus

$$\begin{pmatrix} t_1 & t_1 & t_1 & t_1 \\ t_1 & t_2 & t_2 & t_2 \\ t_1 & t_2 & t_3 & t_3 \\ t_1 & t_2 & t_3 & t_4 \end{pmatrix}. \tag{9.121}$$

Consider the distribution of $X(t_2)$ conditional on $X(t_1)$ and $X(t_4)$. From the definition of the Brownian movement we can write

$$X(t_2) = m(t_2) + \sigma(t_2) Y(t_2)$$
$$= m(t_2) + z(t_2), \tag{9.122}$$

where

$$m(t) = \frac{(t_4 - t) X(t_1) + (t - t_1) X(t_4)}{t_4 - t_1} \tag{9.123}$$

$$\sigma^2(t) = \frac{(t - t_1)(t_4 - t)}{(t_4 - t_1)}, \tag{9.124}$$

and $Y(t_2)$ is a random variable which is independent of $X(t_1)$ and $X(t_4)$, and normally distributed with zero mean and unit variance.

Similarly we can write

$$X(t_3) = m(t_3) + \sigma(t_3) Y(t_3)$$
$$= m(t_3) + z(t_3), \tag{9.125}$$

where $Y(t_3)$ is a similar normal variate independent of $X(t_1)$ and $X(t_4)$ but not of $X(t_2)$. We want the correlation of $z(t_2)$ and $z(t_3)$ and since we already know their variances, we need only to find their covariance.

From (9.125)

$$X(t_3) = \frac{t_4 - t_3}{t_4 - t_2} X(t_2) + \frac{t_3 - t_2}{t_4 - t_2} X(t_4) + z_1,$$

where z_1 is normally distributed with zero mean independently of $X(t_1)$,

$X(t_2)$, and $X(t_4)$. Then $z(t_3)$ is the sum of a function of $X(t_1)$ and $X(t_4)$ and

$$z_1 + \frac{t_4 - t_3}{t_4 - t_2}\{m(t_2) + z(t_2)\}.$$

Since its mean is zero,

$$E\{z(t_2)z(t_3)\} = \frac{t_4 - t_3}{t_4 - t_2}\text{var}\{z(t_2)\}$$

$$= \frac{(t_4 - t_3)(t_2 - t_1)}{t_4 - t_1}. \tag{9.126}$$

Dividing by the variances, the correlation coefficient of $z(t_2)$ and $z(t_3)$ is

$$\rho\{z(t_2), z(t_3)\} = \left|\frac{(t_4 - t_3)(t_2 - t_1)}{(t_4 - t_2)(t_3 - t_1)}\right|^{\frac{1}{2}}. \tag{9.127}$$

This could be described as the 'partial' correlation of $X(t_2)$ with $X(t_3)$ when the effects of $X(t_1)$ and $X(t_4)$ are removed. The theory of partial correlation is treated in books on mathematical statistics (e.g. Kendall and Stuart, Vol. I, 1958), and (9.127) can also be easily derived from (9.121) by using it.

If we write ρ for the correlation between $z(t_2)$ and $z(t_3)$,

$$\rho^2 = \left(\frac{t_4 - t_3}{t_4 - t_2}\right)\bigg/\left(\frac{t_1 - t_3}{t_1 - t_2}\right), \tag{9.128}$$

and this is one of the 'cross-ratios' of the four points with coordinates (t_1, t_2, t_3, t_4). It is therefore, as is easily proved, unaltered by a homographic transformation of these points, i.e. a transformation of the form

$$t' = \frac{at + b}{ct + d}. \tag{9.129}$$

Notice, however, that it is necessary to restrict the possible such transformations to those which either preserve the order $t_1 < t_2 < t_3 < t_4$ or reverse it.

If we consider $X(t_1)$, $X(t_2)$, $X(t_3)$, where $0 < t_1 < t_2 < t_3$, and define Y by

$$X(t_2) = m(t_2) + \sigma_2 Y, \tag{9.130}$$

with $m(t)$ and σ_2 given by (9.83) and (9.84), then Y, considered as a function of t_2 in the interval (t_1, t_3), will remain invariant under a homographic transformation, and, in particular, a transformation of the form $t' = t^{-1}$. From this, for example, we can conclude that the random set of points in the interval $[0, a]$, at which $X(t) = 0$, has similar properties to the transform, by the transformation $t' = at^{-1}$, of the set of points in $[a^{-1}, \infty]$ at which $X(t) = 0$.

9.14. First passage and absorption probabilities

We follow the notation and theory given by Lévy (1948a). Let $X(0) = 0$, and consider $X(\tau)$ in the interval $(0, t)$. Write

$$M(t) = \max_{0 \leqslant \tau \leqslant t} X(\tau), \tag{9.131}$$

$$m(t) = \min_{0 \leqslant \tau \leqslant t} X(\tau), \tag{9.132}$$

$$Y_0(t) = |X(t)|, \tag{9.133}$$

$$Y_1(t) = M(t) - X(t), \tag{9.134}$$

$$Y_2(t) = X(t) - m(t). \tag{9.135}$$

Consider first the distribution of $M(t)$. Its distribution could be found by solving the diffusion equation with a single absorbing boundary. This can be done by the method of images in just the same way that we solved the gambler's ruin problem in § 3.7, in the case where one of the players has an infinite capital. In fact the Brownian motion is an approximation to such a random walk.

We shall instead follow Lévy's method (1948a, p. 210) and state the result as follows.

THEOREM 9.12. *The distribution of $M(t)$ is given by*

$$\text{pr}\{M(t) < x\} = \left(\frac{2}{\pi t}\right)^{\frac{1}{2}} \int_0^x e^{-u^2/2t} \, du. \tag{9.136}$$

Proof. (Compare Theorem 9.10.) Consider the probability that $M(t) \geqslant x$, and remember that $X(\tau)$ ($0 \leqslant \tau \leqslant t$) is continuous with probability one. If $M(t) \geqslant x$, $X(\tau)$ must attain the value x for the first time at some point t_0. Then $X(t) - X(t_0)$ is symmetrically distributed about zero in a continuous distribution so that

$$\text{pr}\{M(t) \geqslant x\} = 2\,\text{pr}\{M(t) \geqslant x, X(t) \geqslant x\}$$

$$= 2\,\text{pr}\{X(t) \geqslant x\}$$

$$= \text{pr}\{|X(t)| \geqslant x\} \tag{9.137}$$

$$= \left(\frac{2}{\pi t}\right)^{\frac{1}{2}} \int_x^\infty e^{-u^2/2t} \, du. \tag{9.138}$$

This proves (9.136) and also the following theorem.

THEOREM 9.13. *The random variables $M(t)$, $-m(t)$, and $|Y_0(t)|$ all have the same distribution.*

From (9.136) it follows that

$$E(M(t)) = \left(\frac{2}{\pi t}\right)^{\frac{1}{2}} \int_0^\infty u e^{-u^2/2t} \, du$$

$$= \left(\frac{2}{\pi}\right)^{\frac{1}{2}} t^{\frac{1}{2}}. \tag{9.139}$$

Thus $E\{M(t)\}$ and $E\{M(t)-m(t)\}$ increase as the square root of t. Consider a sum

$$S_n = X_1 + \ldots + X_n,$$

where the X_i are independent random variables with the same distribution having zero mean and a finite variance. Then 'in the large', i.e. for n large, S_n will behave like a Brownian movement and we can expect

$$E\left(\max_{1 \leqslant i \leqslant n} S_i\right)$$

to increase as $n^{\frac{1}{2}}$. We return to this problem in § 10.12 and § 10.19.

Theorem 9.12 enables us to find the 'first passage distribution' for the Brownian movement. It is easy to see that if $x > 0$, there is probability one that sooner or later $X(t)$ will exceed x. The distribution of T, the time at which this happens is easily obtained from (9.138).

THEOREM 9.14

$$\text{pr}(T \leqslant t) = \left(\frac{2}{\pi t}\right)^{\frac{1}{2}} \int_x^\infty e^{-u^2/2t} \, du. \tag{9.140}$$

Proof. $T \leqslant t$ if and only if $M(t) \geqslant x$, and thus (9.140) follows. Putting $u^2 = x^2 t^{-1}$ where $0 \leqslant \tau \leqslant t$, we get the equivalent form

$$\text{pr}(T \leqslant t) = \frac{x}{(2\pi)^{\frac{1}{2}}} \int_0^t \tau^{-\frac{3}{2}} e^{-x^2/2\tau} \, d\tau. \tag{9.141}$$

This is the Gaussian first passage distribution studied in § 7.23.

Now consider the distribution of $Y_1(t)$. Since

$$Y_1(t) = \max_{0 \leqslant \tau \leqslant t} \{X(\tau) - X(t)\},$$

this has the same distribution as

$$\max_{0 \leqslant \tau \leqslant t} \{X(\tau) - X(0)\} = M(t),$$

because the process is symmetric in time. Thus $Y_1(t)$ and $Y_2(t)$ have the distribution defined by (9.138).

Lévy (1948a) has also obtained the joint distributions of $\{M(t), X(t)\}$, $\{M(t), Y_1(t)\}$, and $\{X(t), m(t), M(t)\}$. Daniels (1941) has given the distribution of $M(t)-m(t)$ as an approximation to an analogous discrete random walk problem.

We now consider the zeros of a Brownian movement when it is known that $X(0) = 0$. With probability one, $X(t)$ is a continuous function of t, and therefore the set of zeros of $X(t)$ almost certainly forms a closed set. The values of t for which $X(t) \neq 0$ must therefore form a set of open intervals.

Suppose $X(0) = x > 0$, and consider the probability that there is no zero in the interval $(0, t_0)$. This is the probability that $m(t_0) > -x$, or that $M(t_0) < x$, and must therefore equal

$$\left(\frac{2}{\pi t_0}\right)^{\frac{1}{2}} \int_0^x e^{-u^2/2t_0} du. \tag{9.142}$$

As $x \to 0$, this tends to zero so that the probability of there being a zero in $(0, t_0)$ tends to one, as we would expect from the very irregular behaviour of $X(t)$ near a zero. However, we can calculate the conditional probability that there is no zero in $(0, t_1)$ $(t_1 > t_0)$ given that there is no zero in $(0, t_0)$, but that $X(0) = 0$.

This is defined as the limit as $x \to 0$ of the ratio

$$\frac{\left(\dfrac{2}{\pi t_1}\right)^{\frac{1}{2}} \displaystyle\int_0^x e^{-u^2/2t_0} du}{\left(\dfrac{2}{\pi t_0}\right)^{\frac{1}{2}} \displaystyle\int_0^x e^{-u^2/2t_1} du},$$

and must equal $t_0^{\frac{1}{2}} t_1^{-\frac{1}{2}}$. Thus if we consider the open intervals on the t-axis at which there are no zeros, and choose any one at random, without reference to its length, the probability that it is larger than t_1, known that it is larger than t_0, must be $t_0^{\frac{1}{2}} t_1^{-\frac{1}{2}}$.

We continue to assume $X(0) = 0$, and calculate the probability that there is no zero in (t_0, t_1), where $0 < t_0 < t_1$.

THEOREM 9.15. *If $0 < t_0 < t_1$ and $X(0) = 0$, the probability that there is no zero in (t_0, t_1) is*

$$1 - 2\pi^{-1} \cos^{-1}(t_0/t_1)^{\frac{1}{2}}. \tag{9.143}$$

Proof. Consider $X(t_0)$. The probability that it is zero is zero, and since the process is symmetric we suppose that $X(t_0) > 0$, and multiply our results by 2. The probability that there is no zero in (t_0, t_1) is then

$$1 - 2 \int_0^\infty \mathrm{pr}\{M(t_1 - t_0) < x, X(t_0) = x\} dx.$$

Using (9.141) this is equal to

$$1 - 2 \int_0^\infty (2\pi t_0)^{-\frac{1}{2}} e^{-x^2/2t_0} x (2\pi)^{-\frac{1}{2}} \int_0^{t_1 - t_0} \tau^{-\frac{3}{2}} e^{-x^2/2\tau} d\tau \, dx$$

$$= 1 - 2(2\pi t_0^{\frac{1}{2}})^{-1} \int_0^{t_1 - t_0} \tau^{-\frac{3}{2}} \int_0^\infty e^{-(x^2/2t_0) - (x^2/2\tau)} x \, dx \, d\tau$$

$$= 1 - 2(2\pi t_0^{\frac{1}{2}})^{-1} \int_0^{t_1 - t_0} \frac{t_0 \tau^{\frac{1}{2}}}{t_0 + \tau} d\tau.$$

In this integral put $\tau = t_0 u^2$ and

$$a = \left(\frac{t_1 - t_0}{t_0}\right)^{\frac{1}{2}}.$$

The probability then becomes

$$1 - 2\pi^{-1} \int_0^a \frac{du}{1 + u^2} = 1 - 2\pi^{-1} \tan^{-1}\left(\frac{t_1 - t_0}{t_0}\right)^{\frac{1}{2}}$$

$$= 1 - 2\pi^{-1} \cos^{-1}\left(\frac{t_0}{t_1}\right)^{\frac{1}{2}}. \tag{9.144}$$

This completes the proof.

If t is any given point, there is unit probability that $X(t) \neq 0$, so that we can define a random quantity T_0 such that $X(T_0) = 0$ is the nearest zero to t on the left. Since $X(0) = 0$, T_0 is distributed on the interval $(0, t)$ and using the above result we have

$$\text{pr}(T_0 < t_0) = 2\pi^{-1} \sin^{-1}\left(\frac{t_0}{t}\right)^{\frac{1}{2}},$$

a distribution with a density (a beta distribution)

$$\{\pi t_0^{\frac{1}{2}}(t - t_0)^{\frac{1}{2}}\}^{-1}. \tag{9.145}$$

This can be expressed by writing $T_0 = t \sin^2 \theta$, where θ is a random variable uniformly distributed between 0 and $\frac{1}{2}\pi$, or alternatively T_0 can be regarded as the projection on the segment $(0, t)$ of a point which is distributed uniformly on the circumference of a circle which has $(0, t)$ as one of its diameters. Note the rather unexpected fact that T_0 is distributed on $(0, t)$ symmetrically about the point $\frac{1}{2}t$. It will be seen in the next chapter that the above results have important analogues in the theory of discrete random walks.

9.15. A more realistic model of observed Brownian motion

Since $X(t)$ in the above model has a zero probability of having a derivative at any point it does not provide a very good model for the position of a particle which physically must have a velocity. A number of different attempts have been made to rectify this situation (an exact theory with bibliographic references is given by Doob 1942). Consider the probability distribution of the velocity, $u(t)$, at the time t. $\{u(t)\}$ is a random process and if its realizations can be integrated, the distribution of the position $X(t)$ (supposing that $X(0) = 0$) can be found by integration.

We begin by supposing that the equation of motion of the particle can be written

$$\frac{du(t)}{dt} = -\beta u(t) + A(t). \tag{9.146}$$

Here β is a frictional coefficient (divided by the mass) which expresses the frictional resistance to motion as a constant multiple of the velocity. $A(t)$ is a term representing the external force which is here due to the impacts of molecules, and hence $A(t)$ is some kind of a process of 'shocks'.

Equation (9.146) is the approach used by Uhlenbeck and Ornstein (1930) (see also Wang and Uhlenbeck 1945). However, on solving this equation heuristically it is found that $u(t)$, also, is a random process with no derivative and hence (9.146) cannot be strictly true.

Doob writes (9.146) in the form

$$du(t) = -\beta u(t)\, dt + dB(t) \qquad (9.147)$$

and tries to give the differentials in this equation a meaning.

Clearly everything will depend on the meaning given to $B(t)$, the process of random shocks, $dB(t)$ is to be, in some sense, the differential of a random process. We shall assume that $B(t)$ is the random variable of an additive homogeneous process with a finite second moment. The mean value of

$$B(t+s) - B(t) \quad (s > 0)$$

is therefore linearly proportional to s and it is convenient to take it as zero. In fact since the frictional term, $-\beta u(t)$, is itself the mean effect of molecular shocks, the mean of $dB(t)$ is zero by definition if the particle is not to have a 'drift' with time. If we write

$$\sigma(s)^2 = E\{B(t+s) - B(t)\}^2, \qquad (9.148)$$

$\sigma(s)^2$ will be independent of t and linearly proportional to s, so that we can write

$$\sigma(s)^2 = \sigma^2 s. \qquad (9.149)$$

We can now introduce the idea of a 'stochastic integral'. Let $f(t)$ be a continuous function. Then

$$\int_a^b f(t)\, dB(t) \qquad (9.150)$$

can be easily defined in a manner analogous to the Riemann integral. We divide the interval $\langle a, b \rangle$ by points t_i so

$$a = t_0 < t_1 < \ldots < t_{n-1} < t_n = b$$

and consider the sum

$$S = \sum_{i=1}^{n} f(\tau_i)\{B(t_i) - B(t_{i-1})\}, \qquad (9.151)$$

where the τ_i are any numbers satisfying $t_{i-1} \leqslant \tau_i \leqslant t_i$. If we let

$$\delta = \max(t_{i+1} - t_i)$$

tend to zero we can show that the random variable S converges in quadratic

29

mean (and therefore in probability) to a random variable which we take as the definition of (9.150). This can be done in various ways. One is to divide $[a, b]$ according to a definite scheme of division such as is obtained by taking

$$t_i = a + (b-a)i2^{-n} \quad (i = 0, \ldots, 2^n),$$

for which it is then easy to show that the sum S converges in quadratic mean to a random variable as n increases. Having done this the continuity of $f(t)$ makes it easy to show that any other division results in a sum which converges to the same random variable as $\delta \to 0$. Alternatively the method described by Doob in the above reference may be used. This starts by considering functions $f(t)$ which are equal to some constant C in an interval (t_1, t_2) contained in $[a, b]$ and zero elsewhere. The integral with respect to such a function is defined to be

$$C\{B(t_2) - B(t_1)\}.$$

We can then define successively the integral of finite sums of such functions, infinite sums of such functions and the limits of such infinite sums, each successive extension of the definition being easily shown to be unique. A method of definition of some such kind is necessary because (9.150) cannot be defined as the ordinary Riemann–Stieltjes integral of $f(t)$ with respect to a realization of $B(t)$, the latter not being, in general, of bounded variation.

From the method of definition we can now immediately deduce formulae for the expectation of such stochastic integrals since, under the restriction that $f(t)$ is continuous on $[a, b]$, it is obvious that the expectation can be taken under the integral sign. We therefore have

$$E \int_a^b f(t) \, dB(t) = 0, \tag{9.152}$$

$$E \int_a^b f(t_1) \, dB(t_1) \int_a^b g(t_2) \, dB(t_2) = \sigma^2 \int_a^b f(t)g(t) \, dt. \tag{9.153}$$

We interpret (9.147) in an integral form by supposing that it means that for any continuous function $f(t)$, and any $a, b > 0$,

$$\int_a^b f(t) \, du(t) = -\beta \int_a^b f(t)u(t) \, dt + \int_a^b f(t) \, dB(t), \tag{9.154}$$

the integrals being stochastic Riemann–Stieltjes integrals defined in probability in the above manner. Using the particular function

$$f(t) = e^{\beta t},$$

we have, with unit probability,

$$\int_a^b e^{\beta t} \, du(t) = -\beta \int_a^b e^{\beta t} u(t) \, dt + \int_a^b e^{\beta t} \, dB(t).$$

Since integration by parts is allowable, and putting $a = 0$, $b = t$,

$$u(t) = u(0)e^{-\beta t} + e^{-\beta t} \int_0^t e^{\beta \tau} \, dB(\tau). \tag{9.155}$$

We now make further assumptions regarding the $B(t)$ process. We assume that it is a Brownian motion process of the type discussed in §9.9, and thus that $B(t)$ is normally distributed with mean zero and variance $\sigma^2 t$. We further assume that $B(t)$, for $t > 0$, is independent of $u(0)$. In actual fact the effect of $B(t)$ is that of a large number of shocks resulting from the impact of the particle with the molecules. So long as these shocks are independent and have bounded variance, for a sufficiently coarse scale of time their overall effect will be practically that of a Brownian motion since their sum over any small (but not too small) interval will be practically normally distributed. Taking $B(t)$ to be an additive normal process is thus really equivalent to assuming that $u(t)$ will not change very much in an interval of time which is still large enough for the change in $B(t)$ to be practically normally distributed. This is a consequence of the fact that m, the mass of the particle, is much larger than that of the molecules which surround it.

If we now consider (9.155) and let t increase, the effect of the initial value, $u(0)$, will gradually vanish and the distribution of $u(t)$ will become normal, whatever that of $u(0)$. Now by the principle of equipartition of energy the distribution of $u(t)$ must tend to a normal distribution with zero mean and variance kTm^{-1}, where k is Boltzmann's constant, T is the absolute temperature in degrees, and m is the mass of the particle.

We therefore obtain a stationary distribution if we assume that $u(0)$ is also normally distributed with mean zero and variance kTm^{-1}. Using (9.153), the last term in (9.155) will have a normal distribution with mean zero and variance

$$\frac{\sigma^2(1 - e^{-2\beta t})}{2\beta}. \tag{9.156}$$

Adding the variances of the terms on the right-hand side of (9.155) we have

$$kTm^{-1} = kTm^{-1}e^{-2\beta t} + \frac{\sigma^2(1 - e^{-2\beta t})}{2\beta}$$

and hence

$$\sigma^2 = 2\beta kTm^{-1}.$$

From the above results we have obtained not only the distribution of $u(t)$, but also the transition probabilities since

$$u(s+t) = u(s)e^{-\beta t} + e^{-\beta(s+t)} \int_s^{s+t} e^{\beta \tau} \, dB(\tau), \tag{9.157}$$

and this defines the Ornstein–Uhlenbeck process as a solution of the stochastic differential equation (9.147).

Since $u(t)$ is continuous with probability one it can be integrated to give the displacement, $x(t)$. Clearly

$$x(t) = x(0) + \beta^{-1}(1 - e^{-\beta t})u(0) + \beta^{-1}\int_0^t \{1 - e^{-\beta(t-\tau)}\} dB(\tau). \quad (9.158)$$

Thus $x(t)$ is normally distributed, and using (9.152) and (9.153) again, we see that it has mean $x(0)$, and variance

$$2kTm^{-1}\beta^{-2}(e^{-\beta t} - 1 + \beta t). \quad (9.159)$$

Thus for large t the variance is proportional to t and the Brownian motion is a good approximation, but for small t the variance is proportional to t^2 so that the velocity exists.

Doob (1942) gives a penetrating analysis of the above process, and the method may be easily generalized to give the distribution of the fluctuations of a suspended galvanometer mirror. This requires the addition of another force, the restoring force, proportional to x (see also Moyal 1949, and the references there given) and thus to a generalization of the stochastic differential equation (9.147).

9.16. Stable distributions

Much research has been done on distributions which are 'stable' in some sense, meaning thereby that they are such that the sum of two independent random variables following such distributions also has the same distribution after rescaling. Lukacs (1960), following Gnedenko and Kolmogorov (1954) (and also Loève 1963) defines a distribution, $F(x)$, as stable if given any two independent random variables X_1, X_2 following this distribution, and any positive numbers c_1, c_2, there exist finite real numbers A, B (where $B > 0$) such that

$$\frac{c_1 X_1 + c_2 X_2 - A}{B} \quad (9.160)$$

has the same distribution as X and Y. This is therefore not so much a property of the distribution $F(x)$ as of the whole class of distributions, $F(A + Bx)$, obtained from $F(x)$ by rescaling and relocating. Such a class may be described as a distribution type. Lévy has a similar definition but omits the possibility of relocation, i.e. insists that A in (9.160) is always zero. The distributions satisfying the condition we have given above he calls quasi-stable. It seems that the least confusion will be caused if we follow the nomenclature of Lukacs as above, and call Lévy's stable distributions 'stable in the restricted sense'.

If $\phi(t)$ is the characteristic function of a stable distribution the above condition can be put in the form that given $c_1 > 0$, $c_2 > 0$, there exist numbers A and B $(B > 0)$, such that

$$\phi(c_1 t)\phi(c_2 t) = e^{iAt}\phi(Bt). \quad (9.161)$$

It follows from this that for any positive integer n, there exists numbers A_n and B_n $(B_n > 0)$ such that

$$\phi(t)^n = \exp(iA_n t)\phi(B_n t), \tag{9.162}$$

and this can be rewritten

$$\phi(t)^{1/n} = \phi(tB_n^{-1})\exp(-iA_n tB_n^{-1}n^{-1}), \tag{9.163}$$

which shows that the distribution is necessarily infinitely divisible. Starting from this fact we can now prove the following important theorem.

THEOREM 9.16. $F(x)$ *is a stable distribution if and only if its characteristic function can be written in the form*

$$\phi(t) = \exp[iat - c|t|^\alpha\{1 + ibt|t|^{-1}\omega(|t|, \alpha)\}] \tag{9.164}$$

where a, b, c, α *satisfy the inequalities*

$$-\infty < a < \infty, \quad |b| \leqslant 1,$$

$$c \geqslant 0, \quad 0 < \alpha \leqslant 2,$$

and

$$\omega(|t|, \alpha) = \tan \tfrac{1}{2}\pi\alpha, \quad \text{if } \alpha \neq 1,$$

$$= 2\pi^{-1}\log|t|, \quad \text{if } \alpha = 1.$$

Proof (Gnedenko and Kolmogorov 1954). Since $F(x)$ is infinitely divisible we have

$$\log \phi(t) = iat - \tfrac{1}{2}\sigma^2 t^2 + \int_{-\infty}^{0-0} \left(e^{itu} - 1 - \frac{itu}{1+u^2}\right) dM(u)$$

$$+ \int_{0+0}^{\infty} \left(e^{itu} - 1 - \frac{itu}{1+u^2}\right) dN(u) \tag{9.165}$$

where $M(u)$, $N(u)$ satisfy the conditions of Theorem 9.5. Apply this to equation (9.161) after taking logarithms, and we get, from the uniqueness of the representation (Theorem 9.6),

$$\sigma^2 B^2 = \sigma^2(c_1^2 + c_2^2), \tag{9.166}$$

$$M(uB^{-1}) = M(uc_1^{-1}) + M(uc_2^{-1}), \tag{9.167}$$

$$N(uB^{-1}) = N(uc_1^{-1}) + N(uc_2^{-1}). \tag{9.168}$$

The derivation of (9.167) and (9.168) is not quite obvious, and is obtained as follows. On replacing t in $\log \phi(t)$ by $c_1 t$ we get integrals such as

$$\int_{-\infty}^{\infty} \left(e^{ic_1 tu} - 1 - \frac{ic_1 tu}{1+u^2}\right) dM(u) = \int_{-\infty}^{\infty} \left(e^{itu} - 1 - \frac{itu}{1+u^2 c_1^{-2}}\right) dM(uc_1^{-1}). \tag{9.169}$$

However, the integral

$$it \int_{-\infty}^{\infty} \left(\frac{u}{1+u^2 c_1^{-2}} - \frac{u}{1+u^2}\right) dM(uc_1^{-1})$$

is convergent and this difference can be absorbed into the term iat in (9.165) which corresponds only to a location of the distribution. Thus integrals such as (9.169) can be replaced by

$$\int_{-\infty}^{\infty} \left(e^{itu} - 1 - \frac{itu}{1+u^2}\right) dM(uc_1^{-1})$$

and (9.167), (9.168) follow from the uniqueness of a Fourier–Stieltjes representation.

Changing notation slightly and remembering that $B^{-1} > 0$, it follows that for any finite set of positive numbers $\alpha_1, \ldots, \alpha_n$, there exists a positive number α_0 such that

$$N(\alpha_0 u) = N(\alpha_1 u) + \ldots + N(\alpha_n u) \quad \text{(all } u > 0\text{)}.$$

Thus we can write

$$N(\alpha_0 u) = nN(u),$$

where α_0 is a function of n. Alternatively, we can write this as

$$n^{-1}N(u) = N\left\{\frac{u}{\alpha_0(n)}\right\}.$$

Let $r = mn^{-1}$ be a positive rational number so that m, n are integers. Then it follows from the above that to every such r there is a positive number $A(r)$ such that

$$rN(u) = N\{A(r)u\}. \tag{9.170}$$

If $N(u)$ is not identically equal to zero it is easy to see that $A(r)$ is non-increasing. Then for any $\lambda > 0$, $A(\lambda - 0)$ and $A(\lambda + 0)$ must exist, the limits being taken through rational values. From (9.170) it also follows that $A(r)$ must be continuous and non-increasing. Hence for every real number $\lambda > 0$,

$$\lambda N(u) = N\{A(\lambda)u\}. \tag{9.171}$$

It is easy to verify that since

$$\lim_{u \to 0} N(u) = 0,$$

$$\lim_{\lambda \to \infty} A(\lambda) = 0,$$

and

$$\lim_{\lambda \to 0} A(\lambda) = \infty.$$

The function $y = A(\lambda)$ thus has an inverse, $\lambda = B(y)$, where y lies in the interval $[0, \infty]$ and (9.171) can then be rewritten in the form

$$N(yu) = B(y)N(u) \tag{9.172}$$

for all $u > 0$, $y > 0$. This is a functional equation which we must solve.

Suppose that $N_1(u)$ and $N_2(u)$ are two functions which satisfy (9.172) for the given $B(y)$ (which is itself unknown). Furthermore, suppose that $N_2(u) \neq 0$.

Write

$$R(u) = \frac{N_2(u)}{N_1(u)}.$$

Then

$$R(yu) = \frac{N_2(yu)}{N_1(yu)} = \frac{B(y)N_2(u)}{B(y)N_1(u)} = R(u). \tag{9.173}$$

Since y is arbitrary this means that $R(u)$ must be a constant independent of u, and hence all solutions of (9.172) must be constant multiples of each other.

In (9.172) fix $y \neq 1$. Then the sequence of values $N(u)$, $N(yu)$, $N(y^2 u), \ldots$ must form a geometric sequence with ratio $B(y)$, and furthermore since $B(y)$ is continuous, $N(u)$ can only be of the form $K|u|^{-\alpha}$. Since $N(u)$ is non-decreasing and $N(\infty) = 0$, K must be negative $(= -A_2$, say) and $0 < \alpha < 2$ to make (9.165) converge. We write α_2 for the value of α.

Similarly it follows that $M(u)$ must be of the form

$$A_1|u|^{-\alpha_1} \quad (u < 0),$$

where $A_1 > 0$, $0 < \alpha_1 < 2$.

If $\sigma^2 \neq 0$ it follows from (9.166) that

$$B^2 = c_1^2 + c_2^2. \tag{9.174}$$

On the other hand, from (9.167) it follows that

$$A_1|B|^{\alpha_1} = A_1(|c_1|^{\alpha_1} + |c_2|^{\alpha_1}), \tag{9.175}$$

and $\alpha_1 < 2$, so that (9.174), (9.175) together imply that $A_1 = 0$. Similarly $A_2 = 0$. Conversely it follows that if $A_1 > 0$ or $A_2 > 0$ we must have $\sigma^2 = 0$.

If c_1 and c_2 are both non-zero we must have $\alpha_1 = \alpha_2$. This follows from (9.167) and (9.168) on putting $c_1 = c_2 = 1$, for we then get $B^{\alpha_1} = B^{\alpha_2} = 2$, so that $\alpha_1 = \alpha_2 = \alpha$, say. Thus the characteristic function $\phi(t)$ must satisfy the equation

$$\log \phi(t) = iat - \tfrac{1}{2}\sigma^2 t^2 + A_1 \int_{-\infty}^{0} \left(e^{itu} - 1 - \frac{itu}{1+u^2}\right)|u|^{-1-\alpha}\, du$$

$$+ A_2 \int_{0}^{\infty} \left(e^{itu} - 1 - \frac{itu}{1+u^2}\right)|u|^{-1-\alpha}\, du,$$

where $A_1, A_2 \geqslant 0$, and $\sigma^2 = 0$ if $A_1 + A_2 > 0$, or $A_1 + A_2 = 0$ if $\sigma^2 > 0$. It now remains to evaluate these integrals. To do this we consider separately the three cases $0 < \alpha < 1$, $\alpha = 1$, $1 < \alpha < 2$.

If $0 < \alpha < 1$ suppose first that $t > 0$. The parts of the integrals involving $itu(1+u^2)^{-1}|u|^{-1-\alpha}$ are finite, and can be absorbed in the constant a. We now have

$$\int_{0}^{\infty} (e^{itu} - 1)|u|^{-1-\alpha}\, du = t^\alpha \int_{0}^{\infty} (e^{iv} - 1)v^{-1-\alpha}\, dv,$$

and integrating around a positive quadrant of radius R we get, after letting R tend to infinity,

$$\int_0^{i\infty} (e^{iv} - 1)v^{-1-\alpha}\, dv = i^{-\alpha} \int_0^\infty (e^{-v} - 1)y^{-1-\alpha}\, dy$$

$$= e^{-\frac{1}{2}i\alpha\pi} \int_0^\infty (e^{-v} - 1)y^{-1-\alpha}\, dy.$$

Similarly

$$\int_0^\infty (e^{-iv} - 1)v^{-1-\alpha}\, dv = e^{\frac{1}{2}i\alpha\pi} \int_0^\infty (e^{-v} - 1)y^{-1-\alpha}\, dy,$$

so that we can write

$$\log \phi(t) = ita + t^\alpha L(\alpha)\{(A_1 + A_2)\cos\tfrac{1}{2}\pi\alpha + i(A_1 - A_2)\sin\tfrac{1}{2}\pi\alpha\},$$

where $L(\alpha)$ is a function of α alone. Since $\cos\tfrac{1}{2}\pi\alpha > 0$ we can write this as

$$\log \phi(t) = ita - ct^\alpha(1 + ib\tan\tfrac{1}{2}\pi\alpha),$$

where $-1 \leqslant b \leqslant 1$, $c > 0$, and $t > 0$. From the properties of characteristic functions we have, for $t < 0$,

$$\log \phi(t) = \log \overline{\phi(-t)} = ita - c|t|^\alpha(1 + ibt|t|^{-1}\tan\tfrac{1}{2}\pi\alpha),$$

so that for all real t,

$$\log \phi(t) = ita - c|t|^\alpha(1 + ibt|t|^{-1}\tan\tfrac{1}{2}\pi\alpha). \tag{9.176}$$

For $\alpha = 1$ we have to evaluate integrals such as

$$\int_0^\infty \left(e^{itu} - 1 - \frac{itu}{1+u^2}\right)\frac{du}{u^2},$$

where $t > 0$. This is equal to

$$\int_0^\infty u^{-2}(\cos tu - 1)\, du + i\int_0^\infty u^{-2}\left(\sin tu - \frac{ut}{1+u^2}\right) du$$

$$= -\tfrac{1}{2}\pi t + i\lim_{\delta\to 0}\left\{\int_\delta^\infty u^{-2}\sin tu\, du - t\int_\delta^\infty \frac{du}{u(1+u^2)}\right\}$$

$$= -\tfrac{1}{2}\pi t - i\lim_{\delta\to 0}\left[t\int_\delta^{t\delta} u^{-2}\sin u\, du - t\int_\delta^\infty \left\{u^{-2}\sin u - \frac{1}{u(1+u^2)}\right\} du\right]$$

$$= -\tfrac{1}{2}\pi t - it\log t + itK,$$

where

$$K = \int_0^\infty \left\{\frac{\sin u}{u^2} - \frac{1}{u(1+u^2)}\right\} du$$

which is a convergent integral. Using a similar result for the integral over the range $(-\infty, 0)$, we can write $\log\phi(t)$ in the form

$$ita - \tfrac{1}{2}(A_1 + A_2)\pi t - i(A_1 - A_2)t\log t,$$

and arguing as before, if $t < 0$,

$$\log \phi(t) = ita - \tfrac{1}{2}(A_1 + A_2)\pi|t| - i(A_1 - A_2)t\log|t|.$$

This is easily transformed into the form

$$ita - c|t|\{1 + 2ibt|t|^{-1}\pi^{-1}\log|t|\}, \qquad (9.177)$$

where $c \geqslant 0$, $-1 \leqslant b \leqslant 1$. Notice that for $b = 0$ this is the second characteristic function of the Cauchy distribution which has already been proved to be stable in §7.19.

Finally, we have the case $1 < \alpha < 2$. This is evaluated in just the same way as the case $0 < \alpha < 1$.

9.17. Properties of stable distributions

We have already mentioned the two particular cases of the normal distribution and the Cauchy distribution which are obviously stable. If we take $a = 0$, $c = 1$, $b = -1$, $\alpha = \tfrac{1}{2}$, we get the probability distribution on $(0, \infty)$ with density function

$$f(x) = (2\pi)^{-\frac{1}{2}}x^{-\frac{3}{2}}e^{-\frac{1}{2}x^{-1}}. \qquad (9.178)$$

This is the first passage time distribution for the Brownian movement and has been studied in §7.23 and §9.14. The fact that it is stable is obvious by scale considerations on the Brownian movement, together with the fact that realizations of the latter are continuous with probability one. Notice that a similar argument does not hold for the first passage time distribution of non-normal additive homogeneous processes because the realizations of such processes are not continuous.

No other explicit representation of stable distributions in terms of elementary functions are known although certain expansions of their probability densities in terms of higher transcendental functions have been obtained (Zolotarev 1954) for particular values of α and b (see also Pollard 1946 and Bergström 1952). The fact that all stable distributions have absolutely continuous distribution functions follows at once from the fact that

$$\phi(t) = o(e^{-|t|^\alpha}) \quad (\alpha > 0).$$

Other work on the analytic properties of stable distributions is described in Lukacs (1960).

The most illuminating feature of the theory of stable laws arises from the study of their domains of attraction which are defined in analogy with the domain of attraction of the normal distribution which we have studied in §8.20. If $X_1, X_2, \ldots$ are independent random variables having a distribution $F(x)$, and if there exist numbers A_n, B_n, such that

$$B_n^{-1} \sum_1^n X_i - A_n \qquad (9.179)$$

has a distribution which converges to a distribution function $G(x)$, $F(x)$ is said to belong to the domain of attraction of $G(x)$. Thus all stable distributions have non-empty domains of attraction since they belong to their own domains of attraction. Furthermore, if a distribution is not stable it obviously has no domain of attraction. From this it follows that a stable distribution cannot have a finite variance if it is not a normal distribution.

From the above results we have the following theorem.

THEOREM 9.17. *The necessary and sufficient condition that a distribution be the limiting distribution of a sum of the form* (9.179), *where the X_i are independently distributed in the same distribution, is that it be stable.*

If the X_i are normally and independently distributed with variance σ^2, the sum, $\sum_1^n X_i$, has a variance $n\sigma^2$ which increases as n. If the X_i were similarly distributed in the Cauchy distribution the sum has the Cauchy distribution with a scale factor which increases as n, in contrast to the normal case in which it increases as $n^{\frac{1}{2}}$. Formula (9.164) shows the similar behaviour for the general case of stable distributions. Ignoring the case where $\alpha = 1$, $b \neq 0$, for the moment, the first term, iat, represents only a shift in position, and the term $|t|^\alpha$ shows that the dispersion of the sum increases as $n^{\alpha^{-1}}$, so that the rate of increase can take any form such as n^β with $\beta \geqslant \frac{1}{2}$. When $b \neq 0$ and $\alpha = 1$, the effect of the term $w(|t|, 1) = 2\pi^{-1}\log|t|$ is simply to alter the location.

In §8.20, we have determined a necessary and sufficient condition that a distribution belongs to the domain of attraction of the normal distribution. A similar theorem has been given for distributions belonging to the domain of attraction of any given stable law. This is due to Gnedenko and Doeblin (see Gnedenko and Kolmogorov 1954, p. 175, for references and especially the note by the translator, K. L. Chung).

As we have said before, Lévy defined stability in a more restricted sense by insisting that in the definition using (9.160), A should be zero. Clearly any distribution stable in this sense is also stable in the general sense used above and must have a characteristic function of the form required by Theorem 9.16.

Suppose that X and Y are independent random variables whose distribution has the characteristic function, (9.164),

$$\exp[iat - c|t|^\alpha\{1 + ibt|t|^{-1}w(|t|, \alpha)\}].$$

Then $(c_1 X + c_2 Y)B^{-1}$ $(B > 0)$ will have the characteristic function

$$\exp[\![iaB^{-1}(c_1 t + c_2 t) - cB^{-\alpha}(c_1^\alpha + c_2^\alpha)|t|^\alpha[1 + ibt|t|^{-1}w\{B^{-1}(c_1 + c_2)|t|, \alpha\}]\!]].$$

If we insist that $(c_1 X + c_2 Y)B^{-1}$ have the same distribution we have $a = 0$ and $w = \text{const}$. We then have the result that the second characteristic functions of all distributions stable in the restricted sense are of the form

(9.164) above with $a = 0$, and $b = 0$ when $\alpha = 1$. Thus they can be written as

$$-c|t|^\alpha(\cos\gamma + it|t|^{-1}\sin\gamma), \tag{9.180}$$

where $c > 0$, $0 < \alpha \leqslant 2$, $\cos\gamma > 0$, and γ is determined by the relations

$$\cos\gamma = K^{-1}\cos\tfrac{1}{2}\pi\alpha,$$

$$\sin\gamma = -bK^{-1}\sin\tfrac{1}{2}\pi\alpha,$$

$$K^2 = \cos^2\tfrac{1}{2}\pi\alpha + b^2\sin^2\tfrac{1}{2}\pi\alpha.$$

9.18. Further results on the decomposability of distributions

After the above long excursus on the theory and application of infinitely divisible distributions we can now return to consider the decomposition of distributions and the various manners in which they can be factorized. For a detailed survey of this subject see Lukacs (1960) and Dugué (1957).

To deal with problems of this kind it is convenient, following Lukacs, to write

$$N(\phi) = -\int_0^a \log|\phi(t)|\,dt, \tag{9.181}$$

where $\phi(t)$ is a characteristic function, and a is some suitably chosen positive constant. Since $\phi(t)$ is continuous we can choose a so that $\phi(t) \neq 0$ in the closed interval $[0, a]$. We shall use $N(\phi)$ as a measure of the departure of the distribution corresponding to $\phi(t)$ from the degenerate distribution in which the probability is concentrated at one point. We can do this because $N(\phi) \geqslant 0$, and equals zero only if the distribution is degenerate.

Let $F(x)$ be the distribution whose characteristic function is $\phi(t)$. Let X_1, X_2 be independent random variables having this distribution. Then the distribution of $X_1 - X_2$ is symmetrical and has the characteristic function $|\phi(t)|^2$. Let this distribution be denoted by $F_1(x)$. From Theorem 5.22 the function of concentration, $Q_1(l)$ say, of $X_1 - X_2$ is not greater than the function of concentration of $F(x)$. Let $a > 0$. Then by Theorem 6.26,

$$\frac{a}{7}\int_{|x|\geqslant a^{-1}} dF_1(x) \leqslant \int_0^a \{1 - |\phi(t)|^2\}\,dt$$

$$\leqslant 2\int_0^a \{1 - |\phi(t)|\}\,dt$$

$$\leqslant -\int_0^a \log|\phi(t)|\,dt = N(\phi). \tag{9.182}$$

Then

$$Q_1(2a^{-1}) \geqslant 1 - \frac{a}{7} \int_{|x| \geqslant a^{-1}} dF_1(x)$$

$$\geqslant 1 + \int_0^a \log |\phi(t)| \, dt.$$

If $\{\phi_n\}$ is a sequence of characteristic functions such that $N(\phi_n) \to 0$ for some particular value of $a > 0$, then the corresponding values of $Q_1(a^{-1})$ must tend to unity.

Furthermore we have (Theorem 6.25),

$$1 - |\phi(2t)|^2 \leqslant 4\{1 - |\phi(t)|^2\}.$$

Hence using this repeatedly we see that if $N(\phi_n) \to 0$ for some $a > 0$, it does so also for $2a, 4a, \ldots$ and therefore for all a, so that $Q(2a^{-1})$ tends to zero for all $a > 0$. Hence the distribution, $F_1(x)$, tends to a degenerate distribution, and so also must $F(x)$ after relocation if necessary. Thus there exists a_n real so $e^{ia_n t}\phi_n(t) \to 1$ uniformly in every finite t-interval.

Finally, if $\phi(t) = \phi_1(t)\phi_2(t)$, and a is chosen so that $|\phi(t)| > 0$ in $[0, a]$, we must have

$$N(\phi) = N(\phi_1) + N(\phi_2). \tag{9.183}$$

We can now prove the following theorem due to Khintchine.

THEOREM 9.18. *If a characteristic function has no indecomposable factor it is infinitely divisible.*

Proof. Let $\phi(t)$ be the characteristic function. Since it has no indecomposable factor we can, for every positive integer n, write it as a product of characteristic functions,

$$\phi(t) = \phi_1(t) \ldots \phi_n(t).$$

Choosing a so that $\phi(t) \neq 0$ for $0 \leqslant t \leqslant a$, and defining $N(\phi)$ with this value of a in (9.181), let d be the greatest lower bound of

$$\max_i N(\phi_i),$$

for all such decompositions with $n = 1, 2, \ldots$. We first show $d = 0$. If this were not so we could extract from all the decompositions a sequence $\phi^{(1)}(t), \phi^{(2)}(t), \ldots$ such that

$$N(\phi^{(i)}) \geqslant d, \quad \text{all } i, \ N(\phi^{(i)}) \to d,$$

and $\phi^{(i+1)}(t)$ is a factor of $\phi^{(i)}(t)$, and $\phi(t)$. $\phi^{(i)}(t)$ is then a convergent sequence of characteristic functions which must converge to a characteristic function, $\Phi(t)$ say, which is a factor of $\phi(t)$, and $N(\Phi) = d$. There cannot be more than a finite number of such Φ's and since, by hypothesis, ϕ contains no indecomposable factor, each can itself be decomposed into characteristic functions with N's smaller than d. This is a contradiction and thus $\phi(t)$ is infinitely divisible.

The converse of this theorem is false and it is possible for an infinitely divisible distribution to contain an indivisible factor. A variety of examples to illustrate this are known and we give one simple example here. Consider the probability generating function

$$\frac{1-a}{1-az},$$ (9.184)

where $0 < a < 1$. This is the generating function of the geometric distribution and is obviously infinitely divisible. It can also be written in the form

$$\left(\frac{1+az}{1+a}\right)\left(\frac{1+a^2z^2}{1+a^2}\right)\left(\frac{1+a^4z^4}{1+a^4}\right)\cdots,$$ (9.185)

which is an infinite product of factors each of which is the generating function of a binomial distribution and is indecomposable.

To prove this equality consider the product

$$(1+a)(1+a^2)\ldots(1+a^{2^n}) = \left(\frac{1-a^2}{1-a}\right)\left(\frac{1-a^4}{1-a^2}\right)\cdots\left(\frac{1-a^{2^{n+1}}}{1-a^{2^n}}\right)$$

$$= \frac{1-a^{2^{n+1}}}{1-a}.$$

Letting n tend to infinity and using the same result with a replaced by az we easily show that (9.185) is equal to (9.184).

We are now in a position to prove the main theorem on the decomposition of distributions which is also due to Khintchine.

THEOREM 9.19. *Every characteristic function, $\phi(t)$, is the product of at most two non-degenerate characteristic functions, such that one does not have any indecomposable factors, and the other is a finite or convergent infinite product of indecomposable factors.*

Proof. Choose a such that $\phi(t) \neq 0$ in the closed interval $[0, a]$. We can suppose that $|\phi(t)|$ is not identically unity in this interval as otherwise the distribution would be degenerate. As before define $N(\phi)$ by (9.181). Then $\phi(t)$ may have an indecomposable factor $\phi_1(t)$ such that $N(\phi_1) > \frac{1}{2}N(\phi)$. If this is so we can write

$$\phi(t) = \phi_1(t)\Phi(t),$$

where $N(\Phi) < \frac{1}{2}N(\phi)$. If this is not possible, there may be indecomposable factors $\phi_i(t)$ (at most three of these) such that $N(\phi_i) > \frac{1}{4}N(\phi)$, and then

$$\phi(t) = \phi_1(t)\phi_2(t)\phi_3(t)\Phi(t),$$

where $N(\Phi) < \frac{1}{4}N(\Phi)$. Carrying on the argument in a similar way we can write

$$\phi(t) = \phi_1(t)\ldots\phi_n(t)\Phi(t),$$

where $N(\phi_i) > 2^{-k}N(\phi)$, $N(\Phi) < 2^{-k}N(\phi)$, for each positive integer k, and n depends on k.

This procedure can come to an end for some k if $\Phi(t)$ has no indecomposable factor and the conclusion of the theorem would then hold.

Suppose that the procedure can be continued indefinitely by applying the same argument at each stage to $\Phi(t)$ thus getting an infinite sequence $\phi_i(t)$. Let the sequence of ϕ_i obtained in this way be $\{\phi_i(t)\}$, $i = 1, 2,...$, and write

$$\Phi_n(t) = \phi(t)\{\phi_1(t)... \phi_n(t)\}^{-1}.$$

The two functions $\{\phi_1(t)... \phi_n(t)\}$ and $\Phi_n(t)$ are characteristic functions and we wish to prove that as n increases they converge to characteristic functions. This would be simple if they were real for then $\{\phi_1(t)... \phi_n(t)\}$ would be non-increasing in n for each real t and $\Phi_n(t)$ would then be non-decreasing. However, they are not real in general and it is necessary to consider their convergence more carefully.

From the mode of construction of the $\phi_i(t)$ it follows that

$$\sum_1^\infty N(\phi_i)$$

is convergent. Then using the above argument involving Theorem 6.26 we see that it is possible to relocate the distributions to which the ϕ_i correspond in such a way that

$$\prod_{i=n}^{n+m} \phi_i(t)$$

converges to unity in every finite t interval as n increases uniformly in $m > 0$. This implies that the infinite product

$$\prod_{i=1}^\infty \phi_i(t)$$

is convergent and is therefore a characteristic function. Since

$$\Phi_n(t) = \Phi(t)\left\{\prod_{i=1}^n \phi_i(y)\right\}^{-1}$$

then also converges as n increases, its limit is also a characteristic function $\Phi(t)$. $\Phi(t)$ is a factor of $\Phi_n(t)$ and contains no indecomposable factor whose corresponding N is positive, and therefore no indecomposable factor at all. This completes the proof. As we have mentioned before (§9.1), a decomposition of the above kind is not in general unique.

There are a number of other special results on the decomposition of distributions which deserve mention. These results are all of a special character and no general theory of divisibility yet exists. We have already (§9.2) constructed an absolutely continuous distribution which is indecomposable. Cramer (1947) proposed the problem of finding an absolutely continuous indecomposable distribution whose range was a finite interval, and this was solved by Lévy in 1952 (Dugué 1957, p. 38). Let $l_0,\ldots, l_{n-1}$ be n positive quantities less than unity, which are mutually incommensurable, and let $\phi_i(t)$ be the characteristic function of the uniform distribution on the interval $(0, l_i)$. Then

$$\phi_i(t) = \frac{\sin l_i t}{l_i t}.$$

These characteristic functions have no common factor. To prove this we first observe that any common factor of the $\phi_i(t)$ can have no zeros in the finite part of the complex plane, for all the $\phi_i(t)$ have zeros at different points. Furthermore, since the range of any component is finite, its characteristic function must be an integral function of order one. However, an integral function of order one which has no finite zeros is necessarily of the form ae^{bz} and hence cannot be the characteristic function of a non-degenerate distribution.

Now take a probability generating function,

$$P(z) = p_0 + p_2 z^2 + \ldots + p_{2n-2} z^{2n-2},$$

which is indecomposable. From §9.1 a suitable example is

$$P(z) = n^{-1}(1 + z^2 + \ldots + z^{2n-2}),$$

where n is prime. Put

$$\phi(t) = p_0 \phi_0(t) + p_2 e^{2it}\phi_1(t) + \ldots + p_{2n-2} e^{2(n-1)it}\phi_{2n-2}(t). \qquad (9.186)$$

This is the characteristic function of a mixture of distribution which is absolutely continuous, and confined to a finite interval. We prove that this mixture is indecomposable.

If it were decomposable, suppose that it is the distribution of $Z = X + Y$, where X and Y are independent random variables with non-degenerate distributions. By relocation we can suppose that their regions of support (which must be confined to finite intervals) have extreme left-hand points equal to zero.

Write

$$Z = Z_1 + Z_2,$$
$$X = X_1 + X_2,$$
$$Y = Y_1 + Y_2,$$

where Z_1, X_1, Y_1 are the integral parts of Z, X, Y. Z_1 can only take even values and hence $Z_1 = X_1 + Y_1$, and thus $Z_2 = X_2 + Y_2$. Z_1 is a discrete variable with a probability generating function

$$n^{-1}(1 + z^2 + \ldots + z^{2n-2}),$$

which is indecomposable. Thus one of X_1, Y_1 must have a degenerate distribution concentrated at the value zero. Suppose that this is Y_1 so that $Y_1 = 0$. Then the characteristic function of X_2 must divide each of the $\phi_i(t)$. Since this is impossible, (9.186) is indecomposable.

We have already seen in §9.1 that it is possible to have independent random variables X_1, X_2, X_3, X_4, having different indecomposable distributions such that $X_1 + X_2$ has the same distribution as $X_3 + X_4$. Dugué (1951b) has shown that the same result is possible when X_1 and X_2 have the same distribution.

9.19. Some particular cases

A more special class of problems is concerned with the existence of factors of particular types of distributions such as the normal and Poisson.

Consider first the case of a normal component (Fisher and Dugué 1948). The expression

$$\left(\frac{2}{3\pi}\right)^{\frac{1}{2}} (1 - \tfrac{8}{9}x^2)^2 \, e^{-2x^2/3} \tag{9.187}$$

is non-negative, and its integral over the range $(-\infty, \infty)$ can be easily verified to be unity so that it can be regarded as the density of a probability distribution. By considering the equation

$$e^{-\frac{3}{8}t^2} = \left(\frac{2}{3\pi}\right)^{\frac{1}{2}} \int_{-\infty}^{\infty} e^{itx - \frac{3}{4}x^2} \, dx,$$

and differentiating, it is easy to see that the characteristic function of (9.187) is

$$e^{-\frac{3}{8}t^2}(1 - t^2 + \tfrac{1}{4}t^4) = e^{-\frac{3}{8}t^2}(1 - \tfrac{1}{2}t^2)^2. \tag{9.188}$$

If we multiply this by

$$e^{-\frac{1}{8}t^2},$$

which is the characteristic function of a normal distribution, we obtain

$$e^{-\frac{1}{2}t^2}(1 - \tfrac{1}{2}t^2)^2 = \{e^{-\frac{1}{4}t^2}(1 - \tfrac{1}{2}t^2)\}^2. \tag{9.189}$$

From §9.2 we see that this is the square of the characteristic function of an indecomposable distribution, and thus the distribution of the sum of two independent random variables with indecomposable distributions may have a distribution with a proper normal component.

A similar result can be established with a Poisson component in the convolution of two distributions without Poisson components (Lukacs 1960). To show this it is simplest to use generating functions. Consider the two probability generating functions

$$P_1(z) = \tfrac{2}{3} + \tfrac{1}{3}z, \tag{9.190}$$

$$P_2(z) = \left(\frac{3}{4 - z^2}\right)^{\frac{1}{2}}. \tag{9.191}$$

$P_1(z)$ is the generating function of a simple binomial distribution which is indecomposable. $P_2(z)$ is the generating function of a negative binomial (on the values 0, 2, 4,...) which we have shown before to be infinitely divisible. However, it cannot have a Poisson component, because if it did, its points of support could not be the even integers alone. However,

$$P_1(z)P_2(z) = \left(\frac{1}{3}\right)^{\frac{1}{2}} \left(\frac{1+\frac{1}{2}z}{1-\frac{1}{2}z}\right)^{\frac{1}{2}}$$

$$= (\tfrac{1}{3})^{\frac{1}{2}} \exp \tfrac{1}{2}\{\log(1+\tfrac{1}{2}z) - \log(1-\tfrac{1}{2}z)\}$$

$$= (\tfrac{1}{3})^{\frac{1}{2}} \exp\left(z + \frac{z^3}{3} + \frac{z^5}{5} + ...\right), \qquad (9.192)$$

which has a Poisson component

$$\exp(-\tfrac{1}{2} + \tfrac{1}{2}z)$$

since if this is divided into (9.198) we get

$$(\tfrac{1}{3})^{\frac{1}{2}} \exp\left(\tfrac{1}{2} + \tfrac{1}{2}z + \frac{z^3}{3} + ...\right),$$

which has an expansion in positive powers.

The above results show the complexity of the problem of divisibility in general. Many further theorems can be obtained for the decomposability of characteristic function which are analytic, i.e. analytic in a neighbourhood of $t = 0$. An account of these is given in Lukacs (1960).

The non-uniqueness of decomposition into prime factors recalls the similar situation in some algebraic rings. Dugué (1957) suggests the tool used to deal with this situation in algebra, i.e. the theory of ideals, might have an analogue in the theory of probability but no such theory yet exists.

Bibliographical notes

For the general theory of the subjects contained in this chapter see Lévy (1937a, 1948a), Gnedenko and Kolmogorov (1954), Linnik (1964), and Lukacs (1960). Many relevant papers will be found in the Russian journal *Theory of Probability* of which an English translation now appears in the U.S.A. The reader should also consult La Salle (1962). The statistical problem of estimating the structure of an additive homogeneous process from an observed realization is considered by Rubin and Tucker (1959), and first-passage distributions for such processes are discussed by Baxter and Donsker (1957) and by Kac (1959c).

30

Exercises

Exercise 9.1. Show that the converse of Theorem 9.1 is false, i.e. that there exist distributions $F_1(x)$, $F_2(x)$, such that $F_1 * F_2$ has at least one point of increase which cannot be written as $x_1 + x_2$, where x_1 is a point of support of $F_1(x)$, and x_2 a point of support of $F_2(x)$.

Exercise 9.2. Show that the Cauchy distribution whose characteristic function is $\exp -|t|$ can be represented in the Lévy representation (9.30) where $a = 0$, $\sigma = 0$, $M(x) = -(\pi x)^{-1}$ $(x < 0)$, $N(x) = -(\pi x)^{-1}$ $(x > 0)$.

Exercise 9.3. Show that the second characteristic function of the distribution whose characteristic function is given by (9.69) can be represented in the form (9.52) when $f(x)$ is bounded and absolutely integrable over the interval $(0, \infty)$.

10. The Random Walk

10.1. THE study of random walks is the study of the sums of random variables, these variables varying in complexity from simple independent distributions in one dimension to random variables of a much more complicated character with their sums subject to further conditions. In this book we have already considered a number of problems of this kind. Thus in Chapter 8 we proved a number of theorems on the asymptotic distribution, and overall behaviour, of sums of the form

$$S_n = X_1 + \ldots + X_n, \tag{10.1}$$

where the X_n are independent random variables.

In this chapter we shall consider further problems of this kind and we begin by studying random variables whose values are confined to the integers 0, $\pm 1,\ldots$. In Chapter 3 we studied random walks in which the individual steps took only the values 0, ± 1, whilst the sum S_n was restricted to a finite set of integers. Such a set may be called a 'lattice' and we use this word similarly in higher dimensions.

What happens in this case is then dependent on what conditions are imposed at the boundaries $S_n = 0$, N. We have already discussed such problems in some detail for reflecting and absorbing boundaries and considered some of the resulting absorption and passage-time distributions.

When the individual steps are no longer confined to ± 1 the problem becomes much more complicated and we begin by considering techniques for dealing with this case which is of importance in statistical theory in connection with 'sequential' sampling.

10.2. The general random walk on a one-dimensional lattice

Suppose that the X_n are independent random variables with the distribution

$$\operatorname{pr}(X = j) = p_j \quad (j = 0, \pm 1,\ldots). \tag{10.2}$$

Let the walk be confined to the range $0, 1,\ldots, N$ and start at i. We suppose that as soon as

$$S_n = i + X_1 + \ldots + X_n \geqslant N$$

or

$$\leqslant 0, \tag{10.3}$$

the process stops. A variety of questions may then be asked about S_n such as its distribution at 'time' n conditional on the walk not having stopped, the probability distribution of the time at which the walk stops, and so on. Consider the probability of ever stopping at the boundary 0. Suppose that this is P_i, i being the initial position. It is easy to prove that if the distribution (10.2) is not concentrated at $X = 0$, the probability that S_n never satisfies the inequalities (10.3) is zero, and thus $1 - P_i$ is the probability that the walk stops with $S_n \geqslant N$.

It is convenient to define the conventional values

$$P_i = 1, \quad i \leqslant 0$$

$$= 0, \quad i \geqslant N. \tag{10.4}$$

Then, considering what happens at the first step, we have

$$P_i = \sum_{j=-\infty}^{\infty} p_j P_{i+j} \tag{10.5}$$

$$= \sum_{j=1-i}^{N-i-1} p_j P_{i+j} + \sum_{j=-\infty}^{-i} p_j. \tag{10.6}$$

For $i = 1,\ldots,\ N-1$ these provide $N-1$ equations for the probabilities P_i, and it is easy to show that these equations are sufficient to determine the P_i uniquely. However, for a general distribution, $\{p_j\}$, and N not very small their solution can be very awkward. When the distribution, $\{p_j\}$, is confined to a small range it is, however, sometimes possible to find an explicit solution in terms of the roots of the generating functions of $\{p_j\}$.

If the range of the distribution $\{p_n\}$ is finite and small compared with N it is also possible to obtain a good approximation by using the idea of a 'martingale' as described in § 3.7. This is equivalent to a method of approximation used by Feller for this purpose (Feller 1957, p. 333). Suppose that the range of the distribution $\{p_n\}$ is from $-a$ to b, where a, b are positive integers, and suppose also that the mean of X_n is zero, so that

$$\sum_{-a}^{b} np_n = 0.$$

Now modify the definition of the process to allow the state to consist of all values from $-a$ to $N+b$, and suppose that the states $-a$, $1-a,\ldots$, 0, N, $N+1,\ldots$, $N+b$ are all absorbing states so that once the process lands in one of them further X's are zero. Let P be the probability that the process ultimately lands in one of the states N, $N+1,\ldots$, $N+b$. Then $1-P$ will be the probability that it ultimately lands in one of the states $-a$, $1-a,\ldots$, 0. Whatever the value of n, and the initial state i,

$$E(S_n) = i$$

so that the sequence $\{S_n\}$ is a martingale. From the limiting behaviour we

have
$$(N+b)P \geqslant \lim_n E(S_n) \geqslant NP - a(1-P).$$

From these two inequalities we have
$$i(N+b)^{-1} \leqslant P \leqslant (1+a)(N+a)^{-1}.$$

When the mean of $\{p_n\}$ is not zero this method can be extended by using the roots of the equation obtained by equating the generating function to unity.

This method of argument can also be regarded as a special case of the use of a general theorem due to Wald which we will consider later (§ 10.13), and which, in suitable cases, also gives the distribution of the time of arrival at the absorbing state.

10.3. The diffusion approximation

Under very wide conditions on the distribution of $\{X_n\}$ we know that the distribution of sums of the form (10.1) are well approximated by normal distributions. This suggests that, 'in the large', problems of random walk can be approximated by problems on the Brownian motion discussed in the previous chapter, and if this is so we expect that the distribution of S_n, with given boundary conditions, could be approximated by the solutions of a diffusion equation. We shall do this here in a purely heuristic manner following the argument for the Brownian movement given in §9.10. The exact theory involves rather elaborate arguments (see Watterson 1962). We shall always assume that var(X_t) is finite. If this is not true the Brownian approximation need not be valid as is illustrated by the stable distributions considered in §9.16 and §9.17.

It is necessary to rescale both the variable S_n and 'time' (as measured by n), both of which are here discrete. Let S_n be the state of the process at the nth step and write $Z_t = N^{-1}S_n$, where $t = N^{-1}n$ is the new 'time'. We consider the behaviour with time of Z_t when N is large. At time t, Z_t will approximate by a continuous distribution whose probability density we write as $p(z, t)$.

Suppose that
$$E(X_n) = m_1 = \Sigma n p_n. \tag{10.7}$$

Then except near the boundaries
$$E(S_{N(t+h)} - S_{Nt}) = Nhm_1, \tag{10.8}$$

$$E(Z_{t+h} - Z_t) = hm_1,$$

and
$$\lim_{h \to 0} h^{-1} E(Z_{t+h} - Z_t) = m_1. \tag{10.9}$$

This is a measure of the 'drift' of the process, considered in the large.

Now suppose that the second moment of X_n is

$$E(X_n^2) = \sum_n n^2 p_n = Nm_2, \tag{10.10}$$

where m_2 is fixed as N increases. Then except near the boundaries

$$E(S_{N(t+h)} - S_{Nt})^2 = N^2 h m_2 + Nh(Nh-1)m_1^2 \tag{10.11}$$

so that

$$E(Z_{t+h} - Z_t)^2 = hm_2 + h^2 m_1^2 + O(N^{-1}),$$

and

$$\lim_{h \to 0} h^{-1} E(Z_{t+h} - Z_t)^2 = m_2. \tag{10.12}$$

We also assume that the absolute third moment of X_n is finite.

Taking h as small compared with unity, and N^{-1} small compared with h, we consider the distribution $p(z, t+h)$, at time $t+h$, and express it in terms of $p(z, t)$, the distribution at time t. Assuming that z is not near the boundaries and writing $r(z, h)$ for the density of a continuous probability distribution approximating to the distribution of $Z_{t+h} - Z_t$ (which consists of the sum of many terms such as X_n after rescaling) we have

$$p(z, t+h) = \int p(z-u, t) r(u, h) \, du.$$

Here the range of integration is taken as effectively infinite. If $p(z-u, t)$ can be expanded in a Taylor series about the point z we have

$$p(z, t+h) = \int \{p(z, t) - u p_z(z, t) + \tfrac{1}{2} u^2 p_{zz}(z, t) + O(u^3)\} r(u, h) \, du$$
$$= \int \{p(z, t) r(u, h) - u p_z(z, t) r(u, h)$$
$$+ \tfrac{1}{2} u^2 p_{zz}(z, t) r(u, t)\} \, du + O(h^3). \tag{10.13}$$

From above we have

$$\int r(u, h) \, du = 1,$$
$$\int u r(u, h) \, du = h m_1 + o(h), \tag{10.14}$$
$$\int u^2 r(u, h) \, du = h m_2 + o(h). \tag{10.15}$$

Putting these in (10.13) we get

$$p(z, t+h) = p(z, t) - h m_1 p_z(z, t) + \tfrac{1}{2} h m_2 p_{zz}(z, t) + o(h).$$

Dividing by h, and proceeding to the limit,

$$\frac{\partial p(z, t)}{\partial t} = -m_1 \frac{\partial p(z, t)}{\partial z} + \tfrac{1}{2} m_2 \frac{\partial p(z, t)}{\partial z^2}. \tag{10.16}$$

This is the required diffusion equation. To establish that the solution of this equation for prescribed initial and boundary conditions exists, is unique, and that after rescaling the solution of the random walk converges to it, requires a lengthy argument which we will not give here, and we refer the reader to Watterson (1962) who gave the first known proof.

In the absence of any boundaries the solution of (10.16) is any multiple of the normal distribution

$$p(z, t) = (2\pi m_2 t)^{-\frac{1}{2}} \exp\left\{-\frac{(z - m_1 t)^2}{2m_2 t}\right\}. \tag{10.17}$$

Clearly the random walk is usually well approximated by this when the steps have finite variance. Similarly the random walk with absorbing or reflecting boundaries can be approximated by the solutions of (10.16) with the appropriate boundary conditions and we have discussed such solutions in a previous chapter. In this way we can not only obtain probabilities of absorption but also approximations to the first passage-time distributions in random walks. The 'gambler's ruin' problem considered in Chapter 3 can be dealt with in this way and a numerical example of such an approximation is given by Bartlett (1955).

An interesting problem of this type which can also be approximated by the use of the theory of Brownian motion is that of the 'extent' of the random walk S_n (Daniels 1941). Suppose for simplicity that the X_n can take only the values ± 1 with probabilities $\frac{1}{2}$. Then the extent of the random walk of n steps is defined as

$$\max_{0 \leqslant m \leqslant n} S_n - \min_{0 \leqslant m \leqslant n} S_n,$$

where we put $S_0 = 0$.

Daniels obtains an explicit solution of the problem of finding the distribution of the extent and also considers the Brownian motion approximation which is equivalent to considering the distribution of

$$\max_{0 \leqslant u \leqslant t} X(u) - \min_{0 \leqslant u \leqslant t} X(u),$$

where $X(u)$ is the random variable defining a simple Brownian movement without drift.

10.4. Distribution of the zeros

Consider first the random walk in which $X_n = \pm 1$ with probabilities $\frac{1}{2}$, and with no other restrictions on the sums S_n. We study the distributions of the zeros of the successive sums S_n, i.e. the values of n at which $S_n = 0$. Clearly we can expect this problem to be closely mimicked by the distributions of the zeros of the Brownian movement which we have studied in §9.14. However, the discrete problem has a very considerable interest of its own besides giving an introduction to an important class of combinatorial relations which hold for more general problems.

THEOREM 10.1. *Putting* $S_0 = 0$, *the probability that* $S_{2n} = 0$ *at the step* $2n$ *for the first time after* S_0 *(i.e.* $S_1 \neq 0, \ldots, S_{2n-1} \neq 0$*) is*

$$p_{2n} = (2n-1)^{-1}\binom{2n-1}{n-1}2^{1-2n}. \tag{10.18}$$

Proof. To prove the result we use generating functions and a recurrence principle. To do this observe that a return to zero ($S_k = 0$) can only occur when k is even. If, in an unrestricted walk, $S_{2n} = 0$, S_{2k} may have been zero for some step $2k < 2n$. Let P_{2n} be the probability that $S_{2n} = 0$. This is obviously

$$P_{2n} = \binom{2n}{n} 2^{-2n}. \tag{10.19}$$

Then enumerating the various possibilities we have

$$P_{2n} = \sum_{k=1}^{n} p_{2k} P_{2n-2k}, \tag{10.20}$$

where we put $P_0 = 1$. Introducing the generating functions

$$P(z) = \sum_{k=0}^{\infty} P_{2k} z^{2k}$$

$$= \sum_{k=0}^{\infty} \binom{2k}{k} 2^{-2k} z^{2k}$$

$$= (1-z^2)^{-\frac{1}{2}}, \tag{10.21}$$

and

$$p(z) = \sum_{k=0}^{\infty} p_{2k} z^{2k}, \tag{10.22}$$

we get from (10.20)

$$P(z) - 1 = p(z) P(z)$$

so that

$$p(z) = 1 - (1-z^2)^{\frac{1}{2}}$$

from which we find

$$p_{2n} = \frac{\frac{1}{2}\left(-\frac{1}{2}\right)\left(-\frac{3}{2}\right)\cdots\left(-\frac{2n-3}{2}\right)}{n!}(-1)^{n+1}$$

$$= (2n-1)^{-1}\binom{2n-1}{n-1}2^{1-2n},$$

which checks with (10.18). This approach is easily generalized to the case where $X_n = \pm 1$ with probabilities p and $q = 1-p$.

We can look at this distribution in another way. Since the first step must be either $+1$ or -1, and the process is symmetric, p_{2n}, as we have said above, gives the probability of a first passage from $+1$ to 0 (and therefore from 0 to -1) in $2n-1$ steps. Hence the result can be found using the reflection principle described in § 3.7 and formula (3.93). Hence

$$\left\{\frac{1-(1-z^2)^{\frac{1}{2}}}{z}\right\}^N \tag{10.23}$$

is the probability generating function of the first passage from 0 to $-N$ (or $+N$), as we have already shown in §3.7. The evaluation of the coefficients of (10.23) is analytically awkward and these probabilities are best obtained directly by the reflection argument used in §3.7 where it was shown that the probability of a first passage from 0 to $N > 0$ at the stage $2n - N$ is

$$\frac{N}{2n-N}\binom{2n-N}{n}2^{N-2n}. \qquad (10.24)$$

By the argument above this is also equal to the probability that starting from zero, the Nth return occurs at the stage $2n$.

We now compare these formulae with those of the passage-time distributions in the theory of the Brownian motion. Consider (10.24) and suppose N and n large. In fact let M be a large number and put $N = xM^{\frac{1}{2}}$, $2n = tM$, so that t measures 'time' in units of M steps. The possible values of $2n$ are even numbers so that t increases in jumps of amount $2M^{-1}$. Using this and the approximation (§2.9) to the binomial distribution we find for the probability that a Brownian motion starting from $X(0) = 0$ reaches $X(t) = x$ for the first time in the interval $(t, t+dt)$ the probability

$$(2\pi)^{-\frac{1}{2}}\frac{x}{t^{\frac{3}{2}}}e^{-x^2/2t}\,dt, \qquad (10.25)$$

which is just what we obtained before in (9.141). In just the same way we can obtain approximations for the distribution of

$$\max_{1\leqslant i\leqslant n} S_i = \max_{1\leqslant i\leqslant n}(X_1+\ldots+X_i), \qquad (10.26)$$

as in (9.136) (Theorem 9.12). These approximate results, like the ones in the next section, can be applied to much more general walks in which the steps have any distribution with zero mean and finite variance.

10.5. The inverse sine laws

Following Feller (1957b) we now prove some further important results for the simple random walk with $X_i = \pm 1$ with probabilities $\frac{1}{2}$. Consider the set of values of S_i for $0 \leqslant i \leqslant 2n$, and join them to form a path. The line between S_{i-1} and S_i is called a 'side' of the path and we say the side is positive if at least one of S_{i-1} and S_i is positive. Notice that because of the distribution of the X_i it is not possible for a 'side' to be horizontal. Consider the probability that a specified number of steps occur in which the resulting side is positive, during the first $2n$ steps starting from $S_0 = 0$. We could have considered this probability for an odd number of steps but it is slightly more convenient to take an even number. The number of steps out of the first $2n$ for which the sides are positive is necessarily even because returns to zero can only take place at an even number of steps and so we take it as $2k$. We therefore write $p_{2k,2n}$ for the required probability.

THEOREM 10.2

$$p_{2k,2n} = \binom{2k}{k}\binom{2n-2k}{n-k}2^{-2n}. \tag{10.27}$$

When $k = n$, $p_{2k,2n}$ is the probability that every successive pair of vertices is such that at least one of them is positive and non-zero. Since two successive vertices must correspond to different values of S this is the same as the probability that $S_k \geqslant 0$ for $k = 1,\ldots, 2n$, and is therefore equal to

$$1 - p_2 - p_4 - \ldots - p_{2n} = \binom{2n}{n}2^{-2n} = P_{2n}, \tag{10.28}$$

by the proof of the previous theorem, since this is the coefficient of z^{2n} in $\{1 - p(z)\}(1 - z^2)^{-1} = P(z)$. Thus (10.27) is correct for $k = n$, and similarly it is correct for $k = 0$.

Suppose $1 \leqslant k \leqslant n-1$. Then the path must pass through zero and we use again the standard recurrence argument. Let $2r$ be the first value for which $S = 0$. Then the path stays always on the positive side until $2r$, an event with probability $\frac{1}{2}p_{2r}$, or it stays always on the negative side, an event with the same probability.

Summing over all cases we have for $1 \leqslant k \leqslant n-1$,

$$p_{2k,2n} = \frac{1}{2}\sum_{r=1}^{k} p_{2r}p_{2k-2r,2n-2r} + \frac{1}{2}\sum_{r=1}^{n-k} p_{2k}p_{2k,2n-2r}. \tag{10.29}$$

Assume that (10.27) holds for values of n less than the value on the left of (10.29). We can write (10.27) in the form

$$p_{2k,2n} = P_{2k}P_{2n-2k}, \tag{10.30}$$

by using (10.28). Inserting this in (10.29) we get

$$p_{2k,2n} = \frac{1}{2}P_{2n-2k}\sum_{r=1}^{k} p_{2r}P_{2k-2r} + \frac{1}{2}P_{2k}\sum_{r=1}^{n-k} p_{2k}P_{2n-2r-2k}$$

$$= P_{2k}P_{2n-2k},$$

on using (10.20). Since (10.30) is true for $n = 1$, it is true generally. Thus (10.27) is the required distribution.

Using Stirling's formula on (10.27) we see that $p_{2k,2n}$ is asymptotically equal to

$$\pi^{-1}k^{-\frac{1}{2}}(n-k)^{-\frac{1}{2}} \tag{10.31}$$

when k and $n-k$ are large. If we take fixed constants x, y satisfying $0 < x < y < 1$, the sum of $p_{2k,2n}$ over values of k such that

$$nx < k < ny$$

will converge to

$$\pi^{-1} \int_x^y u^{-\frac{1}{2}} (1-u)^{-\frac{1}{2}} \, du,$$

so that asymptotically we have a beta-type distribution, with parameters $\frac{1}{2}, \frac{1}{2}$. For these values the integral can be evaluated and we have

$$\pi^{-1} \int_0^x u^{-\frac{1}{2}} (1-u)^{-\frac{1}{2}} \, du = 2\pi^{-1} \sin^{-1} x^{\frac{1}{2}}. \tag{10.32}$$

This is the origin of the name 'inverse-sine law' which is applied to this distribution.

This result is a particular case of much more general asymptotic results which require only that the distribution of the individual steps has zero mean and finite standard deviation, and in fact an analogous theorem holds for the simple Brownian motion.

Returning again to the simple walk with unit steps, consider the probability that $\max_{0 \leqslant k \leqslant 2n} S_k$ occurs at the point k for the first time. Let this be $\pi_{k,2n}$. We show that

$$\pi_{k,2n} = p_{k,2n}, \qquad \text{if } k \text{ is even,}$$

$$= p_{k-1,2n}, \quad \text{if } k \text{ is odd.} \tag{10.33}$$

Consider first the case $k = 0$. For this to happen we must have $S_k \leqslant 0$ for $k = 1, \ldots, 2n$, an event which has probability P_{2n}. This is also the probability that a random walk of length $2n-1$ attains its maximum at $k = 0$. The probability that a walk of length $2n$ or $2n+1$ attains its maximum for the first time at $k = 2n$ ($k = 2n+1$ respectively) is obtained by reversing the direction of time and is the probability that $S_k < 0$ for $k = 1, \ldots, 2n$ ($2n+1$ respectively). It is therefore $\frac{1}{2} P_{2n}$.

For a path of length $2n$ to attain its maximum at k for the first time it must consist of two parts, a walk of length k with the first maximum at k and a walk of length $2n-k$ with $S_i \leqslant 0$ for $i = 1, \ldots, 2n-k$. The probabilities of these depend on whether k is even or odd. If we write $k = 2l$, or $2l+1$, and insert the corresponding probabilities we see that the probability of the maximum being attained at k for the first time is

$$\tfrac{1}{2} P_{2l} P_{2n-2l} = p_{2l,2n} \tag{10.34}$$

where $k = 2l$ or $2l+1$. This is very nearly the distribution of the amount of time spent on the positive side, the difference arising from the fact that in that problem the amount of time must be even. Approximating in the same way we obtain the inverse-sine law for the position at which the maximum is first obtained. The remarkable fact about this result is that the distribution of k is very nearly symmetrical about the point $\frac{1}{2}n$ so that the distribution of the first maximum is very nearly the same as that of the last maximum. This is due to the fact that the probability of there being more than one maximum is very small.

10.6. The distribution of max S_k

For an unrestricted random walk with $X_k = \pm 1$ with probabilities $\frac{1}{2}$, consider the distribution of

$$T_n = \max_{1 \leqslant k \leqslant n} (0, S_k). \tag{10.35}$$

This can be obtained in various ways. We have

$$v_{x,n} = \binom{n}{\frac{1}{2}(x+n)} 2^{-n} \tag{10.36}$$

for the probability that $S_n = x$. Here x and n must both be odd or both even. Using the reflection principle described in §3.6 we have

$$u_{x,n}^{(t)} = \mathrm{pr}\!\left(S_n = x, \ \max_{1 \leqslant k \leqslant n} S_k < t \right)$$

$$= v_{x,n} - v_{2t-x,n}, \tag{10.37}$$

where x and n are integers which are both odd or both even, and t is any positive integer greater than x. For convenience make the convention that

$$v_{x,n} = 0,$$

if x and n are not both even, or both odd. Then the probability that

$$\max_{1 \leqslant k \leqslant n} S_k < t \quad (t = 0, 1, \ldots, n)$$

must be equal to

$$\sum_{\substack{\text{all} \\ x < t}} v_{x,n} - \sum_{\substack{\text{all} \\ x < t}} v_{2t-x,n},$$

and therefore

$$\pi_t = \mathrm{pr}\!\left(\max_{1 \leqslant k \leqslant n} S_k = t \right)$$

$$= \sum_{\substack{\text{all} \\ x < t+1}} v_{x,n} - \sum_{\substack{\text{all} \\ x < t+1}} v_{2t+2-x,n} - \sum_{\substack{\text{all} \\ x < t}} v_{x,n} + \sum_{\substack{\text{all} \\ x < t}} v_{2t-x,n}$$

$$= v_{t,n} + v_{t+1,n}. \tag{10.38}$$

Notice that one of the two terms in (10.38) must be zero. Consider the generating function of this distribution. If $n = 2m$, and $m > 0$, we find

$$\pi_n(z) = \sum_{t=0}^{n} \pi_t z^t$$

$$= \binom{2m}{m} 2^{-2m} + \sum_{s=1}^{m} \binom{2m}{m-s} 2^{-2m} z^{2s} (1 + z^{-1}), \tag{10.39}$$

whilst if $n = 2m+1$,

$$\pi_n(z) = \sum_{t=0}^{n} \pi_t z^t$$

$$= \sum_{s=0}^{m} \binom{2m+1}{m-s} 2^{-2m-1} z^{2s}(1+z). \tag{10.40}$$

Multiplying $\pi_n(z)$ by w^n and summing over all values of n from zero to infinity we obtain a generalized generating function which we can write

$$\pi(w,z) = \sum_{n=0}^{\infty} \pi_n(z) w^n$$

$$= (1-w^2)^{-\frac{1}{2}} + \sum_{m=1}^{\infty} \sum_{s=1}^{m} \binom{2m}{m-s} 2^{-2m} z^{2s}(1+z^{-1}) w^{2m}$$

$$+ \sum_{m=0}^{\infty} \sum_{s=0}^{m} \binom{2m+1}{m-s} 2^{-2m-1} z^{2s}(1+z) w^{2m+1}. \tag{10.41}$$

The diffusion approximation to (10.38), the distribution of max S_k, can be easily verified to be that already obtained in the last chapter (§ 9.14) for the distribution of the maximum in a Brownian motion.

The mean of the distribution of max S_k is also of interest. Write

$$Z_i = 1 \quad \text{if} \quad S_i > S_j \quad \text{for} \quad 1 \leqslant j < i,$$

$$= 0, \quad \text{otherwise.}$$

Then the mean of the distribution of max S_k $(1 \leqslant k < n)$ is

$$\sum_{1}^{n} E(Z_i),$$

and by what has been proved above, $E(Z_i)$ is the probability that the first maximum occurs at the ith place, which is $\frac{1}{2} P_{2m}$ when $i = 2m$, or $i = 2m+1$. Thus the mean is

$$\tfrac{1}{2} P_0 + P_2 + P_4 + \ldots + \tfrac{1}{2} P_{2[\frac{1}{2}n]}.$$

Since by using Stirling's formula we easily see that

$$P_{2n} \sim (\pi n)^{-\frac{1}{2}},$$

the above series is asymptotically equal to

$$\pi^{-\frac{1}{2}} \sum_{m=1}^{\frac{1}{2}n} m^{-\frac{1}{2}},$$

which in turn is asymptotically equal to

$$(2\pi)^{-\frac{1}{2}} n^{\frac{1}{2}}. \tag{10.42}$$

This clearly links up with the fact that the expectation of maximum of a Brownian movement increases as $t^{\frac{1}{2}}$, where t is the time. The above results on the distribution of the maxima of partial sums can be widely generalized as we shall see later in connexion with the Pollaczek–Spitzer theorem (§ 10.18). Results for first-passage distributions for the very special case where $X_k = \pm 1$ can also be easily obtained when the probabilities of $X = 1$, $X = -1$ are no longer $\frac{1}{2}$ but p and $q = 1-p$.

10.7. Random walk on a two-dimensional lattice

A number of interesting results of a different kind can be obtained when the random walks are no longer one-dimensional but take place on the points of an integral-valued lattice in two or more dimensions. Thus in two dimensions we suppose that the walk takes place on the points $m, n = 0, \pm 1, \pm 2, \ldots$. We suppose that it begins at the point $(0,0)$ and that at any point the next step can be to any of the four closest points, each such step having probability $\frac{1}{4}$. Then the probability of being at the point (m, n) after N steps, if there are no further restrictions on the walk, is

$$\Sigma \binom{N}{k}\binom{k}{\frac{1}{2}(k-m)}\binom{N-k}{\frac{1}{2}(N-k-n)}4^{-N}, \qquad (10.43)$$

where the sum is taken over all values of k such that $m \leqslant k \leqslant N-n$, and such that $k-m$ and $N-k-n$ are both even or both odd. Then N and $m+n$ must be also both even or both odd. We shall later consider as before the probabilities of such walks under various boundary conditions but before doing this we find the probability of ever returning to the origin in an infinitely long walk. In one dimension the analogous probability is obviously unity but the two and higher dimensional analogues are not obvious and were first discussed by Pólya (1921).

THEOREM 10.3. *In the unrestricted two-dimensional walk as defined above the probability of ever returning to the origin is unity.*

Proof. Write p_n for the probability of returning to the origin at the nth step. Then n must be even so we only consider p_{2n} which from (10.43) is equal to

$$p_{2n} = \sum_{k=0}^{n}\binom{2n}{2k}\binom{2k}{k}\binom{2n-2k}{n-k}4^{-2n}$$

$$= \sum_{k=0}^{n}\binom{2n}{n}\binom{n}{k}^{2}4^{-2n}.$$

The sum

$$\sum_{k=0}^{n}\binom{n}{k}^{2}$$

can be evaluated by observing that it is the coefficient of z^0 in

$$(1+z)^n(1+z^{-1})^n = z^{-n}(1+z)^{2n},$$

and is therefore equal to

$$\binom{2n}{n}.$$

Thus

$$p_{2n} = 4^{-2n}\binom{2n}{n}^2. \tag{10.44}$$

Using Stirling's formula again we see that for large n, p_{2n} is approximately equal to $(\pi n)^{-1}$.

We now use a recurrence argument. Let q_{2n} be the probability of returning to the origin at step $2n$ for the first time. Then

$$p_{2n} = \sum_{k=1}^{n} q_{2k} p_{2n-2k} \quad (n = 1, 2, \ldots). \tag{10.45}$$

Introducing the generating functions (with $p_0 = 1$)

$$P(z) = \sum_{n=0}^{\infty} p_{2n} z^{2n}, \tag{10.46}$$

$$Q(z) = \sum_{n=1}^{\infty} q_{2n} z^{2n}, \tag{10.47}$$

we get

$$Q(z) = \frac{P(z)-1}{P(z)}. \tag{10.48}$$

$Q(z)$ is the generating function of a probability distribution which is proper (i.e. $\sum q_{2n} = 1$) if $Q(1) = 1$. Since the series for $P(z)$ consists of non-negative terms and is divergent for $z = 1$, we must necessarily have $Q(1) = 1$, and the theorem is proved. As we shall see later this result is false for lattices in three or more dimensions. See also Domb (1954), Foster and Good (1953), and Montroll (1956, 1964).

From the obvious generalization of the Central Limit Theorem of Chapter 8, with finite variances, it follows that for large N the distribution of the point (m, n) is approximately a circular normal distribution. Its two means are clearly zero, and the variances are obviously both equal to $\frac{1}{2}n$. Thus·the expected square of the distance from the origin is n.

10.8. Two-dimensional lattice with boundaries

Suppose again that we have a random walk on the points of a two-dimensional lattice with all four directions equally probable. We consider only a set of points, D, containing $(0, 0)$ and such that each point in D is connected with

(0, 0) by a path lying entirely in D. The 'interior' points of D are defined to be those points whose four neighbouring points belong to D. The 'boundary' of D consists of all points of D which are not interior points. Then most problems of interest arise when we suppose that the boundary points are 'absorbing', i.e. that no move from them is allowed, and we ask for the probability that a random walk beginning from (0, 0), or from any other interior point of D, will end at any specified boundary point. By adding such probabilities we can obtain the probability that the walk ends on any specified subset of the boundary.

Let P be a boundary point, and $p(x, y)$ the probability of absorption at P if the walk begins at an interior point (x, y), i.e. this is the probability of arriving at P before any other boundary point. Then for every interior point (x, y) we must have

$$4p(x, y) = p(x+1, y) + p(x-1, y) + p(x, y+1) + p(x, y-1). \quad (10.49)$$

A function satisfying such an equation is said to be a 'discrete harmonic function' in analogy with functions of two continuous variables, $f(x, y)$ say, for which the value of $f(x, y)$ is equal to the mean of its values on any small circle with centre (x, y). A detailed study of such discrete harmonic functions, without reference to probability, has been made by Heilbronn (1949).

Suppose that the domain is finite. We then have the two following results corresponding to the usual theory of harmonic functions.

THEOREM 10.4. $p(x, y)$ *either is constant inside the domain or attains its maximum at a boundary point.*

THEOREM 10.5. *If $p(x, y)$ is a discrete harmonic function and is specified on the boundary (and in the present probability case we put $p(x, y) = 1$, when (x, y) is the specified boundary point, and $p(x, y) = 0$ for other boundary points) the solution of the above set of equations (10.49) is unique.*

Proofs of Theorems 10.4 *and* 10.5. Let M be the maximum of $p(x, y)$ on the domain and its boundary. If M is attained at an interior point it must also be attained at the four neighbours of this point and continuing the argument for these, $p(x, y)$ must equal M everywhere. This proves 10.4. If $p(x, y)$ were not determined uniquely by the boundaries values, the difference of two solutions would have all boundaries values equal to zero but some interior point with $p(x, y) \neq 0$. This is impossible by Theorem 10.4 and thus Theorem 10.5 is proved.

For an infinite domain on a two-dimensional lattice Heilbronn shows that (10.49) has a unique solution when values are prescribed at given boundary points. For a three-dimensional lattice this is no longer true but it is never-the-less obvious that the probability problem has a unique solution.

For particular sets of boundaries $p(x, y)$ has been evaluated explicitly by McCrea and Whipple (1940), and by Barnett (1963). They approach the problem in a slightly different manner in that they begin by determining a function, $F(x, y)$, which is the expected number of times the particle leaves the point (x, y) when the walk begins at some specified point $((0, 0)$ say) and is absorbed at the boundary. The probability of being absorbed at the boundary (m, n) is then $\frac{1}{4}$ of the sum of the $F(x, y)$ at points neighbouring to (x, y). In the cases considered the domain is a rectangle, an infinite strip, a half-plane, or infinite quadrant. Barnett solves these problems also for the more general case when the probabilities of a walk in the four directions north, south, east, and west (say), are p, q, r, and s, where $p+q+r+s = 1$.

Using their results it is possible to obtain again the result of Theorem 10.3 that the probability of ever returning to the point $(0, 0)$ in an unrestricted random walk is unity when $p = q = r = s = \frac{1}{4}$. Barnett goes further than this and shows that for general p, q, r, s, the necessary and sufficient condition that the probability of return is unity is that $p = q$, $r = s$. He also obtains a general method of solution for arbitrarily placed absorbing points.

Return problems for unrestricted random walks on other types of lattice, such as one composed of equilateral triangles, have also been considered by various writers (Domb 1954).

Other problems have also been considered. Thus one may ask how many different points the particle passes through in the course of the first n steps. Dvoretzky and Erdös (1951) show that the expected number of such points is

$$\frac{\pi n}{\log n} + O\left\{\frac{n \log \log n}{(\log n)^2}\right\}, \tag{10.50}$$

and they give similar results for walks in higher dimensions.

Hammersley (1956) considers a random walk with $p = q = r = s = \frac{1}{4}$, and defines the 'area' enclosed by this walk to consist of the number of points not belonging to the walk which are inaccessible to a path from infinity which has no common points with the path of the walk. He obtains inequalities for the expectation of the area.

Another generalization arises when we assume that the steps are no longer independent. Thus after a step in some direction one could assume that the path must turn to the left or to the right with probabilities $\frac{1}{2}(1+\alpha)$, $\frac{1}{2}(1-\alpha)$, any movement in the same direction as the previous move being prohibited. The generating function for the position after n steps can be evaluated explicitly (Goldstein 1951, Gillis 1960, Seth 1963) and the probability of ever returning to the origin is again unity.

Random walks on lattices have been related to the theory of electrical networks by Nash–Williams (1959).

10.9. The self-avoiding random walk

A much more difficult problem is that of the random walk with the restriction that the path never passes through the same point twice. This is

31

said to be a 'self-avoiding' walk and has been extensively studied because it is a simplified version of a problem important in theoretical chemistry, that of finding the statistical distribution of the length of polymer molecules (see § 10.22). Very few theoretical results in such problems are known.

It is convenient to consider frequencies rather than probabilities. Let C_N be the number of self-avoiding walks with N steps, and $C_N(m, n)$ the number of such walks which start from $(0,0)$ and end at (m, n). Then the probability distribution of the end of the walk could be taken as $C_N(m, n)C_N^{-1}$, although

TABLE 10.1

Number of self-avoiding walks on a square lattice

N	C_N	N	C_N
1	4	9	16 268
2	12	10	44 100
3	36	11	120 292
4	100	12	324 932
5	284	13	881 500
6	780	14	2 374 444
7	2172	15	6 416 596
8	5916	16	17 245 332

this is in some ways not a natural definition. Most of the known results have been concerned with obtaining an estimate of C_N as a function of N, and not with $C_N(m, n)$.

No formula for C_N as a function of N is known. Enumeration quickly becomes difficult as N increases but with the aid of a computer and a special programme it is possible to simplify the work. Fisher and Sykes (1959) have given results for $N = 1(1)\ 16$ (Table 10.1).

J. L. Martin (1962) devised a computer programme for this purpose, and gives the results for other types of lattice.

Hammersley and Morton (1954) have proved that there exists a constant, $k > 0$, such that $n^{-1}\log C_n$ converges to k. In order to do this write

$$K(n) = n^{-1}\log C_n. \tag{10.51}$$

We know that $1 < C_n \leqslant 4^n$ so that $0 < K(n) \leqslant \log 4$. Consider the set of all random walks which commence with a self-avoiding walk of length m, and then have a self-avoiding walk of length n which may intersect the first part. The number of such walks is $C_m C_n$ and is not less than C_{m+n}. Thus

$$0 < (m+n)K(m+n) \leqslant mK(m)+nK(n) \leqslant (m+n)\log 4. \tag{10.52}$$

$K(n)$ is bounded and we have to show that

$$\underline{\lim} K(n) = \overline{\lim} K(n). \tag{10.53}$$

To prove this write

$$k = \underline{\lim} K(n) \tag{10.54}$$

and, given $\varepsilon > 0$, choose r so that

$$K(r) < k + \varepsilon.$$

If n is any integer greater than r we can write $n = tr + s$ where t and s are integers, and $0 \leqslant s < r$. Then using the inner inequality in (10.52) repeatedly we have

$$nK(n) \leqslant rtK(rt) + sK(s)$$
$$\leqslant rtK(r) + sK(s)$$
$$\leqslant nK(r) + rK(s)$$
$$\leqslant n(k + \varepsilon) + r \log 4,$$

so that

$$K(n) \leqslant k + \varepsilon + \frac{r}{n} \log 4.$$

Letting n tend to infinity we get

$$\overline{\lim} K(n) \leqslant k + \varepsilon,$$

so that (10.53) is proved.

k has not been evaluated theoretically but can be estimated from Table 10.1, and strict inequalities can be given for it. Using $c_{15} = 6\,416\,596$ and $c_{16} = 17\,245\,332$ we get e^k approximately equal to 2·688 but a better estimate by Fisher and Sykes gives $e^k = 2·639 \pm 0·003$. Another method of estimating k is to do a sampling or Monte Carlo experiment and this has been investigated by Wall and Erpenbeck (1959). This subject has a large literature, much of it concerned with self-avoiding walks on other types of lattices in both two and three dimensions. A survey with an extensive bibliography will be found in the above-mentioned paper of Fisher and Sykes (see also Domb 1964, Lehman and Weiss 1958).

10.10. The random walk on higher dimensional lattices

Consider a three-dimensional lattice defined by coordinates (x, y, z) each of which can take the integral values $0, \pm 1, \pm 2, \ldots$. This is known as a 'cubic' lattice. The walk starts at $(0, 0, 0)$ and takes unit steps along the positive or negative directions of the three coordinate axes. There are six of these and it is supposed that they have equal probabilities $\frac{1}{6}$. We then have the following result.

THEOREM 10.6. *In the above three-dimensional random walk the probability of ever returning to the origin is less than unity and is about* 0·3405.

Proof. We argue in the same manner as in the proof of Theorem 10.3. Let p_n be the probability of returning to $(0,0,0)$ at the nth step. Then $p_n = 0$ if n is odd, as before. If n is even and equal to $2m$ we have

$$p_{2m} = 6^{-2m} \sum_{i,j} \frac{(2m)!}{(i!)^2(j!)^2\{(m-i-j)!\}^2},\tag{10.55}$$

where the sum is taken over all i, j such that $i \geqslant 0$, $j \geqslant 0$, $i+j \leqslant m$. We can rewrite this as

$$p_{2m} = \frac{1}{2^{2m}}\binom{2m}{m}\sum_{i,j}\left(\frac{1}{3^m}\frac{m!}{i!\,j!\,m-i-j!}\right)^2.\tag{10.56}$$

As in the proof of Theorem 10.3 put q_{2m} for the probability of returning to $(0,0,0)$ for the first time at the step $2m$, and write

$$P(z) = \sum_{m=0}^{\infty} p_{2m} z^{2m},$$

where $p_0 = 1$, and

$$Q(z) = \sum_{m=1}^{\infty} q_{2m} z^{2m}.$$

Then arguing as before we again get the equation

$$Q(z) = \frac{P(z)-1}{P(z)},\tag{10.57}$$

so that the probability of ultimate return is

$$Q(1) = \frac{P(1)-1}{P(1)} < 1,\tag{10.58}$$

if $P(1)$ is finite. To show that $P(1)$ is finite consider (10.56). The sum is not greater than

$$\sum_{i,j}^{m}\left(\frac{1}{3^m}\frac{m!}{i!\,j!\,m-i-j!}\right)\max_{i,j}\left(\frac{1}{3^m}\frac{m!}{i!\,j!\,m-i-j!}\right)$$

$$= \max_{i,j}\frac{1}{3^m}\frac{m!}{i!\,j!\,m-i-j!}.\tag{10.59}$$

For the binomial distribution whose typical term is

$$\binom{n}{k}p^k q^{n-k}$$

it is known that the largest term is that for which k is the nearest integer to np. It might be suspected that a similar result holds for a multinomial in the sense that given the distribution

$$\frac{n!}{k_1!\dots k_s!}p_1^{k_1}\dots p_s^{k_s}\quad(\Sigma p_s = 1, \Sigma k_s = n)$$

the largest term should have $|k_i - np_i| \leqslant 1$. Curiously enough, this is false. However, by considering the terms neighbouring to the maximum and their ratio to the latter it is easy to show that

$$np_i - 1 \leqslant k_i \leqslant (n+s-1)p_i, \tag{10.60}$$

for the maximum term, and thus $|k_i - np_i| < s$.

We know from §1.12 that

$$n! = (2\pi)^{\frac{1}{2}} n^{n+\frac{1}{2}} e^{-n+r_n},$$

where $0 < r_n < (12n)^{-1}$. Using this we see that the maximum term in (10.59) is not greater than

$$(2\pi)^{-1} e^{1/12m} \frac{m^{m+\frac{1}{2}}}{x^{x+\frac{1}{2}} y^{y+\frac{1}{2}} (m-x-y)^{m-x-y+\frac{1}{2}}}, \tag{10.61}$$

where $x > 0$, $y > 0$ are real numbers such that $x + y < m$, and both differ from $\frac{1}{3}m$ by quantities not greater than 3. Thus the above expression is not greater than Km^{-1} where K is a positive constant so that $P(z)$ is convergent for $z = 1$ because p_{2m} is of order $m^{-\frac{3}{2}}$, and the probability of return is less than unity. Its evaluation is awkward but Montroll (1956, 1964) has given it as 0·340537330, and it can also be found in terms of a closed elliptic integral (McCrea and Whipple 1940). Starting from $(0,0,0)$ the number of subsequent returns therefore clearly has a geometric distribution. Similar calculations can be carried out for higher dimensional cubic lattices and the probability of ultimate return is again less than unity.

Dvoretsky and Erdös (1951) have studied the way in which the path tends to infinity in three or more dimensions, and McCrea and Whipple show how to use the method of images to obtain absorption probabilities when there are planar boundaries.

Random walks on other types of three-dimensional lattices have been considered (Lehman 1951, Montroll in Beckenbach 1964) and the same sort of result obtained.

Self-avoiding random walks in three-dimensional lattices of various kinds are considered in the papers referred to in connexion with the two-dimensional case.

10.11. Time-continuous random walks on lattices

Instead of assuming that the steps of the random walk take place at integral values of time we may suppose time to be continuous. Let the walk take place on the one-dimensional lattice with values 0, ± 1, ± 2,... and begin at the point k. In any small interval of time, δt, we suppose that the chance of moving one step to the right $(k \to k+1)$ is

$$\lambda \delta t + o(\delta t), \tag{10.62}$$

and similarly the chance of moving one step to the left ($k \to k-1$) is

$$\mu\delta t + o(\delta t). \tag{10.63}$$

Let the distribution of the position after a time t be $P_{k,n}(t)$, where n takes integral values. The determination of the distribution $P_{k,n}(t)$, either for an unrestricted walk, or with reflecting or absorbing boundaries, has been considered by a number of writers (a unified theory with a bibliography is given by Heathcote and Moyal 1959).

Consider first the unrestricted walk. Then using the forward Kolmogorov equations derived in § 3.11 we have

$$\frac{dP_{k,n}(t)}{dt} = -(\lambda+\mu)P_{k,n}(t) + \lambda P_{k,n-1}(t) + \mu P_{k,n+1}(t). \tag{10.64}$$

These are similar to equations (3.136) but here the possible states are all the positive and negative integers and not just the non-negative ones only. To obtain the solution it is more convenient to use the backward equations (3.120), and these are

$$\frac{dP_{k,n}(t)}{dt} = -(\lambda+\mu)P_{k,n} + \lambda P_{k+1,n}(t) + \mu P_{k-1,n}(t). \tag{10.65}$$

For the unrestricted walk (10.64) and (10.65) are the same equations since it is obvious that $P_{k,n}(t)$ is a function of $n-k$ and t only. However, when boundary conditions are imposed this is no longer true and, as is often the case, the backward equations are simpler to solve.

We next have to consider the uniqueness of the solutions of this system of equations. This uniqueness does not follow from the criteria given in Chapter 3, but follows from the general theory of discontinuous processes given by Moyal (1957) whether there exists a boundary or not (if there exist two boundaries the state of the system is bounded and the uniqueness follows from the usual theory of finite sets of differential equations). From a probability point of view the fact that the system must lead to a unique probability solution follows from the fact that the number of changes of state in an interval, $(0, T)$, is a random variable with a Poisson distribution with mean $(\lambda+\mu)$, and thus using Tchebychev's inequality we see that no probability can escape to infinity.

The most systematic way of solving such systems is to introduce a generating function to collect all the $P_{k,n}(t)$ for different values of n into one function and then use a Laplace transform on the time variable. Thus we write, following Heathcote and Moyal,

$$G_k(z, t) = \sum_n P_{k,n}(t)z^n, \tag{10.66}$$

so that

$$\frac{dG_k(z, t)}{dt} = -(\lambda+\mu)G_k(z, t) + \lambda G_{k+1}(z, t) + \mu G_{k-1}(z, t), \tag{10.67}$$

and then put

$$g_k(z, s) = \int_0^\infty e^{-st} G_k(z, t)\, dt, \tag{10.68}$$

so that

$$(s + \lambda + \mu) g_k(z, s) = \lambda g_{k+1}(z, s) + \mu g_{k-1}(z, s) + z^k. \tag{10.69}$$

To ensure the convergence of the integral we insist that the real part of s be positive. We then have to solve (10.69) with whatever boundary conditions are given. This is done by Heathcote and Moyal with one or two reflecting or absorbing boundaries. Here we consider only the simplest case of no boundaries. Then the transition probabilities must depend only on $k - n$, so that

$$G_k(z, t) = z^k G_0(z, t),$$

and

$$g_k(z, s) = z^k g_0(z, s). \tag{10.70}$$

Putting (10.70) in (10.69) we get

$$g_0(z, s) = (s + \lambda + \mu - \lambda z - \mu z^{-1})^{-1}. \tag{10.71}$$

We expand this in a Laurent series convergent in a ring $0 < \alpha < |z| < \beta$ where $0 < \alpha < 1 < \beta$. Let $u_1(s)$, $u_2(s)$ be the two roots of the equation

$$s + \lambda + \mu - \lambda z - \mu z^{-1} = 0. \tag{10.72}$$

These roots are

$$u_1(s) = (2\lambda)^{-1}[s + \lambda + \mu - \sqrt{\{(s + \lambda + \mu)^2 - 4\lambda\mu\}}],$$

$$u_2(s) = (2\lambda)^{-1}[s + \lambda + \mu + \sqrt{\{(s + \lambda + \mu)^2 - 4\lambda\mu\}}], \tag{10.73}$$

and are therefore real for $s > 0$. The Laurent expansion of (10.70) is therefore

$$g_0(z, s) = \lambda^{-1}\{u_2(s) - u_1(s)\}^{-1} \left[\sum_{r=0}^\infty \left\{ \frac{u_1(s)}{z} \right\}^r + \sum_{r=1}^\infty \left\{ \frac{z}{u_2(s)} \right\}^r \right]. \tag{10.74}$$

Picking out the powers of z and inverting the Laplace transforms (as given, for example, in Erdelyi 1954) we get

$$P_{0,r}(t) = (\lambda\mu^{-1})^{\frac{1}{2}r} e^{-(\lambda+\mu)t} I_{|r|}(2t\lambda^{\frac{1}{2}}\mu^{\frac{1}{2}}), \tag{10.75}$$

where $I_r(x)$ is the modified Bessel function of the first kind.

This result is easily obtained by a simple probabilistic argument. Consider the positive and negative jumps during the time interval $(0, t)$. Since λ and μ, the probability density of positive and negative jumps, are not dependent on n, the numbers of positive and negative jumps during $(0, t)$ are independent Poisson variates with means λt and μt. Their algebraic sum is r which is therefore distributed as the difference between two Poisson variates, and we have derived this distribution in §2.13, agreeing with (10.75).

When reflecting boundaries are introduced the system can be interpreted in terms of queueing theory. If absorbing barriers are used the probabilities of absorption can be found by considering the corresponding embedded chain. Notice that it is not difficult to obtain approximate solutions to problems of this kind by using an approximating diffusion equation.

Similar processes can be studied for higher dimensional lattices and probabilities of absorption and of ultimate return to the initial point are again given by those of the corrresponding embedded process with discrete time.

10.12. General random walks in one dimension

We now drop the condition that the individual steps are lattice variates but return to discrete time and therefore consider sums of the form

$$S_n = X_1 + \ldots + X_n,$$

where the X_n are independent random variables with the same distribution. Many of the problems associated with such sums, in the absence of boundary restrictions, have already been considered in Chapter 8.

When the variance of the distribution of the X_i is finite the distribution of S_n tends to normality as n increases and if we rescale the time variable n, and the scale in which S_n is measured we can show that, 'in the large', the statistical properties of the path, $(0, S_1, \ldots, S_n)$, of the walk are therefore closely approximated by the path of a Brownian movement process. In particular, therefore, we can expect that if $E(X_i) = 0$, and $S_0 = 0$,

$$E\left(\max_{0 \leqslant k \leqslant n} S_k\right) \tag{10.76}$$

will increase proportionally to $n^{\frac{1}{2}}$, when n is large.

It is of some interest to calculate this in special cases. We have already done this in § 10.6 for the case where $X_i = \pm 1$ with probabilities $\frac{1}{2}$. The case where the X_i are normally distributed is also interesting. This was first obtained by Anis and Lloyd (1953) who used a complicated argument to show that (10.76) is exactly equal to

$$(2\pi^{-1})^{\frac{1}{2}} \sum_{r=1}^{n} r^{-\frac{1}{2}}. \tag{10.77}$$

We shall later derive this as a special case of a corollary to Spitzer's theorem. Clearly (10.77) is one half of the expectation of

$$T_n = E\left(\max_{0 \leqslant k \leqslant n} S_k - \min_{0 \leqslant k \leqslant n} S_k\right), \tag{10.78}$$

and Anis (1955, 1956) also obtained higher moments of this quantity. These higher moments are, however, more easily obtained by using an extension of the Pollaczek–Spitzer theorem (§ 10.18). (10.77) is asymptotically equal to a multiple of $n^{\frac{1}{2}}$ which again checks with the Brownian motion approximation on putting in the variance of the individual steps.

Expressions such as T_n were considered by Hurst (1956) in a study of the operation of a dam fed by annual random inputs of water. Suppose that the dam is initially partly filled, and that each year an amount of water is released equal to the mean input. Then the rate at which T_n grows with n can be used to throw some light on how large the dam should be in order to provide a reliable source of water.

In fact Hurst used a slightly different quantity. He supposed that the inputs, X_i, to the dam all had the same distribution on $(0, \infty)$, and considered the expected range of the sequence of successive sums

$$S_i = \sum_1^i (X_i - M), \tag{10.79}$$

where M was the observed mean. For values of i not near the total observed number this should behave in nearly the same way (see Feller 1951).

Returning to (10.78), if the X_i have zero means and finite variances we see, by using the Brownian motion approximation, that T_n should ultimately increase as a multiple of $n^{\frac{1}{2}}$. Hurst carried out a large number of numerical experiments with sequences of statistical data such as streamflows, rainfalls, temperatures, etc. These showed that T_n varied as n^σ where σ has values between 0·69 and 0·80 with a mean of 0·72, whereas the theoretical value should be 0·5. This phenomenon can be explained as we shall see later (§ 10.19).

10.13. Wald's identity

A number of boundary problems on random walk in one dimension with discrete time steps can be solved approximately, or sometimes exactly, by using a remarkable identity discovered by A. Wald (1944, 1946). Suppose that the walk starts at the point 0 and makes successive independent steps $X_1, X_2, \ldots$ having the same distribution. Let $a > 0$, $b < 0$, be finite constants, and let the walk end as soon as

$$S_n = X_1 + \ldots + X_n$$

satisfies $S_n \geqslant a$, or $S_n \leqslant b$.

We first observe that the probability of this happening after a finite number of steps must be unity, unless $\mathrm{var}(X_i) = 0$. For, by Tchebychev's inequality, if $\mathrm{var}(X_i) > 0$ we can choose n so large that

$$\mathrm{pr}(|S_n| \leqslant a + |b|) < \tfrac{1}{2}, \quad \text{say.}$$

Consider the successive independent sums S_n, $S_{2n} - S_n$, $S_{3n} - S_{2n}, \ldots$. The probability that all of these do not exceed $a + |b|$ is zero, and if any one of them does, the walk must have moved outside the interval (b, a). In fact by carrying this argument a little further we see that the probability that $|S_n| \leqslant a+|b|$, for all n in $1 \leqslant n \leqslant N$, must ultimately tend to zero at least as fast as $e^{-\alpha N}$ where α is a positive constant.

THEOREM 10.7 (Wald's identity). *Suppose that the moment generating function,*

$$M(t) = \int_{-\infty}^{\infty} e^{tx} \, dF(x), \tag{10.80}$$

where $F(x)$ is the distribution function of the X_i, exists for all real t in some finite or infinite interval, and let n be the first value for which $S_n \geqslant a$ or $S_n \leqslant b$. Then for every real or complex value of t for which

$$1 \leqslant |M(t)| < \infty, \tag{10.81}$$

$$E\{(\exp t S_n)M(t)^{-n}\} = 1. \tag{10.82}$$

Notice that (10.82) would be identically true if there were no barriers a and b, and n was a fixed integer. The interest of the result arises from the fact that n is a random variable depending on the sequence $S_1, S_2, \ldots$.

Proof of Theorem 10.7. Suppose that the walk is unrestricted so that it does not end when $S_n \geqslant a$, or $S_n \leqslant b$. Then for any positive integers n, N, we have

$$E(\exp t S_N) = M(t)^N$$
$$= E[\exp\{t S_n + t(S_N - S_n)\}], \tag{10.83}$$

where n may be greater or less than N. Now define n to be the first value for which $S_n \geqslant a$ or $S_n \leqslant b$. Thus n is now a random variable depending on the sequence $S_1, S_2, \ldots$ but (10.83) remains true.

Let P_N be the probability that $n \leqslant N$, and E_N, E_N^* be expectations conditional on the events $n \leqslant N$, $n > N$. From what we have seen above

$$1 - P_N < K_\alpha N^{-\alpha}, \tag{10.84}$$

where K_α is a constant and α any fixed positive integer. We can now write (10.83) in the form

$$E(\exp t S_N) = P_N E_N[\exp\{t S_n + t(S_N - S_n)\}]$$
$$+ (1 - P_N)E_N^*[\exp\{t S_n + t(S_N - S_n)\}]. \tag{10.85}$$

When $n \leqslant N$, $S_N - S_n$ is the sum of $N - n$ random variables independent of S_n, and the first term in (10.85) can therefore be written

$$P_N E_N\{(\exp t S_n)M(t)^{N-n}\}.$$

Thus

$$P_N E_N\{(\exp t S_n)M(t)^{-n}\} + M(t)^{-N}(1 - P_N)E_N^*(\exp t S_N) = 1. \tag{10.86}$$

In the second term the expectation involves the condition $b < S_N < a$, and is therefore bounded independently of N. If t is any real or complex number such that $M(t)$ exists and $|M(t)| > 1$ we can let N tend to infinity and obtain (10.82), since $1 - P_N \to 0$.

THEOREM 10.8 (Wald 1946). *Under the restriction that $M(t)$ exists for all real t, the identity* (10.82) *can be differentiated under the expectation sign any number of times at all points at which* $|M(t)| \geqslant 1$.

Proof. Since $M(t)$ exists everywhere it is easy to see that its derivatives of all finite orders exist for all real t. Thus all derivatives of

$$(\exp t S_n) M(t)^{-n}$$

exist for all t such that $|M(t)| \geqslant 1$, and any set of finite derivatives of $M(t)$ are bounded in any given finite interval in t.

Consider a finite interval in the range of values of real t for which $M(t)| \geqslant 1$. We show that for $k = 1, 2, \ldots$

$$E\left| \frac{d^k}{dt^k} (\exp t S_n) M(t)^{-n} \right| \tag{10.87}$$

is uniformly bounded in this interval. This is the expectation of the modulus of a finite sum of terms of the form

$$(S_n^r \exp t S_n) f(t)$$

where r, n are non-negative integers and $f(t)$ is bounded in any finite interval. These are bounded above by

$$|S_n|^r (\exp t_0 |S_n|) |f(t)|$$

where t_0 is an upper bound of t in the interval. But since $M(t)$ exists for all real t this clearly has a bounded expectation when integrated over the whole distribution of unrestricted values of S_n.

Taking the expectation of E is equivalent to integrating over a certain set and we have shown above that the integrand is bounded on this set, uniformly for t in the given finite interval, by a quantity with a finite expectation. Thus differentiation under the integral is legitimate for any value of t in the *closed* interval. In particular this is legitimate at $t = 0$, since the fact that $M''(0) > 0$ implies that either $[0, \delta]$ or $[-\delta, 0]$, $\delta > 0$ and sufficiently small, is contained in the set where $|M(t)| \geqslant 1$.

Differentiating (10.82) once and putting $t = 0$, we have

$$E(S_n) = E(X_i)E(n). \tag{10.88}$$

Notice that in this formula S_n and n are jointly distributed random variables, S_n having the value of the sum at the point at which it first meets or traverses the boundaries.

When $E(X_i) = 0$, a further result is useful. Differentiating (10.82) twice we obtain

$$E[(\exp t S_n) M(t)^{-n} \{ S_n^2 - 2n S_n M'(t) M(t)^{-1}$$
$$+ n(n+1) M'(t)^2 M(t)^{-2} - n M''(t) \}] = 0. \tag{10.89}$$

Putting $t = 0$, and using the facts that $M'(0) = E(X_i) = 0$, $M''(0) = E(X_i^2)$, we get

$$E(S_n^2) = E(n)E(X_i^2). \tag{10.90}$$

In the above theory we have assumed a and b to be finite. We can make one of them infinite ($b = -\infty$, say) provided we impose a condition that makes the probability that $S_n \geqslant a$ for some $n \leqslant N$ tend to unity as $N \to \infty$. In order to do this we assume that $E(X_i) > 0$ so that $M'(0) > 0$, and $M(t) > 1$ for all $t > 0$.

10.14. Consequences of Wald's identity

Formulae (10.88) and (10.90) are useful in estimating $E(n)$, the expected duration of the random walk. To do this we need to know $E(S_n)$ or $E(S_n^2)$, and these can also be obtained, at least approximately, from Wald's identity. To do this we first consider the properties of a more restricted class of moment generating functions and we prove the following lemma (Wald 1944).

LEMMA 10.1. *Suppose that the distribution of the* Z_i *is such that*

(1) $E(X_i) \neq 0$,   i.e. $M'(0) \neq 0$;

(2) $\mathrm{var}(X_i) \neq 0$,   i.e. $M''(0) - M'(0)^2 > 0$;

(3) *there exists* $\delta > 0$, *such that*

$$\mathrm{pr}(\exp X_i < 1 - \delta) > 0,$$
$$\mathrm{pr}(\exp X_i > 1 + \delta) > 0. \tag{10.91}$$

This means that the distribution of X_i *does not lie entirely in the non-positive or non-negative parts of the line.*

(4) $M(t)$ *exists for all real* t.

Then there exists one and only one value of t, t_0 *say, such that* $t_0 \neq 0$, *and*

$$M(t_0) = 1. \tag{10.92}$$

Proof. Since

$$M(t) = \int(\exp tx)\,dF(x)$$

exists for all real t, it follows that

$$E(X_i \exp tX_i) = M'(t), \tag{10.93}$$
$$E(X_i^2 \exp tX_i) = M''(t), \tag{10.94}$$

the expectations on the left-hand side being finite for all real finite t.

Then $M''(t) > 0$ for all t, and from (10.91), $M(t) \to \infty$ as $t \to +\infty$ or $-\infty$. Thus $M(t)$ has a strict minimum which occurs at a finite point, t_1, say. At this point $M(t_1) < 1$ and there must therefore exist another point $t_0 \neq 0$, at which $M(t_0) = 1$. Clearly zero and t_0 are the only such points, which proves the lemma.

We now suppose the random walk ends when $S_n \geqslant a$, or $S_n \leqslant b$, and we try to find the probabilities, P_a and $P_b = 1 - P_a$ say, of these events. We first assume that the distribution of the X_i satisfies the conditions of Lemma 10.1. Let $E_1(\exp t_0 S_n)$ be the expectation of $\exp t_0 S_n$ conditional on $S_n \geqslant a$, and n being the first step for which one of the boundaries is reached. $E_2(\exp t_0 S_n)$ is the similar expectation given that $S_n \leqslant b$. Then since the probability of reaching a boundary in a finite number of steps is unity,

$$P_a E_1(\exp t_0 S_n) + P_b E_2(\exp t_0 S_n) = E(\exp t_0 S_n) = 1. \qquad (10.95)$$

Thus

$$P_a = \frac{1 - E_2(\exp t_0 S_n)}{E_1(\exp t_0 S_n) - E_2(\exp t_0 S_n)}. \qquad (10.96)$$

The difficulty in using this formula is in obtaining the two expectations, for with a general distribution of the X_i, S_n will not be exactly equal to a (or b) when it first passes the boundary (two special cases in which this is true will be considered in § 10.15). To obtain an approximation we argue as follows (Wald 1944). Suppose that the distribution of the X_i satisfies the conditions of Lemma 10.1, and that $t_0 > 0$. If $t_0 < 0$, the argument goes through similarly.

Then since $a > 0$, $b < 0$, we have

$$E_1(\exp t_0 S_n) \geqslant \exp t_0 a > 1 > \exp t_0 b \geqslant E_2(\exp t_0 S_n). \qquad (10.97)$$

Under the restriction that $S_{n-1} < a$, $S_n \geqslant a$, $\exp t_0 S_{n-1}$ must have some distribution on the range $(\exp t_0 b, \exp t_0 a)$. Write

$$\exp t_0 S_{n-1} = \zeta \exp t_0 a,$$

where ζ has some distribution on the interval $\{\exp t_0(b-a), 1\}$ with cumulative distribution function $P(\zeta)$. Then

$$E_1(\exp t_0 S_n) = \int \zeta \exp t_0 a E(\exp t_0 X_n | \exp t_0 X_n > \zeta^{-1}) \, dP(\zeta).$$

As we do not know the distribution of $\exp t_0 S_{n-1}$ it is usually not possible to obtain this exactly but it sometimes happens that we can put a bound on

$$E(\exp t_0 X_n | \exp t_0 X_n > \zeta^{-1}),$$

as, for example, if the distribution of X_n has a finite range. Thus if X_n has a finite range R,

$$E_1(\exp t_0 S_n) < \exp t_0(a+R).$$

In this way better bounds for (10.96) can be found than those obtained by using (10.97) which are

$$0 \leqslant P_a \leqslant \exp -t_0 a. \qquad (10.98)$$

When the range of X_n is so small that it can be neglected a good estimate

P_a is thus

$$P_a = \frac{1-\exp t_0 b}{\exp t_0 a - \exp t_0 b}. \qquad (10.99)$$

These inequalities have been obtained under the assumption that the distribution of the X_i satisfies the conditions of Lemma 10.1. When condition (2) is not satisfied $\mathrm{var}(X_i) = 0$ and the walk never reaches the boundary. If condition (3) is not satisfied the walk will reach the boundary a with unit probability if the distribution lies entirely to the right of zero, and similarly will reach b with unit probability if it lies entirely to the left.

If condition (1) is not satisfied we have $E(X_i) = 0$, so that

$$P_a E_1(S_n) + (1 - P_a) E_2(S_n) = 1, \qquad (10.100)$$

and

$$P_a = \frac{1 - E_2(S_n)}{E_1(S_n) - E_2(S_n)}. \qquad (10.101)$$

This may be used in a similar way.

When condition (4) is not satisfied the situation is more complicated.

10.15. Some special cases

The above approximations are clearly related to the method of finding absorption probabilities described in §3.6. Consider again the case of the simple random walk on a one-dimensional lattice in which $X_1 = +1, -1,$ with probabilities p and q. Take $a > 0, b < 0$, as integers.

Then

$$M(t) = p e^t + q e^{-t}. \qquad (10.102)$$

If this is to equal unity we must have

$$p u^2 - u + q = 0, \qquad (10.103)$$

where

$$u = e^t.$$

One root is $u = 1$ and the other is $t_0 = qp^{-1}$. Suppose $p < q$ so that $u > 1$, $t_0 > 1$. For this random walk there is no overshoot of the boundary so that $E_1(\exp t_0 S_n) = \exp t_0 a$, and similarly for b. Then

$$P_a = \frac{1 - \exp t_0 b}{\exp t_0 a - \exp t_0 b} = \frac{1 - (q/p)^b}{(q/p)^a - (q/p)^b}$$

in agreement with (3.72). The case $p = q$ is even simpler.

A more interesting case is obtained by assuming that the distribution of the X_i is normal with mean m and standard deviation σ. Then the moment generating function of the distribution of the X_i is

$$M(t) = \exp(mt + \tfrac{1}{2}\sigma^2 t^2). \qquad (10.104)$$

This satisfies the conditions of Lemma 10.1 and we have to look for the non-zero root of $M(t) = 1$, which is

$$t_0 = -2m\sigma^{-2}. \tag{10.105}$$

Using this in (10.96) we obtain, as an approximation, the value of P_a given by (10.99).

This case links up with the Brownian motion process when we rescale X_i and time. Thus if a and b are large compared with m and σ, the error in (10.96) becomes negligible and we get

$$P_a = \frac{1-e^\alpha}{e^\beta - e^\alpha}, \tag{10.106}$$

where α and β are the limiting values of $-2mb\sigma^{-2}$ and $-2ma\sigma^{-2}$.

10.16. The distribution of *n* in general

Wald's identity can be used not only to find the moments of n, the suffix of the first n for which $S_n \geqslant a$ or $S_n \leqslant b$, but also to find the generating function of its distribution.

Consider first the case where $b = -\infty$. Then n is the smallest integer for which $S_n \geqslant a$, and in order to obtain a proper distribution we assume that $E(X_i) > 0$. Then $M(t) > 1$ when $t > 0$ because $M'(0) > 0$.

Write $\bar{S}_n = a$ if $S_n \geqslant a$ and $U_n = S_n - \bar{S}_n$ where n is the first value of n for which $S_n > 0$. U_n is a random variable whose expectation we can often bound in the manner described above, and for the case of $X_i = \pm 1$ is identically zero.

Write $\Phi(z)$ for the moment generating function of n. Neglecting U_n and using Wald's identity, we have $S_n = a$, and

$$E\{M(t)^{-n}\} = \Phi\{-\log M(t)\}$$

$$= \exp at, \tag{10.107}$$

for all values of t such that $|M(t)| > 1$, and in particular for all $t > 0$. Write $t(\tau)$ for a root of the equation

$$\tau = -\log M(t), \tag{10.108}$$

with a non-negative real part. Then

$$\Phi(\tau) = \exp at(\tau), \tag{10.109}$$

which is the required generating function.

When b is finite write $\bar{S}_n = a$ if $S_n \geqslant a$ for the first time at n, and $\bar{S}_n = b$ if $S_n \leqslant b$ for the first time. Let p_n be the probability that $\bar{S}_n = a$, given that n is the first value for which the boundaries are exceeded, and $\{p_n^*\}$ the probability distribution of n. In the above notation this means that

$$\sum p_n^* z^n = \Phi(-\log z). \qquad (10.110)$$

Then for values of t such that $|M(t)| \geq 1$ we have

$$\sum_{n=1}^{\infty} p_n^* M(t)^{-n} \{p_n \exp at + (1-p_n) \exp bt\} = 1. \qquad (10.111)$$

Write

$$\Phi(z) = P_a \Phi_1(z) + (1-P_a)\Phi_2(z), \qquad (10.112)$$

where $\Phi_1(z)$, $\Phi_2(z)$, are the characteristic functions of n conditional on $S_n \geq a$, $S_n \leq b$, respectively.

If we can neglect $U_n = S_n - \bar{S}_n$, we have, from (10.99),

$$P_a = \frac{1 - \exp t_0 b}{\exp t_0 a - \exp t_0 b}, \qquad (10.113)$$

where t_0 is the root of $M(t) = 1$ other than $t = 0$, the conditions of Lemma 10.1 being assumed. Furthermore, we have

$$\Phi_1(\log z) = \sum_{n=1}^{\infty} p_n p_n^* z^n \left(\sum_{n=1}^{\infty} p_n p_n^*\right)^{-1},$$

so that

$$\Phi_1\{-\log M(t)\} = \frac{\sum_{n=1}^{\infty} p_n p_n^* M(t)^{-n}}{\sum_{n=1}^{\infty} p_n p_n^*}$$

$$= P_a^{-1} \sum_{n=1}^{\infty} p_n p_n^* M(t)^{-n}, \qquad (10.114)$$

and similarly

$$\Phi_2\{-\log M(t)\} = (1-P_a)^{-1} \sum_{n=1}^{\infty} (1-p_n)p_n^* M(t)^{-n}. \qquad (10.115)$$

Then

$$\Phi\{-\log M(t)\} = \sum_{n=1}^{\infty} p_n^* M(t)^{-n}$$

$$= \sum_{n=1}^{\infty} (p_n + 1 - p_n)p_n^* M(t)^{-n}$$

$$= P_a \Phi_1\{-\log M(t)\} + (1-P_a)\Phi_2\{-\log M(t)\}.$$

If $\tau = -\log M(t)$, this can be written

$$P_a \Phi_1(\tau) + (1-P_a)\Phi_2(\tau) = \Phi(\tau). \qquad (10.116)$$

We known that the equation $M(t) = 1$ has two real roots $t = 0$, t_0. In the neighbourhood of these roots $M'(t) \neq 0$, and $M(t)$ is an analytic function. Thus for τ small and real, the equation

$$M(t) = e^{-\tau}$$

will have two roots $t_1(\tau)$, $t_0(\tau)$ such that $t_1(\tau) \to 0$ and $t_0(\tau) \to t_0$ as $\theta \to 0$.

Using these two values in (10.111) we get

$$P_a\Phi_1(\tau)\exp at_1(\tau)+(1-P_a)\Phi_2(\tau)\exp bt_1(\tau) = 1. \qquad (10.117)$$

$$P_a\Phi_1(\tau)\exp at_2(\tau)+(1-P_a)\Phi_2(\tau)\exp bt_2(\tau) = 1. \qquad (10.118)$$

These are two equations for the functions $\Phi_1(\tau)$ and $\Phi_2(\tau)$, and after solving them we can use (10.116) to obtain $\Phi(\tau)$ for τ small.

10.17. The distribution of n in special cases

Consider first the case where $X_i = +1$, -1, with probabilities p and $q = 1-p$, and a and b are integers. P_a has been found above. We suppose $p \neq q$ (the case $p = q = \frac{1}{2}$ can be treated similarly). Since $M(t) = pe^t + qe^{-t}$ we have to find the roots of the equation $M(t) = e^{-\tau}$ which can be written

$$pw^2 - e^{-\tau}w + q = 0,$$

where

$$w = e^t.$$

These roots are

$$w_{0,1} = \tfrac{1}{2}p^{-1}\{e^{-\tau} \pm \sqrt{(e^{-2\tau} - 4pq)}\}, \qquad (10.119)$$

so that we can write

$$e^{t_{1,2}} = (2p)^{-1}e^{-\tau}\{1 \pm \sqrt{(1 - 4pqe^{2\tau})}\}.$$

Inserting these values in the two equations (10.117) and solving for $\Phi_1(\tau)$, we get

$$\Phi_1(\tau) = \frac{w_2^b - w_1^b}{P_a(w_1^a w_2^b - w_1^b w_2^a)}. \qquad (10.120)$$

Using the fact that $w_1 w_2 = -qp^{-1}$, this is equal to $Q_a^{-1}Q_i(z)$ of equation (3.88) after changing the notation so that $N = a-b$, $i = -b$, $z = e^\tau$, and remembering that Q_a^{-1} is involved since $\Phi_1(\tau)$ is a conditional characteristic function conditional on the boundary attained being the upper boundary, whereas $Q_i(z)$ of (3.88) is a generating function without this condition.

The case where the X_i are normally distributed is also interesting. $M(t)$ is given by (10.104) and the equation $M(t) = e^{-\tau}$ becomes

$$mt + \tfrac{1}{2}\sigma^2 t^2 = -\tau$$

so that

$$t = \sigma^{-2}\{-m \pm \sqrt{(m^2 - 2\sigma^2\tau)}\}. \qquad (10.121)$$

Now suppose that $b = -\infty$ and $m \geqslant 0$. The probability of ultimately reaching the barrier a is unity so that we put $P_a = 1$. Then

$$\Phi(\tau) = \Phi_1(\tau) = \exp - at_1(\tau). \qquad (10.122)$$

The root t_1 required here is the one with $\tau < 0$ and $t_1 > 0$, which is

$$t_1 = \sigma^{-2}\{-m + \sqrt{(m^2 - 2\sigma^2\tau)}\}.$$

Thus

$$\Phi(\tau) = \exp a\sigma^{-2}\{m - \sqrt{(m^2 - 2\sigma^2\tau)}\} \quad (\tau < 0). \tag{10.123}$$

This result ignores the size of the overshoot. More interesting is the deduction of the limiting distribution of n, suitably scaled, when a is large compared with σ. This must clearly be the moment generating function of the first passage-time distribution for the Brownian movement. To verify this put $\sigma^2 = 1$, $a = AN^{\frac{1}{2}}$, $m = MN^{-\frac{1}{2}}$, $n = tN$, $\tau = \theta N^{-1}$. The moment generating function of t, as N becomes large, is then

$$\psi(\theta) = E(\exp t\theta)$$

$$= \lim_N E(\exp \tau n)$$

$$= \exp AM[1 - \sqrt{\{1 - (2\theta/M^2)\}}]. \tag{10.124}$$

Referring back to equation (7.175) we see that this is the correct moment generating function for the first passage distribution of the Brownian movement.

10.18. The Pollaczek–Spitzer identity

We now prove a remarkable theorem on the distribution of the maxima of successive sums of random variables which has many implications in probability theory.

THEOREM 10.9. *Suppose that* $X_1, X_2,\ldots$ *is a sequence of independent random variables with the same distribution. Write*

$$S_n = X_1 + \ldots + X_n, \quad S_0 = 0,$$

$$T_n = \max_{0 \leqslant k \leqslant n} S_k, \tag{10.125}$$

$$S_n^+ = \max(0, S_n). \tag{10.126}$$

Let $\phi_n(t)$ *be the characteristic function of* T_n, *and* $\psi_n(t)$ *that of* S_n^+. *Then*

$$\sum_0^\infty \phi_n(t)z^n = \exp \sum_1^\infty n^{-1}\psi_n(t)z^n, \tag{10.127}$$

the two series being obviously convergent for $|z| < 1$.

This very striking result has occasioned a large body of research and a number of generalizations and alternative proofs. It was first stated in this form by Spitzer (1956) (see also Pollaczek 1952), who arrived at it by a generalization of earlier work of E. Sparre Andersen. Most published proofs are based on combinatorial ideas and are somewhat complicated. We give here a slight modification of a relatively simple proof due to Baxter (1961).

Proof. By the continuity theroem for characteristic functions it is sufficient to prove the result when the distribution of X_i is a discrete distribution on the finite set of positive and negative integers

$$-r, \ 1-r,..., \ -1, \ 0, \ 1, \ 2,..., \ s.$$

We shall denote by

$$\int (...) \, dP \qquad (10.128)$$

integration over the joint distribution of as many of the X_i as are necessary to specify the expression inside the brackets. Since the distribution of X_i is discrete, this integration is in fact a summation. Write

$$f_+(z, w) = \exp \sum_{n=1}^{\infty} \frac{z^n}{n} \int_{S_n > 0} w^{S_n} \, dP$$

$$= \sum_{0}^{\infty} z^n a_n(w), \quad \text{say}, \qquad (10.129)$$

and similarly

$$f_-(z, w) = \exp \sum_{n=1}^{\infty} \frac{z^n}{n} \int_{S_n \leqslant 0} w^{S_n} \, dP$$

$$= \sum_{0}^{\infty} z^n b_n(w), \quad \text{say}. \qquad (10.130)$$

The expressions $a_n(w)$ and $b_n(w)$ will be products of powers of terms of the form

$$\int_{S_n > 0} w^{S_n} \, dP$$

and

$$\int_{S_n \leqslant 0} w^{S_n} \, dP$$

respectively. Instead of considering $\phi(t)$, write $\pi(w)$ for the probability generating function of the X_i, and $\Phi_n(w)$ for the probability generating function of T_n, which must also have a discrete distribution. Thus

$$\Phi_n(w) = \int w^{T_n} \, dP.$$

Then the result (10.127) can be written

$$\sum_{n=0}^{\infty} z^n \int w^{T_n} \, dP = f_+(z, w) f_-(z, 1). \qquad (10.131)$$

Write L_n for the first suffix, k, in the sequence $(S_0, S_1,..., S_n)$ which is such that $S_k = T_n$. Then L_n is itself a random variable and $L_n = 0$ if $S_1 \leqslant 0$, $S_2 \leqslant 0,..., S_n \leqslant 0$. We now prove

$$f_+(z, w) = \sum_{n=0}^{\infty} z^n \int_{L_n = n} w^{S_n} \, dP \qquad (10.132)$$

and

$$f_{-}(z, w) = \sum_{n=0}^{\infty} z^n \int_{L_n=0} w^{S_n} dP. \tag{10.133}$$

Write

$$p_n(w) = \int_{L_n=n} w^{S_n} dP, \tag{10.134}$$

$$q_n(w) = \int_{L_n=0} w^{S_n} dP, \tag{10.135}$$

so that we want to prove

$$p_n(w) = a_n(w), \quad q_n(w) = b_n(w).$$

We clearly have

$$\pi(w)^n = \int w^{S_n} dP$$

$$= \sum_{k=0}^{n} \int_{L_n=k} w^{S_n} dP. \tag{10.136}$$

Suppose that $L_n = k$. This is equivalent to asserting the two conditions,

$$(S_k > 0, S_k > S_1, ..., S_k > S_{k-1})$$

and

$$(S_k \geqslant S_{k+1}, S_k \geqslant S_{k+2}, ..., S_k \geqslant S_n).$$

The first of these depends only on the values of $X_1, ..., X_k$, and the second only on $X_{k+1}, ..., X_n$. Then

$$\int_{L_n=k} w^{S_n} dP = \int_{L_k=k} w^{S_k} dP \int_{L_{n-k}=0} w^{S_n-S_k} dP$$

$$= p_k(w)q_k(w). \tag{10.137}$$

Putting this in (10.136) we get

$$\pi(w)^n = \sum_{k=0}^{n} p_k(w)q_{n-k}(w). \tag{10.138}$$

If we introduce the generating functions

$$P(z, w) = \sum_{n=0}^{\infty} z^n p_n(w), \tag{10.139}$$

$$Q(z, w) = \sum_{n=0}^{\infty} z^n q_n(w), \tag{10.140}$$

we have

$$\{1 - z\pi(w)\}^{-1} = \sum_{n=0}^{\infty} z^n \pi(w)^n$$

$$= P(z, w)Q(z, w). \tag{10.141}$$

We now use the following simple lemma.

LEMMA 10.2. *Suppose $R(w)$ is a finite polynomial in positive and negative powers of w, and let $\{p_k(w)\}$, $\{q_k(w)\}$, $k = 0, 1,\ldots$ be similar finite polynomials with the additional restriction that the $p_k(w)$ have only powers w^s with $s \geqslant 1$, and $q_k(w)$ has only powers w^s such that $s < 1$. Furthermore suppose that*

$$p_0(w) = q_0(w) = 1$$

and

$$R^n(w) = \sum_{k=0}^{n} p_k(w)q_{n-k}(w)$$

for all $n > 0$.

Then $R(w)$ uniquely defines the $p_k(w)$ and the $q_k(w)$.

Proof. This is true for $n = 0$, 1. Suppose it is true for values $0, 1,\ldots, n-1$. Then

$$R(w)^n = p_n(w) + q_n(w) + \sum_{k=1}^{n-1} p_k(w)q_{n-k}(w)$$

so that

$$p_n(w) + q_n(w)$$

is uniquely defined. Since $p_n(w)$ and $q_n(w)$ have no powers of w in common they are themselves uniquely defined which proves the lemma by induction.

Now we already have

$$f_+(z, w)f_-(z, w) = \sum_{n=0}^{\infty} z^n \sum_{k=0}^{n} a_k(w)b_{n-k}(w)$$

$$= \exp \sum_{n=1}^{\infty} \frac{z^n}{n} \int w^{S_n} dP$$

$$= \exp - \log\{1 - z\pi(w)\}$$

$$= \{1 - z\pi(w)\}^{-1}$$

$$= \sum_{n=0}^{\infty} z^n \pi(w)^n. \tag{10.142}$$

Using the lemma, $\pi(w)$ uniquely defines $a_k(w)$ and $b_k(w)$, which must therefore be respectively equal to the $p_k(w)$ and $q_k(w)$ of (10.139), and (10.140). Thus (10.132) and (10.133) are true. We can now write

$$\int w^{T_n} dP = \sum_{k=0}^{n} \int_{L_n=k} w^{T_n} dP$$

$$= \sum_{k=0}^{n} \int_{L_n=k} w^{T_k} dP \int_{L_{n-k}=0} w^{T_{n-k}} dP$$

$$= \sum_{k=0}^{n} p_k(w)q_{n-k}(1),$$

since $T_{n-k} = 0$, if $L_{n-k} = 0$. Taking generating functions on both sides we get (10.127) as required.

Identities of this type can be obtained for a number of other problems concerning cumulative sums of random variables. For these see Baxter (1961) and the references given in the bibliographical notes at the end of this chapter. A continuous analogue of Theorem 10.9 is given by Baxter and Donsker (1957).

10.19. Some consequences of the Pollaczek–Spitzer theorem

Differentiating (10.127) with respect to t, and putting $t = 0$, we can obtain the lower moments of T_n. Thus

$$\sum_1^\infty \phi_n'(0)z^n = \left\{ \sum_{n=1}^\infty n^{-1}\psi_n'(0)z^n \right\} \exp \sum_1^\infty n^{-1}\psi_n(0)z^n$$

$$= (1-z)^{-1} \sum_{n=1}^\infty n^{-1}\psi_n'(0)z^n.$$

Then since

$$\phi_n'(0) = E(T_n),$$

$$\psi_n'(0) = E(S_n^+),$$

we have

$$E(T_n) = \sum_1^n k^{-1}E(S_k^+). \tag{10.143}$$

Similarly, differentiating twice, we obtain

$$\sum_1^\infty \phi_n''(0)z^n = (1-z)^{-1} \sum_{n=1}^\infty n^{-1}\psi_n''(0)z^n + (1-z)^{-1}\left\{ \sum_1^\infty n^{-1}\psi_n'(0)z^n \right\}^2, \tag{10.144}$$

from which we get $(n \geqslant 2)$

$$E(T_n^2) = \sum_{k=1}^n k^{-1}\{E(S_k^+)^2\} + \sum_{k=2}^n \sum_{l=1}^{k-1} l^{-1}(k-l)^{-1}E(S_l^+)E(S_{k-l}^+). \tag{10.145}$$

Consider the special case in which the X_i are distributed normally with mean zero and unit standard deviation. Then S_n is distributed normally with mean zero and standard deviation $n^{\frac12}$. From this

$$E(S_n^+) = n^{\frac12}(2\pi)^{-\frac12} \int_0^\infty x e^{-\frac12 x^2}\, dx$$

$$= n^{\frac12}(2\pi)^{-\frac12},$$

so that

$$E(T_n) = (2\pi)^{-\frac12} \sum_{k=1}^n k^{-\frac12}, \tag{10.146}$$

which is the result of Anis and Lloyd (1953). In a similar way we may obtain the higher moments given by Anis (1955, 1956).

Clearly if the X_i have any distribution with zero mean and a finite second moment the distribution of $n^{-\frac{1}{2}}S_n$ will tend to normality with second moment σ^2 and it can then be easily proved by dominated convergence that the expectation of S_n^+ will be asymptotically equal to

$$\sigma n^{\frac{1}{2}}(2\pi)^{-\frac{1}{2}}.$$

Thus $E(T_n)$ will increase asymptotically as $n^{\frac{1}{2}}$ as has already been conjectured above (§ 10.6) from the Brownian motion analogy.

Similarly in this case

$$E\left(\max_{1\leqslant k\leqslant n} S_n - \min_{1\leqslant k\leqslant n} S_n\right) \tag{10.147}$$

will increase asymptotically as $n^{\frac{1}{2}}$ multiplied by a constant.

Hurst's calculations (§ 10.12), using the modified sums (10.79), did not show this behaviour, a fact which has caused a good deal of puzzlement but can be explained in the following manner (Moran 1964). In § 9.16 we have seen that $\exp-|t|^\alpha$ $(0 < \alpha \leqslant 2)$ is the characteristic function of a stable distribution and if $1 < \alpha < 2$ this distribution has a first but not a second moment. Thus if the X_i follow such a distribution, $E(S_n^+)$ will be a multiple of $n^{\alpha^{-1}}$, and $E(T_n)$ will be a multiple of

$$\sum_{k=1}^{n} k^{\alpha^{-1}-1}$$

which is asymptotically equal to $Kn^{\alpha^{-1}}$. Hence $E(T_n)$, and similarly

$$E\left(\max_{1\leqslant k\leqslant n} S_n - \min_{1\leqslant k\leqslant n} S_n\right),$$

can be made to increase asymptotically as n^ρ, where ρ is any number satisfying $\frac{1}{2} \leqslant \rho < 1$.

The random variables used by Hurst all must have finite second moments. However, if a distribution is such that $E(X^2)$ is very much larger than $\{E(|X|)\}^2$, we may expect that for smaller values of n, $E(S_n^+)$ will increase faster than $n^{\frac{1}{2}}$. In particular it can be shown that for the gamma-type distribution whose probability density is

$$\Gamma(\alpha)^{-1}x^{\alpha-1}e^{-x}, \tag{10.148}$$

$E(S_n^+)$ (after X is corrected to have zero mean) can be made to increase approximately as n, provided n is not too large, and if α is very small. Thus if $\alpha = 0.001$, $E(S_n^+)$ will vary roughly as n for $n \leqslant 500$.

The behaviour of $E(T_n)$ as a function is therefore related to the nature of the tail of the distribution, and for distributions, symmetric or not, for which the second moment about the mean is large compared with the square of the absolute first moment about the mean we may expect $E(T_n)$ to increase faster than $n^{\frac{1}{2}}$ for the smaller values of n.

10.20. Further remarks on the one-dimensional random walk

We have seen above that Wald's theorem can be used to obtain the distribution of first passage times for both one-sided and two-sided boundaries on a random walk. The chief disadvantage of this method is that with the exception of the cases where (1) t is discrete and the X_i take the values $0, \pm 1$, (2) t is continuous and $X_i(t)$ is the Brownian movement, the result is not exact.

For one-sided boundaries, the Spitzer–Pollaczek theorem enables us, at least in principle, to obtain the first passage-time distribution exactly because if N is the first passage-time to a boundary $a \geqslant 0$,

$$\mathrm{pr}(N > n) = \mathrm{pr}(T_n < a). \tag{10.149}$$

This type of dual relationship between two distributions has already occurred a number of times in this book. Since we can, in principle, find the right-hand side from its characteristic function as given by (10.127) we can also obtain the left-hand side. This relationship deserves further study.

An interesting class of random walk problems which is mathematically related to transmission line theory in physics has been studied by Redheffer (1961). A particle moves with a constant velocity on the real axis $(-\infty < x < \infty)$. The direction of motion may be either to the right or to the left. In any interval $(x, x+dx)$ there is a probability, $r_1(x)\,dx$, of the particle reversing its direction if it moves into this interval from the left, and similarly $r_2(x)\,dx$, if it moves into this interval from the right. From this specification it is then possible to set up differential equations for the two functions $r(x, y)$, $t(x, y)$, the total probabilities that a particle approaching the interval (x, y) $(-\infty < x < y < \infty)$ from the left is reflected by the interval back into $(-\infty, x)$ or transmitted into (y, ∞).

By symmetry similar results can be obtained for a particle approaching from the right. These ideas can clearly be generalized in various directions and in particular if generalized to motion in three dimensions the problem becomes that of neutron diffusion, a problem with a very large literature. For a detailed mathematical analysis of the one-dimensional case see also Mycielski and Paszkowski (1956), Paszkowski (1956), and Redheffer (1960).

10.21. The isotropic random walk in two and three dimensions

Suppose that a random walk starts from the origin $(0, 0)$ in the plane and makes successive steps of lengths $r_1, r_2, \ldots$ whose directions are independently distributed uniformly, i.e. the angles $\theta_1, \theta_2, \ldots$ which they make with some fixed direction are uniformly and independently distributed on the interval $(0, 2\pi)$. We look for the distribution of the end of the nth step. In the case where all the steps have the same length this problem was apparently first proposed by K. Pearson in 1905 and the first solutions obtained by Kluyver (1906) and Pearson (1906). It was later studied by Markov (1912) and Rayleigh (1919). Some of the results seem to be rediscovered at regular intervals. Similarly in three or more dimensions there is a corresponding problem in which the steps have distributions which are uniform in all directions.

Let X_i be the vector representing the ith step whether considered in two, three, or more dimensions. The distribution of X_i is to be circularly or spherically symmetric about the origin and if this distribution has a finite second moment (and under even less restrictive conditions) the distribution of the end of the nth step, which is given by

$$S_n = X_1 + \ldots + X_n,$$

will converge to normality as n increases, with a second moment along each axis equal to $n\sigma^2$, where σ^2 is the expectation of the square of the component of X_i along any fixed axis. The main interest is to determine the distribution of S_n for finite n and given assumptions about the distribution of the X_i. Apart from the obvious results when X_i has an isotropic normal distribution, the only case which has been worked out in detail is that where the X_i have a constant length and a uniformly random direction.

It is now convenient to consider the two- and three-dimensional cases separetely and we begin with the former. The distribution of S_n being symmetric, we may write the probability that the end point lies in the interval $(x, x+dx; y, y+dy)$ as

$$\frac{1}{\pi}\phi_n(r^2)\,dx\,dy = \{\phi_n(r^2)\,d(r^2)\}\frac{d\theta}{2\pi}$$

where $r^2 = x^2 + y^2$. Thus $\phi_n(r^2)$ is the probability density of the distribution of r_n^2.

If the components of S_n along the x- and y-axes are U, V, we clearly have

$$E(U) = E(V) = 0,$$

and

$$E(|S_n|^2) = E(U^2) + E(V^2)$$

$$= 2nE(u_i^2),$$

where u_i is the component of X_i along the x-axis. Then

$$E(u_i^2) = l^2 E(\cos^2\theta_i)$$

$$= \frac{l^2}{2\pi}\int_0^{2\pi}\cos^2\theta\,d\theta$$

$$= \tfrac{1}{2}l^2,$$

and thus

$$E(|S_n|^2) = nl^2.$$

This is useful for checking calculations of the exact distribution of $|S_n|$.

For $n = 1$ there is no probability density since the distribution is singular. Suppose that all steps have the same length l. Let θ be the angle between the first and second steps. Then $r^2 = 2l^2(1 - \cos \theta)$. θ is uniformly distributed over the interval $(0, 2\pi)$ and thus the distribution of r^2 is

$$\frac{1}{2\pi l^2 \sin \theta} = \frac{1}{2\pi l r (1 - \frac{1}{4} r^2 l^{-2})^{\frac{1}{2}}}.$$

Then

$$\phi_2(r^2) = \frac{1}{\pi r \sqrt{(4l^2 - r^2)}} \quad (r \leqslant 2l). \tag{10.150}$$

The distribution for higher values of n can be found by using a recurrence relation. Suppose that r_n and r_{n+1} are the distances of the nth and $(n+1)$th point from the origin, and $\phi_n(r^2)$, $\phi_{n+1}(r^2)$ the probability densities of the distributions of r_n^2 and r_{n+1}^2. Since we can write

$$r_{n+1}^2 = r_n^2 + l^2 - 2r_n l \cos \theta,$$

where θ is uniformly distributed on the range $(0, \pi)$, we have

$$\phi_{n+1}(r^2) = \frac{1}{\pi} \int_0^\pi \phi_n(r^2 + l^2 - 2rl \cos \theta) \, d\theta, \tag{10.151}$$

where $\phi_n(x) = 0$ for $x < 0$, $x > n^2 l^2$. It can be verified that a solution of this recurrence relation which agrees with $\phi_2(r^2)$ is given by

$$\phi_n(r^2) = \frac{1}{2\pi} \int_0^\infty u J_0(ur) J_0(ul)^n \, du. \tag{10.152}$$

For $n = 3$ this can be expressed in terms of elliptic integrals, and for $n = 4(1)\ 7$, $\phi_n(r^2)$ has been calculated numerically by Pearson and Blakeman (1906). For $n > 7$ and not large enough for the normal approximation to be useful, Pearson gives a series expansion. He gives numerical solutions for $n = 2(1)\ 7$ and considers a number of other problems to which these solutions can be applied. Further developments are given in Durand and Greenwood (1955, 1957). These results have been used in electrical theory.

In three dimensions the resulting distributions turn out to be simpler and here it is definitely better to use characteristic functions. Let the position after the nth step be given by a vector $\mathbf{R} = (X_n, Y_n, Z_n)$, and the individual steps be vectors $\mathbf{r}_1, \ldots, \mathbf{r}_n$ of length l and directions uniformly distributed. Then $\mathbf{R} = \mathbf{r}_1 + \ldots + \mathbf{r}_n$. We suppose that the components of $\mathbf{r}_i$ are (x_i, y_i, z_i) whose distribution function is $F(x_i, y_i, z_i)$. This is a singular distribution concentrated on the surface of the sphere $x_i^2 + y_i^2 + z_i^2 = l^2$. Using the same kind of argument as above it is easy to show that $E(|R|^2) = nl^2$.

Write $G(X_n, Y_n, Z_n)$ for the distribution function of (X_n, Y_n, Z_n). This must be singular for $n = 1$, but we can see directly and also verify from what follows that for $n \geqslant 2$ it has a probability density. Defining the characteristic functions to be

$$\phi(u, v, w) = \iiint \{\exp i(ux + vy + wz)\} \, dF(x, y, z), \qquad (10.153)$$

and

$$\Phi_n(u, v, w) = \iiint \{\exp i(uX_n + vY_n + wZ_n)\} \, dG(X_n, Y_n, Z_n), \qquad (10.154)$$

we see immediately that

$$\Phi_n(u, v, w) = \phi(u, v, w)^n. \qquad (10.155)$$

When $G(X_n, Y_n, Z_n)$ has a probability density it will be given by

$$g(X_n, Y_n, Z_n) = \frac{\partial^3}{\partial X_n \, \partial Y_n \, \partial Z_n} G(X_n, Y_n, Z_n)$$

$$= (2\pi)^{-3} \iiint \{\exp - i(uX_n + vY_n + wXZ_n)\} \Phi(u, v, w) \, du \, dv \, dw. \qquad (10.156)$$

To evaluate $\phi(u, v, w)$ we have

$$\phi(u, v, w) = (4\pi l^2)^{-1} \int_0^\pi \int_{-\pi}^\pi \{\exp il(u^2 + v^2 + w^2)^{\frac{1}{2}} \cos \theta\} 2\pi l^2 \sin \theta \, d\psi \, d\theta,$$

where (ψ, θ) are polar coordinates. This is equal to

$$\frac{\sin l(u^2 + v^2 + w^2)^{\frac{1}{2}}}{l(u^2 + v^2 + w^2)^{\frac{1}{2}}}, \qquad (10.157)$$

and thus

$$g(X_n, Y_n, Z_n) = (2\pi)^{-3} \iiint \{\exp - i(uX_n + vY_n + wZ_n)\}$$

$$\times \left(\frac{\sin l(u^2 + v^2 + w^2)^{\frac{1}{2}}}{l(u^2 + v^2 + w^2)^{\frac{1}{2}}} \right)^n \, du \, dv \, dw. \qquad (10.158)$$

We now introduce polar coordinates for u, v, w, taking the axis along the direction of (X_n, Y_n, Z_n). Put $R = (X_n^2 + Y_n^2 + Z_n^2)^{\frac{1}{2}}$, $r = (u^2 + v^2 + w^2)^{\frac{1}{2}}$. Then

$$g(X_n, Y_n, Z_n) = (2\pi^2 R)^{-1} \int_0^\infty \sin(|R|r) \left(\frac{\sin lr}{lr} \right)^n r \, dr, \qquad (10.159)$$

which is the required solution. This integral clearly exists for $n \geqslant 3$. It also exists for $n = 2$, but this is slightly more complicated to prove. By straight-forward but tedious algebra it is possible to evaluate it for any value of $n \geqslant 3$.

For $n = 1$ the probability is uniformly concentrated in the surface of a sphere of radius l. For $n = 2$ the probability density can be easily found directly.

For $n = 3$ we expand the expression

$$\sin(|R|r) \sin^3 lr$$

in a series of cosines and obtain

$$g_n(X_n, Y_n, Z_n) = (8\pi l^3)^{-1}, \quad 0 \leqslant |R| \leqslant l,$$
$$= (16\pi l^3 |R|)^{-1}(3l - |R|), \quad l \leqslant |R| \leqslant 3l,$$
$$= 0, \quad |R| \geqslant 3l. \tag{10.160}$$

Rayleigh (1919) (and Chandrasekhar 1943) similarly give the results for $n = 4$, $n = 6$. Vincenz and Bruckshaw (1960) obtain a recurrence relation by which they calculate $g_n(X_n, Y_n, Z_n)$ for $n = 3(1)\ 8$.

For higher values the distribution tends to normality fairly rapidly and it is not difficult to obtain an expansion in an infinite series giving the corrections to the normal approximation in the same manner as that given by Pearson for the two-dimensional case.

The tendency to normality implies that if l is made small and n large the process will tend to be that of a three-dimensional Brownian motion, and the approximating normal distribution will satisfy a diffusion equation. In this way approximations may be found for probabilities of absorption in various boundaries, and the corresponding first passage distributions.

An important theory of random walks with spherical symmetry in two or more dimensions is given by Kingman (1963a).

10.22. Stiff chains

When the individual steps are no longer independent we get a more difficult and more interesting class of problems which may be exemplified by the problem of determining the distribution of the distance between the two end carbon atoms in a rubber molecule. To a first approximation such a molecule may be regarded as consisting of $n+1$ atoms of negligible size which are joined to one another in a chain, the distance between successive atoms being fixed and equal. We take these distances as the unit of length and suppose that the direction of the join from the nth to the $(n+1)$th atom is represented by a unit vector, $\mathbf{p}_n$. We fix the position of the first atom at the origin of coordinates and suppose that $\mathbf{p}_1$ is uniformly distributed over all directions. We suppose that successive links are such that they make a constant angle α with the previous link but are otherwise uniformly distributed, i.e. the azimuthal angle is uniformly distributed over the interval $(0, 2\pi)$.

The position of the $(n+1)$th carbon atom can be represented by a vector, S, so that

$$\mathbf{S}_n = \mathbf{p}_1 + \ldots + \mathbf{p}_n, \tag{10.161}$$

and we wish to consider the asymptotic distribution of $\mathbf{S}_n$. Suppose that it has coordinates (X_n, Y_n, Z_n). From the spherical symmetry of the distribution of p_1, and therefore of S_n, we have

$$E(X_n) = E(Y_n) = E(Z_n) = 0,$$

and

$$E(S_n^2) = 3E(X_n^2). \tag{10.162}$$

Let $u_1, \ldots, u_n$ be the lengths of the projections, with correct sign, of $p_1, p_2, \ldots, p_n$, on the x-axis. It is easy to see that u_i is distributed in the rectangular distributions on the interval $(-1, 1)$ so that $E(u_i) = 0$, $E(u_i^2) = \frac{1}{3}$.

Write $c = \cos \alpha$, $s = \sin \alpha$, and consider p_{i+1}. This can be written as

$$p_{i+1} = cp_i + sr_{i+1}, \tag{10.163}$$

where r_{i+1} is a unit vector which is constrained to be at right angles to p_i but is otherwise uniformly distributed. For carbon atoms in a rubber molecule $c = -\frac{1}{3}$ and α is about $107\cdot46°$. The expectation of the projection of r_{i+1} on the x-axis is zero, whether $p_1, \ldots, p_i$ be fixed or not. Using (10.163) repeatedly we have

$$p_2 = cp_1 + sr_2,$$

$$p_3 = c^2 p_1 + csr_2 + sr_3,$$

$$p_n = c^{n-1} p_1 + c^{n-2} sr_2 + \ldots + csr_{n-1} + sr_n. \tag{10.164}$$

Let $v_2, \ldots, v_n$ be the projections of $r_2, \ldots, r_n$ on the x-axis. Then

$$(1-c)X_n = (1-c^n)u_1 + (1-c^{n-1})sv_2 + \ldots + (1-c^2)sv_{n-1} + (1-c)sv_n. \tag{10.165}$$

u_1 and the v_i are not independent, but since $E(v_i) = 0$, whatever the values of $u_1, v_2, \ldots, v_{i-1}$, we have

$$E(u_1 v_i) = 0, \quad i > 1,$$

$$E(v_i v_j) = 0, \quad i \neq j. \tag{10.166}$$

Furthermore, the initial vector p_1 is uniformly distributed in all directions, and therefore, considered by themselves, $r_2, r_3, \ldots, r_n$ are also uniformly distributed in all directions. Thus

$$E(u_1^2) = E(v_i^2) = \frac{1}{3}. \tag{10.167}$$

It follows that

$$E(X_n^2) = \left(\frac{1-c^n}{1-c}\right)^2 E(u_1^2) + s^2 \left(\frac{1-c^{n-1}}{1-c}\right)^2 E(v_2^2) + \ldots + s^2 E(v_n^2)$$

$$= \frac{1}{3(1-c)^2} \{n(1-c^2) - 2c + 2c^{n+1}\}. \tag{10.168}$$

This result was first obtained by Eyring (1932).

As a certain degree of randomness is introduced at each step, links which are far apart will be almost independent, and we expect that the distribution of (X_n, Y_n, Z_n) will tend to a trivariate normal distribution with zero means, and variances given by (10.168). As successive links are not independently distributed the ordinary central limit theorems do not apply. By using the extension to non-independent components due to Bernstein (1926–7) it is, however, possible to prove this rigorously (Moran 1948a). Another proof is given by Kac (1959a). The exact distribution for $n = 3$, 4, 5 is given by Treloar (1943).

A much more detailed study of this problem has been made by Daniels (1952) who also considers the two-dimensional analogue, and the limiting case when α is small and n large. In physical applications to long-chain molecules each atom must occupy a non-zero amount of space and hence unrestricted random walks of the above type cannot be exact models. No satisfactory method of dealing with this 'excluded volume' problem has yet been found. The similar situation for walks on lattices has been considered above in § 10.9.

If the atoms and their links are assumed to occupy no volume we may suppose that the direction of the first link is fixed, and try to find the distribution of the direction of the nth link. Let the distribution of the direction of $\mathbf{p}_n$ be represented by a probability measure, $P_n(E)$, on the surface of a unit sphere, with a polar direction parallel to p_1. Setting up polar coordinates, (θ, ϕ), where θ is the angle from the pole and ϕ is the azimuth we see that $P_2(E)$ is concentrated uniformly on the circle $\theta = \alpha$, whilst $P_3(E)$ is a distribution with a probability density spread out over the region $0 \leqslant \theta \leqslant 2\alpha$. We thus have a situation analogous to Pearson's random walk problem considered above but for which the 'state space' is the surface of a sphere instead of the plane. This problem has been considered by Goudsmit and Saunderson (1940) using spherical harmonics. When the steps (measured by α) are small the problem can be approximated by a diffusion process, or Brownian motion, on the surface of the sphere and has been discussed in this form by Perrin (1928). In this case there is a partial differential equation on the surface of the sphere which is analogous to the diffusion equation (see also Klein 1952 and Roberts and Ursell 1959–60).

10.23. Random walk on groups

In the previous sections we have considered random walks on the plane, in space, on the circle, and on the sphere. For the general random walk on a sphere we have seen (§ 10.22) that if all azimuthal angles are equally probable the distribution of the nth step will tend to uniformity on the surface of the sphere whatever the distribution of the polar angle with the sole exception of the case where all the probability is concentrated in the direction of the pole. On the other hand, for a random walk on the circumference of a circle, necessary and sufficient conditions for the distribution to tend to uniformity are not known. For the problem of random walks on more general algebraic structures see Grenander (1963) and for the particular case of a finite Abelian group Good (1951).

Bibliographical notes

A very detailed study of first passage distributions in random walks has been made by Kemperman (1961) who studies in detail the Wald and Pollaczek–Spitzer identities. The applications of Wald's identity to sequential analysis are considered in Wald (1947). Spitzer (1964) is the best reference on the relationship between random walk and potential theory. A generalization of Wald's identity to random walks in two dimensions has been given by Barnett (1965).

Exercises

Exercise 10.1. Show that for a random walk on a rectangular periodic lattice (e.g. wrapped on a torus) in which there are non-zero probabilities of moving from a given point to each of its four neighbours, the mean recurrence time is equal to the total number of points.

Exercise 10.2. Two particles perform independent random walks on a square lattice starting from the origin and having probabilities $\frac{1}{4}$ of moving to each of the four nearest neighbours. Prove that they meet at least once, with probability one.

Exercise 10.3. Using (10.109) derive the Borel–Tanner distribution considered in § 3.17

Exercise 10.4. Let V_1, V_2,... be a sequence of independent random variables identically distributed on $(0, \infty)$. Using the Pollaczek–Spitzer identity prove that

$$nE \max(1, V_1, V_1 V_2,..., V_1 ... V_n)$$

$$= \sum_{j=0}^{n-1} E \max(1, V_1,..., V_1 ... V_j) E \max(1, V_1 ... V_{n-j}),$$

where the first term in the sum is taken as $E \max(1, V_1 ... V_n)$.

BIBLIOGRAPHY

AEPPLI, A. 1924. Zur Theorie verketetter Wahrscheinlichkeiten. Thesis, Zürich.

AITCHISON, J. 1963. Inverse distributions and independent gamma-distributed products of random variables. *Biometrika* **50**, 505–8.

—— and BROWN, J. A. C. 1957. *The logarithmic normal distribution with special reference to its uses in economics. Department of applied economics monographs* No. 5. Cambridge University Press.

AITKEN, A. C. 1939. *Statistical mathematics.* Oliver and Boyd, Edinburgh.

—— 1950. On the statistical independence of quadratic forms in normal variates. *Biometrika* **37**, 93–6.

ANDERSEN, E. SPARRE 1949. On the number of positive sums of random variables. **32**, 27–36.

—— 1950. On the frequency of positive partial sums of a series of random variables. *Mat. Tidsskr.* B 33–5.

—— 1953a. On sums of symmetrically dependent random variables. *Skand. Aktuartidsskr.* **36**, 123–38.

—— 1953b, 1954. On the fluctuations of sums of random variables. *Mathematica scand.* **1**, 263–85; **2**, 195–223.

ANDERSON, T. W. 1958. *An introduction to multivariate analysis.* Wiley, New York.

ANIS, A. A. 1955. The variance of the maximum of the partial sums of a finite number of independent normal variates. *Biometrika* **42**, 96–101.

—— 1956. On the moments of the maximum of the partial sums of a finite number of independent normal variates. *Biometrika* **43**, 79–84.

—— and LLOYD, E. H. 1953. On the range of partial sums of a finite number of independent normal variates. *Biometrika* **40**, 35–42.

ANSCOMBE, F. J. 1950. Sampling theory of the negative binomial and logarithmic series distributions. *Biometrika* **37**, 358–82.

ARBOUS, A. G. and KERRICH, J. C. 1951. Accident statistics and the concept of accident proneness. *Biometrics* **7**, 340–432.

—— and SICHEL, H. S. 1954. New techniques for the analysis of absenteeism data. *Biometrika* **41**, 77–90.

ARLEY, N. 1943. *On the theory of stochastic processes and their application to the theory of cosmic radiation.* Gads Forlag, Copenhagen.

AROIAN, L. A. 1947. The probability function of the product of two normally distributed variables. *Ann. math. Statist.* **18**, 265–71.

—— 1950. On the levels of significance of the incomplete beta function and the *F*-distributions. *Biometrika* **37**, 219–23.

BAGNOLD, R. A. 1941. *The physics of blown sand and desert dunes.* Methuen, London.

BAHADUR, R. R. 1960. Some approximations to the binomial distribution function. *Ann. math. Statist.* **31**, 43–54.

BAILEY, N. T. J. 1951. On estimating the size of mobile populations from recapture data. *Biometrika* **38**, 293–306.

—— 1957. *The mathematical theory of epidemics.* Griffin, London.

33

BAILEY, N. T. J. 1964. *The elements of stochastic processes with applications to the natural sciences.* Wiley, New York.

BARNARD, G. A. 1951. The theory of information. *Jl R. statist. Soc.* B **13**, 46–64.

BARNETT, V. D. 1963. Some explicit results for an asymmetric two-dimensional random walk. *Proc. Camb. phil. Soc. math. phys. Sci.* **59**, 451–62.

—— 1965. Wald's identity and absorption probabilities for two-dimensional random walks. *Proc. Camb. phil. Soc. math. phys. Sci.* **61**, 747–62.

BARTLETT, M. S. 1937. Note on the derivation of fluctuation formulae for statistical assemblies, *Proc. Camb. phil. Soc. math. phys. Sci.* **33**, 390–3.

—— 1938. The characteristic function of a conditional statistic. *J. Lond. math. Soc.* **13**, 62–7.

—— 1945. Negative probability. *Proc. Camb. phil. Soc. math. phys. Sci.* **41**, 71–3.

—— 1955. *Stochastic processes.* Cambridge University Press.

—— 1963. Statistical estimation of density functions. *Sankhyā* A **25**, 245–54.

—— and KENDALL, D. G. 1951. On the use of the characteristic functional in the analysis of some stochastic processes occurring in physics and biology. *Proc. Camb. phil. Soc. math. phys. Sci.* **47**, 65–76.

BARTON, D. E. 1958. The matching distributions: Poisson limiting forms and derived methods of approximation. *Jl R. statist. Soc.* B **20**, 73–92.

—— and DAVID, F. N. 1958. Runs in a ring. *Biometrika* **45**, 572–8.

BATEMAN, G. 1948. On the power function of the longest run as a test for randomness in a sequence of alternatives. *Biometrika* **35**, 97–112.

BATEMAN, H. 1932. *Partial differential equations.* Cambridge University Press.

BAXTER, G. 1961. An analytic approach to finite fluctuation problems in probability. *J. Analyse math.* **9**, 31–70.

—— and DONSKER, M. D. 1957. On the distribution of the supremum functional for processes with stationary independent increments. *Trans. Am. math. Soc.* **85**, 73–87.

BAYES, T. 1763. An essay towards solving a problem in the doctrine of chances. *Phil. Trans. R. Soc.* **53**, 370–418. Reprinted with introduction by G. A. Barnard in *Biometrika* **45**, 293–315 (1958).

BEALL, G. and RESCIA, R. R. 1953. A generalization of Neyman's contagious distributions. *Biometrics* **9**, 354–86.

BECKENBACH, E. F. 1964. *Applied combinatorial mathematics.* Wiley, New York.

BELL, D. A. 1960. *Electrical noise.* Van Nostrand, London.

BERGSTRÖM, H. 1952. On some expansions of stable distributions. *Arch. math.* **2**, 375–8.

—— 1963. *Limit theorems for convolutions.* Almqvist and Wiksell, Stockholm.

BERNSTEIN, S. 1926–7. Sur l'extension du théorème limité du calcul des probabilités aux sommes des quantités dépendantes. *Math. Annln* **97**, 1–59.

—— 1943. Return to the problem of evaluating the approximation in the limiting formula of Laplace (in Russian). *Izv. Akad. Nauk SSSR.* Serija matematiceskaja **7**, 3–16.

BERRY, A. 1941. The accuracy of the Gaussian approximation to the sum of independent variates. *Trans. Am. math. Soc.* **49**, 122–36.

BETHE, H. A. 1937. Nuclear physics, B; Nuclear dynamics, theoretical. *Rev. mod. Phys.* **9**, 120.

BHARUCHA-REID, A. T. 1960. *Elements of the theory of Markov processes and their applications.* McGraw-Hill, New York.

BIENAYMÉ, J. 1853. Considérations à l'appui de la découverte de Laplace sur la loi des probabilités, dans la méthode des moindres carrés. *C. r. hebd. Séanc. Acad. Sci., Paris* **37**, 309–24.

BIRKHOFF, G. 1937. On product integration. *J. Math. Phys.* **16**, 104–32.

BIRNBAUM, Z. W. 1942. An inequality for Mill's ratio. *Ann. math. Statist.* **13**, 245–6.

—— and MARSHALL, A. W. 1961. Some multivariate Chebyshev inequalities with extensions to continuous parameter processes. *Ann. math. Statist.* **32**, 687–703.

—— RAYMOND, J. and ZUCKERMANN, H. S. 1947. A generalization of Tshebyshev's inequality to two dimensions. *Ann. math. Statist.* **18**, 70–9.

BISLEY, M. T. L. 1957. *Probability: an intermediate textbook*. Cambridge University Press.

BLANC-LAPIERRE, A. and FORTET, R. 1953. *Théorie des fonctions aléatoires*. Masson, Paris.

BOAS, R. P. 1949. Representation of probability distributions by Charlier series. *Ann. math. Statist.* **20**, 376–92.

BOCHNER, S. 1932. *Vorlesungen über Fouriersche Integrale*. Leipzig, Akademische Verlagsgesellschaft (Chelsea reprint 1948).

BOREL, E. 1909. Sur les probabilités dénombrables et leurs applications arithmétiques. *Circolo mat.* **26**, 247–71.

—— 1942. Sur l'emploi du théorème de Bernoulli pout faciliter le calcul d'un infinité de coefficients. Application au problème de l'attente à un guichet. *C. r. hebd. Séanc. Acad. Sci., Paris* **214**, 452–6.

—— and CHÉRON, A. 1940. *Théorie mathématique du bridge à la portée de tous*. Gauthier–Villars, Paris.

BOURGET, J. 1871. Des permutations. *Nouv. Annls Math.* (2) **10**, 254–68.

BOX, G. E. P. and MULLER, M. E. 1958. A note on the generation of random normal deviates. *Ann. math. Statist.* **29**, 610–11.

BREIMAN, L. 1961. Optimal gambling systems for favourable games. *Proceedings of the fourth Berkeley Symposium on mathematical statistics and probability*, Vol. I, pp. 65–78. University of California Press, California.

BREITENBERGER, E. 1963. Analogues of the normal distribution on the circle and the sphere. *Biometrika* **50**, 81–8.

BROCKMEYER, E., HALSTRØM, H. L., and JENSEN, A. 1948. *The life and works of A. K. Erlang. Trans. Dan. Acad. tech. Sci.* No. 2.

BRODERICK, T. S. 1937. On some symbolic formulae in probability theory. *Proc. R. Ir. Acad. A* **44**, 19–28.

BROWN, G. W. 1951. History of RAND's random digits—summary, Monte Carlo method. *Appl. Math. Ser.* **12**, 31–3.

BROWN, R. 1828. A brief account of microscopical observations made in the months of June, July, and August 1827, on the particles contained in the pollen of plants; and on the general existence of active molecules in organic and inorganic bodies. *Phil. Mag.* (2nd series) **4**, 161–73.

CADWELL, J. H. 1951. The bivariate normal integral. *Biometrika* **38** 475–9.

CAMP, B. H. 1948. Generalization to *N* dimensions of inequalities of the Tchebycheff type. *Ann. math. Statist.* **19**, 568–74.

CAMPBELL, N. R. 1909. The study of discontinuous phenomena. *Proc. Camb. phil. Soc. math. phys. Sci.* **15**, 117–36.

CANTELLI, F. P. 1917. Sulla probabilità come limite della frequenza. *Atti Accad. naz. Lincei Rc. Sed. solen. s. 5,* **26**, 39–45.

CARLEMAN, T. 1926. *Les fonctions quasi-analytiques*. Gauthier–Villars, Paris.

CHANDRASEKHAR, S. 1943. Stochastic problems in physics and astronomy. *Rev. mod. Phys.* **15**, 1–89.

CHAPMAN, D. G. 1954. The estimation of biological populations. *Ann. math. Statist.* **25**, 1–15.

CHAUNDY, T. W. 1935. *The differential calculus.* Clarendon Press, Oxford.

CHENG, T. T. 1949. The normal approximation to the Poisson distribution and a proof of a conjecture of Ramanujan. *Bull. Am. math. Soc.* **55**, 396–401.

CHU, J. T. 1955. On bounds for the normal integral. *Biometrika* **42**, 263–5.

CHUNG, K. L. 1941. On the probability of the occurrence of at least *m* events among *n* arbitrary events. *Ann. math. Statist.* **12**, 328–38.

—— 1942. On mutually favourable events. *Ann. math. Statist.* **13**, 338–49.

—— 1943a. Generalization of Poincaré's formula in the theory of probability. *Ann. math. Statist.* **14**, 63–5.

—— 1943b. On fundamental systems of probabilities of a finite number of events. *Ann. math. Statist.* **14**, 123–33.

—— 1943c. Further results on probabilities of a finite number of events. *Ann. math. Statist.* **14**, 234–7.

—— 1960. *Markov chains with stationary transition probabilities.* Grundlehr. math. Wiss. **104**.

—— and HSU, L. C. 1945. A combinatorial formula and its application to the theory of probability of arbitrary events. *Ann. math. Statist.* **16**, 91–5.

COCHRAN, W. G. 1934. The distribution of quadratic forms in a normal system, with applications to the analysis of covariance. *Proc. Camb. phil. Soc. math. phys. Sci.* **30**, 178–91.

—— 1951. Improvement by means of selection. *Proceedings of the second Berkeley symposium on mathematical statistics and probability,* pp. 449–70. University of California Press, California.

CONDON, E. U. and BREIT, G. 1936. The energy distribution of neutrons slowed by elastic impacts. *Phys. Rev.* **49**, 229–31.

COURANT, R., FRIEDRICHS, K., and LEWY, H. 1928. Differenzengleichingen der mathematischen Physik. *Math. Annln* **100**, 32–74.

COX, D. R. 1955. A use of complex probabilities in the theory of stochastic processes. *Proc. Camb. phil. Soc. math. phys. Sci.* **51**, 313–19.

—— and MILLER, H. D. 1965. *The theory of stochastic processes.* Methuen, London.

—— and SMITH, W. L. 1961. *Queues.* Methuen, London.

CRAIG, A. T. 1943. Note on the independence of certain quadratic forms. *Ann. math. Statist.* **14**, 195–7.

CRAIG, C. C. 1936. On the frequency function of *xy*. *Ann. math. Statist.* **7**, 1–15.

CRAMÉR, H. 1939. On the representation of a function by certain Fourier integrals. *Trans. Am. math. Soc.* **46**, 191–201.

—— 1947. Problems in probability theory. *Ann. math. Statist.* **18**, 165–93.

—— 1962. *Random variables and probability distributions,* 2nd edn. *Cambridge Mathematical Tracts,* No. 36. Cambridge University Press.

—— and WOLD, H. 1936. Some theorems on distribution functions. *J. Lond. math. Soc.* **11**, 290–4.

CSASZAR, A. 1955. Sur la structure des espaces de probabilité conditionnelle. *Acta math. hung.* **6**, 337–61.

CURNOW, R. N. 1961. Optimal programmes for varietal selection. *Jl R. statist. Soc.* B **23**, 282–318.

DABONI, L. 1959. A property of Poisson distributions. *Boll. Un. mat. ital.* **14**, 318–20.

DANIELS, H. E. 1941. The probability distribution of the extent of a random chain. *Proc. Camb. phil. Soc. math. phys. Sci.* **37**, 244–61.

—— 1952. The statistical theory of stiff chains. *Proc. R. Soc. Edinb.* **63**, 290–311.

—— 1954. Saddlepoint approximations in statistics. *Ann. math. Statist.* **25**, 631–50.

—— 1956. The approximate distribution of serial correlation coefficients. *Biometrika* **43**, 169–85.

DARLING, D. A. 1957. The Kolmogorov–Smirnov, Cramér–von Mises tests. *Ann. math. Statist.* **28**, 823–38.

DARMOIS, G. 1928. *Statistique mathèmatique.* Octave Doin, Paris.

—— 1951. Sur les diverses propriétés charactéristiques de la loi de probabilité de Laplace–Gauss. *Bull. Inst. int. Statist.* **33**, 79–82.

DARWIN, J. H. 1960. An ecological distribution akin to Fisher's logarithmic distribution. *Biometrics* **16**, 51–60.

DAS, S. C. 1956. The numerical evaluation of a class of integrals. II. *Proc. Camb. phil. Soc. math. phys. Sci.* **52**, 442–8.

DAVENPORT, W. B. and ROOT, W. L. 1958. *An introduction to the theory of random signals and noise.* McGraw-Hill, New York.

DAVID, F. N. 1950. Two combinatorial tests of whether a sample has come from a given population. *Biometrika* **37**, 97–110.

—— 1955. Studies in the history of probability and statistics. I. Dicing and gaming. *Biometrika* **42**, 1–15.

—— 1962. *Games, gods and gambling.* Griffin, London.

—— and BARTON, D. E. 1962. *Combinatorial chance.* Griffin, London.

—— and FIX, E. 1961. Rank correlation and regression in a non-normal surface. *Proceedings of the fourth Berkeley symposium on mathematical statistics and probability*, Vol. I, pp. 177–97. University of California Press, California.

—— and KENDALL, M. G. 1949, 1951, 1953, 1955, 1958. Tables of symmetric functions. *Biometrika* **36**, 431–49; **38**, 435–62; **40**, 427–46; **42**, 223–42 (correction, **45**, 292).

—— and MALLOWES, C. L. 1961. The variance of Spearman's rho in normal samples. *Biometrika* **48**, 19–28.

DAVID, S. T., KENDALL, M. G., and STUART, A. 1951. Some questions of distribution in the theory of rank correlation. *Biometrika* **38**, 131–40.

DAVIES, R. O. and LEECH, J. W. 1954. The statistics of scaled random events. *Proc. Camb. phil. Soc. math. phys. Sci.* **50**, 575–80.

DE CASTRO, G. 1952. Notes on the difference of binomial and Poisson variates. *Port. math.* **11**, 173–5.

DELTHEIL, R. 1926. *Probabilités géometriques. Traité de calcul des probabilités et de ses applications.* Gauthier–Villars, Paris.

DEMING, W. E. and GLASSER, G. J. 1959. On the problem of matching lists by samples. *J. Am. statist. Ass.* **54**, 403–15.

DIANANDA, P. H. 1953. Some probability limit theorems with statistical applications. *Proc. Camb. phil. Soc. math. phys. Sci.* **49**, 239–46.

—— 1954. The central limit for *m*-dependent random variables asymptotically stationary to the second order. *Proc. Camb. phil. Soc. math. phys. Sci.* **50**, 287–92.

—— 1955. The central limit theorem for *m*-dependent variables. *Proc. Camb. phil. Soc. math. phys. Sci.* **51**, 92–5.

DINGLE, R. B. 1949. The Bose–Einstein statistics of particles with special reference to the case of low temperatures. *Proc. Camb. phil. Soc. math. phys. Sci.* **45**, 275–85.

DOMB, C. 1952. On the use of a random parameter in combinatorial problems. *Proc. phys. Soc.* A **65**, 305–9.

—— 1954. On multiple returns in the random walk problem. *Proc. Camb. phil. Soc. muth. phys. Sci.* **50**, 586–91.

—— 1964. Some statistical problems connected with crystal lattices (with discussion). *Jl R. statist. Soc.* B **26**, 367–97.

DOOB, J. L. 1942. The Brownian movement and stochastic equations. *Ann. Math.* **43**, 351–69.

—— 1953. *Stochastic processes.* Wiley, New York.

DORFMAN, R. 1943. The detection of defective members of large populations. *Ann. math. Statist.* **14**, 436–40.

DUARTE, F. J. 1927. *Nouvelle tables de log n! depuis n* = 1 *jusqu'à n* = 3000. Kundig, Geneva; Index Generalis, Paris.

DUGUÉ, D. 1951*a*. Analyticité et convexité des fonctions charactéristiques. *Annls Inst. Poincaré* **12**, 45–56.

—— 1951*b*. Sur certains examples de décomposition en arithmétique des lois de probabilité. *Annls Inst. Poincaré* **12**, 159–69.

—— 1957. *Arithmétique des lois de probabilités. Mém. Sci. math.* **137**. Gauthier-Villars, Paris.

DURAND, D. and GREENWOOD, J. A. 1955. The distribution of length and components of *n* random unit vectors. *Ann. math. Statist.* **26**, 233–46.

—— —— 1957. Random unit vectors. II. Usefulness of Gram–Charlier and related series in approximating distributions. *Ann. math. Statist.* **28**, 978–86.

DVORETZKY, A. and ERDÖS, P. 1951. Some problems on random walk in space. *Proceedings of the second Berkeley symposium on mathematical statistics and probability*, pp. 353–67. University of California Press, California.

—— and WOLFOWITZ, J. 1951. Sums of random integers reduced modulo *m. Duke math. J.* **18**, 501–7.

DYNKIN, E. B. 1960. *Theory of Markov processes.* Pergamon Press, Oxford.

DYSON, F. J. 1953. Fourier transforms of distribution functions. *Can. J. Math.* **5**, 554–8.

EGGENBERGER, F. and PÓLYA, G. 1923. Über die Statistik verketteter Vorgänge. *Z. angew. Math. Mech.* **3**, 279–89.

EHRENFEST, P. and T. 1960. Ueber eine Aufgabe aus der Wahrscheinlichkeitsrechnung, die mit der kinetischen Deutung der Entropievermehrung zusammenhängt. *Math.-naturw. Bl.* Nos. 11 and 12 (1906). (Also in collected papers of P. Ehrenfest.)

EINSTEIN, A. 1905. Über die von der molekular-kinetischen Theorie der Wärme geforderte Bewegung von in ruhenden Flüssigkeiten suspendierten Teilchen. *Annln Phys.* **17**, 549–60.

—— 1906*a*. Zur Theorie der Brownschen Bewegung. *Annln Phys.* **19**, 371–81.

—— 1906*b*, 1911. Eine neue Bestimmung der Molekuldimensionen. *Annln Phys.* **19**, 289–306 (also **34**, 591–2).

—— 1907. Theoretische Bemerkungen über die Brownsche Bewegung. *Z. Elektrochem.* **13**, 41–2.

—— 1908. Elementare Theorie der Brownschen Bewegung. *Z. Elektrochem.* **14**, 235–9.

—— 1956. *Investigations on the theory of the Brownian movement* (ed. R. Fürth, trans. by A. D. Cowper). Dover, New York.

EPSTEIN, B. 1947. The mathematical description of certain breakage mechanisms leading to the logarithmic-normal distribution. *J. Franklin Inst.* **244**, 471–7.

EPSTEIN, B. 1948. Some applications of the Mellin transform in statistics. *Ann. math. Statist.* **19**, 370–9.

—— 1954. Tables for the distribution of exceedances. *Ann. math. Statist.* **25**, 762–8.

ERLANG, A. K. 1929. Problem 15. *Mat. Tidsskr.* B, p. 36. (See also Brockmeyer, Halstrøm, and Jensen 1948.)

ESSEEN, G. 1945. Fourier analysis of distribution functions. A mathematical study of the Laplace–Gaussian law. *Acta Math., Stockh.* **77**, 1–125.

EWENS, W. J. 1963. The mean time for absorption in a process of genetic type. *J. Aust. math. Soc.* **3**, 375–83.

EYRING, H. 1932. The resultant electric moment of complex molecules. *Phys. Rev.* **39**, 746–8.

FEDERIGHI, E. T. 1959. Extended tables of the percentage points of Student's *t*-distribution. *J. Am. statist. Ass.* **54**, 683–8.

FELLER, W. 1936. Über den zentralen Grenzwertsatz der Wahrscheinlichkeitsrechnung. *Math. Z.* **40**, 521–59.

—— 1937a. Über den zentralen Grenzwertsatz der Wahrscheinlichkeitsrechnung. *Math. Z.* **42**, 301–12.

—— 1937b. Zur Theorie der stochastischen Prozesse (Existenz- und Eindeutigkeitssätze). *Math. Annln* **113**, 113–60.

—— 1943a. The general form of the so-called law of the iterated logarithm. *Trans. Am. math. Soc.* **54**, 373–402.

—— 1943b. On a general class of 'contagious' distributions. *Ann. math. Statist.* **14**, 389–400.

—— 1945a. The fundamental limit theorems in probability. *Bull. Am. math. Soc.* **51**, 800–32.

—— 1945b. A note on the law of large numbers and 'fair' games. *Ann. math. Statist.* **16**, 301–4.

—— 1945c. On the normal approximation to the binomial distribution. *Ann. math. Statist.* **16**, 319–29 (correction **21**, 301).

—— 1946. The law of the iterated logarithm for identically distributed random variables. *Ann. Math.* **47**, 631–8.

—— 1951. The asymptotic distribution of the range of sums of independent random variables. *Ann. math. Statist.* **22**, 427–32.

—— 1957a. On boundary conditions for the Kolmogorov differential equations. *Ann. Math.* **65**, 527–70.

—— 1957b. An introduction to probability theory and its applications, Vol. I, 2nd edn. (Vol. II, 1966).

FERGUSON, T. S. 1962. A representation of the symmetric bivariate Cauchy distribution. *Ann. math. Statist.* **33**, 1256–66.

FINNEY, D. J. 1941. On the distribution of a variate whose logarithm is normally distributed. *Jl R. statist. Soc.* Suppl. **7**, 155–61.

—— 1962. Cumulants of truncated multinormal distributions. *Jl R. statist. Soc.* B **24**, 535–6.

—— 1964. Screening processes: problems and illustrations. *Contributions to statistics presented to Professor P. C. Mahalanobis.* Pergamon, Oxford.

FISHER, M. E. and SYKES, M. F. 1959. The excluded volume problems and the Ising model of ferromagnetism. *Phys. Rev.* **114**, 45–58.

FISHER, R. A. and WISHART, J. 1927. On the distribution of the error of an interpolated value and on the construction of tables. *Proc. Camb. phil. Soc. math. phys. Sci.* **23**, 912–21.

FISHER, R. A. 1930. *The genetical theory of natural selection.* Clarendon Press, Oxford.

—— 1935. The mathematical distributions used in the common tests of significance. *Econometrica* **3**, 353–65.

—— 1949. *The theory of inbreeding.* Oliver and Boyd, Edinburgh.

—— 1953. Dispersion on a sphere. *Proc. R. Soc.* A **217**, 295–305.

—— 1958. *Statistical methods for research workers.* Oliver and Boyd, Edinburgh.

—— CORBET, A. S., and WILLIAMS, C. B. 1943. The relation between the number of species and the number of individuals in a random sample of an animal population. *J. Anim. Ecol.* **12**, 42–57.

—— and DUGUÉ, D. 1948. Un résultat assez inattendu d'arithmétique des lois du probabilité. *C. r. hebd. Séanc. Acad. Sci., Paris* **227**, 1205–6.

—— and YATES, F. 1938. *Statistical tables for biological, agricultural and medical research.* Oliver and Boyd, London (and subsequent editions).

FISZ, M. 1963. *Probability theory and mathematical statistics,* 3rd edn. Wiley, New York.

FIX, E. 1949. Tables of non-central χ^2. *Univ. Calif. Publs Statist.* **1**, 15–19.

FLETCHER, A., MILLER, J. C. P., ROSENHEAD, L., and COMRIE, L. J. 1962. *Index of mathematical tables,* 2 vols. Blackwell, Oxford.

FORTET, R. 1951. Random functions from a Poisson process. *Proceedings of the second Berkeley symposium on mathematical statistics and probability,* pp. 373–85. University of California Press, California.

FOSTER, F. G. 1951. Markov chains with an enumerable number of states and a class of cascade processes. *Proc. Camb. phil. Soc. math. phys. Sci.* **47**, 77–85.

—— 1953. On the stochastic matrices associated with certain queueing processes. *Ann. math. Statist.* **24**, 355–60.

—— and GOOD, I. J. 1953. On a generalization of Pólya's random walk theorem. *Q. Jl Math.* **4**, 120–6.

—— and REES, D. H. 1957. Upper percentage points of the generalized beta distribution. *Biometrika* **44**, 237–47.

FOWLER, R. H. 1936. *Statistical mechanics,* 2nd edn. Cambridge University Press.

FOX, C. 1957. Some applications of Mellin transforms to the theory of bivariate statistical distributions. *Proc. Camb. phil. Soc. math. phys. Sci.* **53**, 620–8.

FRANCIS, V. J. 1946. On the distribution of the sum of n sample values drawn from a truncated normal distribution. *Jl R. statist. Soc.* Suppl. **8**, 223–32.

FRANEL, J. 1917. *Vjschr. naturf. Ges. Zurich* **62**, 285–95.

FRÉCHET, M. 1937, 1938. *Recherches théoriques modernes sur la théorie des probabilités.* 1^r Livre. *Généralités sur les probabilités. Variables aléatoires.* 2e Livre. *Méthode des fonctions arbitraires. Théorie des événements en chaînés dans le cas d'un nombre fini d'états possibles.* Gauthier–Villars, Paris.

—— 1940, 1943. *Les probabilités associées à un système d'événements compatibles et dépendants. Actualités scientifiques et industrielles.* Nos. 859 and 942. Hermann, Paris.

FREEMAN, H. 1952. *Mathematics for actuarial students,* Part II. Cambridge University Press.

FREUND, J. E. and POZNER, A. N. 1956. Some results in restricted occupancy theory. *Ann. math. Statist.* **27**, 537–9.

FRIEDMAN, B. 1949. A simple urn model. *Communs pure appl. Math.* **2**, 59–70.

FRISCH, R. 1926. Sur les semi-invariants et moments employés dans l'étude des distributions statistiques. *Skr. norske Vidensk-Akad.* Mat-naturv. Kl. No. 3.

FRY, T. C. 1928. *Probability and its engineering uses.* Van Nostrand, New York.

FURRY, W. H. 1937. On fluctuation phenomena in the passage of high energy electrons through lead. *Phys. Rev.* **52**, 569–81.

GALTON, F. 1873. Problem 4001. *Mathl. Quest. Solut.* **19**, 103–5.

—— 1889. *Natural inheritance.* London.

GARTI, Y. and CONSOLI, T. 1954. Sur la densité de probabilité du produit de variables aléatoires de Pearson de type III. *Studies math. Mech. presented to R. v. Mises*, pp. 301–9. Academic Press, New York.

GAUSS, C. F. 1821. *Theoria combinationis observationum.*

GEARY, R. C. 1930. The frequency distribution of the quotient of two normal variables. *Jl R. statist. Soc.* **93**, 442–6.

—— 1936. The distribution of 'Student's' ratio for non-normal samples. *Jl R. statist. Soc.* Suppl. **3**, 178–84.

—— 1944. Extension of a theorem of Harald Cramér on the frequency distribution of the quotient of two variables. *Jl R. statist. Soc.* **107**, 56–7.

GEIRINGER, H. 1938. On the probability theory of arbitrarily linked events. *Ann. math. Statist.* **9**, 260–71.

GENERAL ELECTRIC COMPANY, U.S.A. 1962. *Tables of the individual and cumulative terms of the Poisson distribution.* Van Nostrand, Princeton.

GILL, A. 1962. Synthesis of probability transformers. *J. Franklin Inst.* **274**, 1–19.

GILLIS, J. 1960. A random walk problem. *Proc. Camb. phil. Soc. math. phys. Sci.* **56**, 390–2.

GNEDENKO, B. V. and KOLMOGOROV, A. N. 1954. *Limit distributions for sums of independent random variables* (trans. by K. L. Chung). Addison-Wesley, Cambridge, Mass.

—— KOROLUK, V. S., and SKOROKHOD, A. V. 1961. Asymptotic expansions in probability theory. *Proceedings of the fourth Berkeley symposium on mathematical statistics and probability*, Vol. II, pp. 153–70. University of California Press, California.

GODWIN, H. J. 1955. On generalizations of Tchebychef's inequalities. *J. Am. statist. Ass.* **50**, 923–45.

—— 1964. *Inequalities on distribution functions. Griffin's statistical monographs* No. 16. Griffin, London.

GOLDSTEIN, S. 1951. On diffusion by continuous movement, and on the telegraph equation. *Q. Jl Mech. appl. Math.* **4**, 129–56.

GOLOMB, S. W. 1961. Permutations by cutting and shuffling. *SIAM Rev.* **3**, 293–7.

GOOD, I. J. 1949. The number of individuals in a cascade process. *Proc. Camb. phil. Soc. math. phys. Sci.* **45**, 360–3.

—— 1950a. *Probability and the weighing of evidence.* Griffin, London.

—— 1950b. A proof of Liapounoff's inequality. *Proc. Camb. phil. Soc. math. phys. Sci.* **46**, 353.

—— 1951. Random motion on a finite Abelian group. *Proc. Camb. phil. Soc. math. phys. Sci.* **47**, 756–62 (Corrigenda **48**, 368).

—— 1953. The population frequencies of species and the estimation of population parameters. *Biometrika* **40**, 237–64.

—— 1961. An asymptotic formula for the differences of the powers of zero. *Ann. math. Statist.* **32**, 249–56.

GOUDSMIT, S. and SAUNDERSON, J. L. 1940. Multiple scattering of electrons. *Phys. Rev.* **57**, 24–9.

GRAB, E. L. and SAVAGE, I. R. 1954. Tables of the expected values of $1/X$ for positive Bernoulli and Poisson variable. *J. Am. statist. Ass.* **49**, 169–77 (also p. 906).

34

GRAVES, L. M. 1946. *The theory of functions of real variables*. McGraw-Hill, New York.

GREENWOOD, J. A. and HARTLEY, H. O. 1962. *Guide to tables in mathematical statistics*. Princeton University Press.

GREENWOOD, M. 1931. On the statistical measure of infectiousness. *J. Hyg., Camb.* **31**, 336–51.

—— 1950. Accident proneness. *Biometrika* **37**, 24–9.

—— and YULE, G. U. 1920. An inquiry into the nature of frequency-distributions of multiple happenings, etc. *Jl R. statist. Soc.* **83**, 255–79.

GRENANDER, U. 1951. On empirical spectral analysis of stochastic processes. *Ark. Mat.* **1**, 503–31.

—— 1963. *Probabilities on algebraic structures*. Almqvist and Wiksell, Stockholm, and New York.

—— POLLAK, H. O., and SLEPIAN, D. 1959. The distribution of quadratic forms in normal variates: a small sample theory with applications to spectral analysis. *J. Soc. ind. appl. Math.* **7**, 374–401.

GRIMM, H. 1962. Tafeln der negativen Binomialverteilung. *Biometr. Z.* **4**, 239–62.

GRUNDY, P. M. 1951. The expected frequencies in a sample of an animal population where the abundancies of species are log normal. *Biometrika* **38**, 427–34.

GUMBEL, E. J., GREENWOOD, J. A., and DURAND, D. 1953. The circular normal distribution: theory and tables. *J. Am. statist. Ass.* **48**, 131–52.

—— and VON SCHELLING, H. 1950. The distribution of the number of exceedances. *Ann. math. Statist.* **21**, 247–62.

GUPTA, H. 1950. Tables of differences of zero. *Research Bulletin of the East Punjab University*, No. 2.

GURLAND, J. 1957. Some interrelations among compound and generalized distributions. *Biometrika* **44**, 265–8.

—— 1958. A generalized class of contagious distributions. *Biometrics* **14**, 229–49.

HACATUROV, A. A. 1954. Determination of the value of the measure for a region of n-dimensional Euclidean space from its values for all half-spaces. *Usp. mat. Nauk* **9**, 205–12.

HADEN, H. G. 1947. A note on the distribution of the different orderings of n objects. *Proc. Camb. phil. Soc. math. phys. Sci.* **43**, 1–9.

HAIGHT, F. A. 1961. Index to the distributions of mathematical statistics. *J. Res. natn. Bur. Stand.* **B65**, 23–60.

—— and BREUER, M. A. 1960. The Borel–Tanner distribution. *Biometrika* **47**, 143–50.

HAJNAL, J. 1956. The ergodic properties of non-homogeneous finite Markov chains. *Proc. Camb. phil. Soc. math. phys. Sci.* **52**, 67–77.

—— 1958. Weak ergodicity in non-homogeneous Markov chains. *Proc. Camb. phil. Soc. math. phys. Sci.* **54**, 233–46.

HALDANE, J. B. S. 1940. The cumulants and moments of the binomial distribution and the cumulants of χ^2 for an $n \times 2$ table. *Biometrika* **31**, 392–6.

—— 1948. Note on the median of a multivariate distribution. *Biometrika* **35**, 414–15.

HALL, P. 1927. The distribution of means for samples of size N drawn from a population in which the variate takes values between 0 and 1, all such values being equally probable. *Biometrika* **19**, 240–5.

HALMOS, P. R. 1944. Random alms. *Ann. math. Statist.* **15**, 182–9.

—— 1950. *Measure theory*. Van Nostrand, New York.

HALTON, J. H. 1962. Sequential Monte Carlo. *Proc. Camb. phil. Soc. math. phys. Sci.* **58**, 57–78.

HAMMERSLEY, J. M. 1952. Tauberian theory for the asymptotic forms of statistical frequency functions. *Proc. Camb. phil. Soc. math. phys. Sci.* **48**, 592–9 (see also **49**, 735).

—— 1956. The area enclosed by Polyá's walk. *Proc. Camb. phil. Soc. math. phys. Sci.* **52**, 78–87.

—— 1957. Percolation processes. II. The connective constant. *Proc. Camb. phil. Soc. math. phys. Sci.* **53**, 642–5.

—— and HANDSCOMB, D. C. 1964. *Monte Carlo methods.* Methuen, London.

—— and MORTON, K. W. 1954. Symposium on Monte Carlo methods: poor man's Monte Carlo. *Jl R. Statist. Soc.* B **16**, 23–38.

HANNAN, E. J. 1960. *Time series analysis. Methuen's monographs.* Methuen, London.

—— 1965. Group representations and applied probability. *J. app. Probability* **2**, 1–68. Also published in Methuen's *Supplementary review series in applied probability*, Vol. 3 (1965).

HARDY, G. H. and ROGOSINSKI, W. W. 1944. *Fourier series, Cambridge tracts in mathematics and mathematical physics* No. 38. Cambridge University Press.

HARRIS, T. E. 1951. Some mathematical models for branching processes. *Proceedings of the second Berkeley symposium on mathematical statistics and probability.* pp. 305–28. University of California Press, California.

—— 1963. *The theory of branching processes.* Springer, Berlin.

HARTER, H. L. 1964. *New Tables of the incomplete gamma-function ratio and of percentage points of the chi-square and beta distributions.* Government Printing Office, Washington.

HARTMAN, P. and WINTNER, A. 1941. On the law of the iterated logarithm. *Am. J. Math.* **63**, 169–76.

HARVARD COMPUTATION LABORATORY 1955. *Tables of the cumulative binomial probability distribution. Annals of the computation laboratory.* Harvard University Press, Cambridge, Mass.

HASTINGS, C. 1955. *Approximations for digital computers.* Princeton University Press, Princeton.

HAWKINS, D. and ULAM, S. 1944. Theory of multiplicative processes. I. Los Alamos declassified document 265.

HEATHCOTE, C. R. and MOYAL, J. E. 1959. The random walk (in continuous time) and its application to the theory of queues. *Biometrika* **46**, 400–11.

HEILBRONN, H. A. 1949. On discrete harmonic functions. *Proc. Camb. phil. Soc. math. phys. Sci.* **45**, 194–206.

HELLY, E. 1912. Ueber lineare Funktionaloperationen. *Sber. Akad. Wiss. Wien.* **121**, 265–97.

HEYDE, C. C. 1963. On a property of the lognormal distribution. *Jl R. statist. Soc.* B **25**, 392–2.

HODGES, J. L. 1955. On the non-central beta-distribution. *Ann. math. Statist.* **26**, 648–53.

HOEFFDING, W. and ROBBINS, H. 1948. The central limit theorem for dependent random variables. *Duke math. J.* **15**, 773–80.

—— and SHRIKHANDE, S. S. 1955. Bounds for the distribution function of a sum of independent, identically distributed random variables. *Ann. math. Statist.* **26**, 439–49.

HULL, T. E. and DOBELL, A. R. 1962. Random number generators. *SIAM Rev.* **4**, 230–54.

HURST, H. E. 1956. Methods of using long-term storage in reservoirs. *Proc. Instn civ. Engrs*, Part I, **5**, 519–90.

HUYGENS, C. 1957. *De ratiociniis in ludo aleae.*

IMHOF, J. P. 1961. Computing the distribution of quadratic forms in normal variables. *Biometrika* **48**, 419–26.

IRWIN, J. O. 1927. On the frequency-distribution of the means of samples from a population having any law of frequency with finite moments. *Biometrika* **19**, 225–39.

—— 1937. The frequency distribution of the difference between two independent variates following the same Poisson distribution. *Jl R. statist. Soc.* **100**, 415–16.

—— 1941. Discussion of a paper by Chambers and Yule. *Jl R. statist. Soc.* Suppl. **7**, 101–7.

—— 1943. A table of the variance of $\sqrt{x}$ when x has a Poisson distribution. *Jl R. statist. Soc.* **106**, 143–4.

—— 1946. On the characteristic function of the distribution of the product of two normal variates. *Proc. Camb. phil. Soc. math. phys. Sci.* **42**, 82–4.

—— 1954. A distribution arising in the study of infectious diseases. *Biometrika* **41**, 266–8.

—— 1955. A unified derivation of some well-known frequency distributions of interest in biometry and statistics. *Jl R. statist. Soc.* **118**, 389–404.

—— 1963. The place of mathematics in medical and biological statistics. *Jl R. statist. Soc.* A **126**, 1–44.

—— 1964. The personal factor in accidents—a review article. *Jl R. statist. Soc.* A **127**, 438–51.

JAMES, G. S. 1952. Notes on a theorem of Cochran. *Proc. Camb. phil. Soc. math. phys. Sci.* **48**, 443–6.

JAMES, W. and STEIN, C. 1961. Estimation with quadratic loss. *Proceedings of the fourth Berkeley symposium on mathematical statistics and probability*, Vol. II, pp. 361–79. University of California Press, California.

JAMES-LEVY, G. E. 1956. A nomogram for the integral law of Student's distribution. *Theory of probability and its applications* **1**, 246–8.

JEFFREYS, H. 1951. *Theory of probability*, 3rd edn. Cambridge University Press.

JENSEN, A. 1950. *Mòe's principle.* K.T.A.S., Copenhagen.

JOHN, P. W. M. 1961. A note on the quadratic birth process. *J. Lond. math. Soc.* **36**, 159–60.

JOHNSON, N. L. 1949a. Systems of frequency curves generated by methods of translation. *Biometrika* **36**, 149–76.

—— 1949b. Bivariate distributions based on simple translation systems. *Biometrika* **36**, 297–304.

——1959. On an extension of the connection between Poisson and χ^2 distributions. *Biometrika* **46**, 352–63.

—— 1960. An approximation to the multinomial distribution: some properties and applications. *Biometrika* **47**, 93–102.

—— and YOUNG, D. H. 1960. Some applications of two approximations to the multinomial distribution. *Biometrika* **47**, 463–9.

JOHNSON, W. E. 1921–4. *Logic*, 3 Vols. Cambridge University Press.

JOSEPH, A. W. 1959. A note on the paper by D. D. Kosambi and U. V. Ramamohana Rao on 'the efficiency of randomization by card shuffling'. *Jl R. statist. Soc.* A **122**, 373–4.

KAARSEMAKER, L. and VAN WIJNGAARDEN, A. 1953. Tables for use in rank correlation. *Statistica Neerlandica* **7**, 41–54.

KABE, D. G. 1958. Some applications of Meijer *G*-functions to distribution problems in statistics. *Biometrika* **45**, 578–80.

KAC, M. 1945. A remark on the independence of linear and quadratic forms involving independent Gaussian variates. *Ann. math. Statist.* **16**, 400–1.

—— 1947. Random walk and the theory of Brownian motion. *Am. math. Mon.* **54**, 369–91.

—— 1959*a*. *Probability and related topics in physical sciences*. Interscience, London.

—— 1959*b*. *Statistical independence in probability, analysis and number theory. Carus Mathematical Monographs* No. 12. Mathematical Association of America (Wiley).

—— 1959*c*. Some remarks on stable processes with independent increments. *Probability and statistics: the Harald Cramér volume*. Wiley, New York.

KAMAT, A. R. 1958*a*. Hypergeometric expansion for incomplete moments of the bivariate normal distribution. *Sankhyā* **20**, 317–20.

—— 1958*b*. Incomplete moments of the trivariate normal distribution. *Sankhyā* **20**, 321–2.

KAMKE, E. 1932. *Einführung in die Wahrscheinlichkeitsrechnung*. Hirzel, Leipzig.

KARLIN, S. and MCGREGOR, J. L. 1957*a*. The differential equations of birth and death processes, and the Stieltjes moment problem. *Trans. Am. math. Soc.* **85**, 489–546.

—— —— 1957*b*. The classification of birth and death processes. *Trans. Am. math. Soc.* **86**, 366–400.

—— —— 1958. Linear growth, birth and death processes. *J. Math. Mech.* **7**, 643–62.

KATTI, S. S. and GURLAND, J. 1961. The Poisson–Pascal distribution. *Biometrics* **17**, 527–38.

KATZ, L. 1952. The distribution of the number of isolates of a social group. *Ann. math. Statist.* **23**, 271–6.

KAWADA, Y. 1950. Independence of quadratic forms in normally correlated variables. *Ann. math. Statist.* **21**, 614–15.

KEMENY, J. G. and SNELL, J. L. 1960. Finite Markov chains. van Nostrand, Princeton, New Jersey.

KEMP, C. D. and KEMP, A. W. 1956. Generalized hypergeometric distributions. *Jl R. statist. Soc.* B **18**, 202–11.

KEMPERMAN, J. H. B. 1961. The passage problem for a stationary Markov chain. University of Chicago Press, Chicago.

KENDALL, D. G. 1948. On some modes of population growth leading to R. A. Fisher's logarithmic series distribution. *Biometrika* **35**, 6–15.

—— 1949. Stochastic processes and population growth. *Jl R. statist. Soc.* **11**, 230–64.

—— 1950*a*. An artificial realization of a simple 'birth and death' process. *Jl R. statist. Soc.* **12**, 116–19.

—— 1950*b*. Random fluctuations in the age-distribution of a population whose development is controlled by the simple birth-and-death process. *Jl. R. statist. Soc.* B **12**, 278–85.

—— 1951. Some problems in the theory of queues. *Jl R. statist. Soc.* **13**, 151–85.

KENDALL, M. G. 1938. A new measure of rank correlation. *Biometrika* **30**, 81–93.

—— 1941. Proof of relations connected with the tetrachoric series and its generalizations. *Biometrika* **32**, 196–8.

—— 1949. On the reconciliation of theories of probability. *Biometrika* **36**, 101–16.

—— 1955. *Rank correlation methods*, 2nd edn. Griffin, London.

KENDALL, M. G. 1956. Studies in the history of probability and statistics. II. The beginnings of a probability calculus. *Biometrika* **43**, 1–14.

—— 1957. Studies in the history of probability and statistics. V. A note on playing cards. *Biometrika* **44**, 260–2.

—— 1961. Studies in the history of probability and statistics. XII. The book of fate. *Biometrika* **48**, 220–2.

—— 1963. Studies in the history of probability and statistics. XIII. Isaac Todhunter's history of the theory of probability. *Biometrika* **50**, 204–5.

—— and BABINGTON-SMITH, B. 1938. Randomness and random sampling numbers. *Jl. R. statist. Soc.* **101**, 147–66.

—— —— 1939*a*. Second paper on random sampling numbers. *Jl R. statist, Soc.* Suppl. **6**, 51–61.

—— —— 1939*b*. *Tables of random sampling numbers. Tracts for computers* No. 24. Cambridge University Press.

—— KENDAL, S. F. H. and BABINGTON-SMITH, B. 1939. The distribution of Spearman's coefficient of rank correlation in a universe in which all rankings occur an equal number of times. *Biometrika* **30**, 251–73.

—— and MORAN, P. A. P. 1963. *Geometrical probability.* Griffin, London.

—— and MURCHLAND, J. D. 1964. Statistical aspects of the legality of gambling. *Jl R. statist. Soc.* A **127**, 359–91.

KERRIDGE, D. A. 1965. A probability derivation of the non-central χ^2 distribution. *Aust. J. statist.* **7**, 37–9.

KEYNES, J. M. 1921. *A treatise on probability.* Macmillan, London.

KHAMIS, S. H. 1964. New tables of the chi-squared integral. *Bull. Inst. int. statist.* **40**, Book 2, 799–823.

KHINTCHINE, A. YA. 1933. *Asymptotische Gesetze der Wahrscheinlichkeitsrechnung.* Springer, Berlin (reprinted by Chelsea, New York).

—— 1937. Contribution à l'arithmétique des lois de distribution. *Bull. math. Univ. Moscou.* Sect. A **1**, 6–17.

—— 1938. Theory of fading spontaneous effects. *Izv. Acad. Nauk SSSR* Sér Math. No. 3, 313–22.

—— 1950. *Analytical foundations of physical statistics.* Akad. Nauk SSSR (trans. into English and published by the Hindustani Publishing Corporation, Delhi, India 1961).

—— 1957. *Mathematical foundations of information theory.* Dover, New York.

KIMBALL, A. W. and LEACH, E. 1959. Approximate linearization of the incomplete β-function. *Biometrika* **46**, 214–18.

KINGMAN, J. F. C. 1962. The imbedding problem for finite Markov chains. *Z. Wahrscheinlichkeitstheorie* **1**, 14–24.

—— 1963*a*. Random walks with spherical symmetry. *Acta math. Stockh.* **109**, 11–53.

—— 1963*b*. On inequalities of Tchebychev type. *Proc. Camb. phil. Soc. math. phys. Sci.* **59**, 135–46.

—— 1963*c*. Ergodic properties of continuous time Markov processes and their discrete skeletons. *Proc. Lond. math. Soc.* **13**, 593–604.

—— and TAYLOR, S. J. 1966. *Introduction to measure and probability.* Cambridge University Press.

KIRKHAM, W. J. 1935. Moments about the arithmetic mean of a binomial frequency distribution. *Ann. math. Statist.* **6**, 96–101.

KLEIN, G. 1952. Mean first passage times of Brownian motion and related problems. *Proc. R. Soc.* A **211**, 431–43.

KLUYVER, J. C. 1906. A local probability problem. *Proc. K. ned. Akad. Wet.* **8**, 341–50.

KOLMOGOROV, A. N. 1928, 1930. Ueber die Summen durch den Zufall bestimmter unabhängigen Grossen. *Math. Annln* **99**, 309–19; **102**, 484–8.

—— 1929. Ueber das Gesetz des iterierten Logarithmus. *Math. Annln* **101**, 126–35.

—— 1931. Über die analytische Methoden in der Wahrscheinlichkeitsrechnung. *Math. Annln* **104**, 415–58.

—— 1932. Sulla forma generale di una processo stochastico omogeneo. *Atti Accad. naz. Lincei. Cl. Sci. Fis. Mat. Nat.* **15**, 805–8, 866–9.

—— 1933. Grundbegriffe der Wahrscheinlichkeitsrechnung. *Ergebnisse der Mathematik.* Springer, Berlin. (English edition, Chelsea, New York, 1950).

—— 1936. Anfangsgründe der Markoffschen Ketten mit unendlich vielen möglichen Zuständen. *Mat. Sb.* **1**, 607–10.

—— 1941. Über das logarithmisch normale Verteilungsgesetz der Dimensionen der Teilchen bei Zerstückelung. *Dokl. Akad. Nauk SSSR* **31**, 99–101.

KOSAMBI, D. D. and RAO, U. V. R. 1958. The efficiency of randomization by card shuffling. *Jl R. statist. Soc.* A **121**, 223–33.

KOSTELYANEC, P. O. and RESETNYAK, YU. G. 1954. The determination of a completely additive function by its values on half-spaces. *Usp. mat. Nauk* **9**, 135–40.

KRASNER, M. and RANULAC, B. 1937. Sur une propriété des polynômes de la division du cercle. *C. r. hebd. Séanc. Acad. Sci., Paris* **204**, 397–9.

KRISHNA IYER, P. V. 1954. Some distributions arising in matching problems. *J. Indian Soc. agric. Statist.* **6**, 5–29.

KULLBACK, S. 1936. The distribution laws of the difference and quotient of variables independently distributed in Pearson type III laws. *Ann. math. Statist.* **7**, 51–3.

—— 1959. *Information theory and statistics.* Wiley, New York.

—— KUPPERMAN, M., and KU, H. H. 1961. An application of information theory to the analysis of contingency tables, with a table of $2n \log n$, $n = 1(1) 10,000$. *J. Res. natn. Bur. Stand.* **66**B, 217–43.

LAHA, R. G. 1954. On a characterization of the gamma distribution. *Ann. math. Statist.* **25**, 784–7.

—— 1959. On the laws of Cauchy and Gauss. *Ann. math. Statist.* **30**, 1165–74.

LAMPERTI, J. 1959. Problem 4841. *Am. math. mon.* **66**, 317.

LANCASTER, H. O. 1954. Traces and cumulants of quadratic forms in normal variables. *Jl R. statist. Soc.* B **16**, 247–54.

—— 1960. The characterization of the normal distribution. *J. Aust. math. Soc.* **1**, 368–83.

LAPLACE, P. S. 1812. *Théorie analytique des probabilités.* Paris. (Four supplements published later—see collected works 1847, 1886).

LARGUIER, E. H. 1936. On a method for evaluating the moments of a Bernoulli distribution. *Ann. math. Statist.* **7**, 191–5.

LASALLE, J. P. 1962. *Recent Soviet contributions to mathematics.* Macmillan, New York.

LEDERMANN, W. 1950. On the asymptotic probability distribution for certain Markov processes. *Proc. Camb. phil. Soc. math. phys. Sci.* **46**, 581–94 (correction **47**, 626).

—— and REUTER, G. E. H. 1954. Spectral theory for differential equations of simple birth and death processes. *Phil. Trans. R. Soc.* A **246**, 321–69.

LEHMAN, R. S. 1951. A problem on random walk. *Proceedings of the second Berkeley symposium on mathematical statistics and probability*, pp. 263–8. University of California Press, California.

LEHMAN, R. S., and WEISS, G. H. 1958. A study of the restricted random walk. *J. Soc. indust. appl. Math.* **6**, 257–78.

LÉVY, P. 1937a. *Théorie de l'addition des variables aléatoires.* Paris.

—— 1937b. Sur les exponentielles des polynômes. *Annls scient. Ec. norm., Paris* sup. **54**, 231–92.

—— 1939a. Sur certains processus stochastiques homogènes. *Compositio math.* **7**, 283–339.

—— 1939b. L'addition des variables aléatoires définies sur un circonférence. *Bull. de Soc. math. de Fr.* **67**, 1–41.

—— 1940. Le mouvement brownien plan. *Am. J. Math.* **62**, 487–550.

—— 1944a. Une propriété d'invariance projective dans le mouvement brownien. *C. r. hebd. Séanc. Acad. Sci., Paris* **219**, 378–9.

—— 1944b. Un théorème d'invariance projective relatif au mouvement brownien. *Comment. math. helvet.* **16**, 242–8.

—— 1948a. *Processus stochastiques et mouvement brownien.* Gauthier–Villars, Paris.

—— 1948b. The arithmetical character of the Wishart distribution. *Proc. Camb. phil. Soc. math. phys. Sci.* **44**, 295–7.

LIAPOUNOV, A. 1901. Nouvelle forme du théorème sur la limite de la probabilité. *Mém. Acad. Sci. St. Petersburg* **12**, 1–24.

LIEBERMANN, G. J. and OWEN, D. B. 1961. *Tables of the hypergeometric distribution.* Stanford University Press.

LINDEBERG, J. W. 1922. Eine neue Herleitung des Exponentialgesetz in der Wahrscheinlichkeitsrechnung. *Math. Z.* **15**, 211–25.

LINNIK, YU. V. 1956. A remark on Cramér's theorem on the decomposition of the normal law. *Theory of probability and its applications* **1**, 435–6.

—— 1957. On factorizing the composition of a Gaussian and a Poissonian law. *Theory of probability and its applications* **2**, 31–57.

—— 1961. On the probability of large deviations for the sums of random variables. *Proceedings of the fourth Berkeley symposium on mathematical statistics and probability*, Vol. II, pp. 289–306. University of California Press, California.

—— 1964. *Decompositions of probability distributions* (ed. S. J. Taylor). Oliver and Boyd, London.

LITTLEWOOD, J. E. 1944. *Theory of functions.* Oxford University Press.

LOÉVE, M. 1950. Fundamental limit theorems of probability theory. *Ann. math. Statist.* **21**, 321–338.

—— 1963. *Probability theory*, 3rd edn. van Nostrand, New York.

LOMNICKI, Z. A. and ZAREMBA, S. K. 1957. A further instance of the central limit theorem for dependent random variables. *Math. Z.* **66**, 490–4.

LORD, R. D. 1954a. The use of the Hankel transform in statistics. I. General theory and examples. *Biometrika* **41**, 44–55.

—— 1954b. The use of the Hankel transform in statistics. II. Methods of computation. *Biometrika* **41**, 344–50.

LOTKA, A. J. 1931. The extinction of families. *J. Wash. Acad. Sci.* **21**, 377–80, 453–59.

—— 1939. A contribution to the theory of self-renewing aggregates, with special reference to industrial replacement. *Ann. math. Statist.* **10**, 1–25.

—— 1945. Population analysis as a chapter in the mathematical theory of evolution. In *Essays in growth and form presented to D'Arcy Wentworth Thompson* (ed. W. E. Le Gros Clark and P. B. Medawar). Clarendon Press, Oxford.

LOWAN, A. N. and LEDERMANN, J. 1939. On the distribution of errors in nth tabular differences. *Ann. math. Statist.* **10**, 360–4.

LUKACS, E. 1942. A characterization of the normal distribution. *Ann. math. Statist.* **13**, 91–3.

—— 1955. A characterization of the gamma distribution. *Ann. math. Statist.* **26**, 319–24.

—— 1958. Some extensions of a theorem of Marcinkiewicz. *Pacif. J. Math.* **8**, 487–501.

—— 1960. *Characteristic functions. Griffin's statistical monographs* No. 5, Griffin, London.

—— 1961. Recent developments in the theory of characteristic functions. *Proceedings of the fourth Berkeley symposium on mathematical statistics and probability,* Vol. II, pp. 307–35. University of California Press, California.

—— and KING, E. P. 1954. A property of the normal distribution. *Ann. math. Statist.* **25**, 389–94.

—— and LAHA, R. G. 1964. *Applications of characteristic functions. Griffin's statistical monographs* No. 14. Griffin, London.

McCREA, W. H. 1936. A problem on random paths. *Mathl Gaz.* **20**, 311–17.

—— and WHIPPLE, F. J. W. 1940. Random paths in two and three dimensions. *Proc. R. Soc. Edin.* **60**, 281–98.

MacDONALD, D. K. C. 1952. Information theory and its application to taxonomy. *J. appl. Phys.* **23**, 529–31.

MACEDA, E. C. 1948. On the compound and generalized Poisson distributions. *Ann. math. Statist.* **19**, 414–16.

McFADDEN, J. A. 1960. Two expansions for the quadrivariate normal integral. *Biometrika* **47**, 325–33.

McKENDRICK, A. G. 1914. Studies in the theory of continuous probabilities with special reference to its bearing on natural phenomena of a progressive nature. *Proc. Lond. math. Soc.* (2) **13**, 401–16.

—— 1926. Applications of mathematics to medical problems. *Proc. Edinb. Math. Soc.* **44**, 98–130.

MacMAHON, P. A. 1915–16. *Combinatory analysis,* 2 Vols. Cambridge University Press.

MALLOWS, C. L. 1956. Generalizations of Tchebycheff's inequalities. *Jl R. statist. Soc.* B **18**, 139–76.

—— 1963. A generalization of the Chebyshev inequalities. *Proc. Lond. math. Soc.* **13**, 385–412.

MARCINKIEWICZ, J. 1938. Sur une propriété de la loi de Gauss. *Math. Z.* **44**, 612–18.

MARITZ, J. S. 1952. Note on a certain family of discrete distributions. *Biometrika* **39**, 196–8.

MARKOV, A. A. 1884. On certain applications of algebraic continued fractions (thesis in Russian). St Petersburg.

—— 1912. *Wahrscheinlichkeitsrechnung.* Taubner, Leipzig.

MARSAGLIA, G. 1954. Iterated limits and the central limit theorem for dependent variables. *Proc. Am. math. Soc.* **5**, 987–91.

—— 1961*a*. Expresssing a random variable in terms of uniform random variables. *Ann. math. Statist.* **32**, 894–8.

—— 1961*b*. Generating exponential random variables. *Ann. math. Statist.* **32**, 899–900.

MARSHALL, A. W. and OLKIN, I. 1960. Multivariate Chebyshev inequalities. *Ann. math. Statist.* **31**, 1001–14.

MARTIN, J. L. 1962. The exact enumeration of self-avoiding walks on a lattice. *Proc. Camb. phil. Soc. math. phys. Sci.* **58**, 92–101.

MAULDON, J. G. 1956. Characterizing properties of statistical distributions. *Q. Jl Math.* (2) **7**, 155–60.

—— 1959. A generalization of the beta-distribution. *Ann. math. Statist.* **30**, 509–20.

MERRINGTON, M. and THOMPSON, C. M. 1943. Tables of percentage points of the inverted beta (*F*) distribution. *Biometrika* **33**, 73–88.

MEYER, H. A. (ed.) 1956. *Symposium on Monte Carlo methods.* Wiley, New York.

MIDAS. 1964. *How to win at this and that.* Collins, Fontana books.

MIDDLETON, D. 1960. *An introduction to statistical communication theory.* McGraw-Hill, New York.

MILLER, J. C. P. 1950. Errors in Fisher and Yates tables. *Mathl. Tabl. natn. Res. Coun., Wash.* **4**, 27.

—— (ed.) 1954. *Tables of binomial coefficients. Royal Society mathematical tables,* 3. Cambridge University Press.

MIRSKY, L. 1955. *An introduction to linear algebra.* Clarendon Press, Oxford.

MOLINA, E. C. 1930. The theory of probability: some comments on Laplace's Théorie Analytique. *Bull. Am. math. Soc.* **36**, 369–92.

—— 1942. *Poisson's exponential binomial limit.* van Nostrand, New York.

MONTMORT, P. R. 1708. *Essai d'analyse sur les jeux de hasard.* Paris.

MONTROLL, E. W. 1956. Random walks in multi-dimensional spaces, especially on periodic lattices. *J. Soc. indust. appl. Math.* **4**, 241–60.

—— 1964. Random walks on lattices. *Proceedings of symposia in applied mathematics,* Vol. 16, pp. 193–220. American Mathematical Society.

MOOD, A. M. 1940. The distribution theory of runs. *Ann. math. Statist.* **11**, 367–92.

MORAN, P. A. P. 1948*a*. The statistical distribution of the length of a rubber molecule. *Proc. Camb. phil. Soc. math. phys. Sci.* **44**, 342–4.

—— 1948*b*. Rank correlation and product-moment correlation. *Biometrika* **35**, 203–6.

—— 1951. A mathematical theory of animal trapping. *Biometrika* **38**, 307–11.

—— 1952. A characteristic property of the Poisson distribution. *Proc. Camb. phil. Soc. math. phys. Sci.* **48**, 206–7.

—— 1954. The dilution assay of viruses. *J. Hyg., Camb.* **52**, 189–93.

—— 1956. The numerical evaluation of a class of integrals. *Proc. Camb. phil. Soc. math. phys. Sci.* **52**, 230–3.

—— 1958. Approximate relations between series and integrals. *Mathl Tabl. natn. Res. Coun., Wash.* **12**, 34–7.

—— 1961. Entropy, Markov chains and Boltzmann's H-theorem. *Proc. Camb. phil. Soc. math. phys. Sci.* **57**, 833–42.

—— 1962*a*. *The statistical processes of evolutionary theory.* Clarendon Press, Oxford.

—— 1962*b*. Polysomic inheritance and the theory of shuffling. *Sankhyā* **24**, 63–72.

—— 1964. On the range of cumulative sums. *Ann. Inst. statist. Math.* **16**, 109–12.

—— 1966. Accurate approximations for t-tests. *Essays presented to J. Neyman.* Wiley, London.

MORLAT, G. 1952. Sur une généralisation de la loi de Poisson. *C. r. hebd. Séanc. Acad. Sci., Paris* **235**, 933–5.

MOSES, L. E. and OAKFORD, R. V. 1963. *Tables of random permutations.* Stanford University Press.

MOSIMAN, J. E. 1962. On the compound multinomial distribution, the multivariate β-distribution, and correlations among proportions. *Biometrika* **49**, 65–82.

—— 1963. On the compound negative multinomial distribution and correlations among inversely sampled pollen counts. *Biometrika* **50**, 47–54.

MOYAL, J. E. 1949. Stochastic processes and statistical physics. *Jl R. statist. Soc.* B **11**, 150–210.

—— 1957. Discontinuous Markov processes. *Acta Math., Stockh.* **98**, 221–64.

MÜNSTER, A. 1950. On the theory of grand partition functions. *Proc. Camb. phil. Soc. math. phys. Sci.* **46**, 319–30.

MYCIELSKI, J. and PASZKOWSKI, S. 1956. Sur un problème du calcul de probabilité. I. *Studia math.* **15**, 188–200.

NABEYA, S. 1961. Absolute and incomplete moments of the multivariate normal distribution. *Biometrika* **48**, 77–84.

NASH-WILLIAMS, C. ST J. 1959. Random walk and electric current in networks. *Proc. Camb. phil. Soc. math. phys. Sci.* **55**, 181–94.

NATIONAL BUREAU OF STANDARDS 1950. Tables of the binomial probability distribution. *Appl. math. Ser.* **6**.

—— 1952. A guide to tables of the normal probability integral. *Appl. math. Ser.* **21**.

—— 1959. Tables of the bivariate normal distribution and related functions. *Appl. math. Ser.* **50**.

NEYMAN, J. 1939. On a new class of 'contagious' distributions applicable in entomology and bacteriology. *Ann. math. Statist.* **10**, 35–57.

NICHOLSON, C. 1943. The probability integral for two variables. *Biometrika* **33**, 59–72.

NICHOLSON, W. L. 1954. A computing formula for the power of analysis of variance tests. *Ann. math. Statist.* **25**, 607–10.

—— 1956. On the normal approximation to the hypergeometric distribution. *Ann. math. Statist.* **27**, 471–83.

—— 1961. Occupancy probability distribution critical points. *Biometrika* **48**, 175–80.

OLKIN, I. and PRATT, J. W. 1958. A multivariate Tchebycheff inequality. *Ann. math. Statist.* **29**, 226–34.

Ordnance Corps, U.S. Army 1952. *Tables of the cumulative binomial probabilities.*

OTTER, R. 1949. The multiplicative process. *Ann. math. Statist.* **20**, 206–24.

OWEN, D. B. 1957. *The bivariate normal probability distribution.* Sandia Corporation.

—— 1962. *Handbook of statistical tables.* Addison-Wesley, Mass.

PACKER, L. R. 1950. The distribution of the sum of n rectangular variates. I. *J. Inst. act. Stud. Soc.* **10**, 52–61.

PALM, C. 1947, 1954. *Tables of the Erlang loss formula.* Fritz, Stockholm.

PARZEN, E. 1962. On estimation of a probability density function and mode. *Ann. math. Statist.* **33**, 1065–76.

PASZKOWSKI, S. 1956. Sur un problème du calcul de probabilité. II. *Studia math.* **15**, 273–98.

PATIL, G. P., KAMAT, A. R., and WANI, J. K. 1964. *Certain studies on the structure and statistics of the logarithmic series distribution and related tables.* Office of Aerospace Research, United States Air Force.

—— and SESHADRI, V. 1964. Characterization theorems for some univariate probability distributions. *Jl R. statist. Soc.* B **26**, 286–92.

PATNAIK, P. B. 1949. The non-central χ^2 and F-distributions and their applications. *Biometrika* **36**, 202–32.

PEARSON, E. S. and HARTLEY, H. O. 1956. *Biometrika tables for statisticians*, Vol. I. Cambridge University Press.

PEARSON, K. 1922. *Tables of the incomplete gamma function.* Biometrika Office, University College, London.

PEARSON, K. 1930, 1931. *Tables for statisticians and biometricians*, Part I, 3rd edn.; Part II. Cambridge University Press.

—— 1948. *Tables of the incomplete beta function*. Biometrika Office. University College, London.

—— and BLAKEMAN, J. 1906. Mathematical contributions to the theory of evolution. XV. A mathematical theory of random migration. *Draper's company research memoirs*, Biometric Series, III.

PERRIN, F. 1928. Etude mathématique du mouvement brownien de rotation. *Annls scient. Ec. norm. sup., Paris* **45**, 1–51.

PETERS, J. 1922. *Zehnstellige Logarithmentafel*. Reichsamt für Landesaufnahme, Berlin. Reprinted in 1957 by Ungar, New York.

PIERCE, J. A. 1940. A study of the universe of n finite populations with application to moment-function adjustments for grouped data. *Ann. math. Statist.* **11**, 311–34.

PITT, H. R. 1963. *Integration, measure and probability*. Oliver and Boyd, Edinburgh.

PITMAN, E. J. 1956. On the derivatives of a characteristic function at the origin. *Ann. math. Statist.* **27**, 1156–60.

PLACKETT, R. L. 1954. A reduction formula for normal multivariate integrals. *Biometrika* **41**, 351–60.

—— 1958. Studies in history of probability and statistics. VII. The principle of the arithmetic mean. *Biometrika* **45**, 130–5.

—— 1960. *Principles of regression analysis*. Clarendon Press, Oxford.

POLLACZEK, F. 1952. Fonctions charactéristiques de certaines répartitions définies au moyen de la notion d'ordre. Applications à la théorie des attentes. *C. r. hebd. Séanc. Acad. Sci., Paris* **234**, 2334–6.

POLLARD, H. 1946. The representation of $e^{-x^{\lambda}}$ as a Laplace integral. *Bull. Am. math. Soc.* **52**, 908–10.

POLLARD, S. 1921. The Stieltjes integral and its generalizations. *Q. Jl Math.* **49**, 73–138.

PÓLYA, G. 1921. Über eine Aufgabe der Wahrscheinlichkeitsrechnung betreffend die Irrfahrt in Strassennetz. *Math. Annln* **84**, 149–60.

—— 1930. Sur quelques points de la théorie des probabilités. *Annls Inst. h. Poincaré.* **1**, 117–61.

—— 1949a. Remarks on computing the probability integral in one and two dimensions. *Proceedings of the first Berkeley symposium on mathematical statistics and probability*, pp. 63–78. University of California Press, California.

—— 1949b. Remarks on characteristic functions. *Proceedings of the first Berkeley symposium on mathematical statistics and probability*, pp. 115–23. University of California Press, California.

—— and SZEGÖ, G. 1925. *Aufgaben und Lehrsätze aus den Analysis*. Vols. 1 and 2. *Grundlehren der mathematischen Wissenschaften*, Band XIX, XX. Springer, Berlin.

PRESTON, F. W. 1948. The commonness, and rarity, of species. *Ecology* **29**, 254–83.

PROKHOROV, YU. V. 1966. Characterization theorems. *Proceedings of the fifth Berkeley symposium on mathematical statistics and probability*. University of California Press, California.

QUENOUILLE, M. H. 1949a. The evaluation of probabilities in a normal multivariate distribution with special reference to the correlation ratio. *Proc. Edinb. math. Soc.* (2) **8**, 95–100.

—— 1949b. A relation between the logarithmic, Poisson, and negative binomial distributions. *Biometrics* **5**, 162–4.

QUENOUILLE, M. H. 1959. Tables of random observations from standard distributions. *Biometrika* **46**, 178–202.

RAIKOV, D. A. 1938. On the decomposition of the Gaussian and Poisson laws. *Izv. Akad. Nauk SSSR* Ser. Math. 91–124.

—— 1939. On the composition of analytic distribution functions. *Dokl. Akad. Nauk. SSSR* **23**, 511–14.

RAJSKI, C. 1961. A metric space of discrete probability distributions. *Inf. Control* **4**, 371–7.

RAMSEY, F. P. 1931. Truth and probability (1926). Posthumous paper published in *The foundations of mathematics and other logical essays* by F. P. Ramsey. Kegan Paul, London.

RAND CORPORATION. 1955. *A million random digits with* 100,000 *normal deviates.* The Free Press, Glencoe, Illinois.

RAO, C. R. 1961. Generation of random permutations of given number of elements using random sampling methods. *Sankhyā* A **23**, 305–7.

RAO, K. S. and KENDALL, D. G. 1950. Second limit theorem in the calculus of probabilities. *Biometrika* **37**, 224–40.

RAYLEIGH, LORD 1919. On the problem of random vibrations, and of random flights in one, two, or three dimensions. *Phil. Mag.* (6th Series) **37**, 321–47.

REDHEFFER, R. 1960. The Mycielski–Paszkowski diffusion problem. *J. Math. Mech.* **9**, 607–21.

—— 1961. Difference equations and functional equations in transmission line theory. *Modern mathematics for the engineer*, second series. McGraw-Hill, New York.

RENYI, A. 1955. On a new axiomatic theory of probability. *Acta math. hung.* **6**, 285–335.

—— 1962. *Wahrscheinlichkeitsrechnung.* Deutscher Verlag der Wissenschaften, Berlin.

RICE, S. O. 1944, 1945. Mathematical analysis of random noise. *Bell Syst. tech. J.* **23**, 282–332; **24**, 46–156.

RICHTER, H. 1956. *Wahrscheinlichkeitsrechnung.* Springer, Berlin.

RIDER, P. R. 1962. Expected values and standard deviations of the reciprocal of a variable from a decapitated negative binomial distribution. *J. Am. statist. Ass.* **57**, 439–45.

RIORDAN, J. 1958. *An introduction to combinatorial analysis.* Wiley, New York.

ROBBINS, H. 1948a. Convergence of distributions. *Ann. math. Statist.* **19**, 72–6.

—— 1948b. The asymptotic distribution of the sum of a random number of random variables. *Bull. Am. math. Soc.* **54**, 1151–61.

—— 1948c. The distribution of a definite quadratic form. *Ann. math. Statist.* **19**, 266–70.

—— 1952. A note on gambling systems and birth statistics. *Am. math. Mon.* **59**, 685–6.

—— 1955. A remark on Stirling's formula. *Am. math. Mon.* **62**, 26–9.

—— and PITMAN, E. J. G. 1949. Application of the method of mixtures to quadratic forms in normal variates. *Ann. math. Statist.* **20**, 552–60.

ROBERTS, P. H. and URSELL, H. D. 1959–60. Random walk on a sphere and on a Riemannian manifold. *Phil. Trans. R. Soc.* A **252**, 317–56.

ROMIG, H. C. 1953. 50–100 *Binomial tables.* Wiley, New York.

ROSENBAUM, S. 1961. Moments of a truncated bivariate normal distribution. *Jl R. statist. Soc.* B **23**, 405–8.

ROSENBLATT, M. 1956. Remarks on some non-parametric estimates of a density function. *Ann. math. Statist.* **27**, 832–7.

—— 1961. Independence and dependence. *Proceedings of the fourth Berkeley symposium on mathematical statistics and probability*, Vol. II, pp. 431–43. University of California Press, California.

—— 1962. *Random processes*. Oxford University Press, New York.

ROWLAND, E. N. 1936. The theory of the mean square variation of a function formed by adding known functions with random phases, and applications to the theories of the shot effect and of light. *Proc. Camb. phil. Soc. math. phys. Sci.* **32**, 580–97.

—— 1937. The theory of the shot effect. II. *Proc. Camb. phil. Soc. math. phys. Sci.* **33**, 344–58.

ROY, S. N. 1957. Some aspects of multivariate analysis. Wiley, New York.

ROYDEN, H. L. 1953. Bounds on a distribution function when its first n moments are given. *Ann. math. Statist.* **24**, 361–76.

RUBEN, H. 1954. On the moments of order statistics in samples from normal populations. *Biometrika* **41**, 200–27 (correction **41**, (3–4), ix).

—— 1960. The probability content of regions under spherical normal distributions. I. *Ann. math. Statist.* **31**, 598–618.

—— 1961a. A power series expansion for a class of Schläfli functions. *J. Lond. math. Soc.* **36**, 69–77.

—— 1961b. On the numerical evaluation of a class of multivariate normal integrals. *Proc. R. Soc. Edinb.* **65**, 272–81.

—— 1961c. The probability content of regions under spherical normal distributions. III. The bivariate normal integral. *Ann. math. Statist.* **32**, 171–86.

—— 1962a. An asymptotic expansion for a class of multivariate normal integrals. *J. Aust. math. Soc.* **2**, 253–64.

—— 1962b. A new asymptotic expansion for the normal probability integral and Mill's ratio. *Jl R. statist. Soc.* **24**, 177–9.

—— and TUCKER, H. G. 1959. Estimating the parameters of a differential process. *Ann. math. Statist.* **30**, 641–58.

RUSHBROOKE, G. S. and EVE, J. 1959. On non-crossing lattice polygons. *J. Chem. Phys.* **31**, 1333–4.

SAATY, T. L. 1961. *Elements of queueing theory*. McGraw-Hill, New York.

SAKS, S. 1937. *Theory of the integral*. Stechert, New York.

SALZER, H. E. 1951. Tables of $n!$ and $\Gamma(n+\frac{1}{2})$ for the first thousand values of n. *Appl. math. Ser.* **16**. National Bureau of Standards, Washington.

SAMPFORD, M. R. 1953. Some inequalities on Mill's ratio and related functions. *Ann. math. Statist.* **24**, 130–2.

SANDELIUS, M. 1962. A simple randomization procedure. *Jl R. statist. Soc.* B **24**, 472–81.

SAPOGOV, N. A. 1951. The stability problem for a theorem of Cramér (in Russian). *Izv. Akad. Nauk SSSR*, Ser. Mat. **15**, 205–18. (Select translation in Math. Stat. Prob., Vol. I. Am. Math. Soc. and Inst. of Math. Stat. 1961.)

SARKADI, K. 1957a. On the distribution of the number of exceedances. *Ann. math. Statist.* **28**, 1021–3.

—— 1957b. Generalized hypergeometric distributions. *Nagy. Tud. Akad. Mat. Kut. Int. Közl.* **2**, 59–69.

SAVAGE, I. R. 1961. Probability inequalities of the Tchebycheff type. *J. Res. natn. Bur. Stand.* Section B **65B**, 211–22.

SAWKINS, D. T. 1940. Elementary presentation of the frequency distributions of certain statistical populations associated with the normal population. *J. Proc. R. Soc. N.S.W.* **74**, 209–39.

SCHLÄFLI, L. 1858, 1860. *Q. Jl Math.* **2**, 269–301; **3**, 54–68, 97–108.

SCHRODINGER, E. 1915. Zur Theorie der Fall– und Steigversuche an Teilchen mit Brownscher Bewegung. *Phys. Z.* **16**, 289–95.

—— 1957. A combinatorial problem in counting cosmic rays. *Proc. phys. Soc.* A **64**, 1040–2.

SEAL, H. L. 1950–51. Spot the prior reference. *J. Inst. Act. Stud. Soc.* **10**, 255–8.

SEDGWICK, W. F. 1942. On the theory of successive radioactive transformations. *Proc. Camb. phil. Soc. math. phys. Sci.* **38**, 280–9.

SETH, A. 1963. The correlated unrestricted random walk. *Jl R. statist. Soc.* B **25**, 394–400.

SEVERO, N. C. and ZELEN, M. 1960. Normal approximation to the chi-square and non-central F-probability functions. *Biometrika* **47**, 411–16.

SHANNON, C. E. 1948. The mathematical theory of communication. *Bell Syst. tech. J.* **27**, 379–423, 623–56.

—— 1961. Two-way communication channels. *Proceedings of the fourth Berkeley symposium on mathematical statistics and probability*, Vol. I, pp. 611–44. University of California Press, California.

SHENTON, L. R. 1954. Inequalities for the normal integral including a new continued fraction. *Biometrika* **41**, 177–89.

SHEPPARD, W. F. 1898. On the application of the theory of error to cases of normal distributions and normal correlations. *Phil. Trans. R. Soc.* A **192**, 101–67.

SHOHAT, J. A. and TAMARKIN, J. D. 1943. *The problem of moments. Mathematical surveys* No.1. American Mathematical Society, New York.

SKELLAM, J. G. 1946. The frequency distribution of the difference between two Poisson variates belonging to different populations. *Jl R. statist. Soc.* **109**, 296.

—— 1948. A probability distribution derived from the binomial distribution by regarding the probability of success as variable between the sets of trials. *Jl. R. statist. Soc.* B **10**, 257–61.

—— 1952. Studies in statistical ecology. I. Spatial pattern. *Biometrika* **39**, 346–62.

—— 1958. On the derivation and applicability of Neyman's type A distribution. *Biometrika* **45**, 32–6.

—— and SHENTON, L. R. 1957. Distributions associated with random walk and recurrent events. *Jl R. statist. Soc.* B **19**, 64–118.

SKITOVITCH, V. P. 1954. Linear forms of independent random variables and the normal distribution law. *Izv. Akad. Nauk SSSR* Mat ser. **18**, 185–200.

SLUTSKY, E. E. 1925. Über stochastistische Asymptoten und Grenzwerte. *Metron, Rovigo* **5**, 3–89.

—— 1950. Tables for the computation of the incomplete gamma function and the probability function of χ^2 (Ed. A. N. Kolmogorov) (in Russian). Akad. Nauk. Leningrad.

SMIRNOV, S. V. and POTAPOV, M. K. 1957. A nomogram for the incomplete Γ-function and the χ^2 probability function. *Theory of probability and its applications* **2**, 461–4.

—— —— 1961. Nomogram for the probability function χ^2. *Theory of probability and its applications* **6**, 138–40.

SMITH, W. L. 1962. A note on characteristic functions which vanish identically in an interval. *Proc. Camb. phil. Soc. math. phys. Sci.* **58**, 430–2.

SOBOL, I. M. 1958. Pseudo-random numbers for the machine 'Strela'. *Theory of probability and its applications* (transl. from Russian) **3**, 192–7.

SOPER, H. E. 1922. *Frequency arrays*. Cambridge University Press.

SPITZER, F. 1956. A combinatorial lemma and its application to probability theory. *Trans. Am. math. Soc.* **82**, 323–39.

—— 1964. *Principles of random walk*. Van Nostrand, Princeton.

STECK, G. P. 1958*a*. A uniqueness property not enjoyed by the normal distribution. *Ann. math. Statist.* **29**, 604–6.

—— 1958*b*. A table for computing trivariate normal probabilities. *Ann. math. Statist.* **29**, 780–800.

STEFFENSEN, J. F. 1930. Om Sandsynligheden for at Afkommet uddor. *Mat. Tidsskr.* B **1**, 19–23.

—— 1933. Deux problèmes du calcul des probabilités. *Annls Inst. Poincaré* **3**, 319–44.

STEINHAUS, H. and TRYBULA, S. 1959. On a paradox in applied probabilities. *Bull. Acad. pol. Sci. Sér. Sci. math. astr. phys.* **3**, 67–9.

STEPHAN, F. F. 1945. The expected value and variance of the reciprocal and other negative powers of a positive Bernoullian variate. *Ann. math. Statist.* **16**, 50–61.

STEPHENS, M. A. 1962. Exact and approximate tests for directions. I. *Biometrika* **49**, 463–77.

—— 1963. Random walk on a circle. *Biometrika* **50**, 385–90.

STERRETT, A. 1957. On the detection of defective members of large populations. *Ann. math. Statist.* **28**, 1033–6.

STIELTJES, T. J. 1884. Quelque recherches sur les quadratures dites mécaniques. *Annls scient. Ec. norm. sup., Paris* (3) **1**, 409–26. (*Collected Works* **1** (1918) 377–94.)

—— 1894. Sur certaines inégalités dues à M.P. Tchebycheff. *Collected Works* **2** (1918) 586–93.

STONE, M. H. 1936. The theory of representations for Boolean algebras. *Trans. Am. math. Soc.* **40**, 37–111.

STUART, A. 1950. The cumulants of the first n natural numbers. *Biometrika* **37**, 446.

SUGIYAMA, H. 1961. Some statistical methodologies for epidemiological research and medical sciences. *Bull. Inst. int. Statist.* **38**, 3^e livraison, 137–51.

SWED, F. S. and EISENHART, C. 1943. Tables for testing randomness of grouping in a sequence of alternatives. *Ann. math. Statist.* **14**, 66–87.

SYKES, M. F. 1961. Some counting theorems in the theory of the Ising model and the excluded volume problem. *J. math. Phys.* **2**, 52–62.

SYSKI, R. 1960. *Introduction to congestion theory in telephone systems*. Oliver and Boyd, Edinburgh.

SZEKERES, G. and BINET, F. E. 1957. On Borel fields over finite sets. *Ann. math. Statist.* **28**, 494–8.

TAKACS, L. 1958. On a general probability theorem and its application in the theory of stochastic processes. *Proc. Camb. phil. Soc. math. phys. Sci.* **54**, 219–24.

—— 1960. *Stochastic processes. Methuen's monographs*. Methuen, London.

—— 1962. *Introduction to the theory of queues*. Oxford University Press.

TALLIS, G. M. 1961. The moment generating function of the truncated multi-nomial distribution. *Jl R. statist. Soc.* B **23**, 223–9.

—— 1962. The use of a generalized multinomial distribution in the estimation of correlation in discrete data. *Jl R. statist. Soc.* B **24**, 530–4.

TANG, P. C. 1938. The power function of the analysis of variance tests with tables and illustrations of their use. *Statist. Res. Mem. Univ. Coll. Lond.* **2**, 128–49.

TANNER, J. C. 1953. A problem of interference between two queues. *Biometrika* **40**, 58–69.

TATE, R. F. 1953. On a double inequality of the normal distribution. *Ann. math. Statist.* **24**, 132–4.

TCHEBYCHEV, P. 1874. Sur les valeurs limites des intégrales. *J. Math. pures Appl.* (2) **19**, 157–60.

THATCHER, A. R. 1957. Studies in the history of probability and statistics. VI. A note on the early solutions of the duration of play. *Biometrika* **44**, 515–18.

THIONET, P. 1963. Sur le moment d'ordre (—1) de la distribution binomiale tronquée. Application à l'echantillonnage de Hajek. *Publs Inst. Statist. Univ. Paris* **12**, 93–102.

THOMAS, M. 1949. Generalization of Poisson's binomial limit for use in ecology. *Biometrika* **36**, 18–25.

THOMPSON, C. M. 1941. Percentage points of the incomplete beta function. *Biometrika* **32**, 151–81.

THOMPSON, H. R. 1954. A note on contagious distributions. *Biometrika* **41**, 268–71.

THOMSON, W. E. 1959. ERNIE—a mathematical and statistical analysis. *Jl R. statist. Soc.* A **122**, 301–33.

TIAGO DE OLIVEIRA, J. 1952. Sur le calcul des moments de la réciproque d'une variable aléatoire positive de Bernoulli et Poisson. *Anais Fac. Cienc. Porto* **36**, 165–8.

TIPPETT, L. H. C. 1927. *Random sampling numbers. Tracts for computers* No. 15. Cambridge University Press.

TITCHMARSH, E. C. 1932. *The theory of functions.* Oxford University Press.

TOCHER, K. D. 1954. Symposium on Monte Carlo methods—the application of automatic computers to sampling experiments. *Jl R. statist. Soc.* B **16**, 39–75.

—— 1963. *The art of simulation.* English University Press, London.

TODHUNTER, I. 1949. *A history of the theory of probability from the time of Pascal to that of Laplace.* Chelsea, New York. (First published 1865.)

TOMPKINS, C. B. 1956. Review of the RAND corporations million random digits. *Mathl Tabl. natn. Res. Counc., Wash.* **10**, 39–43.

TORNIER, E. 1936. *Wahrscheinlichkeitsrechnung und allgemeine Integrationstheorie.* Teubner, Leipzig.

TRELOAR, L. R. G. 1943. The statistical length of paraffin molecules. *Proc. phys. Soc.* London **55**, 345–61.

TUKEY, J. W. 1949. Moments of random group size distributions. *Ann. math. Statist.* **20**, 523–39.

TWEEDIE, M. C. K. 1957a. Statistical properties of inverse Gaussian distributions. I. *Ann. math. Statist.* **28**, 362–77.

—— 1957b. Statistical properties of inverse Gaussian distributions. II. *Ann. math. Statist.* **28**, 696–705.

UHLENBECK, G. E. and ORNSTEIN, L. S. 1930. On the theory of the Brownian motion. *Phys. Rev.* **36**, 823–41.

VAN DANTZIG, D. 1951. Une nouvelle généralization de l'inéqualité de Bienaymé. *Annls Inst. Poincaré* **12**, 31–43.

VILLE, J. 1939. *Étude critique de la notion de collectif.* Gauthier–Villars, Paris.

VINCENZ, S. A. and BRUCKSHAW, J. McG. 1960. Note on the probability distribution of a small number of vectors. *Proc. Camb. phil. Soc. math. phys. Sci.* **56**, 21–6.

VOGLER, L. E. 1964. *Percentage points of the beta distribution.* Government Printing Office, Washington.

VON MISES, R. 1918. Über die 'Ganzzahligkeit' der Atomgewichte und verwandte Fragen. *Phys. Z.* **19**, 490–500.

—— 1957. *Probability, statistics, and truth.* Allen and Unwin, London.

WALD, A. 1939. Limits of a distribution function determined by absolute moments and inequalities satisfied by absolute moments. *Trans. Am. math. Soc.* **46**, 280–306.

—— 1944. On cumulative sums of random variables. *Ann. math. Statist.* **15**, 283–96.

—— 1946. Differentiation under the integral sign in the fundamental identity in sequential analysis. *Ann. math. Statist.* **17**, 493–7.

—— 1947. *Sequential analysis.* Wiley, New York.

WALL, F. T. and ERPENBECK, J. J. 1959. New methods for the statistical computation of polymer dimensions. *J. chem. Phys.* **30**, 634–7.

WANG, M. C. and UHLENBECK, G. E. 1945. On the theory of the Brownian motion. II. *Rev. Mod. Phys.* **17**, 323–42.

WATSON, G. S. 1957a. Analysis of dispersion on a sphere. *Mon. Not. R. astr. Soc. Geophys. Suppl.* **7**, 153–9.

—— 1957b. A test for randomness of directions. *Mon. Not. R. astr. Soc. Geophys. Suppl.* **7**, 160–1.

—— and IRVING, E. 1957. Statistical methods in rock magnetism. *Mon. Not. R. astr. Soc. Geophys. Suppl.* **7**, 289–99.

—— and WILLIAMS, E. J. 1956. On the construction of significance tests on the circle and on the sphere. *Biometrika* **43**, 244–53.

WATSON, H. W. (and GALTON, F.) 1874. On the probability of the extinction of families. *J. anthrop. Inst.* **4**, 138–44.

WATTERSON, G. A. 1962. Some theoretical aspects of diffusion theory in population genetics. *Ann. math. Statist.* **33**, 939–57.

WEATHERBURN, C. E. 1946. *A first course in mathematical statistics.* Cambridge University Press.

WEILER, H. 1959. Means and standard deviations of a truncated bivariate normal distribution. *Aust. J. Statist.* **1**, 73–87.

WEISS, L. 1958. Limiting distributions in some occupancy problems. *Ann. math. Statist.* **29**, 878–84.

WELLS, W. T., ANDERSON, R. L., and CELL, J. W. 1962. The distribution of the product of two central or non-central chi-square variates. *Ann. math. Statist.* **33**, 1016–20.

WHITTAKER, E. T. and WATSON, G. N. 1935. *A course of modern analysis.* Cambridge University Press.

WHITTAKER, J. M. 1937. The shot effect for showers. *Proc. Camb. phil. Soc. math. phys. Sci.* **33**, 451–98.

—— 1938. The shot effect with space charge. *Proc. Camb. phil. Soc. math. phys. Sci.* **34**, 158–66.

WHITTLE, P. 1955. Some distribution and moment formulae for the Markov chain. *Jl R. statist. Soc.* B **17**, 235–42.

—— 1957. Curve and periodogram smoothing. *Jl R. statist. Soc.* B **19**, 38–47.

—— 1958. On the smoothing of probability density functions. *Jl R. statist. Soc.* B **20**, 334–43.

WHITWORTH, W. A. 1901. *Choice and Chance*, 5th edn. (Republished by Hafner, New York, 1948.)

—— 1897. *DCC exercises in choice and chance.* Deighton Bell, Cambridge. (Republished by Hafner, New York, 1959.)

WICKSELL, S. D. 1919. A general formula for the moments of the normal correlation function of any number of variates. *Phil. Mag.* **37**, Ser. B, 446–52.

WIDDER, D. V. 1946. *The Laplace transform.* Princeton University Press.

WILKS, S. S. 1942. Statistical prediction with special reference to the problem of tolerance limits. *Ann. math. Statist.* **13**, 400–9.

WILKS, S. S. 1946. *Mathematical statistics*. Princeton University Press, Princeton.

WILLIAMS, J. D. 1946. An approximation to the probability integral. *Ann. math. Statist.* **17**, 363–5.

WINSTEN, C. B. 1946. Inequalities in terms of mean range. *Biometrika* **33**, 283–95.

WINTNER, A. 1947. *The Fourier transforms of probability distributions*. Mimeographed lectures. Baltimore. Edwards, Michigan.

—— 1949. Factorial moments and enumerating distributions. *Skand. Aktuartidsskr.* **32**, 63–8.

—— 1957. Student's distribution and Riemann's elliptic geometry. *Biometrika* **44**, 264–5.

WISE, M. E. 1946. The use of the negative binomial distribution in an industrial sampling problem. *Jl R. statist. Soc.* Suppl. **8**, 202–11.

—— 1950. The incomplete beta function as a contour integral and a quickly converging integral for its inverse. *Biometrika* **37**, 208–18.

—— 1954a. A quickly convergent expansion for cumulative hypergeometric probabilities, direct and inverse. *Biometrika* **41**, 317–329 (correction **42**, 277).

—— 1954b. The ratio of two factorials and some fundamental probabilities. *Proc. K. ned. Akad. Wet.* A **57** (*Indag. Math.* **16**), 513–21.

—— 1960. On normalizing the incomplete beta-function for fitting to dose–response curves. *Biometrika* **47**, 173–5.

—— 1963. Multinomial probabilities and the χ^2 and X^2 distributions. *Biometrika* **50**, 145–54.

WISHART, J. 1927. Approximate quadrature of certain skew curves. *Biometrika* **19**, 1–38, 442.

—— 1947. The cumulants of the z and of the logarithmic χ^2 and t distributions. *Biometrika* **34**, 170–8 (correction **34**, 374).

—— 1949. Cumulants of multivariate multinomial distributions. *Biometrika* **36**, 47–58.

—— 1954. The factorial moments of the distribution of joins between line segments. *Biometrika* **41**, 555–6.

—— 1957. An approximation for the cumulative z-distribution. *Ann. math. Statist.* **28**, 504–10.

—— and HIRSCHFELD, H. O. 1936. A theorem concerning the distribution of joins between line segments. *J. Lond. math. Soc.* **11**, 227–35.

WOODWARD, P. M. 1948. A statistical theory of cascade multiplication. *Proc. Camb. phil. Soc. math. phys. Sci.* **44**, 404–12.

WOUK, A. 1961. On digit distributions for random variables. *J. Soc. ind. appl. Math.* **9**, 597–603.

YAGLOM, A. M. and YAGLOM, I. M. 1959. *Probabilité et information. Monographies Dunod*. Dunod, Paris.

YATES, F. 1934. Contingency tables involving small numbers and the χ^2 test. *Jl R. statist. Soc.* Suppl. **1**, 217–35.

—— 1948. Systematic sampling. *Phil. Trans. R. Soc.* A **241**, 345–77.

YOSIDA, K. and KAKUTANI, S. 1939. Markoff processes with an enumerable infinite number of possible states. *Jap. J. Math.* **16**, 47–55.

YULE, G. U. 1924. A mathematical theory of evolution based on the conclusions of Dr. J. C. Willis, F.R.S. *Phil. Trans. R. Soc.* B **213**, 21–87.

—— 1927. On reading a scale. *Jl R. statist. Soc.* **90**, 570–82.

ZIPF, G. K. 1949. *Human behaviour and the principle of least effort*. Addison-Wesley, Cambridge, Mass.

ZIPPIN, C. 1956. An evaluation of the removal method of estimating animal populations. *Biometrics* **12**, 163–89.

ZOLOTAREV, V. M. 1954. Expression of the density of a stable distribution with exponent α greater than one by means of a density with exponent $1/\alpha$. *Dokl. Akad. nauk SSSR* **98**, 735–8.

—— 1957. Mellin–Stieltjes transforms in probability theory. *Theory of probability and its applications*, **2**, 433–60.

ZYGMUND, A. 1959. *Trigonometric series*, 2 Vols. Cambridge University Press.

AUTHOR INDEX

SUBJECT INDEX